D0929334

MODERN
MACHINING
TECHNOLOGY

Richard Baril

Delmar Publishers Inc.

NOTICE TO THE READER

Publisher does not warrant or guarantee any of the products described herein or perform any independent analysis in connection with any of the product information contained herein. Publisher does not assume, and expressly disclaims, any obligation to obtain and include information other than that provided to it by the manufacturer.

The reader is expressly warned to consider and adopt all safety precautions that might be indicated by the activities described herein and to avoid all potential hazards. By following the instructions contained herein, the reader willingly assumes all risks in connection with such instructions.

The publisher makes no representations or warranties of any kind, including but not limited to, the warranties of fitness for particular purpose or merchantability, nor are any such representations implied with respect to the material set forth herein, and the publisher takes no responsibility with respect to such material. The publisher shall not be liable for any special, consequential or exemplary damages resulting, in whole or in part, from the readers' use of, or reliance upon, this material.

Administrative Editor: Mark W. Huth
Developmental Editor: Marjorie A. Bruce
Production Editor: Patricia O'Connor-Gillivan
Managing Editor: Barbara Christie
Art Director: Ron Blackman
Design Coordinator: John Orozco

For information address Delmar Publishers Inc.,
2 Computer Drive West, Box 15-015,
Albany, New York 12212

Copyright © 1987 by Delmar Publishers Inc.
All rights reserved. No part of this work covered by the copyright hereon may be reproduced or used in any form or by any means — graphic, electronic, or mechanical, including photocopying, recording, taping, or information storage and retrieval systems — without written permission of the publisher.

Printed in the United States of America
Published simultaneously in Canada
by Nelson Canada,
a division of International Thomson Limited

10 9 8 7 6 5 4 3

Library of Congress Cataloging in Publication Data

Baril, Richard.
Modern machining technology.

Includes index.
1. Machine-shop practice.   2. Machining.   I. Title.
TJ1160.B228 1987   671.3'5   87-5325
ISBN 0-8273-2578-9
ISBN 0-8273-2579-7 (instructor's guide)
ISBN 0-8273-2580-0 (study guide)

CONTENTS

CHAPTER ONE

EMPLOYMENT OPPORTUNITIES 1

Introduction 1 • Key Terms 1 • Machine Trades 1 • Job Titles 3 • Selecting a Career 4 • Applying for a Job 5

CHAPTER TWO

SAFETY 7

Introduction 7 • Key Terms 7 • General Rules 7 • Eye Protection 8 • Melting and Heat Treating Furnaces 9

CHAPTER THREE

PRECISION MEASUREMENT 11

UNIT 3–1
INTRODUCTION TO PRECISION MEASUREMENT 12

Key Terms 12 • General 12 • Care of Measuring Tools 13 • Measuring Systems 13

UNIT 3–2
USING RULES (SCALES) 15

Key Terms 15 • Rules 15

UNIT 3–3
USING CALIPERS 18

Key Terms 18 • Calipers 18

UNIT 3–4
MICROMETERS AND VERNIERS 23

Key Terms 23 • The Micrometer Caliper 23 • Vernier Calipers 33 • Gear Tool Vernier 35

UNIT 3–5
TELESCOPING GAGES, PROTRACTORS, AND THICKNESS GAGES 42

Key Terms 42 • Telescoping Gage 42 • The Universal Bevel Protractor 42 • Thickness or Feeler Gage 46

UNIT 3–6
DIAL INDICATORS, DIAL BORE GAGES, AND FIXED GAGES 49

Key Terms 50 • Dial Indicators 50 • Dial Bore Gage 50 • Fixed Gages 50

UNIT 3–7

GAGE BLOCKS, SINE BARS AND SINE TABLES, AND OPTICAL FLATS 56

Key Terms 57 • Gage Blocks 57 • Using Gage Blocks 58 • Working with Optical Flats 62

UNIT 3–8

OPTICAL COMPARATORS AND DIGITAL READOUTS 67

Key Terms 67 • Optical Comparators 67 • Digital Readouts 71

CHAPTER FOUR

LAYOUT OR MARKING OFF 73

The Safe Use of Layout Tools 73

UNIT 4–1

INTRODUCTION TO LAYOUT 74

Key Terms 74 • Laying Out 74 • Surface Preparation 75

UNIT 4–2

LAYOUT TOOLS: SURFACE PLATES, HAMMERS, SCRIBERS, CENTER PUNCHES, AND PRICK PUNCHES 76

Key Terms 76 • Common Layout Tools 76 • Surface Plates 76 • Hammers 77 • Scribers 78 • Center Punches and Prick Punches 79

UNIT 4–3

LAYOUT TOOLS: DIVIDERS AND HERMAPHRODITE CALIPERS 81

Key Terms 82 • Dividers 82 • Hermaphrodite Calipers 82

UNIT 4–4

LAYOUT TOOLS: SQUARES, PROTRACTORS AND DEPTH GAGES, SURFACE GAGES, AND VERNIER HEIGHT GAGES 84

Key Terms 85 • Squares 85 • Protractor and Depth Gage 85 • Surface Gage 85 • Vernier Height Gage 89

UNIT 4–5

LAYOUT TOOLS: TRAMMELS, KEYSEAT CLAMPS, V-BLOCKS, AND ANGLE PLATES 92

Key Terms 92 • Trammels 92 • Keyseat Clamps 92 • V-Blocks 92 • Angle Plates 93

CHAPTER FIVE

BENCHWORK 95

Hand Tool Safety 95

UNIT 5–1

BENCHWORK TOOLS 97

Key Terms 97 • Introduction to Benchwork 97 • Care of Tools 97

UNIT 5–2

USING BENCHWORK TOOLS: HAMMERS, SCREWDRIVERS, AND WRENCHES AND PLIERS 98

Key Terms 98 • Hammers 98 • Screwdrivers 99 • Wrenches and Pliers 100

UNIT 5–3

USING BENCHWORK TOOLS: HACKSAWS, CHISELS, AND SCRAPERS 102

Key Terms 103 • Hacksaw 103 • Chisels 105 • Scrapers 107

UNIT 5–4

USING BENCHWORK TOOLS: FILES 109

Key Terms 110 • Files 110 • Rotary Files and Burs 117

CHAPTER SIX

SAWING 121

Safe Practices in Using Saws 121

UNIT 6–1

BAND MACHINING 122

Key Terms 122 • Cutting Off Stock in a Machine Shop 123 • Contour Band Machining 123 • Band Machines 123 • Feed Systems 123 • Principles of Band Machining 123

UNIT 6–2

THE SAW BLADE 128

Key Terms 128 • Selecting Saw Blades 128 • The Job Selector 129 • Cutting Fluids 132

UNIT 6–3

WELDING SAW BLADES 133

Key Terms 133 • Welding Saw Blades 133 • Making the Weld 136 • Troubleshooting Poor Welds 139

UNIT 6–4

OTHER BAND TOOLS 144

Key Terms 144 • Other Band Tools Available 144 • File Bands and Polishing Bands 145 • Friction Sawing 145 • High-Speed Sawing 145 • Cutting-Off with Band Machines 146

UNIT 6–5

SAWING RECOMMENDATIONS AND TROUBLESHOOTING 149

Key Terms 149 • Organization 149 • General Sawing Tips 149 • Band Machining Operation Recommendations 150 • Troubleshooting 152

CHAPTER SEVEN

DRILLS AND DRILLING OPERATIONS 161

Safe Practices in Drilling 162

UNIT 7–1

DRILLING MACHINES AND TOOL HOLDING METHODS 163

Key Terms 163 • Drilling Machines 164 • Tool Holding Methods 166 • Work Holding Methods 167

UNIT 7–2

DRILLS AND THE USE OF DRILLS 174

Key Terms 174 • Drills 174 • Twist Drill Series and Sizes 177 • Drill Point Angles and Clearances 178 • Drill Sharpening 179 • Drilling Problems 179

UNIT 7–3

DRILLING OPERATIONS 187

Key Terms 187 • Laying Out a Hole for Spot Drilling 188 • Spot Drilling 188 • Chip Formation in Drilling 188 • Pressure on Drill Points 189 • Drilling Speeds and Feeds 191

UNIT 7–4

OTHER DRILLING MACHINE OPERATIONS 198

Key Terms 198 • Introduction 198 • Reamers 200 • Boring 203 • Spot-Facing 203 • Counterboring 203 • Countersinking 203 • Cutting Screw Threads 204 • Tapping Faults 207 • Calculating Tap Drill Sizes 209 • Broken Tap Removal 210 • Threading Dies 210 • Resource Tables for Threading 211 • Emergency Drills 211

CHAPTER EIGHT

THE LATHE AND LATHE OPERATIONS 221

Safe Practices in Lathework 221

UNIT 8–1

THE LATHE 222

Key Terms 223 • Introduction to the Lathe 223 • The Lathe Feed Systems 224 • Types of Headstocks 227 • Size of a Lathe 227 • Spindle Noses 228

UNIT 8–2

THE TOOLS OF THE LATHE 232

Key Terms 233 • The Tools of the Lathe Tool Holders 233 • Lathe Tool Post Systems 234 • Turret Tool Holders 235 • Types of Lathe Chuck 236 • Setting Up a Four-Jaw Chuck 237 • Face Plate 240 • Center or Steady Rests 240 • The Follower Rest 241

UNIT 8–3

LATHE OPERATIONS — FACING 248

Key Terms 248 • Lathe Operations 248 • Facing (Squaring) 249

UNIT 8–4

LATHE OPERATIONS — TURNING 250

Key Terms 250 • Turning on a Lathe 250 • Tool Position 251 • Methods of Turning 252 • Cutting Position 254 • Turning a Shoulder 254 • Grooving 255 •

Cutting Off or Parting 255 • Problems in
Cutting-Off (Parting) 256 • Knurling 258 •
The Knurling Operation 258

UNIT 8–5
LATHE OPERATIONS — BORING 261

Key Terms 261 • General 261 • Boring
Tool Holders, Commercial Types 261 •
Boring Methods 264 • Drilling and
Reaming 264 • Center Drilling
Accurately 269

UNIT 8–6
LATHE OPERATIONS — TAPERING 281

Key Terms 281 • Tapering 281 •
Tapers 281 • Taper Turning Methods 282

UNIT 8–7
LATHE OPERATIONS — THREADING 287

Key Terms 287 • Threading 287 • Screw
Threads 288 • Screw Thread Forms 292 •
Power Threads (Translation Threads) 293 •
Unified Screw Threads 294 • Metric
Threading 294

UNIT 8–8
THREADING TOOLS — COMMERCIAL 297

Key Terms 297 • Setting the Threading
Tool 297 • Setting the Compound: Cutting
a 60-Degree Thread 297

UNIT 8–9
THREAD CUTTING 300

Key Terms 301 • Thread Cutting 301 •
The Thread Cutting Routine 301 • Using
the Threading Dial 302 • Picking Up a
Thread 303 • Grinding Threading Tool
Bits 304 • Cutting Power Threads: Acme
Threads 304 • Grinding the Threading
Tool 304 • Cutting the Acme Thread 305 •
Making Coil Springs 305 • Cutting a
Metric Thread 305 • Cutting Tapered
Threads 306 • Resource Tables for
Threading 306

CHAPTER NINE
CUTTING TOOLS 317

Safe Use of Cutting Tools 317

UNIT 9–1
THE PRINCIPLES OF CUTTING 319

Key Terms 319 • Introduction 319 •
Cutting Tool Action 319 • Basic Forms of
a Cutting Edge 320

UNIT 9–2
CUTTING TOOLS 324

Key Terms 324 • Roughing and Finishing
Cutting Tools 324 • Cutting Tool
Materials 326 • Clearance and Rake 328

UNIT 9–3
CARBIDE TOOLING 341

Key Terms 342 • Introduction 342 • Tool Nomenclature 342 • Basic Tool Types 343 • Machining Recommendations 346 • Troubleshooting Carbide Cutting Tools 353

UNIT 9–4
SPEED, FEED, AND DEPTH OF CUT 356

Key Terms 356 • Introduction 356 • Cutting Speed 356 • Feeds for High-Speed Steel and Carbide Cutting Tools 357 • Sample Problems 358

UNIT 9–5
CUTTING FLUIDS 361

Key Terms 361 • Introduction 361 • Functions of a Cutting Fluid 361 • Basic Types of Cutting Fluids 361 • Uses of Cutting Fluids 363

CHAPTER TEN

MILLING AND MILLING OPERATIONS 367

Milling Machine Safety 367

UNIT 10–1
BASIC MILLING OPERATIONS 368

Key Terms 368 • Introduction 369 • The Parts of the Milling Machine 369 • Milling Cutters 369 • Milling Cutter Holding Devices 375 • Work Holding Methods 377

UNIT 10–2
MILLING MACHINES 382

Key Terms 382 • General Milling Machines 382 • Jig Boring Machines 386 • Suggested Order of Operations 389 • Using a Vertical Milling Machine for Jig Boring 389

UNIT 10–3
MILLING OPERATIONS 394

Key Terms 394 • Basic Milling Operations 394 • Conventional Milling and Climb Milling 395 • Speeds and Feeds 397 • Milling Hints 398

UNIT 10–4
INDEXING 401

Key Terms 401 • Introduction 402 • Direct Indexing 402 • The Universal Dividing Head 403 • Cutting Six Equal Cuts 403 • Cutting a 60-Tooth Gear 404

UNIT 10–5
GEARING 412

Key Terms 413 • Types of Gears 413 • Gear Diameters 414 • Gear Tooth Parts 416 • Cutting a Spur Gear 417

UNIT 10–6
HELICAL GEARING 423

Key Terms 423′ • Terminology 423 • The Helix in Practice 424 • Cutting a Helix 424 • Determining the Direction to Set the Table 424

UNIT 10-7
MEASURING GEAR TEETH 436

Key Terms 436 • Gear Tooth Vernier 436 •
Worm Gearing 437 • The Module System of
Gearing 439 • Bevel Gears 440 •
Procedure for Milling a Bevel Gear 446 •
Involute Gear Cutters 447 • Tooth Sizes 447

CHAPTER ELEVEN

GRINDING 459

Grinding Safety 459

UNIT 11-1
THE GRINDING WHEEL 461

Key Terms 461 • General 461 • Abrasive
Grits and Grains 462 • Grinding Wheel
Components 463 • Factors to Be Considered
in the Selection of the Proper Grinding
Wheel 466 • Grinding Wheel
Specifications 471 • Summary 473

UNIT 11-2
PRECISION GRINDING 474

Key Terms 474 • Introduction 474 •
Surface Grinding 474 • Testing Wheels
before Mounting 475 • Mounting and
Dressing Wheels 475 • Dressing and
Trueing the Wheel 475 • Setting the
Work 477 • Troubleshooting Precision
Surface Grinding 478

UNIT 11-3
OTHER GRINDING OPERATIONS 483

Key Terms 483 • Cylindrical
Grinding 483 • Tool and Cutter
Grinding 484

CHAPTER TWELVE

NUMERICAL CONTROL (N/C), COMPUTER NUMERICAL CONTROL (CNC), AND ROBOTICS 497

UNIT 12-1
NUMERICAL CONTROL (N/C) 498

Key Terms 498 • Introduction 498 •
Tapes and Codes 499 • Programming:
Absolute and Incremental 500 • Axis
Orientation 502 • Machining with N/C 503

UNIT 12-2
COMPUTER NUMERICAL CONTROL (CNC) 509

Key Terms 509 • Introduction 509 •
Computer Graphics 511

UNIT 12-3
ROBOTICS 518

Key Terms 518 • The Advent of the
Robot 518 • Design of an Industrial
Robot 519 • Modes of Operation 521 •
Practical Applications for the Robot 521 •
New Developments in Robots 523

CHAPTER THIRTEEN

SPECIAL-PURPOSE PROCESSES 537

UNIT 13–1
POWDER METALLURGY 538

Key Terms 538 • Introduction 538 • Details in the Process of Powder Metallurgy 538

UNIT 13–2
ELECTRICAL DISCHARGE MACHINING (EDM) 540

Key Terms 540 • Introduction 541 • Conventional EDM 541 • Wire-Cut EDM 544

UNIT 13–3
THERMAL SPRAY PROCESSES 548

Key Terms 548 • Introduction 548 • Electric Arc Spraying 549 • Plasma Spraying 549 • Metallizing 549 • Flame Spraying, Wire Process 550 • Flame Spraying, Thermospray or Powder Process 554

CHAPTER FOURTEEN

METALLURGY 565

UNIT 14–1
THE METALLURGY AND MACHINING OF FERROUS METALS 566

Key Terms 566 • Introduction 566 • The Blast Furnace 567 • Cast Iron 567

UNIT 14–2
HOW STEEL IS MADE 572

Key Terms 573 • Introduction 573 • Metallurgy of Steel 573 • Classification of Steels 575 • Machining Steels 578 • Machining Monel and Nickel Alloys 582

UNIT 14–3
HEAT TREATMENT OF STEELS 583

Key Terms 583 • Forms of Heat Treatment 584 • Using the Letting-Down Process on a Cold Chisel 586 • Hardness Testing 587

UNIT 14–4
THE METALLURGY AND MACHINING OF NONFERROUS METALS 600

Key Terms 600 • Aluminum and Its Alloys 600 • Copper and Its Alloys 602 • Babbitt Bearings 604

CHAPTER FIFTEEN

PLASTICS 607

UNIT 15–1
TYPES OF PLASTICS 608

Key Terms 608 • Thermosets 608 • Thermoplastics 609

UNIT 15–2
MACHINING PLASTICS 609

> Key Terms 609 • Machining
> Recommendations 609 • Annealing 610 •
> Special Notes 611

APPENDIX 1

TABLES 619

APPENDIX 2

GENERAL INFORMATION 649

APPENDIX 3

PLANNING WORK 671

GLOSSARY 681

INDEX 689

INDEX OF TABLES

3–1   Comparison of English (U.S. Customary) and Metric Sizes 14

3–2   Millimeter to Inch Conversions 55

5–1   Hacksaw Pitches for Different Uses 103

5–2   Recommended Speeds in Medium Cut 117

6–1   Blade Specifications 130

6–2   Selecting Cutting Fluids 132

6–3   Blade Selection for Cutting Off Tubing 147

6–4   Blade Selection for Cutting Off Solids 148

6–5   Blade Selection for Cutting Off Structural Shapes and the Preferred Method of Holding Structural Shapes 148

6–6A   Operation Recommendations 150

6–6B   Operation Recommendations 152

7–1   Drill Sizes: Fractional, Wire or Number, Letter and Metric Sizes 180

7–2   Cutting Speeds and RPM (inch) 192

7–3   Cutting Speeds and RPM (metric) 193

7–4A   Suggested Feeds for High-Speed Twist Drills for General-Purpose Work 194

7–4B   Suggested Cutting Speeds for Drilling 194

7–5   Suggested Speeds and Feeds for Reaming Various Materials 203

7–6   Tap Drill Sizes (inches) 213

7–7   Tap Drill Sizes (millimeters) 214

7–8   Cutting Speeds for Taps 215

8–1   Recommended Center Drill Sizes for Different Shaft Diameters 268

8–2   Dimensions of Combined Drills and Countersinks 268

8–3   A Comparison of the Morse Taper and the Browne and Sharpe Taper 285

8–4   Comparing Taper per Foot and Millimeters per Millimeter of Length to Angles in Degrees 286

8–5   Proposed 25 Basic ISO Screw Threads 296

8–6   Acme Threads 306

8–7   Pipe Threads — NPT 307

8–8   Inch to Millimeter Conversions 308

9–1   Suggested Cutting Angles for High-Speed Steel Tool Bits 340

9–2   Tool Holder Identification System 347

9–3   Indexable Inserts Identification System 348

9–4   Cutting Tool Grades and Machining Applications 351

9–5   Machining Recommendations for High-Temperature Alloys 352

9–6   Recommendations for Machining Super Alloys 352

9–7   Suggested Cutting Speeds for High-Speed Steel and Carbide Cutting Tools 357

9–8   Suggested Feeds per Tooth for Milling Cutting Tools 359

10–1   RPM for Different Diameters Based on Cutting Speed 399

10–2   Formulas for Speed and Feed Calculations 400

10–3   A Simple Indexing Chart 405

10–4   Change Gears and Movements for Differential Indexing 408

10–5   Rules and Formulas for Spur Gear Calculations 421

10–6   Leads for Helical Milling (Inch) 427

10–7   Nomograph for Finding Angular Settings of Machine Table for Helical Milling 429

10–8   Leads for Helical Milling (Metric Units) 430

10–9   Rules and Formulas for Helical Gear Calculations 434

10–10A   Rules and Formulas for Worm Wheel Calculations 442

10–10B   Rules and Formulas for Worm Gear Calculations 444

10–11   Basic Rules for the Module System of Gearing 446

10–12   Rules and Formulas for Bevel Gear Calculations 452

10–13   Involute Gear Cutters 454

11–1   Relief or Clearance Tables for Grinding High-Speed Steel Cutting Tools 486

11–2   Grinding Wheel Speeds 487

14–1   Clearance and Rake for High-Speed Steel Tool Bits 571

14–2   Clearance and Rake for Carbide Cutting Tools 571

14–3   Recommended Rakes and Clearances for Cutting Monel and Other Nickel Alloys 582

14–4   Hardness Conversions 595

14–5   Rakes and Clearances for Free Machining Alloys 604

14–6   Rakes and Clearances for Moderately Machinable Alloys 604

14–7   Rakes and Clearances for Difficult to Machine Alloys 604

15–1   Recommended Machining Data for Thermosets 610

15–2   Recommended Machining Data for Thermoplastics 611

A–1   English (U.S. Customary) to Metric Conversions 619

A–2   Metric to English (U.S. Customary) Conversions 620

A–3   Metric and English (U.S. Customary) Conversion Units 621

A–4   Geometry and Circles 622

A–5   Basic Trigonometric Functions 623

A–6   A Comparison of Morse Tapers and Browne and Sharpe Tapers 624

A–7   Tapers and Angles 625

A–8   Double Depth of Screw Threads 626

A–9   Diagonal Measurements of Hexagons and Squares 627

A–10   Cutting Speeds and RPM 628

A–11   Cutting Speeds and RPM of Drills (Fractional) 629

A–12   Cutting Speeds and RPM of Drills (Metric) 629

A–13   Grinding Wheel Speeds 630

A–14   Fractional, Wire or Number, Letter and Metric Drill Sizes 631

A–15   Tap Drill Sizes, English (U.S. Customary) 633

A–16   Tap Drill Sizes, Metric 634

A–17   Unified Screw Threads, Basic Data 635

A–18   American National Pipe Threads, Basic Data 636

A–19   Acme Screw Threads, Basic Data 637

A–20   ISO Metric Threads, Basic Data 638

A–21   Wire Gage Standards Used in the United States 639

A–22   Weights of Steel Bars 640

A–23   Weights of Steel Bars 641

A–24   Weights of Flat Bar Steel Per Linear Foot 642

A–25   Hardness Conversions 642

A–26   Classification of Steels 643

A–27   Steel Compositions 644

A–28   Spark Testing Metals 648

PHOTO GALLERIES

Gallery Six 156
Gallery Seven 218
Gallery Eight 310
Gallery Nine 365
Gallery Ten 456
Gallery Eleven 491
Gallery Twelve 530
Gallery Thirteen 563
Gallery Fifteen 615

TECHNOLOGY UPDATE

Modern Manufacturing — New Trends 44

Cad and Cam in Manufacturing 90

Digital Readouts and Computer Controls 142

Automated Guided Vehicles — Robots in Motion 196

Robots — Loading and Unloading Machine Tool Groups or Cells 230

Robots — Future Trends 278

Laser Beams — Cladding and Surface Alloying 338

Laser Surface Hardening 384

Laser Beam Cutting and Machining 418

Laser Machining on a Metal Lathe 488

Computed Tomography — Testing Finished Parts in the Machine Shop 534

Electrical Discharge Machining (EDM) — New Trends 560

Artificial Machine Vision 592

Manufacturing Automation Protocol (MAP) 613

PREFACE

Modern Machining Technology is a comprehensive text which covers basic machining principles, cutting tool parameters, and machine tool characteristics. Beyond the necessary fundamentals for the beginning machinist or apprentice, the text also presents an exciting look into the future of machining technology.

Firmly believing in the need for the machinist to master the basics, the author's goal was to develop a textbook about machine tools and cutting tools. It is not meant to be a catalog of all available tools. The text deals with those machine tools and cutting tools commonly used by machinists and tool and die makers in their everyday work.

All of the basic operations of machine tools are presented in such a way that the learner will not only know what to do, but will also know why it is being done. The reasons why certain things are done or are not done are also discussed. The consequences of not making the right move are also discussed in detail. Basic operations are described just as they are performed in the trades. The learner is cautioned that there is more than one way of performing a specific operation, but the commonly accepted method is thoroughly described as a starting point upon which the learner can build in advanced courses or on the job. Practical machining hints based on years of shop experience are one of the features of this text.

The heart of machining is the action of the cutting tool. The learner is advised to study carefully the principles governing cutting tool action because these are essential to the proper use of the tools. The text provides a thorough discussion of cutting tool geometry, cutting action, chip formation, tool forming and sharpening, and feeds and speeds. Everything the learner needs to know to make informed decisions about the proper cutting tool for various machining conditions and materials is emphasized throughout the text. The most expensive machine tool will not perform properly, if at all, unless the cutting tool is right for the work being done. Good machinists give a great deal of thought to how and why their cutting tools perform the way they do. Attention is also given to the shape of the tools since the various clearance angles have an impact on how the cutting tool performs.

The discussion of cutting tool speeds and feeds is thorough, including clear explanations of how they affect the performance of the cutting tool. Based upon this information, the numerous recommendations for speeds and feeds for specific jobs and materials, and the guidelines for the selection of the proper cutting tool, the learner will be able to set up a job to achieve small and tightly curled chips. This form of chip is safer for the operator and chip disposal is much less of a problem than is the case with the long, coiled type of chip. Machining has become much faster and better because of this change in the type of chip produced.

Safety is another important area that is stressed in the text, with a separate chapter at the beginning of the text and safety highlights opening the majority of the remaining chapters. The safety precautions are repeated through the text because of their importance in helping the learner develop safe working habits. It is never wise to take safety for granted. Safety is, and always should be, an automatic state of mind to be acted upon immediately and consistently.

The technology of numerical control, computer numerical control and robotics is in widespread use in large-scale manufacturing facilities. It is relatively new to many smaller machine shops, especially jobbing shops. However, more and more shops are discovering the advantages of this technology. The addition of computers to basic numerical control has expanded the capabilities of the manufacturing industries. Continuing development of computer numerical controlled (CNC) machine tools, controls and software will have significant impact on manufacturing in the years to come. The text introduces the technology, demonstrates some of the present capabilities, and provides some insights into future developments.

ORGANIZATION OF THE TEXT

The text is divided into 15 chapters covering major topics. With the exception of chapters 1 and 2, the chapters are divided into smaller units. This division provides necessary information in shorter segments with immediate reinforcement. The opening page of the chapter contains an overview of the chapter content followed by a listing of safety guidelines specific to the machine tool or process being discussed in the chapter. Each unit in the chapter contains specific learning objectives, a list of key terms introduced in the unit, highly illustrated content, tables and charts to summarize information and provide machining recommendations, and review questions to reinforce learning.

At the end of selected chapters, photo galleries are provided to show cutting action and chip formation for specific machining processes. The text also contains special inserts which describe evolving technologies in manufacturing, such as laser cutting, artificial machine vision, automated guided vehicles and manufacturing automated protocol. Listings of these special features are included in the tables of contents under the headings "Photo Galleries" and "Technology Update."

The appendix section contains a large collection of tables of machining data, specialized definitions for grinding and several projects which apply basic machining techniques. An extensive glossary follows the appendixes.

FEATURES

- Controlled reading level to contribute to learner comprehension and retention of technical information.
- All basic machining operations are presented in a logical order and with up-to-date information and clarifying illustrations.
- Each of the basic machining operations is presented so that the learner will not only know what to do, but will also know why it is being done.
- The "do's and don't's" of operations and the consequences of not making the right move are also discussed.
- Cutting tools are covered thoroughly in all phases of machining, with recommendations for selecting the proper cutting tool for the conditions of the job and type of material.
- Learners are advised in detail how to make the proper selection of speeds and feeds in combination with the proper cutting tool for specific operations to achieve small and tightly curled chips (resulting in safer and faster machining, with fewer chip disposal problems).
- There are many trade tips for holding and using tools. Emphasis is on the correct placement,

positioning and use of cutting tools, including sharpening.

- Safety is stressed in a separate chapter. Safety highlights for specific machining operations are emphasized at the beginning of the respective chapter. Wherever necessary, safety precautions are included with machining procedures.
- Each unit begins with objectives stressing the development of safety awareness, proper tool use and care, and accurate machining techniques.
- Measurements are given first in inch dimensions, followed by the metric equivalent in parentheses. The majority of the tables provide both inch and millimeter measurements.
- Many tables throughout the text summarize important information and machining tips for ready reference.
- Step-by-step procedures are presented for machining operations.
- Several troubleshooting guides are provided to help learners determine what is causing a problem with a specific operation and how to correct it.
- Shop math is introduced in detail where appropriate for various operations.
- Detailed coverage of special processes: electrical discharge machining (EDM), metallizing and flame spraying (wire process and thermo spray or powder process).
- Special considerations for machining plastics are included.
- The topics of numerical control, computer numerical control and robotics are described in detail, with many photographs showing state of the art equipment.

- Operational procedures, factors relating to the selection of the proper tool and/or method for a specific job, and troubleshooting guidelines are all presented as numbered lists for ease of use.
- Photo Galleries following selected chapters contain additional photos with explanatory legends highlighting cutting operations and chip formation to reinforce the principles discussed in each chapter.
- Technology Update introduces the learner to new processes, methods and equipment under development or recently introduced to manufacturing.
- Extensive appendix of tables provides machining data for ready reference.
- Comprehensive glossary of machining terms is included.

SUPPLEMENTS

The Student Manual contains a large number of additional questions based upon the text content. The purpose of these questions is to reinforce learner understanding of the machining principles presented.

The Instructor's Guide contains the answers to the review questions in the text and an extensive test covering the context of the text.

For the sake of clarity, some photos show machines and machine operations with guards removed. It must be emphasized that all guards are to be in position when machines are operating.

ACKNOWLEDGMENTS

Appreciation is expressed to the following instructors who reviewed the manuscript for *Modern Machining Technology*. Their thorough evaluations provided the author with valuable guidance in refining the manuscript. David H. Ware, Southwest Oakland Vocational Education Center, Wixom, MI 48096; Eldon Peterson, Rochester Area Vocational Technical Institute, Rochester, MN 55901; Douglas G. Long, Knox County Joint Vocational School, Mount Vernon, OH 43050; Robert C. Eck, Reading Muhlenberg AVTS, Reading, PA 19604; Ed Eaton, Borger High School, Borger, TX 79007; Joseph Kasztejna, Lackawanna County AVTS, Scranton, PA 18504; Dietrich Kanzler, Santa Ana College, Santa Ana, CA 92706; John Pesce, Centennial College of Applied Arts and Technology, Scarborough, Ontario, Canada M1K 5E9.

Author's Acknowledgments

My first acknowledgment must be to Cominco Ltd. of Trail, B.C. where I initially trained as a machinist. Billy Ross, a Scottish fitter, was the first to teach me how to read micrometers. Bill Devitt and Dunc Robertson gave me their vast knowledge of tool bits, much of which is still valid today. Mr. J. Harold Knight was the first to introduce me to apprentice instructing.

Cominco machinists Mike Matteucci and Jim DeBruyn confirmed my own thinking regarding clearances and rakes of modern cutting tools. Jim Ihas was a tremendous help in putting together the chapter on N/C work and CNC work. Jim Greener was responsible for giving me much information on EDM and jigboring work. High school students Steve Hope, Christine McConnachie, Catherine Merlo, and Shelley Morris acted as models in certain photographs, and Steve Robson and Shelly Martin traced many of my sketches.

Special thanks go to John Wilkie and Bill Hindman of the DoAll Company who contributed much technical data and so many photographs. Dave Hoffman of Cincinnati Milacron contributed much technical data and many photographs for the CNC, robotics and milling machine sections. Roy Malcolm and Bob Rutledge of A.J. Forsyth and Co. Ltd. gave me much of their time connecting me with manufacturers, as did Ed Bates of Akhurst Machinery Ltd.

My niece, Leone Gregory, spent many hours typing out the initial manuscript. My wife, Yvonne, has become my best critic, copy editor, and secretary, and has spent many hours reading and typing all of the revisions.

Two school librarians, Bunny Charters, who wrote the first book review for me, and Anne Cossentine, who taught me so much about books and the planning of them, were constantly available to answer my many questions.

Credit must also go to a truly professional group, the Delmar team who put this all together.

I am deeply indebted to my close friend Roy Lewis, Associate Professor at the University of British Columbia, Division of Industrial Education, who encouraged me over the years to write this book.

I would like to thank the following companies for their contributions to this book:

AMERICAN SIP CORPORATION
Elmsford, NY 10523

ARMSTRONG BROS. TOOL CO.
Chicago, IL 60646

ASEA ROBOTICS INC.
New Berlin, WI 53151

AUTOMATIX INC.
Billerica, MA 01821

C-E INDUSTRIAL LASERS, INC.
COMBUSTION ENGINEERING, INC.
Westminster, MD 21157

CLEVELAND TWIST DRILL COMPANY
Cleveland, OH 44101

COMINCO LTD.
Trail, British Columbia, Canada

COOPER TOOL GROUP
Raleigh, NC 27622

DOALL COMPANY
Des Plaines, IL 60016

ERODALL LTD.
Weston, Ontario, Canada M9L 2M8

GATE MACHINERY COMPANY LTD.
London, NW106NY, England

GENERAL ELECTRIC COMPANY
AUTOMATION CONTROLS DEPARTMENT
Charlottesville, VA 22906

HEIDENHAIN CORPORATION
Elk Grove Village, IL 60007

JAPAX INC.
Kawasaki, Kanagawa-Pref 213, Japan

KAR INDUSTRIAL LIMITED
Montreal, Quebec H4S 1C2

KEARNEY & TRECKER CORPORATION
Milwaukee, WI 53214

KENNAMETAL INC.
Latrobe, PA 15650

LEBLOND MAKINO MACHINE TOOL COMPANY
Cincinnati, OH 45208 1386

MAZAK CORPORATION
Florence, KY 41042

METCO INC.
Westbury, NY 11590

MONARCH SIDNEY, DIVISION OF THE
MONARCH MACHINE TOOL COMPANY
Sidney, OH 45365

NEILL TOOLS LTD.
Toronto, Ontario, Canada M8Z 2X3

NEWPORT CORPORATION
Fountain Valley, CA 92728-8020

NORTON COMPANY
Worcester, MA 01606

PRATT BURNERD INTERNATIONAL
West Yorkshire HX15JH, England

SANDVIK CANADA CORPORATION,
DISSTON DIVISION
Vancouver, British Columbia, Canada

SKF & DORMER TOOLS
Mississauga, Ontario, Canada L4W 1S1

SOUTH BEND LATHE INC.
South Bend, IN 46625

STANLEY TOOLS
New Britain, CT 06050

THE DUMORE CORPORATION
Racine, WI 53403

THE L. S. STARRETT COMPANY
Athol, MA 01331

U. S. STEEL CORPORATION
Pittsburgh, PA 15230

CHAPTER ONE

EMPLOYMENT OPPORTUNITIES

OBJECTIVES

After completing this chapter, the student will be able to:

- discuss the importance and extent of machining technology within the manufacturing industry.
- select a particular area of study from the many topics in the field of machining technology.
- plan a personal program of training and retraining that will be required to meet changing opportunities in the various fields of work.

INTRODUCTION

All modern industrialized societies depend upon machine technology. Machine tools, hand tools, power hand tools, and the people trained to use and maintain them are indispensable to our manufacturing industries. These include the fields of transportation, electricity and electronics, household appliances, business machinery, the aerospace industry, and the repair and service industries. All of the people working in these diverse fields are affected either directly or indirectly by the many facets of machine technology.

KEY TERMS

Machine technology	Specialist
Machine tools	Machinist
Hand tools	Tool and die maker
Fitting	Machine operator
Machine work	Instrument mechanic
Fitter	Mechanical technician
Millwright	Mechanical technologist
Benchwork	Apprenticeship
Floorwork	Related fields

MACHINE TRADES

The machine shop trade is very complex in that it is made up of many trades. Essentially, there are two main divisions — fitting and machine work.

Fitting

The fitter, Figure 1–1, works with all of the bench-

1

work tools, including hammers, chisels, files, scrapers, center punches, screwdrivers, wrenches, and so on. The machine tools the fitter uses the most include drill presses, bench grinders, and hydraulic presses. The fitter also makes good use of various power hand tools such as electric or air drills, and electric or air grinders.

The primary job of the fitter is the repair or rebuilding of machinery of all kinds. This job is done in the machine shop. Similar work is done by the

Figure 1–1 A machinist fitter *(Courtesy of Cominco Ltd.)*

millwright who works in the field or in the various plants that make up a company. If a machine breaks down, it is the millwright's job to try to make repairs on the spot. If this fails, the millwright replaces the machine, if possible, and sends the broken machine to the machine shop. There the fitter takes over, dismantling the machine, finding the broken part or parts, making new parts, or seeing to it that they are made, then putting the machine back together again. This class of work is referred to as either benchwork or floorwork. The difference is simply a question of the size of the work: if small, it may be repaired on the bench; if large, it may be repaired on the floor. The same work is done in either case.

It is possible for a fitter to become a specialist. This means that the fitter concentrates on one area of work such as cranes and hoists, or pumps, or compressors. The fitter may also generalize and be responsible for all phases of the fitting field.

Machine Work

Machine work can include all of the drilling operations performed on drill presses and radial drills, and on numerical control (N/C) or computer numerical control (CNC) drilling centers. Lathework, Figure 1–2, includes all of the basic operations done on engine lathes and turret lathes, and on N/C or CNC turning centers. Much flame spraying work (metallizing) is done on engine lathes. There is a great variety of milling machine work performed on vertical milling machines and horizontal milling machines, and on N/C or CNC milling centers, and jig boring on a jig boring machine. Vertical and horizontal boring mills, Figures 1–3 and 1–4, are also used a great deal in larger shops.

As in fitting, a machinist may become a specialist in any one type of machining. A general machinist can be responsible for any or all of the fitting or machining specialties. Tool and die makers, Figure 1–5, considered by many to be the elite of the machine tradesworkers, are responsible for making all tools and dies used in industry. They are required to work to much closer tolerances than the average machinist. A machinist/millwright or an industrial mechanic may be called on to work anywhere in a machine shop, or may be required to go into a plant to make repairs on broken machinery or to install new machinery. An automotive engine machinist may rebuild engines or may build new engines. Similarly, an aero-engine machinist specializes in aircraft engines; a marine machinist trains and works on marine engines (steam, diesel, or gasoline). The aerospace industry has opened up an entirely new, diverse, and exacting field for the machinist. All of the fields listed are highly specialized segments of the machine shop industry.

JOB TITLES

A **machinist** is a highly skilled worker, usually requiring four years of training as a machinist apprentice. If there is a tool and die maker option in the training, the machinist would probably require a fifth year of training.

A **machine operator** is a semiskilled worker, usually requiring one or two years of training. Trained in the operation of a single machine tool, the machine operator is required to set up work and tooling, calculate feeds and speeds for different materials, read and use precision measuring tools, and read sketches and blueprints.

An **instrument mechanic**, Figure 1–6, is highly trained in the basics of machine shop theory and electricity, and electronics. The instrument mechanic generally serves a five-year apprenticeship. He or she is responsible for the construction and calibration of all mechanical and electrical test instruments and gages, etc. The instrument mechanic must learn to install and maintain process control instruments such as flow meters, pressure and temperature recorders, pH meters, and pneumatic and electrical control systems.

There is an increasing need for mechanical technicians and mechanical technologists, especially in the field of computer numerical control (CNC). A **mechanical technician** is someone who is experienced in the techniques of a particular subject such as programming a CNC machine tool or inspecting and

Figure 1–2 Turning a bearing housing *(Courtesy of Cominco Ltd.)*

Figure 1–3 Turning and boring a bronze bushing on a vertical boring mill *(Courtesy of Cominco Ltd.)*

Figure 1-4 Slotting a drive sprocket for a sintering machine with a vertical milling attachment on a horizontal boring mill *(Courtesy of Cominco Ltd.)*

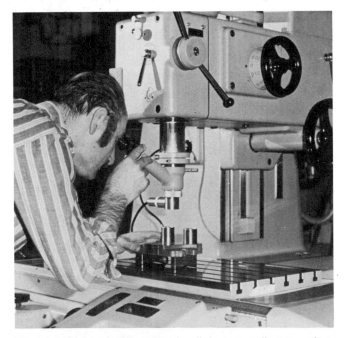

Figure 1-5 A tool and die maker lining up a die set, using a locating microscope on a jig boring machine *(Courtesy of Cominco Ltd. and American SIP Corp.)*

checking parts for accuracy. These jobs do not necessarily require the services of a qualified machinist. A technician usually requires two years of training.

A **mechanical technologist** usually works closely with an engineer in the overall planning of a project. The technologist may also work as a liaison between the engineering office and the machine shop office. He or she requires three to four years of training. The mechanical technologist is essentially a planner, while a technician actually performs a particular procedure.

Any of this type of training can lead in other directions as well. Machinists, for example, can take further schooling to become shop teachers or engineers. There are really no limits to what young people of today can accomplish. All they really need is the will to do it.

SELECTING A CAREER

Machine shops generally fall into two categories — they are either job shops or mass production shops. A mass production shop, for example, may produce only one item, such as an engine or perhaps a carburetor. It may produce these items by the thousands and ship them to wherever they are needed. A job shop, on the other hand, does general machine work and fitting on any type of machinery. The most common type of shop is the job shop.

The training received in secondary schools is designed to help a student make a choice of career, or at least choose a general direction in which to go. If a student decides to become a machinist, for example, he or she should first find out as much as possible about the trade. Talk to machinists, your high school shop teacher, school librarian, and counselor. Ask questions and find out just what a machinist does.

A machinist apprenticeship is the ideal way to get started in the machining industry. This involves a four- to five-year indentured apprenticeship, the length of training time depending upon whether or not a tool and die maker option is offered. If an apprenticeship is not immediately available, look for a pre-apprentice program, which is offered by some schools for a period of six months to one year. This can often lead to an apprenticeship. Always think in terms of enhancing your qualifications with such related courses as mechanical drafting and design, which are valuable no matter which direction you choose. Many prospective employers also prefer to hire someone with first aid and safety training.

APPLYING FOR A JOB

Once a career has been decided upon, a prospective employer must be located and an application must be made for employment. Go to the employment office and ask for an application form. Read the application form thoroughly before attempting to write anything. Fill out the form completely, accurately, and honestly. Take your time and make sure that spelling is correct. Keep in mind that there are more applicants than there are jobs available. As an example, sixty people may be applying for two available apprenticeships, making it very competitive. As many as two-thirds of job applications are thrown out because they are not filled out properly.

A first interview with a prospective employer can be quite nerve wracking, but remember that the interviewer is not your enemy. He or she is simply trying to find out all about you and is trying to decide whether or not you will fit easily into the company work force. At the same time, you are trying to sell your services to the company. During this first look at you, the interviewer will note how you look, how you act, how you move, and how you sit. Both you and your clothes should be clean, neat, and tidy. Sit up straight rather than slouch. Listen closely to what is being asked when you are questioned, and be straightforward in answering. Answer calmly and honestly, and never brag. Show that you are willing to learn, that you are not afraid to work hard, that you are interested and informed about that company, and that you intend to conform to the company's way of doing things. During a first interview, it is poor policy to ask about money and holidays. Finally, show your appreciation and thank the interviewer for the interview. A com-

pany will spend a great deal of money in the training of an apprentice and will look for the best possible candidate. It will take one or two years of training before an apprentice actually starts to earn money for the company. If you are unsuccessful in your first application for an apprenticeship, you might consider trying to get on the company's work force in any capacity. Once you establish a good work record with that company, your chances are greatly increased of getting an apprenticeship the next time one becomes available.

It should be evident that there are many, varied job opportunities in the machine technology field and its related fields. There is no substitute for good training especially now when technology is changing so rapidly. It is always desirable to have a solid base of fundamental knowledge and skills. In addition to this, it is always necessary to rethink your position and to retrain to keep up-to-date in your respective field of

Figure 1–6 An instrument mechanic making adjustments to a pressure transmitter *(Courtesy of Cominco Ltd.)*

work, or even to change your field of work. When you are looking for work, it is absolutely necessary that you have something practical to offer to a prospective employer. Always think of training and retraining as a stepping stone to something better.

REVIEW QUESTIONS

1. Name the main divisions in the machine shop trade.
2. What is the primary job of a fitter?
3. What does machine work include?
4. What is the difference between a mechanical technician and a mechanical technologist?
5. Name two types of machine shop.

CHAPTER TWO

SAFETY

OBJECTIVES

This chapter will prepare the student to:
- practice safety in all phases of machining technology.
- make decisions regarding personal safety while on the job.
- make decisions regarding the safety of other personnel while on the job.

INTRODUCTION

Safety should be uppermost in the minds of all machine technology workers. Unfortunately, safety guidelines are not always put into practice, and people must be constantly reminded about safety. It should become a state of mind so that reactions become automatic in any dangerous situation.

Thought, discipline, and cooperation are required to ensure a safe environment for you and those around you. All accidents are a result of human error. In entering any hazardous situation, your first thought should be, What is the safest way to do this? There are only two possibilities with safety: you are either safe, or you are not safe. If you are not safe, you can be either hurt or killed. If you are to survive, you should practice safety every day of your life.

In setting up safety rules for the machine shop, some rules will be repeated, either for emphasis, or because they apply in more than one place, or because they apply on more than one machine tool.

KEY TERMS

Hazardous situation	Good housekeeping
Safety rules	Horseplay
Safe environment	Eye protection

GENERAL RULES

1. Do not operate any machine before you have received instructions on the correct and safe procedures from your instructor. It is not wise to ask another student for proper procedures.
2. Permission should be obtained to operate machinery, at the instructor's discretion.
3. All machines should have effective guards that should always be in place when machines are operating.
4. Always stop the machine and switch off the power before making any adjustments.

5. Never use your hands or any part of your body to brake a machine to a stop.

6. Only one operator at a time should control a machine.

7. Always keep your hands and body away from all moving parts.

8. Always wear specified eye protection. This is mandatory in all phases of machine shop work.

9. Do not wear loose clothing; roll up sleeves; remove ties, rings, watches, etc.; tie long hair out of the way.

10. Keep tools and other materials in their proper places. Keep the floor clear of chips and objects that could cause a worker to trip. Immediately remove oil or any other spills from the floor. Good housekeeping is always good safety practice.

11. If the instructor leaves the shop for any reason, the power should be shut off.

12. Horseplay, practical jokes, and running should be prohibited in shop areas.

13. Setscrews should be the flush or recessed type.

14. When lifting heavy objects, follow safe lifting procedures. Get help if necessary.

15. Cuts, scratches, bruises, burns, and punctures, no matter how slight, should be reported to the instructor without delay. First aid should be provided immediately to prevent the possibility of infection.

16. Contact lenses should not be worn in any shop or industrial area.

17. In all cases, clean machinery thoroughly when you have finished work.

18. If you have any questions, ask the instructor, not another student.

EYE PROTECTION

Eye protection is important enough to be in a special category of its own.

1. Proper eye protection should be worn in a machine shop under all conditions.

2. In the presence of molten metals, molten case-hardening salts, or sparks of any kind, a proper face shield should be worn.

3. For any operation involving drilling, lathe-work, milling, sawing, grinding, any other machine tool operation, or benchwork, safety glasses must be worn. At times, a proper face shield will be required as well.

4. For heat treating operations or furnace work, a proper face shield must be worn.

5. Gas welding, arc welding, or any other form of welding requires the use of proper welding goggles or face shield and the correct shade of lens in addition to clear glasses for chipping welds.

6. When using any corrosive or explosive materials, enclosed goggles or a proper face shield should be worn.

7. For hygienic reasons, it is best to have your own personal eye protective equipment.

8. If circumstances require sharing goggles or safety glasses, the glasses should be thoroughly cleaned and disinfected after each use for the next person wearing them.

9. Contact lenses should not be worn in the machine shop for the following reasons:
   (a.) Some gases, vapors, and other materials that will harm the eyes can be absorbed by contact lenses.
   (b.) When trapped under contact lenses, dusts and other such materials will harm the eyes or will cause a distraction that could expose the worker to other injury.
   (c.) Electric arcs can cause the contact lenses to stick to the eyes, causing serious injury when trying to remove the lenses.

MELTING AND
HEAT TREATING FURNACES

Furnaces used in conjunction with machine shops can be of two kinds: melting furnaces and heat treating furnaces. Melting furnaces are used mainly for melting bearing metals such as babbitts and bronzes with temperatures varying from 1000 degrees F (537 degrees C) to 2500 degrees F (1370 degrees C). Heat treating furnaces are used mainly for such work as annealing, hardening, and case-hardening, involving temperatures varying from about 1000 degrees F (537 degrees C) to 2500 degrees F (1370 degrees C). Tempering operations, which may or may not involve the use of tempering compounds, involve the use of temperatures ranging from about 350 degrees F (175 degrees C) to as high as 1400 degrees F (760 degrees C). In all cases, you are dealing with hot to molten materials and there are many dangers. The following safety rules should be observed:

1. Permission must be obtained to use the furnaces, at the discretion of the instructor.
2. It is best to start furnaces under the supervision of the instructor. A furnace should be kept under observation while it is operating.
3. Proper and adequate ventilation (exhaust fans) is mandatory, especially immediately above each furnace.
4. Proper protective clothing (gloves, jackets, pants, shoe guards, etc.) is mandatory when working with furnaces and hot or molten materials.
5. Safety glasses should be worn under special metal-screened face shields when dealing with molten materials.
6. Pouring rings or tongs should securely hold and clamp crucibles for pouring.
7. Avoid spilling molten materials on concrete. Concrete can explode when overheated and flying pieces can cause cuts, bruises, and eye injuries. It is best to use cast-iron tables or sand beds for pouring.
8. Avoid breathing fumes from case-hardening pots or tempering pots. Forms of cyanide salts are often used as molten baths. The fumes from them are poisonous.
9. Mark hot items "HOT" with chalk and keep them to one side out of the way.
10. If quenching oils become overheated, they can flare up and become a potential fire hazard.
11. Molten metals should not be poured at temperatures higher than their recommended pouring temperatures. This is particularly true with aluminums, babbitts, and bronzes. Blow holes and cavities can be caused by overheating.
12. Always use the recommended temperatures for heat treatment baths.
13. Moisture and molten materials are an explosive mixture, Figure 2–1. One drop of water, or even unseen moisture, in molten material can explode violently. Be very, very careful to avoid this situation.
14. Preheat any work that is to be lowered into a molten salt bath. Preheating will eliminate any moisture.
15. Preheat containers that are to receive molten materials. Preheating will eliminate any moisture.
16. Treat pickling acids with care. Lower work very slowly into pickling acids. Drain and rinse your work thoroughly after removal. Rinsing should be done in water. Any part of the body that comes into contact with any acid should be washed in water immediately.
17. Rubber gloves should be worn when working with acids.
18. On completion of the work, shut off furnaces, put away equipment, cover all bath and quenching containers, and clean up the area.
19. Materials such as degassing tablets or powders should be stored in airtight containers to keep them dry.

Pay particular attention to numbers 13, 14, 15, 16, and 19.

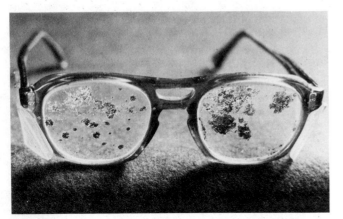

Figure 2–1 These safety glasses and the machinist's eyes survived an explosion of molten bearing metal *(Courtesy of Cominco Ltd.)*

REVIEW QUESTIONS

1. What does a "safe environment" mean?
2. What is a hazardous situation?
3. What are two possibilities with safety?
4. In machine shop practice, what is the best rule regarding eye protection?
5. What is the rule regarding moisture and molten materials?

CHAPTER THREE

PRECISION MEASUREMENT

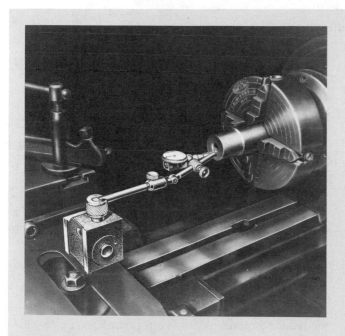

In recent years, especially with the development of space-age technology, the role of machine tools and precision measuring tools is increasing in importance. Machine tools are capable of cutting to much closer tolerances than ever before. Surfaces are flatter and smoother and cutting speeds are faster. Measuring tools have been developed that measure easily and quickly, and with accuracies to millionths of an inch.

Manufacturers produce an endless range of goods that are used throughout the world. Complex manufactured items, such as automobiles, electronic equipment, and airplanes, need parts which are usually manufactured in another part of the country, often in a foreign country. These parts are transported to the assembly site where the goods are assembled. It is essential that the parts be made to such close tolerances that they will fit properly, regardless of where they were made. In other words, the parts must be interchangeable. Precision measurement has a key role to play in the manufacturing process.

UNIT 3-1

INTRODUCTION TO PRECISION MEASUREMENT

OBJECTIVES

After completing this unit, the student will be able to:

- select the proper measuring tool for the job being done.
- explain the concept of accuracy.
- care for measuring tools properly.
- describe the measuring systems in use today.

KEY TERMS

Measurement
Degree of accuracy
Relative
Linear measurement
Pre-set
English measuring system

U. S. Customary measuring system
Metric measuring system
Transition stage
Conversion
Basic dimension

GENERAL

Most machine shop operations require measurements of some kind. These measurements should be made both easily and quickly, and should be within the degree of accuracy required for the work being done. Cutting off stock with a power saw or by hand, for example, demands only the degree of accuracy offered by measurement using a rule (scale) or steel tape. Using a micrometer in this case would be poor prac-

tice. Similarly, when rough turning a piece of stock to a rough diameter, a pair of outside calipers would be accurate enough for measurement and would be much quicker than using a micrometer.

Accuracy is relative to the work being done. It is important to learn from the beginning to use the right measuring tool for the particular work at hand. Just as with poor cutting practices, poor measuring practices can waste valuable time.

Steel rules, calipers, vernier calipers, and micrometers are the measurement tools most commonly needed for general machine shop work. Accuracy for average machine shop work is usually within 0.001 in (0.02 mm). For average tool and die making work, accuracy is usually within 0.0002 in (0.005 mm).

Linear measurement is done on a flat plane or surface by various tools, including rules, steel tapes, dividers, and trammels. When using rules, steel tapes, and slide calipers, sizes can be read directly from graduated scales. Calipers, micrometers, and vernier calipers have legs that are in direct contact with the work. These tools are used to measure diameter or thickness. Sizes can be read directly when micrometers and verniers are used. Calipers are either preset by a rule or micrometer and then compared to the work, or they are set to the work and then checked by rule or micrometer.

CARE OF MEASURING TOOLS

Good measuring tools will last indefinitely if properly cared for. When abused, however, they quickly lose their accuracy. To maintain accuracy, proper storage is mandatory. Tools should always be wiped clean before being put away. Whenever possible, tools should be stored in individual cases. Machinists use tool boxes with narrow, felt-lined drawers. The narrow drawers make it impossible to pile tools on top of one another, which could interfere with their accuracy. The felt lining absorbs moisture, which helps to prevent tools from rusting. When in use, measuring tools should never be placed on or under other objects. It is best to lay them on a firm, flat surface, such as a nearby bench top or stand, where they are readily accessible for use.

MEASURING SYSTEMS

This text deals with both the English and metric measuring systems. The English system is referred to in many schools as the U.S. Customary system. At this time, we are in a transition stage of conversion, somewhere between the two systems. If we do make a total conversion to the metric system, it will not happen quickly and it will not be an easy conversion. Whatever takes place in the years to come, it is the intent of this text to deal with the measurement systems the machinist and student must work with today. You should be prepared to work in both systems, probably for many years.

In the machine shop and the drafting room, the English (U.S. Customary) system deals mainly with the measurement of length in inches and parts of inches. Fractions of an inch are used mainly in halves, quarters, eighths, sixteenths, thirty-seconds, and sixty-fourths of an inch. In decimals, thousandths and ten-thousandths of an inch are the most common

measurements. In the manufacture of precision measuring tools, companies often work in millionths of an inch.

The practical application of the metric system in both the machine shop and the drafting room involves the measurement of length. The length units most used are the meter, the centimeter, and the millimeter. The decimeter is not commonly used.

All machine shop work starts with a drawing or sketch. In mechanical drawings using metric measure, all dimensions are generally in millimeters, no matter how large the dimension. This practice is adopted to avoid misplacing the decimal point or misreading dimensions if other units are used. Drawings in the metric system are not scaled to 1/2, 1/4, or 1/8 as is done in the English (U.S. Customary) system. If they are not drawn full size, they are generally scaled to 1/2.5, 1/5, 1/10, 1/20, 1/100, 1/500, or 1/1000 of full size.

In everyday practice, the machinist or tool and die maker works with metric dimensions in millimeters and fractions of millimeters. To give an idea of the sizes involved, we will compare a few sizes in both measuring systems.

One millimeter is 0.039 in, slightly larger than 1/32 in (0.031 in). Precision dimensions are usually given in hundredths of a millimeter, e.g., 0.02 mm. If you consider that 0.01 mm is equal to 0.0004 in, then you can see that this is close enough for average work in a machine shop.

The basic dimension of the metric system is the meter (m). The multiples and parts of a meter are expressed by adding a prefix representing steps of 1000. As an example, 1000 meters = 1 kilometer (km), one-thousandth of a meter = 1 millimeter (mm). Table 3–1 shows a comparison of metric and English (U.S. Customary) sizes.

When working in more than one measuring system, always keep the following in mind. If work is to be done in metric units, think only in terms of metric. If work is to be done in English (U.S. Customary) units, think only in terms of English (U.S. Customary) units. Do not try to equate the two systems or to think in terms of metric/English (U.S. Customary) conversion unless there is no other way.

Table 3–1 Comparison of English (U.S. Customary) and Metric Sizes

METRIC	METRIC	ENGLISH (U.S. Customary)
One millimeter	1.0 mm	0.03937″
One half millimeter	0.5 mm	0.01969″
One one-hundredth of a millimeter	0.01 mm	0.00039″
Two one-hundredths of a millimeter	0.02 mm	0.00079″
One one-thousandth of a millimeter	0.001 mm	0.00004″
Two one-thousandths of a millimeter	0.002 mm	0.00008″

ENGLISH (U.S. CUSTOMARY)	METRIC
One inch	25.4 mm
One tenth of an inch	2.54 mm
One one-hundredth of an inch	0.254 mm
One one-thousandth of an inch	0.0254 mm
One sixty-fourth of an inch	0.381 mm
One thirty-second of an inch	0.787 mm
One sixteenth of an inch	1.5875 mm
One eighth of an inch	3.175 mm

REVIEW QUESTIONS

1. Accuracy in measurement is relative to the work being done. Explain this statement.
2. List the most commonly used measuring tools for the average run of work.
3. What accuracy is required for average machine shop work?
4. Practical applications in the machine shop involve the use of what metric measurement?
5. What is the basic dimension in the metric system?
6. What accuracy is required for average tool and die making work?
7. What measuring tools are used in linear measurement?
8. What measuring tools are used in measuring diameters on a lathe?

UNIT 3-2

USING RULES (SCALES)

OBJECTIVES

After completing this unit, the student will be able to:

- use the steel rule or scale properly.
- identify inch and metric scales.
- measure work to accuracies of 1/64 in (0.39 mm).

KEY TERMS

Spring tempered rule   Hook rule
Scale

RULES

The most common measuring tool used by machinists is the spring tempered rule, Figure 3–1A. Because of the close connection between drafting and the machine shop, rules are commonly called scales by machinists. This happens because of a convention in drafting where a rule is considered to be a tool used for drawing lines, and a scale is a tool used to measure. The most commonly used rules are the 6-in and 12-in sizes. The most popular combination of divisions on rules is 1/8 in, 1/16 in, 1/32 in, and 1/64 in with an extra 1/2-in scale on each end marked in increments of 1/32 in. Another combination of divisions is tenths and hundredths, which is closely related to measurements in thousandths of an inch as derived from micrometer and vernier scales.

Coming into more common use are the 150-mm and 300-mm rules. One example of a metric rule has two edges in millimeters and two edges in half millimeters, Figure 3–1B. There are many other combinations of divisions in both systems of measure.

A variation of the rule, much more commonly used in a machine shop, is the hook rule. One model has a single, reversible hook, Figure 3–1C. The screw is an eccentric, easily removed and replaced. The hook on the double hook rule, Figure 3–1D, can be shortened or extended on either side, or it can be removed entirely. Again, the 6-in and 12-in or the 150-mm and 300-mm sizes are the most popular with machinists.

Figure 3–1A A spring tempered rule *(Courtesy of The L. S. Starrett Company)*

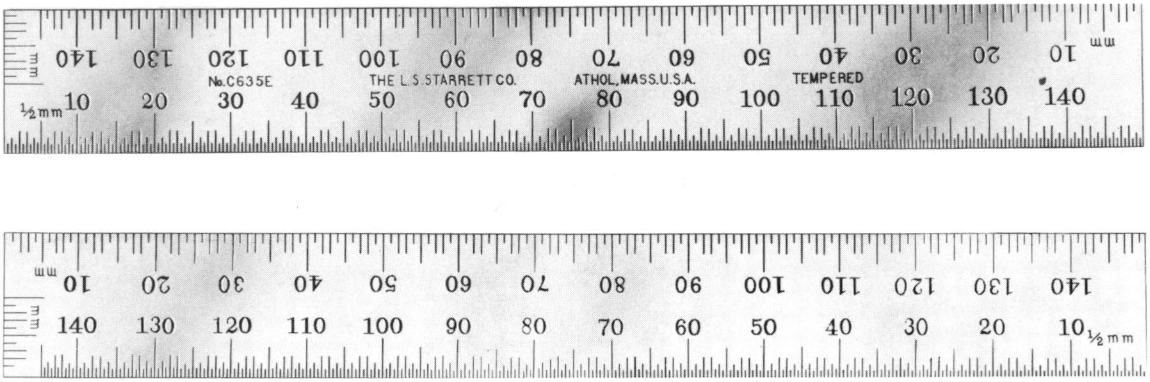

Figure 3–1B A metric rule *(Courtesy of The L. S. Starrett Company)*

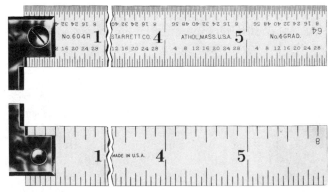

Figure 3–1C A single hook rule *(Courtesy of The L. S. Starrett Company)*

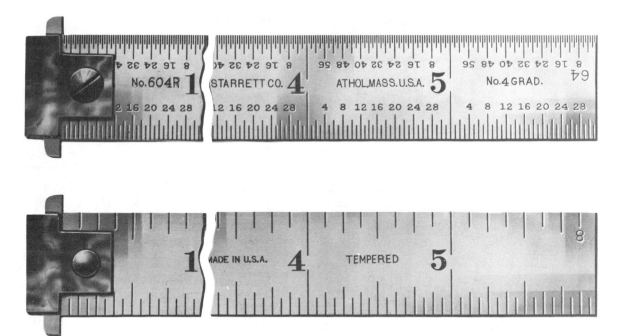

Figure 3–1D A double hook rule *(Courtesy of The L. S. Starrett Company)*

REVIEW QUESTIONS

1. What are the most common units on a 6-in or 12-in machinist scale?
2. What are the most popular sizes of metric rules coming into common use?
3. What forms of the hook rule are used in the machine shop?
4. What are the most popular sizes of hook rules (inch and millimeter)?

UNIT 3-3

USING CALIPERS

OBJECTIVES

After completing this unit, the student will be able to:

- describe the various types of common calipers in use.
- hold the common types of calipers properly.
- practice measuring to accuracies of about 0.005 in (0.13 mm) using the common types of calipers.

KEY TERMS

Spring calipers	Firm joint calipers
Transferring sizes	Lock joint calipers
Toolmaker's calipers	Lock joint transfer calipers

CALIPERS

Spring calipers, Figures 3–2A and 3–2B, are very common measuring tools for machinists. They have either a solid or quick adjusting nut. They are usually used in conjunction with a rule or micrometer, from which the actual measurement is read.

When measuring work with calipers, the best method is to squeeze the legs together by hand to an approximate size. Then the nut is set as a fine adjustment only. This helps to preserve the fine thread of both the nut and screw. The proper use of calipers takes a great deal of practice to develop a fine sense of touch in the fingertips. Figures 3–3A and 3–3B illustrate one method of setting both outside and inside calipers. Figures 3–3C and 3–3D show techniques for measuring with calipers. Figure 3–3E illustrates the very important technique of transferring

sizes from one caliper to another. These are techniques that should be mastered. With practice, it is quite possible to learn to caliper work to within 0.001 in (0.025 mm). For many machinists, it is a matter of pride to be able to set a caliper on a rule to within 0.001 in (0.025 mm). On occasion, you may see someone caliper a moving piece of work on a lathe. This is a poor and dangerous practice and should not be done. Modifications of spring calipers are the toolmaker's calipers, Figures 3–4A and 3–4B. They are much finer and lighter in construction and much more sensitive in use.

Firm joint calipers, Figures 3–5A and 3–5B, have no fine setting. They are rough set by hand and any fine setting is done by lightly tapping a leg against something solid in order to move it slightly. Lock joint calipers, Figures 3–6A and 3–6B, can be set, locked, and then finely adjusted by using the fine adjustment screw.

Lock joint transfer calipers, Figures 3–7A and 3–7B, can be set and locked when measuring recesses, Figures 3–8 and 3–9. In order to remove the calipers from the recess for measuring, the leg is unlocked from the transfer arm. The calipers are then removed from the recess and the leg is returned to its original position and relocked to the transfer arm. The measurement can then be taken.

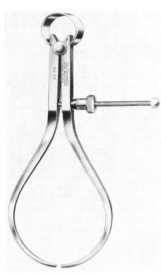

Figure 3–2A Outside spring calipers *(Courtesy of The L. S. Starrett Company)*

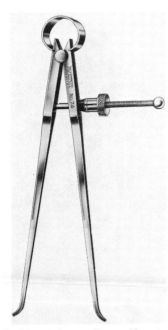

Figure 3–2B Inside spring calipers *(Courtesy of The L. S. Starrett Company)*

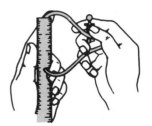

Figure 3–3A Setting an outside spring caliper with a rule *(Courtesy of South Bend Lathe Inc.)*

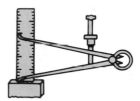

Figure 3–3B Setting an inside spring caliper with a rule *(Courtesy of South Bend Lathe Inc.)*

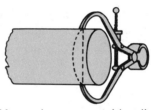

Figure 3–3C Measuring an outside diameter with an outside spring caliper *(Courtesy of South Bend Lathe Inc.)*

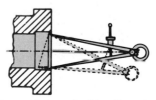

Figure 3–3D Measuring an inside diameter with an inside spring caliper *(Courtesy of South Bend Lathe Inc.)*

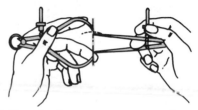

Figure 3–3E Transferring a measurement from an outside caliper to an inside caliper *(Courtesy of South Bend Lathe Inc.)*

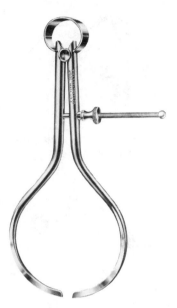

Figure 3–4A Toolmaker's outside calipers *(Courtesy of The L. S. Starrett Company)*

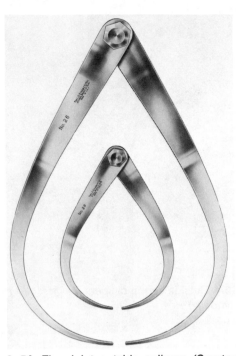

Figure 3–5A Firm joint outside calipers *(Courtesy of The L. S. Starrett Company)*

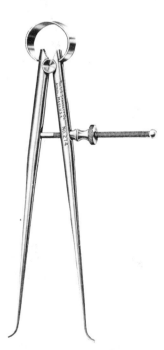

Figure 3–4B Toolmaker's inside calipers *(Courtesy of The L. S. Starrett Company)*

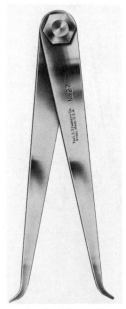

Figure 3–5B Firm joint inside calipers *(Courtesy of The L. S. Starrett Company)*

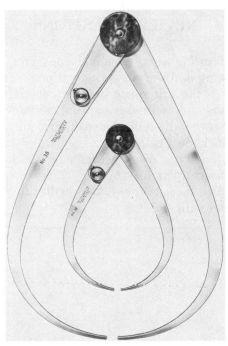

Figure 3–6A Lock joint outside calipers *(Courtesy of The L. S. Starrett Company)*

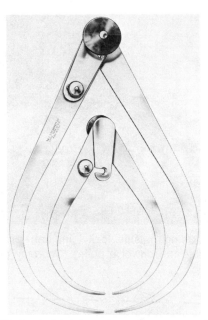

Figure 3–7A Lock joint transfer calipers — outside *(Courtesy of The L. S. Starrett Company)*

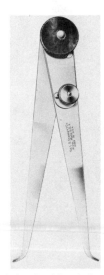

Figure 3–6B Lock joint inside calipers *(Courtesy of The L. S. Starrett Company)*

Figure 3–7B Lock joint transfer calipers — inside *(Courtesy of The L. S. Starrett Company)*

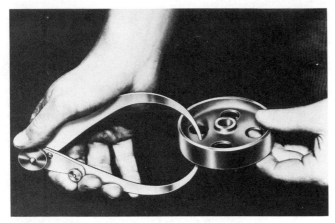

Figure 3-8 Using outside lock joint transfer calipers to measure inside the rim of a pulley *(Courtesy of The L. S. Starrett Company)*

REVIEW QUESTIONS

1. In measuring work, what is the best procedure for adjusting and setting spring calipers?
2. What is the difference between standard spring calipers and toolmaker's spring calipers?
3. What is the difference between firm joint calipers and lock joint calipers?
4. For what type of measurements are lock joint transfer calipers used?
5. How is a measurement taken with the lock joint transfer calipers?

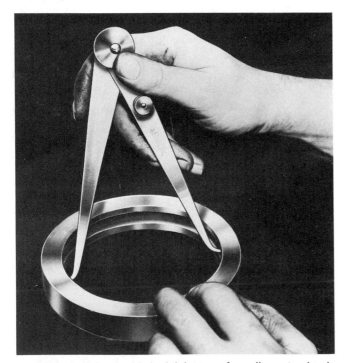

Figure 3-9 Using inside lock joint transfer calipers to check the depth of an inside groove. *(Courtesy of The L. S. Starrett Company)*

UNIT 3-4

MICROMETERS AND VERNIERS

OBJECTIVES

After completing this unit, the student will be able to:

- list the parts of micrometers and verniers.
- hold micrometers and verniers properly.
- measure using both micrometers and verniers.
- measure to accuracies of 0.001 in (0.02 mm) and 0.0001 in (0.002 mm).

KEY TERMS

Micrometer caliper	Vernier scale
Ratchet stop	Micrometer depth gage
Friction thimble	Test rod
Spindle	Vernier caliper
Barrel	Beam
Thou	Vernier plate
Pitch of the thread	Gear tooth vernier

THE MICROMETER CALIPER

A micrometer caliper, more commonly called a mic (pronounced mike), is the most commonly used precision tool for measuring diameters or thicknesses of material. Accuracies of 0.0001 in and 0.002 mm can be achieved.

Part names for micrometers are the same in both measuring systems. The two styles of micrometers illustrated have a ratchet stop, Figure 3–10, or a friction thimble, Figure 3–11.

Reading a Micrometer

The key to reading all micrometers is in the pitch of the thread of the micrometer spindle, Figure 3–12. The spindle is made from a special high carbon steel, hardened, tempered, and precision ground.

On the inch micrometer, the pitch of the screw thread is 1/40 in, that is, there are 40 threads in one inch. In one revolution, the spindle will move 1/40 in which equals 0.025 in or 25 thousandths of an inch. Some machinists use the term "thou" rather than "thousandth." It is simpler to say 25 thou.

- One revolution of the spindle = 1/40 = 0.025 in or 25 thou.
- Two revolutions of the spindle = 2/40 = 0.050 in or 50 thou.
- Three revolutions of the spindle = 3/40 = 0.075 in or 75 thou.
- Four revolutions of the spindle = 4/40 = 0.100 in or 100 thou.
- The barrel has 40 divisions; each one = 0.025 in or 25 thou.

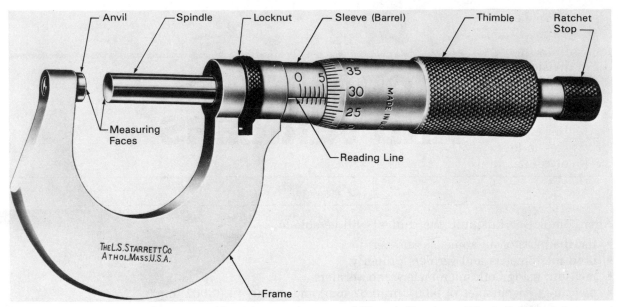

Figure 3–10 A 0–25 mm micrometer with ratchet stop (accuracy 0.01 mm) *(Courtesy of The L. S. Starrett Company)*

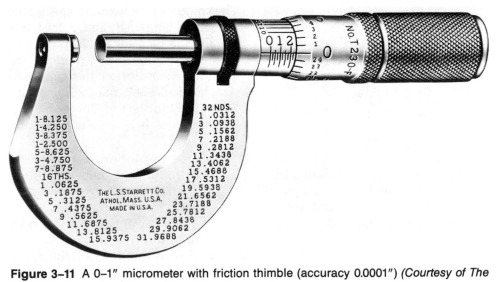

Figure 3–11 A 0–1″ micrometer with friction thimble (accuracy 0.0001″) *(Courtesy of The L. S. Starrett Company)*

- The thimble has 25 divisions; each one = 0.001 in or 1 thou.

To read the micrometer correctly, count the number of divisions you can see on the barrel. Add to that the number of the division on the thimble that lines up with the reading line. That is your reading. Figure 3–13 illustrates the scales of the micrometer. Sample readings are shown in Figure 3–14.

For reading to four decimal places, there is an extra scale, or vernier, on the barrel with lines run-ning parallel to the reading line. The ten divisions of the vernier are numbered 0, 1, 2, 3, 4, 5, 6, 7, 8, 9, and 0, Figure 3–15. When one of these numbers is added to your total micrometer reading, only one of the numbers will match a mark on the thimble, except in the case of a zero reading. In this case, the zeros at each end of the vernier scale will match a thimble mark. Sample readings can be seen in Figure 3–16.

In the metric system, the pitch of the screw thread, or the distance from point to point of the screw thread,

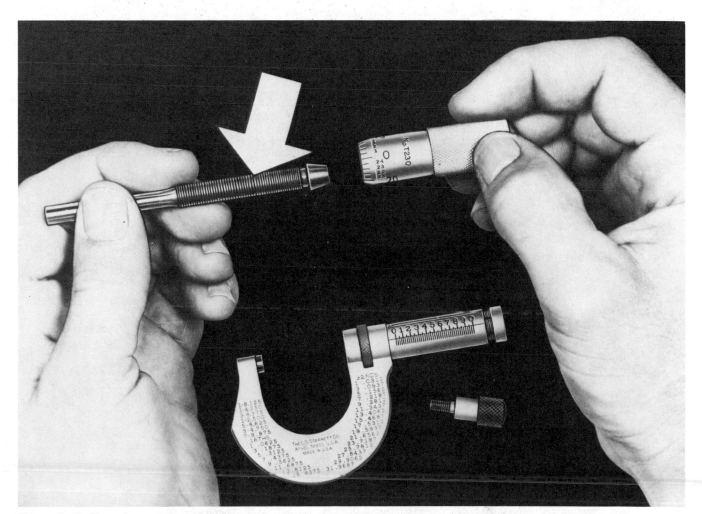

Figure 3–12 The micrometer spindle screw thread *(Courtesy of The L. S. Starrett Company)*

is one-half millimeter (0.5 mm). In other words, one revolution of the thimble will move the spindle 0.5 mm. Two revolutions of the thimble will move the spindle 1.0 mm.

Figure 3–17 illustrates the scales found on the sleeve (barrel) of the micrometer and on the thimble, with an accuracy of 0.01 mm. In the trade, the term *barrel* is used as commonly as the term *sleeve*. Sample readings are shown in Figure 3–18.

To obtain a more accurate reading, an extra scale, or vernier, is added to the micrometer barrel to give an accuracy of 0.002 mm. The vernier scale, Figure 3–19, has five divisions lying parallel to the reading line. Each division is 1/5 of a thimble division, or 1/5 × 0.01 = 0.002 mm. When one of these numbered lines is added to the total micrometer reading, only one numbered line will match a mark on the thimble, except in the case of a zero reading. In this case, the zero at each end of the vernier scale will match a thimble mark. Sample readings can be seen in Figure 3–20. The parts of a micrometer depth gage are illustrated in Figure 3–21 (page 30).

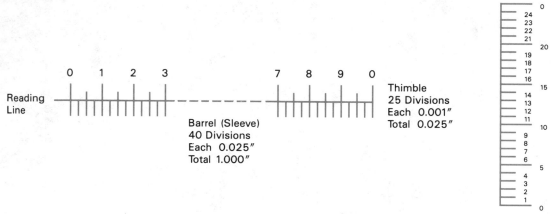

Figure 3–13 The scales of the inch micrometer (accuracy 0.001″)

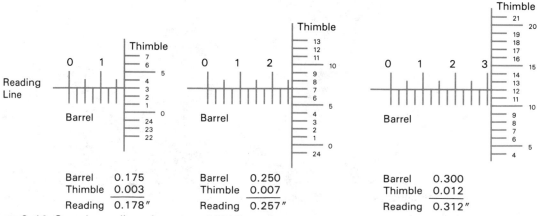

Barrel 0.175	Barrel 0.250	Barrel 0.300
Thimble 0.003	Thimble 0.007	Thimble 0.012
Reading 0.178″	Reading 0.257″	Reading 0.312″

Figure 3–14 Sample readings (accuracy 0.001″)

Inside Micrometers

Inside micrometers, Figure 3–22 (page 30), are used for measuring a variety of inside measurements, Figure 3–23 (page 31), or large hole diameters, Figure 3–24 (page 32). Inside micrometers are read exactly the same as outside micrometers, except that some thimbles will move only 1/2 in. A 1/2-in gage or spacer is added to the set to be positioned between one of the rods and the thimble. In this case, 0.500 in must be added to the micrometer reading to obtain the total reading. Other thimbles will move through 1 in.

In metric sizes, some thimbles will advance only 13 mm. In this case, a 12-mm gage or spacer is added to the set to be positioned between one of the rods and the thimble. The 12 mm must be added to the micrometer reading to obtain the total reading. Other thimbles will advance through 25 mm.

Micrometers are made in many styles such as outside micrometers, inside micrometers, digital

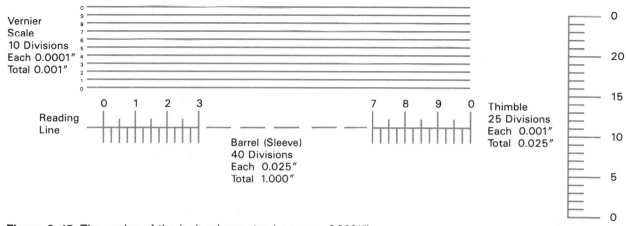

Figure 3–15 The scales of the inch micrometer (accuracy 0.0001″)

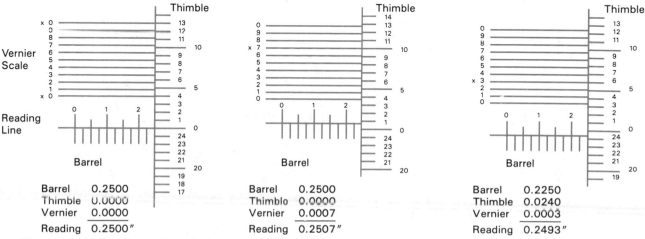

	Barrel	0.2500		Barrel	0.2500		Barrel	0.2250
	Thimble	0.0000		Thimble	0.0000		Thimble	0.0240
	Vernier	0.0000		Vernier	0.0007		Vernier	0.0003
	Reading	0.2500″		Reading	0.2507″		Reading	0.2493″

Figure 3–16 Sample readings (accuracy 0.0001″)

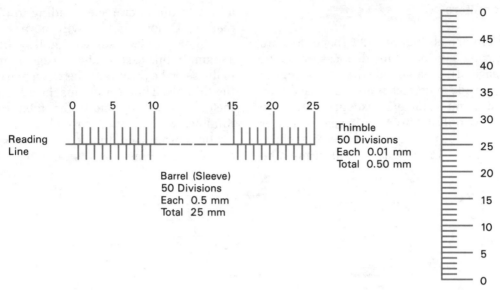

Figure 3–17 The scales of the metric micrometer (accuracy 0.01 mm)

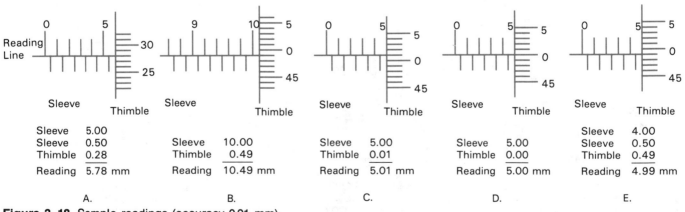

Figure 3–18 Sample readings (accuracy 0.01 mm)

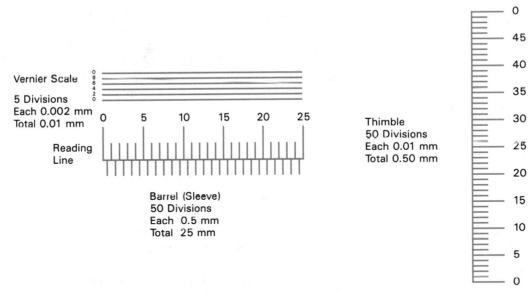

Vernier Scale

5 Divisions
Each 0.002 mm
Total 0.01 mm

Reading
Line

Barrel (Sleeve)
50 Divisions
Each 0.5 mm
Total 25 mm

Thimble
50 Divisions
Each 0.01 mm
Total 0.50 mm

Figure 3–19 The scales of the metric micrometer (accuracy 0.002 mm)

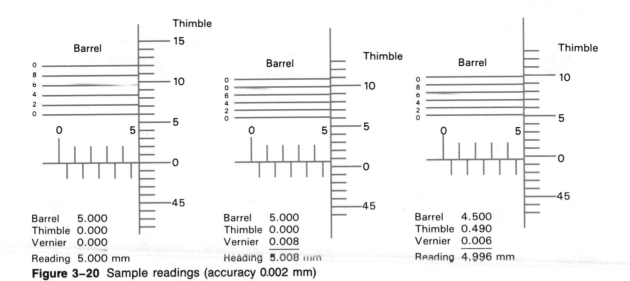

Barrel	5.000
Thimble	0.000
Vernier	0.000
Reading	5.000 mm

Barrel	5.000
Thimble	0.000
Vernier	0.008
Reading	5.008 mm

Barrel	4.500
Thimble	0.490
Vernier	0.006
Reading	4.996 mm

Figure 3–20 Sample readings (accuracy 0.002 mm)

micrometers, blade-type micrometers, screw thread micrometers, and so on. Sizes start at zero and can go up to 120 in on special order. Outside micrometers are made in increments of 1 in, or of 25 mm (in the case of metric sizes). A micrometer is named after its maximum dimension. A 1-in micrometer measures from

0–1 in, a 2-in micrometer measures from 1–2 in, and so on. A 25-mm micrometer measures from 0–25 mm, while a 50-mm micrometer measures from 25–50 mm.

A reading of 0.125 in on a 1-in micrometer is written 0.125 in, but on a 2-in micrometer the same reading is 1.125 in. In metric measurement, a reading of 10.5 mm on a 25-mm micrometer is written as 10.5 mm, but on a 50-mm micrometer, 25 mm is added to make the reading 35.5 mm.

Care of Micrometers

Micrometers should always be checked for accuracy. Never assume that a micrometer is accurate. Check the thimble for play or looseness, and tighten the adjusting nut, Figure 3–25A, if necessary. Also check the micrometer against standards or special test rods. These are made in increments of 1 in starting at a length of 1 in, or in increments of 25 mm starting at 25 mm in length. If it is necessary, adjust the sleeve, Figure 3–25B, to exactly zero.

Always grip micrometers firmly but lightly with your fingertips. Squeezing too tightly decreases the sensitivity in your fingertips. Never spin the micrometer while holding the thimble. This will cause uneven wear on the thread. For fast adjusting, hold the frame and roll the thimble on the hand. Never leave the micrometer tightly closed. An increase in temperature could cause the micrometer to seize. A beginning student should use the ratchet or friction thimble

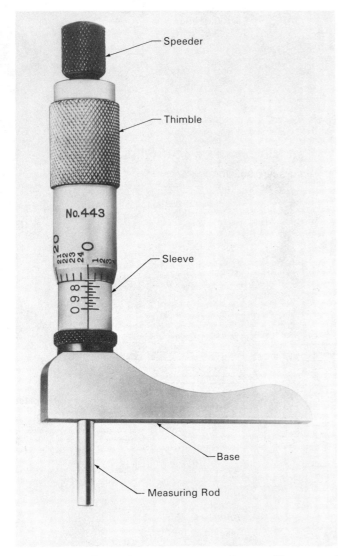

Figure 3–21 Parts of a micrometer depth gage *(Courtesy of The L. S. Starrett Company)*

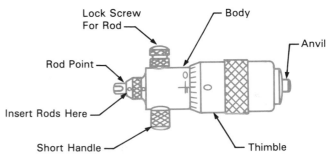

Figure 3–22 An inside micrometer head *(Courtesy of The L. S. Starrett Company)*

when measuring. Remember, a micrometer is a precision instrument, not a G-clamp. Do not measure the work when it is hot. Because of the heat, the work will expand to oversize. The micrometer will also expand and a wrong reading will result. When the work cools, it will shrink and will then be undersize.

Holding and Using Micrometers

The same delicate measuring techniques practiced with calipers should be practiced with micrometers, Figure 3–26A through D. Depending on the size of the micrometer, it can be held in one hand or

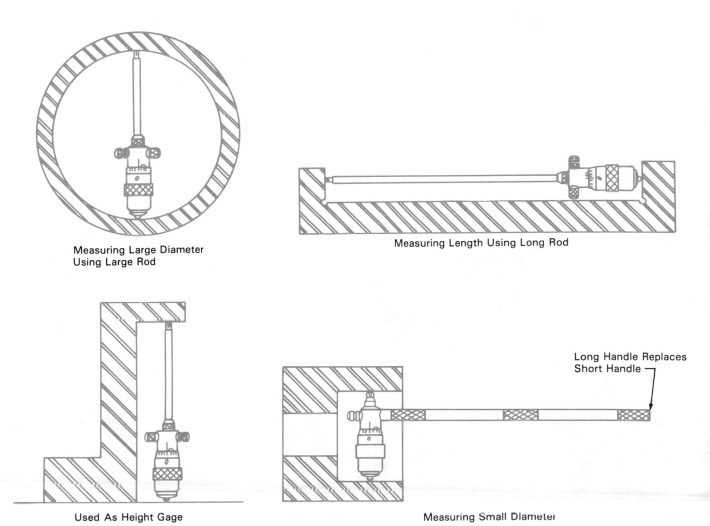

Measuring Large Diameter
Using Large Rod

Measuring Length Using Long Rod

Used As Height Gage

Long Handle Replaces
Short Handle

Measuring Small Diameter

Figure 3–23 Examples of inside measurements *(Courtesy of The L. S. Starrett Company)*

Figure 3–24 A large inside diameter being measured *(Courtesy of The L. S. Starrett Company)*

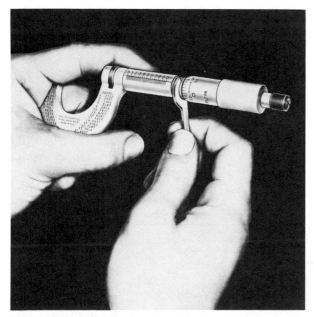

Figure 3–25A Tightening the adjusting nut *(Courtesy of The L. S. Starrett Company)*

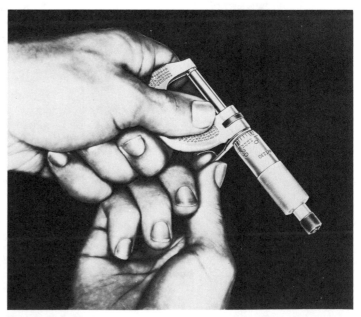

Figure 3–25B Adjusting the sleeve (barrel) *(Courtesy of The L. S. Starrett Company)*

in two, and either vertically or horizontally. Note that the 12-in micrometer, Figure 3–27, is held horizontally instead of vertically. This eliminates any tendency of the micrometer to sag. The same is done with the 24 to 30-in micrometer, Figure 3–28. This same technique should be used with large calipers. Whatever the size of outside micrometers, the thimble will travel only 1 in or 25 mm. Larger outside micrometers, Figure

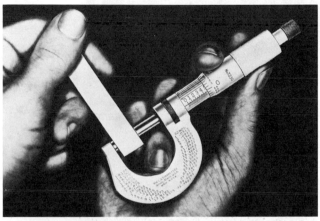

Figure 3–26A Measuring die steel, 0.4995″ *(Courtesy of The L. S. Starrett Company)*

Figure 3–26B Checking a grinding machine operation with a 4–5″ micrometer *(Courtesy of The L. S. Starrett Company)*

3–29 (page 36), often have interchangeable anvils and can be made as large as 120 in on special order.

In Figures 3–26 to 3–28, note how the micrometers are held.

VERNIER CALIPERS

Vernier calipers, Figure 3–30 (page 36), are made in many styles and sizes, but they are essentially similar to read. You first read the beam and then you add what you read on the vernier scale. One graduation on the vernier scale will match a mark on the beam, except in the case of a zero reading. In this case, the zeros at each end of the vernier will match marks on the beam.

Vernier Caliper Readings

In the inch system, vernier plates have 20, 25, and 50 divisions. The number of divisions on the vernier plate tells how much each division is on the beam, in thousandths of an inch. For example, with a 20-division vernier plate, each beam division is 0.020 in. Similarly, a 25-division vernier plate makes each beam division 0.025 in, and a 50-division vernier plate makes each beam division 0.050 in. Keeping this in mind, check the examples in Figure 3–31 (page 37). Since the vernier plates have 25 divisions, each beam division is 0.025 in. Reading the beam in Figure 3–31A first, there is a total of 0.500 in. The zero and 25th vernier divisions line up with a beam division. The total reading is 0.500 in. Figure 3–31B shows an additional sample reading of 0.255 in.

Both of the scales in Figures 3–32 (page 37) and 3–33 (page 37) are in metric. The first is in centimeters and the second is in millimeters. However, the two sizes are identical.

Measuring With Verniers

Figures 3–34 (page 38), 3–35 (page 39), and 3–36 (page 40) illustrate different styles of verniers and how to use them.

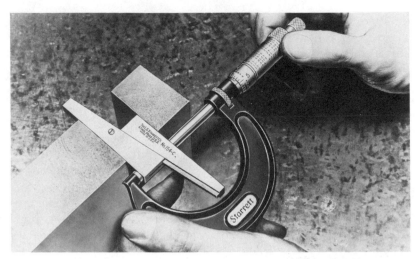

Figure 3–26C Measuring adjustable parallels with a 2″ micrometer *(Courtesy of The L. S. Starrett Company)*

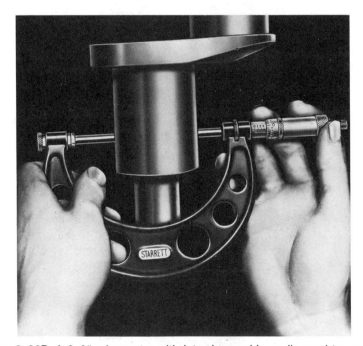

Figure 3–26D A 2–6″ micrometer with interchangeable anvils used to measure the diameter of a crankshaft bearing *(Courtesy of The L. S. Starrett Company)*

Figure 3–27 A 12″ micrometer. Measuring the length of a dovetail base *(Courtesy of The L. S. Starrett Company)*

GEAR TOOTH VERNIER

A gear tooth vernier, Figure 3–37A (page 40), is used to measure the chordal thickness of a gear tooth at the pitch line and also the chordal addendum or corrected addendum of the gear tooth, Figure 3–37B (page 40). Because of the gear tooth terminology, this will be further explained in the gear cutting section under milling.

Figure 3–28 A 24–30″ micrometer used on a turning operation *(Courtesy of The L. S. Starrett Company)*

Figure 3-29 A special large micrometer, 60–66″ *(Courtesy of The L. S. Starrett Company)*

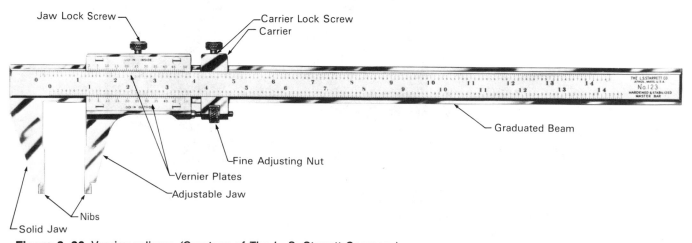

Figure 3-30 Vernier calipers *(Courtesy of The L. S. Starrett Company)*

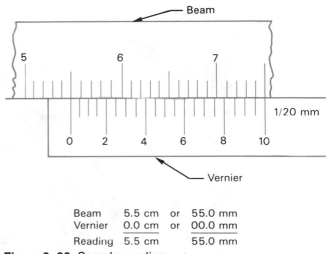

Beam	5.5 cm	or	55.0 mm
Vernier	0.0 cm	or	00.0 mm
Reading	5.5 cm		55.0 mm

Figure 3–32 Sample reading

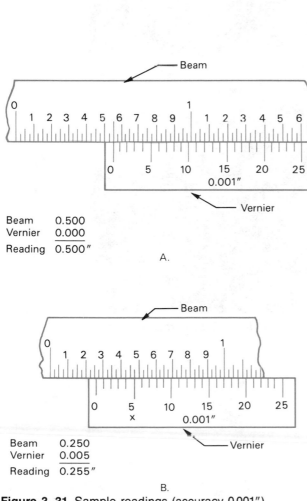

Beam	0.500
Vernier	0.000
Reading	0.500″

A.

Beam	0.250
Vernier	0.005
Reading	0.255″

B.

Figure 3–31 Sample readings (accuracy 0.001″)

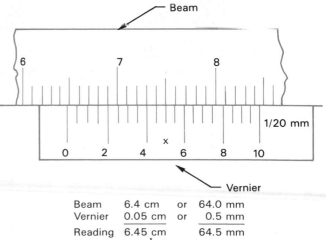

Beam	6.4 cm	or	64.0 mm
Vernier	0.05 cm	or	0.5 mm
Reading	6.45 cm		64.5 mm

Figure 3–33 Sample reading

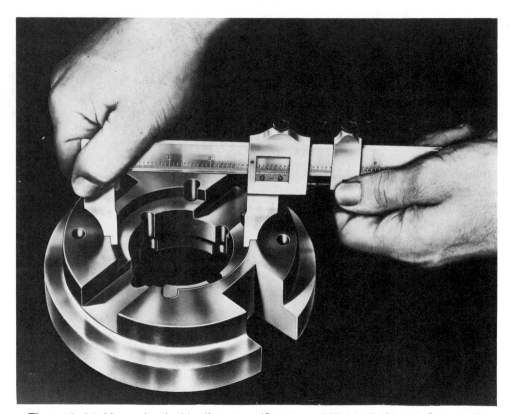

Figure 3–34 Measuring inside diameters *(Courtesy of The L. S. Starrett Company)*

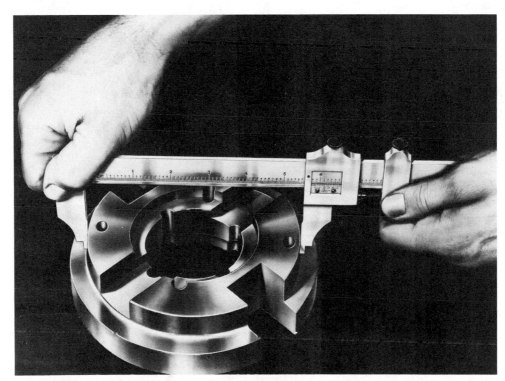

Figure 3–35 Measuring outside diameters *(Courtesy of The L. S. Starrett Company)*

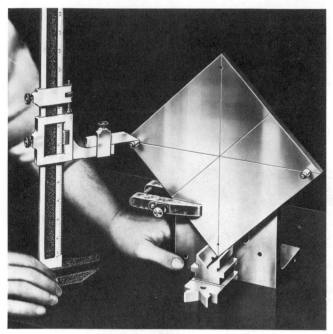

Figure 3–36 Using a vernier height gage *(Courtesy of The L. S. Starrett Company)*

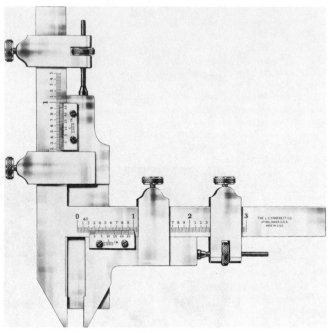

Figure 3–37A A gear tooth vernier *(Courtesy of The L. S. Starrett Company)*

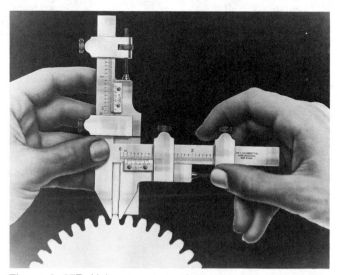

Figure 3–37B Using a gear tooth vernier *(Courtesy of The L. S. Starrett Company)*

REVIEW QUESTIONS

1. What accuracy (in inches and millimeters) can be obtained in measuring using an inch micrometer caliper and a metric micrometer caliper?
2. In one revolution of the inch micrometer, what distance is traveled by the spindle? What distance is traveled in one revolution of the metric micrometer?
3. What precaution should be taken when using a 12-in or larger micrometer as opposed to using a 1-in micrometer?
4. On the scale of an inch micrometer:
   (a.) How many divisions are on the barrel?
   (b.) In decimals, how much is each barrel division?
   (c.) How many divisions are on the thimble?
   (d.) In decimals, how much is each thimble division?
   (e.) On the barrel of some micrometers, there is an extra scale. How many divisions are on that scale?
   (f.) In decimals, how much is each division on that extra scale?
5. List the steps in obtaining a measurement using the inch micrometer.
6. On the metric micrometer:
   (a.) How many divisions are on the barrel?
   (b.) In decimals, how much is each division on the barrel?
   (c.) How many divisions are on the thimble?
   (d.) In decimals, how much is each division on the thimble?
   (e.) If there were an extra scale on the barrel, how many divisions would there be on it?
   (f.) In decimals, how much is each division on the extra scale?
7. List five precautions for the care and use of micrometers.
8. On the scale of some micrometers, there is an extra scale. How many divisions are on that scale?
   (a.) If there are 20 divisions on the vernier scale, how much in decimals is each division on the beam?
   (b.) If there are 25 divisions on the vernier scale, how much in decimals is each division on the beam?
   (c.) If there are 50 divisions on the vernier scale, how much in decimals is each division on the beam?
9. On the metric vernier slide caliper:
   (a.) If there are 25 divisions on the vernier scale, how much is each division on the beam?
   (b.) If there are 50 divisions on the vernier scale, how much is each division on the beam?
10. List the steps in obtaining a measurement using a vernier caliper.

UNIT 3-5

TELESCOPING GAGES, PROTRACTORS, AND THICKNESS GAGES

OBJECTIVES

After completing this unit, the student will be able to:

- set and test sizes with telescoping gages and small hole gages.
- set angles with the different types of protractors to accuracies within 5 minutes.
- use thickness gages to test bearing clearances and gear tooth clearances.

KEY TERMS

Telescoping gage	Universal bevel protractor
Small hole gage	Thickness or feeler gage

TELESCOPING GAGE

A telescoping gage, Figure 3–38, is used to measure inside diameters or hole diameters. The gage consists of a knurled handle, a knurled lock screw, a fixed leg, and a telescoping leg acting against a coil compression spring. The telescoping leg can be locked in position once the hole is measured. The exact size of the hole is obtained by measuring the telescoping gage with a micrometer. Figures 3–39A and 3–39B illustrate the uses of the telescoping gage.

An instrument similar to the telescoping gage is the small hole gage, Figure 3–40 (page 46). The basic difference in the two types is that an adjustment of the knurled screw opens or closes the two halves of the small hole gage. A micrometer is used to obtain an accurate measurement. Figure 3–41 (page 46) illustrates the use of the small hole gage.

THE UNIVERSAL BEVEL PROTRACTOR

The universal bevel protractor, Figure 3–42 (page 47), is used to measure angles to within an accuracy of 5 minutes (1/12 of a degree). The protractor dial and the vernier scale have graduations on both sides of zero. To find the correct angle, it is necessary to read the protractor dial first and then add the vernier reading. The vernier scale must be read in the same direction from zero as the protractor dial, Figure 3–43 (page 47). Figure 3–44 (page 48) illustrates one use of a universal bevel protractor. Figure 3–45 (page 48) shows sample readings.

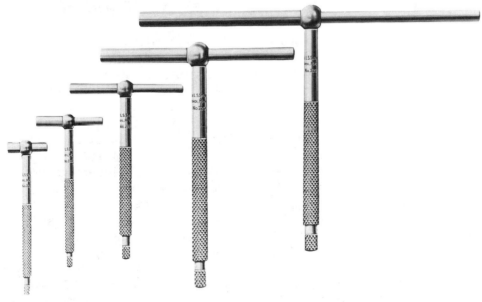

Figure 3–38 A telescoping gage *(Courtesy of The L. S. Starrett Company)*

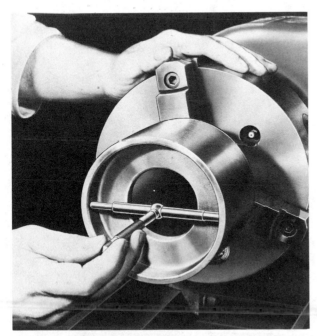

Figure 3–39A Using a large telescoping gage *(Courtesy of The L. S. Starrett Company)*

Figure 3–39B Using a small telescoping gage *(Courtesy of The L. S. Starrett Company)*

MODERN MANUFACTURING – NEW TRENDS

In recent years, factories have been increasingly plagued with multiple problems, including a significant decrease in product quality, high production costs and low or nonexistent company profits. Since manufacturing companies are not in business to lose money, these problems are being evaluated and solutions are being implemented which are changing manufacturing in America. Planning for new factories must address the objectives of producing more and consistently better quality work at lower production costs. To achieve these objectives, factory automation with computer control becomes increasingly important. Many industries have added or are adding to their production facilities computer integrated manufacturing (CIM) systems or flexible manufacturing systems (FMS). These exciting, innovative systems include some or all of the following advanced technology: robotics, coordinate measuring machines, electronic control systems and tool gages, calibration cubes, electronic probes, automatic work and tool changers, automatic machine loading and unloading devices, machine vision and laser tools.

For example, one new FMS for the aerospace industry consists of eight CNC horizontal machining centers, each with a 90-tool magazine, a CNC computer control center, two 10-pallet work changer units, four automated guided vehicles, an inspection station, a wash station, a supervisory computer, and a special chip collection system which separates ferrous and nonferrous materials for reclamation. Plans call for this factory to machine nearly 600 different parts for the aerospace industry. Conventional manufacturing methods would require about 200,000 hours to produce the equivalent volume of work that this FMS can complete in 70,000 hours. The time saved is partly a result of reduced manual involvement by workers. Once work pieces are loaded manually on work pallets, automatic

operation through three work shifts is possible.

In addition, the flexible system permits complete sets of parts for one application to be made in sequence. This system prevents delays at assembly and also helps to keep the inventory of materials, as well as storage requirements, to a minimum. The flexible system will operate on a three-shift basis, normally with two manned shifts and the shift after midnight unmanned. The unmanned shift does not necessarily require lighting, resulting in the conservation of power. Usually a system manager will be on duty in the control room at all times, with one or more operators at the load/unload stations. Future expansion of the system has been planned in advance for the addition of more manufacturing cells. More cells will also require the addition of more automatic storage and retrieval systems.

A Flexible Manufacturing System (FMS) or a Flexible Manufacturing Factory (FMF) is computer controlled from a central location. Such a system contains two or more CNC machine tools

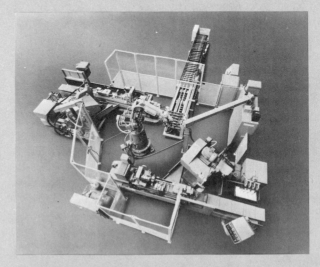

Figure 1 A manufacturing cell with two CNC cylindrical grinders *(Courtesy of Cincinnati Milacron)*

serviced by a robot system. The robot system will load unfinished stock from an incoming conveyor system into a machine tool and unload the finished part from the machine tool to an outgoing conveyor after all machining operations are complete. Moving robots will transfer unfinished stock from the stock room to the incoming conveyor, will transfer finished parts to a storage area, and will also transfer machine tooling to and from the tool room and the machine tool, replacing worn or broken cutting tools.

A Flexible Manufacturing Cell (FMC) containing two unmanned grinding machines serviced by a central fixed robot, Figure 1, is designed for the grinding of cylindrical parts. The two-step grinding machines are positioned offset and back-to-back.

Figure 2 shows a FMC containing a CNC turning center, a vertical CNC machining center, and a horizontal CNC machining center, all serviced by a CNC industrial robot mounted on a car that transports it from machine to machine. A FMC can contain two or more CNC machine tools equipped with a robot system for loading and unloading the machines. A single operator can oversee a FMC, but the cell can also be unmanned for one work shift and might not necessarily be controlled by a central computer. However, the machine tool and the robot, each with its own computer, will be interfaced or coordinated with each other.

Another type of FMC is shown in Figure 3 where two CNC machining centers are serviced by a mobile work-carrying robot and a tool-carrying robot.

Flexible Manufacturing Systems may contain a single machine tool serviced by a robot or they may contain many machine tools serviced by different types of robots. A standard practice is to set up a cell with three or four machine tools serviced by a single robot. This arrangement ensures that the robot will be kept fully occupied at all times. A tool-carrying robot and a work-carrying robot are also a part of the cell. With proper planning, a FMS can be as small or as large as needed. The key is planning to ensure that the system efficiently meets the needs of the individual factory.

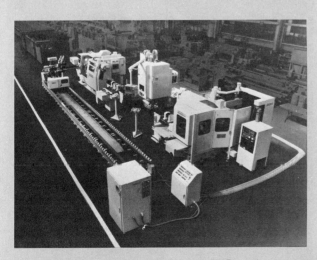

Figure 2 A manufacturing cell with a CNC turning center and two CNC machining centers *(Courtesy of Cincinnati Milacron)*

Figure 3 A manufacturing cell with four CNC machining centers *(Courtesy of Cincinnati Milacron)*

THICKNESS OR FEELER GAGE

A thickness gage, Figure 3–46 (page 49), consists of a number of leaves of hardened and tempered steel. Each leaf is carefully finished to the correct thickness and marked in either inch or millimeter sizes. A typical feeler gage like the one shown may contain eight or more leaves. These eight leaves may vary in thickness from 0.002 in to 0.015 in. A typical metric gage with nine leaves may vary in thickness from 0.04 mm to 0.3 mm. Feeler gages are used in automotive work to check spaces when adjusting tappets in an engine, check gaps between spark plug points, and check the gap between distributor points. In machine shop work, feeler gages are used to check bearing clearances, check play in gears, or fit the gaps between the ends of turned cast iron piston rings.

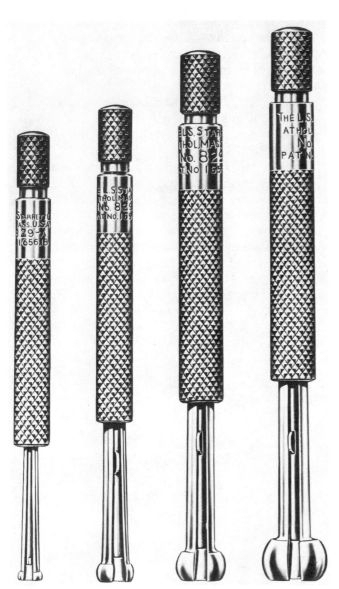

Figure 3–40 A small hole gage *(Courtesy of The L. S. Starrett Company)*

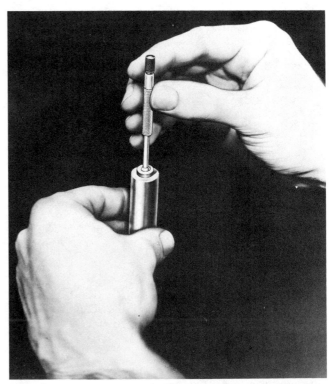

Figure 3–41 Using a small hole gage *(Courtesy of The L. S. Starrett Company)*

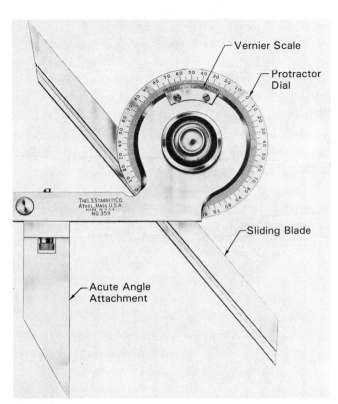

Figure 3–42 A universal bevel protractor *(Courtesy of The L. S. Starrett Company)*

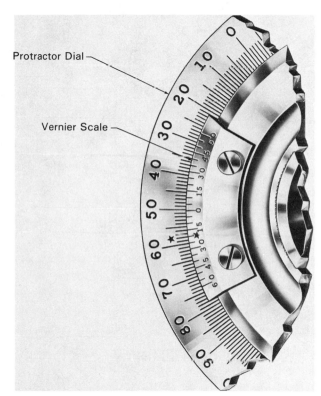

Figure 3–43 The scales of the universal bevel protractor *(Courtesy of The L. S. Starrett Company)*

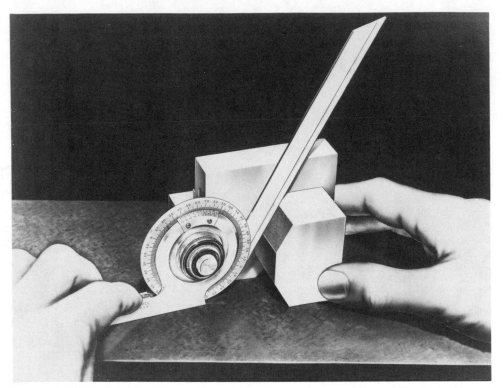

Figure 3–44 Measuring with a universal bevel protractor *(Courtesy of The L. S. Starrett Company)*

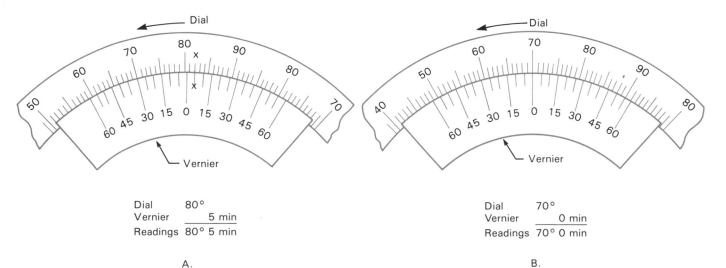

Dial	80°
Vernier	5 min
Readings	80° 5 min

A.

Dial	70°
Vernier	0 min
Readings	70° 0 min

B.

Figure 3–45 Sample readings

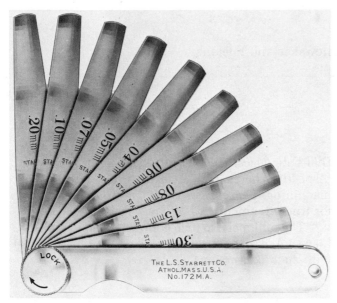

Figure 3–46 A thickness or feeler gage *(Courtesy of The L. S. Starrett Company)*

REVIEW QUESTIONS

1. How is the telescoping gage used to measure inside diameters?
2. In what way does the small hole gage differ from a telescoping gage?
3. What is the accuracy of a universal bevel protractor?
4. List the steps in obtaining a measurement using the universal bevel protractor.
5. Name three uses of thickness gages in the machine shop.

UNIT 3-6

DIAL INDICATORS, DIAL BORE GAGES, AND FIXED GAGES

OBJECTIVES

After completing this unit, the student will be able to:

- read dial indicators to accuracies as close as 0.00005 in and 0.002 mm.
- set and use a dial bore gage.
- use a plug gage to test a turned bore.

- use a ring gage to test a turned shaft.
- define the terms fit, allowance, and tolerance.
- describe the differences between the terms fit, allowance and tolerance.

KEY TERMS

Dial indicator	Tolerance
Mechanical comparator	Cylindrical ring gage
Dial bore gage	Cylindrical plug gage
Fixed gage	Thread ring gage
Running or sliding fit	Thread plug gage
Locational fit	Taper ring gage
Force or shrink fit	Taper plug gage
Allowance	"Go" and "no-go" gage

DIAL INDICATORS

Dial indicators are precision testing dials made in many sizes and in many accuracies. Several dial indicators are shown in Figure 3–47. Accuracies can range from plus or minus 0.001 in to plus or minus 0.00005 in, and from plus or minus 0.01 mm to plus or minus 0.002 mm. Applications of dial indicators range from setting vises square on a milling machine, to trueing work on a lathe.

A universal heavy-duty dial test indicator, Figure 3–48, is probably the most common set that the average machinist will use. With an accuracy of plus or minus 0.001 in, it is used mainly for setting up work accurately on machine tools. Another common use is the accurate straightening of steel shafts between centers. The magnetic base indicator holder, Figure 3–49, is another commonly used tool. The mechanical comparator, Figure 3–50, is used mainly for inspection of duplicate parts, especially in a tool and die shop. Figure 3–51 (page 53) illustrates typical applications of dial indicators.

DIAL BORE GAGE

Dial bore gages, Figure 3–52 (page 54), are used for inspecting bores and checking hole sizes during machining. These gages utilize a three-point contact for true alignment.

The millimeter to inch conversion table, Table 3–2 (page 55), is a very useful tool when you have to deal with such conversions.

FIXED GAGES

Using fixed gages is a quick and easy way to check dimensions of machine parts to see whether or not they are within specified tolerances.

One of the relationships that a machinist must deal with constantly is that of fit, allowance, and tolerance between mating parts. Some machinists unthinkingly use these terms to mean the same thing, but there are differences. **Fit** is the relationship of tightness or looseness between two mating parts, like the diameter of a shaft and the diameter of the mating hole in a bearing. Fits are generally classified as running or sliding fits, locational fits, and force or shrink fits, with more than one classification in each of these categories. Running or sliding fits permit free movement between two parts with proper lubrication. Locational fits are used as a rule where there is no movement between two parts but the parts can be dismantled easily and reassembled. As a rule, force or shrink fits are used when two parts are meant to remain together permanently.

Allowance is the difference in dimensions between the two mating parts, that is, the shaft and the hole. If, for example, the shaft is allowed to be a minimum size of 0.998 in, and the hole is allowed to be a maximum size of 1.000 in, the allowance between the size of the two parts is 0.002 in.

Tolerance is the amount of variation permitted in the size of a single part. If the shaft is allowed to be a maximum size of 0.999 in and a minimum size of 0.998 in, the tolerance permitted in the size of the shaft is 0.001 in. Similarly, if the hole is allowed to be a maximum size of 1.000 in and a minimum size of 0.9995 in, the tolerance permitted in the size of the hole is 0.0005 in.

Examples of fixed gages that are used regularly by machinists are the cylindrical ring, cylindrical plug, thread ring, thread plug, taper ring, and taper plug gages. A wide range of sizes and types is available. A "Go" and "No-Go" gage, for example, is used to check the largest and smallest permissible dimensions. Cylindrical ring gages, Figure 3–53 (page 54), are used to check outside diameters and to set dial bore gages to specific diameters. Cylindrical plug gages, Figure 3–54 (page 55), are used to check bore sizes. Thread ring gages, Figure 3–55 (page 55), are used to check the pitch diameter of external screw threads. Thread plug gages, Figure 3–56 (page 56), are used to check the pitch diameter of internal threads.

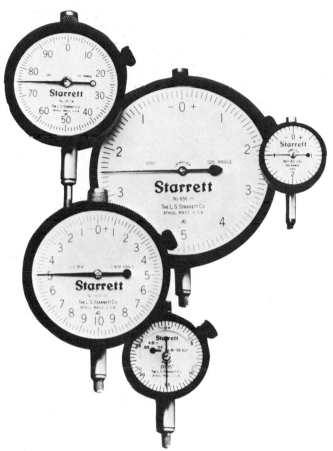

Figure 3–47 A variety of dial indicators *(Courtesy of The L. S. Starrett Company)*

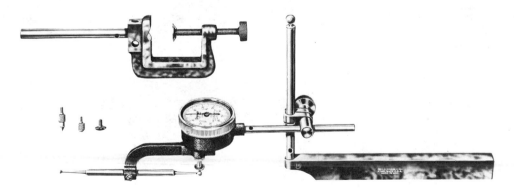

Figure 3–48 A universal heavy-duty dial test indicator *(Courtesy of The L. S. Starrett Company)*

Figure 3–49 A magnetic base indicator holder *(Courtesy of The L. S. Starrett Company)*

Figure 3–50 A mechanical comparator *(Courtesy of The L. S. Starrett Company)*

Figure 3–51A Using dial indicators: testing work for level on a surface grinder *(Courtesy of The L. S. Starrett Company)*

Figure 3–51C Using dial indicators: checking work for runout in a four-jaw chuck *(Courtesy of The L. S. Starrett Company)*

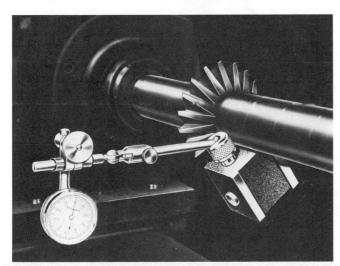

Figure 3–51B Using dial indicators: setting a vise square on a milling machine *(Courtesy of The L. S. Starrett Company)*

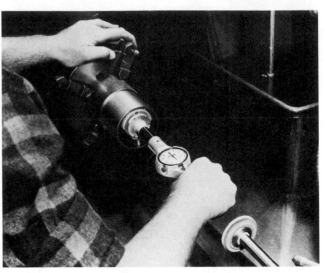

Figure 3–52B Inspecting a bore size during a grinding operation *(Courtesy of The L. S. Starrett Company)*

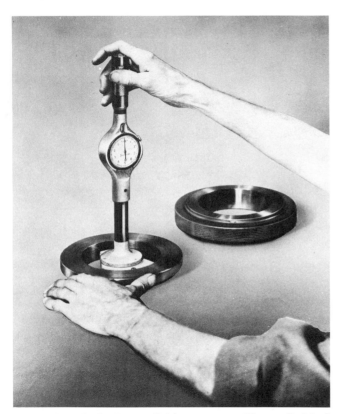

Figure 3–52A Setting a dial bore gage with a cylindrical ring gage *(Courtesy of The L. S. Starrett Company)*

Figure 3–53 A cylindrical ring gage *(Courtesy of DoALL Company)*

Table 3–2 Millimeter to Inch Conversions

mm	Decimal	mm	Decimal	mm	Decimal	mm	Decimal	mm	Decimal
0.01	.00039	0.41	.01614	0.81	.03189	21	.82677	61	2.40157
0.02	.00079	0.42	.01654	0.82	.03228	22	.86614	62	2.44094
0.03	.00118	0.43	.01693	0.83	.03268	23	.90551	63	2.48031
0.04	.00157	0.44	.01732	0.84	.03307	24	.94488	64	2.51969
0.05	.00197	0.45	.01772	0.85	.03346	25	.98425	65	2.55906
0.06	.00236	0.46	.01811	0.86	.03386	26	1.02362	66	2.59843
0.07	.00276	0.47	.01850	0.87	.03425	27	1.06299	67	2.63780
0.08	.00315	0.48	.01890	0.88	.03465	28	1.10236	68	2.67717
0.09	.00354	0.49	.01929	0.89	.03504	29	1.14173	69	2.71654
0.10	.00394	0.50	.01969	0.90	.03543	30	1.18110	70	2.75591
0.11	.00433	0.51	.02008	0.91	.03583	31	1.22047	71	2.79528
0.12	.00472	0.52	.02047	0.92	.03622	32	1.25984	72	2.83465
0.13	.00512	0.53	.02087	0.93	.03661	33	1.29921	73	2.87402
0.14	.00551	0.54	.02126	0.94	.03701	34	1.33858	74	2.91339
0.15	.00591	0.55	.02165	0.95	.03740	35	1.37795	75	2.95276
0.16	.00630	0.56	.02205	0.96	.03780	36	1.41732	76	2.99213
0.17	.00669	0.57	.02244	0.97	.03819	37	1.45669	77	3.03150
0.18	.00709	0.58	.02283	0.98	.03858	38	1.49606	78	3.07087
0.19	.00748	0.59	.02323	0.99	.03898	39	1.53543	79	3.11024
0.20	.00787	0.60	.02362	1.00	.03937	40	1.57480	80	3.14961
0.21	.00827	0.61	.02402	1	.03937	41	1.61417	81	3.18898
0.22	.00866	0.62	.02441	2	.07874	42	1.65354	82	3.22835
0.23	.00906	0.63	.02480	3	.11811	43	1.69291	83	3.26772
0.24	.00945	0.64	.02520	4	.15748	44	1.73228	84	3.30709
0.25	.00984	0.65	.02559	5	.19685	45	1.77165	85	3.34646
0.26	.01024	0.66	.02598	6	.23622	46	1.81102	86	3.38583
0.27	.01063	0.67	.02638	7	.27559	47	1.85039	87	3.42520
0.28	.01102	0.68	.02677	8	.31496	48	1.88976	88	3.46457
0.29	.01142	0.69	.02717	9	.35433	49	1.92913	89	3.50394
0.30	.01181	0.70	.02756	10	.39370	50	1.96850	90	3.54331
0.31	.01220	0.71	.02795	11	.43307	51	2.00787	91	3.58268
0.32	.01260	0.72	.02835	12	.47244	52	2.04724	92	3.62205
0.33	.01299	0.73	.02874	13	.51181	53	2.08661	93	3.66142
0.34	.01339	0.74	.02913	14	.55118	54	2.12598	94	3.70079
0.35	.01378	0.75	.02953	15	.59055	55	2.16535	95	3.74016
0.36	.01417	0.76	.02992	16	.62992	56	2.20472	96	3.77953
0.37	.01457	0.77	.03032	17	.66929	57	2.24409	97	3.81890
0.38	.01496	0.78	.03071	18	.70866	58	2.28346	98	3.85827
0.39	.01535	0.79	.03110	19	.74803	59	2.32283	99	3.89764
0.40	.01575	0.80	.03150	20	.78740	60	2.36220	100	3.93701

Figure 3–54 A cylindrical plug gage *(Courtesy of DoALL Company)*

Figure 3–55 A thread ring gage *(Courtesy of DoALL Company)*

Figure 3–56 A thread plug gage *(Courtesy of DoALL Company)*

REVIEW QUESTIONS

1. What are two uses for a dial indicator?
2. Name the primary use of the mechanical comparator.
3. What is the main purpose of a dial bore gage?
4. What is the purpose of a fixed gage?
5. Define the terms fit, allowance, and tolerance.
6. List the three major types of fits and describe how the fits are applied to parts.
7. Name six fixed gages.
8. For what purpose is a "Go" and "No-Go" gage used?

UNIT 3-7

GAGE BLOCKS, SINE BARS AND SINE TABLES, AND OPTICAL FLATS

OBJECTIVES

After completing this unit, the student will be able to:

- set up stacks of gage blocks to precision sizes.
- use gage blocks, sine bars, and surface plates to check the accuracy of angles (such as tapers).
- Use an optical flat to produce a fringe pattern.

KEY TERMS

Gage block

Wringing

Micrometer height gage

Sine bar

Sine table

Compound sine angle
 plate

Optical flat

Monochromatic light

Wavelength

Interference band

Fringe pattern

GAGE BLOCKS

Gage blocks are the primary standards of the machining industry. One use of gage blocks is to test for the accuracy of such precision measuring tools as micrometers. The blocks are made in a variety of shapes and lengths, in both inch and millimeter sizes. A high grade of tool steel is hardened and tempered. The surfaces of the finished gage are so fine and flat that when they are rubbed together properly, they will actually stick together. This rubbing action is called **wringing**. Another material used is chrome carbide, which is longer wearing and more corrosion resistant than tool steel. These blocks, each of which is accurate to within a few millionths of an inch, are usually called gage blocks, but are sometimes named after the manufacturer. "Jo" blocks are named for the Johanson Co. and "Webber" blocks are named for the Webber division of The L. S. Starrett Co.

Gage blocks come in sets, such as 3 blocks, 28 blocks, 81 blocks, or 111 blocks, to name a few. A common set in a machine shop is an 81-block set. Accuracies in length are usually within 2, 5, or 10 millionths of an inch of the stamped size (0.000027 mm in the metric system). A single 3-in gage block, Figure 3–57, is shown. Figure 3–58 illustrates a mixed set of chrome carbide and tool steel blocks. Figure 3–59 shows a set of inspection-grade angle blocks, accurate to 0.1 second. The wringing of gage blocks together is shown in Figure 3–60 (page 60). A stack of gage blocks, Figure 3–61 (page 61), is being used as a height gage.

A **micrometer height gage** is a column of gage blocks that is set up in 1-in increments and permanently wrung together, Figure 3–62 (page 62). The surfaces are specially treated to prevent corrosion.

Over and under heights can be checked in a single setting since the reference surfaces at the top and bottom of each block and the top and bottom of the adjacent blocks are in the same plane. Readings can also be taken from either the left, center, or right of the gage block column. The gage block column can be moved up or down over a 1-in range by turning the micrometer head. Scales on either side of the gage block column are read to the nearest inch, while the micrometer head can be read to within one ten thousandth of an inch.

Any desired height dimension within the range of the gage can be quickly obtained and transferred to the work.

To make a setting or to take a reading, select a measuring step nearest the dimension required, adjust the micrometer head to the exact setting, and take the reading. The accuracy of the measuring column is plus or minus 0.000060 in (plus or minus 0.0015 mm).

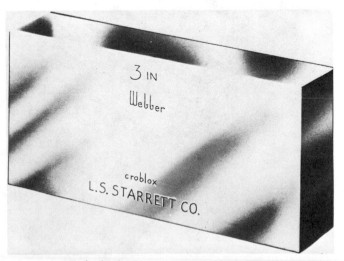

Figure 3–57 A single 3″ gage block *(Courtesy of The L. S. Starrett Company)*

USING GAGE BLOCKS

The length of a gage block is always given as the length between its measuring surfaces, Figure 3–63 (page 62).

When working with gage blocks, it is best to work within the limits of the available set. Use as few blocks as possible, and always work from right to left of the decimal point. If wear blocks are used, their length should be included in the required dimension.

EXAMPLE 1. Set up the dimension 1.4625 in. The set available is an 81-block set, which has an accuracy of plus or minus 0.000001 in.

		Check
1. The dimension required is	1.4625	
2. Eliminate the last number (5)	0.1005	0.1005
	1.3620	
3. Eliminate the last number (2)	0.142	0.142
	1.220	
4. Eliminate the last number (2)	0.120	0.120
	1.100	

Figure 3–58 A mixed set of gage blocks *(Courtesy of The L. S. Starrett Company)*

5. Eliminate the last number (1) <u>0.100</u> 0.100
                                        1.000
6. Eliminate the last number (1) <u>1.000</u> 1.000
                                        0.000 1.4625

Add the check column. This should be the dimension and the gage blocks used in the stack.

EXAMPLE 2. Set up the dimension 3.5625 in. The set available is an 81-block set, which has an accuracy of plus or minus 0.000001 in.

			Check
1. The dimension required is	3.5625		
2. Eliminate the last number (5)	<u>0.1005</u>	0.1005	
	3.462		
3. Eliminate the last number (2)	<u>0.142</u>	0.142	
	3.320		
4. Eliminate the last number (2)	<u>0.120</u>	0.120	
	3.200		
5. Eliminate the last number (2)	<u>0.200</u>	0.200	
	3.000		

6. Eliminate the last number (3) <u>3.000</u> 3.000
                                        0.000 3.5625

Add the check column and use these gage blocks in the stack.

When measuring with blocks in the metric system, the procedure is the same as in the English (U.S. Customary) system.

EXAMPLE 3. Set up the dimension 50.755 mm. The set available is a 45-block set, which has an accuracy of plus or minus 0.00005 mm.

			Check
1. The dimension required is	50.755		
2. Eliminate the last number (5)	<u>1.005</u>	1.005	
	49.750		
3. Eliminate the last number (5)	<u>1.050</u>	1.050	
	48.700		
4. Eliminate the last number (7)	<u>1.700</u>	1.700	
	47.000		

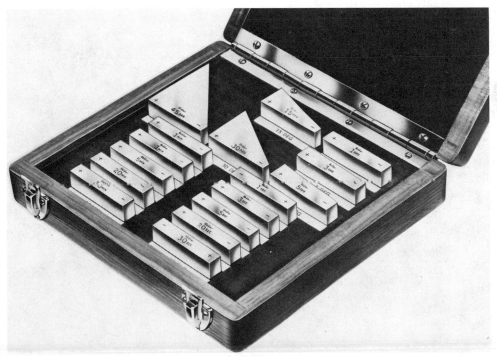

Figure 3–59 A set of inspection-grade angle blocks *(Courtesy of The L. S. Starrett Company)*

5. Eliminate the last number (7)   7.000   7.000
                                                40.000
6. Eliminate the last number (4)   40.000   40.000
                                                00.000   50.755

Add the check column. This should be the dimension and the gage blocks used in the stack. If wear blocks are used, they should be the first numbers eliminated.

Using a Sine Bar With Gage Blocks

A **sine bar**, Figure 3–64A (page 63), is a precision device used for checking the accuracy of tapers of all kinds, Figure 3–64B (page 63). Its length is always known — 5 inches in the case of a 5-in bar, or 10 inches in the case of a 10-in bar.

In trigonometry, the sine of the angle A, Figure 3–65 (page 63), is the length of side a (opposite the angle) divided by the length of side c (hypotenuse):

$$\text{sine angle } A = \frac{\text{opposite}}{\text{hypotenuse}}$$

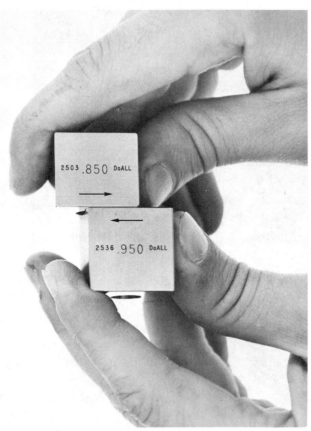

Figure 3–60A Wringing gage blocks, starting position *(Courtesy of DoALL Company)*

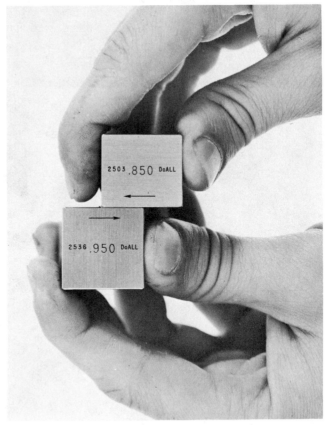

Figure 3–60B Wringing gage blocks, second position *(Courtesy of DoALL Company)*

The sine bar is always the hypotenuse and the stack of gage blocks make up side *a* opposite the angle *A*. The usual problem is to find the angle *A*.

Care of Gage Blocks

Gage blocks must be kept thoroughly clean. It is preferable that the measuring surfaces not be handled, just the sides. Lay the gage blocks on their sides on clean paper, a soft clean cloth, or a clean chamois on the layout table. After use, the blocks should be separated, cleaned, and lightly oiled before being replaced in their case. Some manufacturers will supply their own solvents and oils. As an option, some manufacturers will supply wear blocks, usually 0.020 in, 0.050 in, or 0.100 in (or 1 mm or 2 mm in the metric system). The wear blocks are wrung on each end of a gage block stack. They should be taken into account in the calculation of the dimension.

Sine Tables

A sine table is much wider than a sine bar and has an attached base. Otherwise, in principle its use is the same as that of the sine bar. Some heavy-duty sine plates are rugged enough to hold parts for the machining of angles as well as for the inspection of angles. A compound sine angle plate is two sine plates with hinge lines at right angles to each other. These are used to form compound angles and also to measure them.

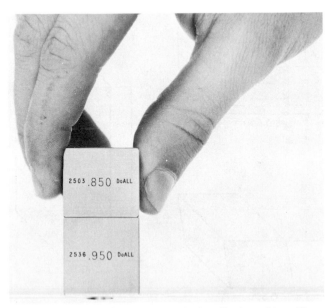

Figure 3–60C Wringing gage blocks, final position *(Courtesy of DoALL Company)*

Figure 3–61 A stack of gage blocks used as a height gage *(Courtesy of DoALL Company)*

WORKING WITH OPTICAL FLATS

The average machinist probably will never use optical flats or light waves in measuring. These are used by precision toolmakers to measure the flatness of a surface in millionths of an inch.

One of the main instruments for measuring with light waves is the optical flat, Figure 3–66. This is a highly polished piece of plate glass, optical glass, or, best of all, fused quartz. One face is polished to almost a perfect flat. This flat face, usually marked with an arrow, is placed carefully on the work surface to be tested and is firmly pressed down until bands appear.

Both the optical flat and the work surface must be absolutely clean, free from grit, oil, fingerprints, or dust. Grain alcohol is typically used to polish each surface with a clean chamois or clean paper. When removing the optical flat from the work surface, lift it carefully. Do not slide it sideways.

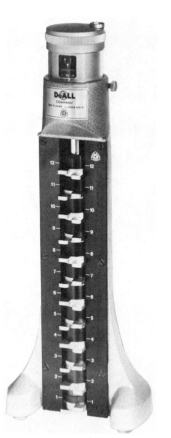

Figure 3–62 Height micrometer *(Courtesy of DoALL Company)*

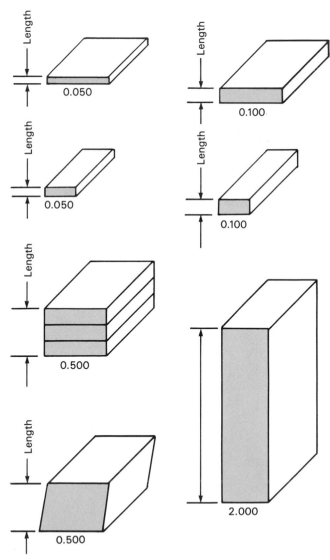

Figure 3–63 The length of a gage block

A monochromatic light source, that is, a light consisting of one wavelength (such as a helium light), is directed down through the optical flat to the work surface, which reflects the light back up through the optical flat to the eye. When the light waves reflected from the work surface interfere with light waves reflected from the underside of the optical flat, a series of alternate light and dark bands or fringes will appear, Figure 3–67. The dark bands are called **interference bands**. It must be remembered that however minute, there will be an air gap or air wedge between the optical flat and the work surface.

Figure 3–64A A granite sine bar *(Courtesy of DoALL Company)*

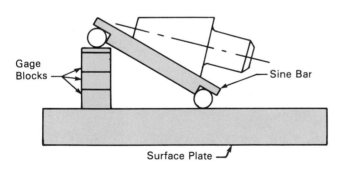

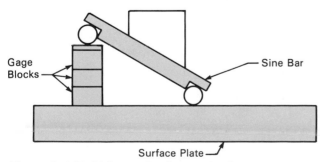

Figure 3–64B Using a sine bar to check tapers

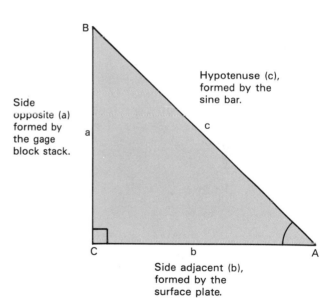

Sine angle A = $\dfrac{\text{Side Opposite}}{\text{Hypotenuse}}$ = $\dfrac{\text{Opp.}}{\text{Hyp.}}$

Sine angle A = $\dfrac{\text{Length of gage block stack}}{\text{Length of sine bar}}$

Height of gage block stack = Length of sine bar × sine angle A

Figure 3–65 Trigonometry calculations involving the sine bar and gage blocks

Fringe Pattern Formations

When interference bands appear, they form different shapes that give an indication of what the surface is like, Figure 3–68A to I. All readings are taken from a reference line (R) or line of contact.

A. The bands do not curve, indicating that the surface is nearly flat along its length.
B. Bands curve toward the reference line, indicating that the surface is convex.
C. Bands curve away from the reference line, indicating that the surface is concave.
D. Part of the surface is flat, but the outer edges drop off.
E. This surface is partly flat but increasingly convex toward the reference line.
F. This surface is partly flat but is increasingly lower toward the bottom left-hand corner.
G. From the lower right to the upper left, this surface is flat, but the slight curvature of the bands away from the reference line indicates it is slightly concave.
H. The surface is flat in the direction the bands run, but diagonally across the center where the bands are widely spread, the surface is higher than at the ends.
I. The two points of contact indicate two high points surrounded by lower areas.

Interpreting Fringe Patterns

When interpreting fringe patterns, Figure 3–69A to J, you must know the line or point of contact or the reference line from which the measurements can be expressed. Dimension lines should pass through the center of the bands. Though the distance between bands varies, the height difference from band to band is always 11.6 millionths of an inch (0.0000116 in), or more commonly, 11.6 mike. This is always counted from the line of contact. The extent of curvature is always measured against the distance between bands. The following items refer to the parts of Figure 3–69.

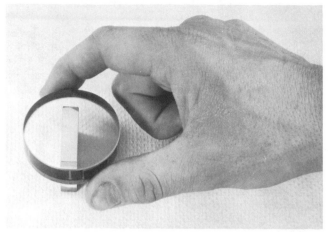

Figure 3–66 An optical flat *(Courtesy of DoALL Company)*

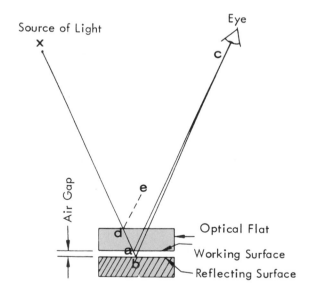

Ignoring refraction and all reflections except the ones we are interested in, the air gap causes two rays to converge at the **eye.**

Figure 3–67 Formation of interference bands *(Courtesy of DoALL Company)*

A. The surface is convex by 1/3 of a band. One-third of 11.6 = 3.866 mike (3.9 mike rounded off) = 3.9 millionths of an inch.

B. The surface is concave by 1/3 of a band. One-third of 11.6 = 3.866 mike (3.9 mike rounded off) = 3.9 millionths of an inch.

C. This surface is convex by 1/2 wavelength high in the center. One-half of 11.6 = 5.8 mike.

D. This surface is convex by 1 wavelength high in the center, which equals 11.6 mike.

E. This surface is also convex, but by 1-1/2 wavelengths high in the center. One and one-half wavelengths = 13.9 mike.

F. This is a very common type of surface. It is flat except at the edges. The edges drop off 1/4 of a wavelength or 2.9 mike.

G. There are two low troughs in this surface with the center and edges at the same height. The ridges are 5/6 of a wavelength high or 5/6 × 11.6 = 9.7 mike high (0.0000097 in).

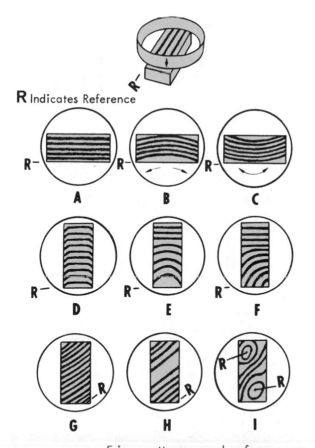

Fringe patterns reveal surface conditions like contour lines on a map.

Figure 3–68 Fringe patterns *(Courtesy of DoALL Company)*

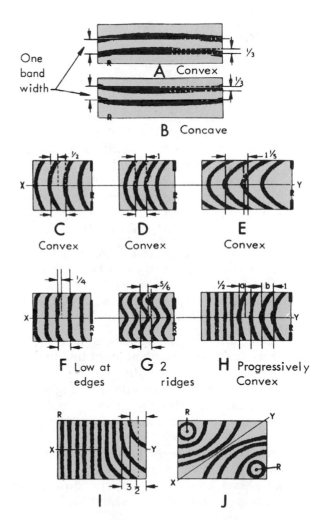

Measurements are the deviation of bands compared to the distance between bands.

Figure 3–69 Interpreting fringe patterns *(Courtesy of DoALL Company)*

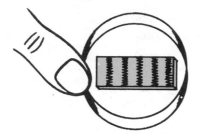

The five fingers represent an air wedge with a height of five half-wave langths. They do not show which end is the open one.

A.

If force at one end does not change the fringe pattern, that end is in contact with the gage block or part.

B.

When force does widen the fringe bands, you know that you are squeezing the air wedge closed. Only slight force is required.

C.

H. Most surfaces are not as uniform as any of the surfaces just described and most change from end to end. Item H starts with a flat surface and becomes progressively more convex toward the right. At "a" the surface is 5.8 mike high, but at "b" the surface is 11.6 mike high.

I. The surface is high near the reference line but rises at the right edge. The top right edge is about 1/2 a wavelength high (5.8 mike). The lower right edge is about 3-1/2 wavelengths high (40.6 mike).

J. There are two high points with a low trough between them. The bottom of this trough is approximated by the line xy. There are four convex bands on each side of the high points. This indicates that the trough is 4-1/2 wavelengths low or 52.3 mike low (0.0000523 in).

The air wedge forms the basic configuration for fringe patterns. This is represented in Figure 3–70. This rule always applies: the fewer the bands, the narrower the angle; the more numerous the bands, the greater the angle.

It is also most important to understand that the number of bands is a measure of height difference, not of absolute height. As in all measurement, there must be a reference from which every length is expressed.

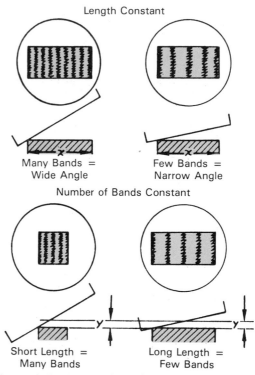

Length Constant

Many Bands = Wide Angle

Few Bands = Narrow Angle

Number of Bands Constant

Short Length = Many Bands

Long Length = Few Bands

These relationships apply to air wedge measurements.

D.

The rule always applies: the fewer the bands, the narrower the angle of the air wedge. The more numerous the bands, the greater the angle of the air wedge.

Figure 3–70 Representing the air wedge *(Courtesy of DoALL Company)*

REVIEW QUESTIONS

1. What does it mean to "wring a gage block"?
2. What is a micrometer height gage? Where are readings taken on this gage?
3. What accuracy is obtained for measurements made with a micrometer height gage?
4. Define the length of a gage block.
5. Set up the following sizes using gage blocks:
   (a.) 2.3626 in
   (b.) 2.3125 in
   (c.) 1.1875 in

6. What is a sine bar and what is its purpose?
7. What is the purpose of wear blocks?
8. State the difference between a sine table and a sine bar.
9. What is the main instrument used in measuring with light waves?
10. Describe the principle of light wave measurement.
11. Define interference bands and fringe patterns.
12. What is the meaning of the term mike when used in light wave measurement?

UNIT 3-8

OPTICAL COMPARATORS AND DIGITAL READOUTS

OBJECTIVES

After completing this unit, the student will be able to:

- describe the operation and use of an optical comparator.
- state the purpose of a digital readout.

KEY TERMS

Optical comparator
Magnification chart
Projected shadow
Chart ring
Angular measuring

Radius and angle chart
Measurement axes
Cartesian coordinates
Deviation

General-purpose chart

Computer-aided design system

OPTICAL COMPARATORS

An optical comparator compares an enlarged shadow of an object with a known chart for size and shape.

Optical comparators, Figure 3–71, are of two types: bench and pedestal. They are identical in operation, but the pedestal type is usually more elaborate. In operation, a light beam projects an enlarged shadow of a part or object being measured through a system of lenses and mirrors onto a magnification chart. The shadow is compared to lines on a chart which correspond to the dimensioning or contours of the part being checked. As an example, a screw thread or a gear tooth may be the part being checked for size or contour. The projected shadow must be sharp and clear. The table on which the part is held can be moved laterally or vertically within an accuracy of 0.0001 in (0.0025 mm). Both movements can be checked on direct reading scales. A chart ring on the projection screen can be rotated for angular measuring to an accuracy of 1 minute.

General-purpose charts, Figure 3–72A, in either clear or matte finish, are useful in the machine shop tool room for inspecting chamfers, form tools, and threads. Radius and angle charts, Figure 3–72B, are used for inspecting and locating holes and radius fillets, and are also useful in inspecting chamfers, threads, and corners.

The latest innovations in optical comparators (or contour projectors) include the use of digital readout equipment and computers, Figures 3–73 (page 71) and 3–74 (page 71). Advanced electronic technology has been coupled with conventional measuring methods to create the control system. Dual axis (X and Y axes) digital readouts display the readings in either English (U.S. Customary) or metric measure, in units of 0.0001 inch (0.002 mm) resolution, and on some comparators, at 0.00005 inch (0.001 mm) resolution. The part coordinate axes are electronically aligned to the worktable measurement axes, virtually eliminating the need for

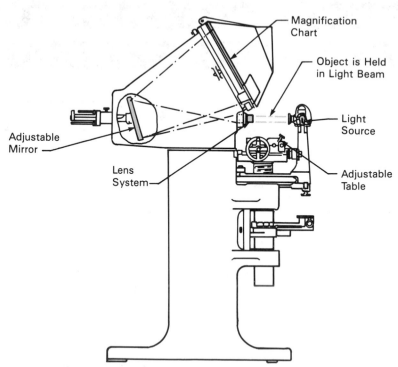

Figure 3–71 Schematic of an optical comparator

physical alignment of the part to the worktable. Display values X and Y can be stored in a special memory for subsequent calculations. Cartesian (X and Y coordinate) deviation of the present X and Y display values can be compared with the previously stored X and Y values. A second push of the Deviation button would provide polar (radius and angle) deviation, if it were required. All of the information can be transferred by use of the Print button to a printer or computer system. It is also possible now to interface (or coordinate) this with a computer-aided design (CAD) system which would provide a faster and easier method of creating an inspection program by graphically simulating the part inspection sequence.

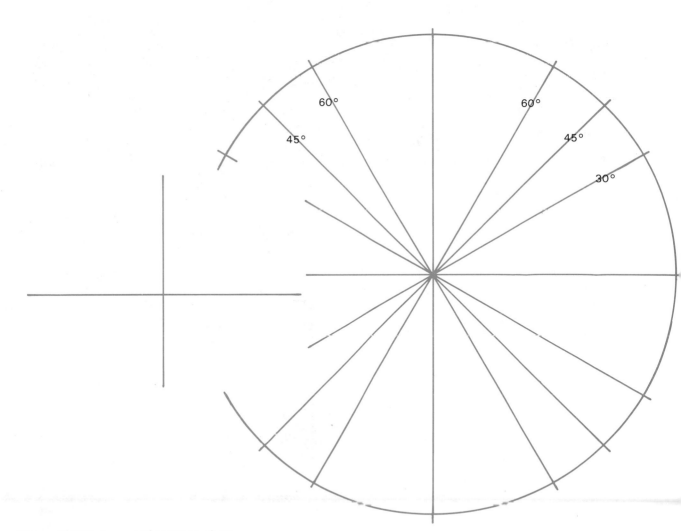

Figure 3–72A A general-purpose chart

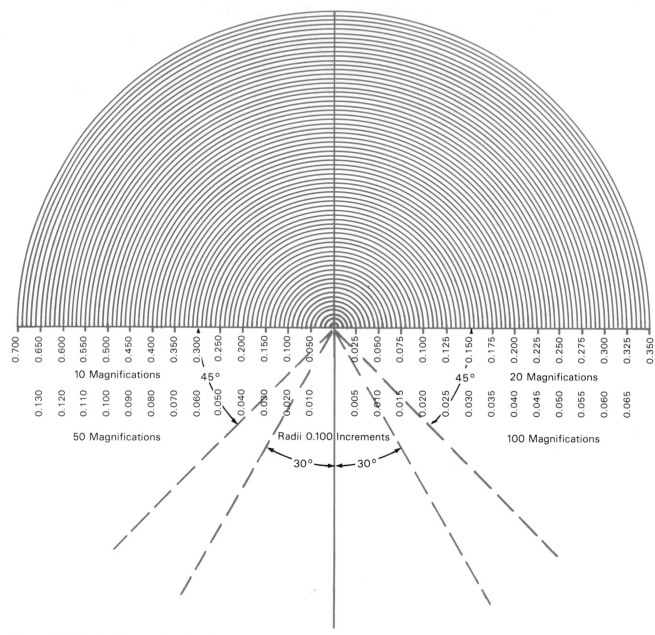

Figure 3–72B A radius and angle chart

DIGITAL READOUTS

Digital readout instruments will give accurate, precise visual readings of a distance moved on an axis. As an example, the movement of a micrometer spindle will activate a small counter that will allow sizes to be read directly on the side of the frame rather than on the scales themselves. Similarly, a digital height gage will allow the same direct vertical readings.

Electrically or electronically, this can be done on a milling machine or on any other machine tool. A simple digital instrument can measure the distance moved by the milling machine table on the x-axis, or a more elaborate digital instrument can measure the distance moved by either the x, y, or z axes. Large, easily readable numbers are flashed on a small screen as movement takes place. Measurements can be made in either English (U.S. Customary) or metric by a flick of a switch.

Figure 3–74 A modern, bench-model, contour projector with a dual-axis, digital, electronic readout *(Courtesy of Optical Gaging Products Inc.)*

Figure 3–73 A modern, floor-model, contour projector with a computer system and printer *(Courtesy of Optical Gaging Products Inc.)*

REVIEW QUESTIONS

1. What is the purpose of an optical comparator?
2. Explain the operating principle of an optical comparator.
3. List four recent innovations in optical comparator design.
4. What does a digital readout instrument actually do?

CHAPTER FOUR

LAYOUT OR MARKING OFF

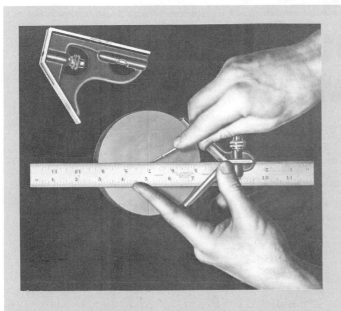

Before actual cutting is done, the machinist must know where the cutting is to be done, how much material is to be removed, what size holes are to be drilled and where, and so on. Gathering this information requires a thorough knowledge of blueprint reading and the ability to plan the work required to do the job. The machinists must also have the ability to transfer the blueprint information to the work.

Layout in the machining process is a very critical step. The machinist must be able to accurately transfer the blueprint information to the work. This transfer is accomplished by marking the necessary guidelines directly on the work. The work must then be set up accurately using the layout lines as a guide. Finally, the work must be cut to exact sizes with the layout lines again being used as a guide.

THE SAFE USE OF LAYOUT TOOLS

Layout tools are also hand tools. Thus, the potential dangers of misuse of hand tools always apply to layout tools as well.
1. When doing layout work, be neat, precise, and clean.
2. Be sure that tools are clean and free of grease or oils.

3. Before starting layout work, be sure that your hands are clean and free of grease or oils.

4. Gather all of the tools you need for a particular project and arrange them neatly where you need them.

5. Be sure that tools such as scribers, center and

prick punches, and so on, are sharp and in good condition at all times.

6. Check punches for mushroomed heads. Keep them dressed properly at all times.
7. Avoid carrying sharp tools in your pockets.
8. Scribers have a single purpose — to mark lines on the work. They are not weapons and should not be used as such, even in fun.
9. Hammers should always be kept in first-class condition. Check for damaged or loosened heads and damaged handles.
10. On completion of work, clean and return tools to their proper places and clean the layout table thoroughly. Use a safety cover if one is available.

UNIT 4-1

INTRODUCTION TO LAYOUT

OBJECTIVES

After completing this unit, the student will be able to:

* decide what lines are necessary for laying out work.
* prepare different surfaces so that layout lines will show up clearly.

KEY TERMS

Laying out or marking off
Scribe
Guidelines
Layout lines
Base line
Copper sulphate solution
Layout die

LAYING OUT

Laying out or marking off, Figure 4–1, is the operation of scribing guidelines on work. Straight lines, curved lines, arcs, and circles act as guides for the machinist to cut to or to set up to, or both. The accuracy of the layout depends on the work to be done. To set layout tools such as dividers accurately, micrometers or vernier calipers are often used.

Layout lines can be marked on either rough or smooth surfaces. A circle may be needed so that a hole can be drilled or bored accurately and on center. A line may be needed on a casting to mark the depth that a machinist will cut to. A **base line** may be needed so the machinist can set up the work. The base line is a line scribed on the work at its base. All measuring and setting up of the work is done from the base line rather than from a surface of the work that may or may not be accurate. The work can also be set level according to the base line.

Figure 4–1 Typical layout work being performed by a journeyman machinist *(Courtesy of Cominco Ltd.)*

Figure 4–2A Brushing on layout dye *(Courtesy of The L. S. Starrett Company)*

SURFACE PREPARATION

Several methods are used to prepare surfaces so that scribed lines will show up clearly. A white solution of latex paint, thinned with water can be painted on any metal surface. This material dries quickly to white. White chalk can also be rubbed on a surface to turn it white. A copper sulphate solution will turn a smooth steel surface to a copper color. Heat will turn a smooth steel surface blue. Blue layout dye thinly brushed or sprayed on any metal surface will dry quickly, Figure 4–2. Of the methods listed, the three most commonly used are the white paint, copper sulphate, and blue layout dye.

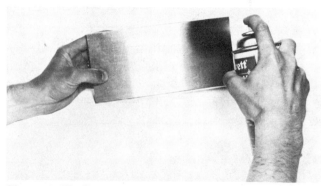

Figure 4–2B Spraying on layout dye *(Courtesy of The L. S. Starrett Company)*

REVIEW QUESTIONS

1. What is the operation of laying out work and why is it necessary?
2. What is a base line?
3. Why is all measuring and setting up of the work done from the base line?
4. Describe five different ways of preparing a surface for layout.
5. List four safety considerations relating to layout work.

UNIT 4-2

LAYOUT TOOLS: SURFACE PLATES, HAMMERS, SCRIBERS, CENTER PUNCHES, AND PRICK PUNCHES

OBJECTIVES

After completing this unit, the student will be able to:

- state why surface plates are valuable in layout work and machining.
- use a good surface plate properly and accurately.
- use hammers, scribers, and punches properly.

KEY TERMS

Surface plate

Reference surface

Black granite surface plate

Master pink granite surface plate

Crystal pink granite surface plate

Striking tool

Ball pein hammer

Pocket scriber

Improved scriber

Center punch

Prick punch

Automatic center punch

COMMON LAYOUT TOOLS

One of the most important layout tools is a good surface plate. Other common layout tools are ball pein hammers, center punches, prick punches, scribers, dividers, hermaphrodite calipers, squares, surface gages, vernier height gages, trammels, keyseat clamps, V-blocks, and angle plates.

SURFACE PLATES

A surface plate is an accurate, flat reference surface from which final dimensions are taken in testing or inspecting work for accuracy. Surface plates are also used for laying out work before machining is done and for testing a machined surface for flatness. Common cast-iron surface plates are well-ribbed to prevent warping. They are hand scraped flat to a high degree of accuracy. These plates are usually finished in sets of three, being scraped in rotation, one against the other.

Another common type of surface plate is the heavy granite plate, Figures 4–3 and 4–4. This type of plate is made with an accuracy ranging from 0.0015 in to 0.00005 in of flatness. Granite plates are made in master pink, crystal pink, and black granite. The master pink plate is probably the best for flatness, wear life, surface finish, and accuracy under load. The life of granite plates is influenced by the percentage of quartz present. Master pink plates contain crystalline quartz. Black granite plates contain no crystalline quartz, but take a fine finish and meet flatness requirements equally well.

HAMMERS

A **hammer** is a striking tool with hardened and tempered ends and a soft central section fitted with a handle. In layout work, the main purpose of the hammer is to strike the center punches or prick punches. It is poor practice to use a hammer directly on a surface plate.

The most common hammer used in the machine shop is the ball pein hammer. The rounded top is called the pein. The most common sizes used for layout work are the 8 oz (227 gm) and the 16 oz (454 gm).

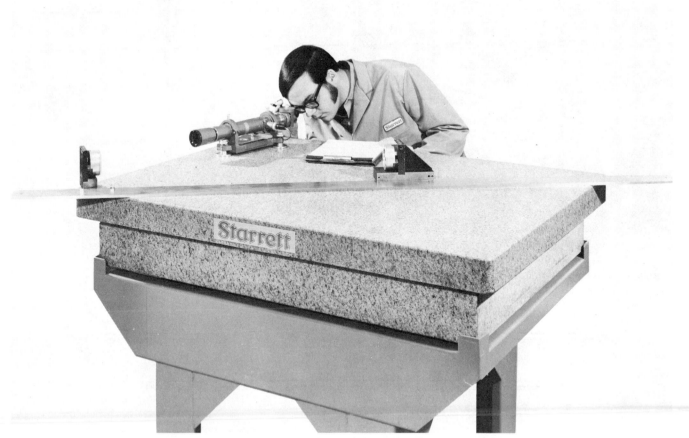

Figure 4–3 Testing a large granite surface plate for flatness *(Courtesy of The L. S. Starrett Company)*

SCRIBERS

Scribers of various styles are used for marking lines on the surfaces of the work. The extremely sharp scriber points are hardened and tempered. Under no circumstances should a scriber ever be struck with a hammer. The pocket scriber, Figure 4–5, can be fitted with either carbide or regular points. The curved and straight points of the improved scriber, Figure 4–6, are replaceable.

Figure 4–5 A pocket scriber *(Courtesy of The L. S. Starrett Company)*

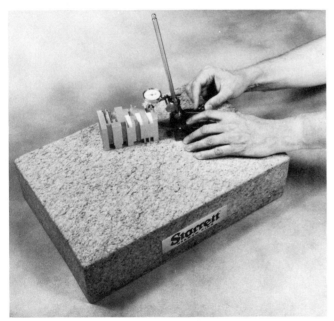

Figure 4–4 Testing a work surface for flatness *(Courtesy of The L. S. Starrett Company)*

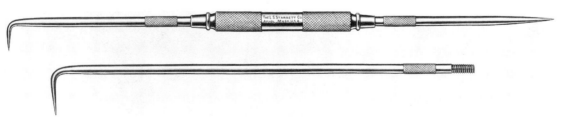

Figure 4–6 The improved scriber *(Courtesy of The L. S. Starrett Company)*

CENTER PUNCHES AND PRICK PUNCHES

Punches have hardened and tempered points. The heads, which are struck with the hammer, are left soft to prevent chipping of hardened metal when two hardened surfaces are struck together. The prick punch has a 30-degree point. Its main use is for light marking of hole centers, hole circles, or lines. The center punch, Figure 4–7, is heavier than the prick punch. Point angles vary from 60 degrees to 90 degrees, depending on use. The most common point angle used in machine shop practice is 60 degrees. The center punch is used for heavier marking. One of the best and easiest ways to sharpen a center punch is to hold it vertically and lean it into the grinding wheel at an angle equal to half of the desired included angle of the point, Figure 4–8. The punch is easier to revolve in this position, the point will be hollow ground, you will be grinding away from the point to prevent a wire edge from forming, and the point will be much neater looking. This is a better method of sharpening a center punch as compared to holding the punch in a horizontal position.

An automatic center punch, Figure 4–9, has an inner spring mechanism which eliminates the need for a hammer. The automatic punch is located on the desired spot and pressed straight down. The releasing mechanism inside allows the punch to snap down, driving it into the work. The force of the blow can be regulated by adjusting the knurled cap to tighten or loosen the spring.

Some of the uses for center punches and prick punches are shown in Figure 4–10, the layout of a hole; Figure 4–11, punching the center; Figure 4–12A, scribing a temporary straight line; and Figure 4–12B, scribing and prick punching a permanent straight line.

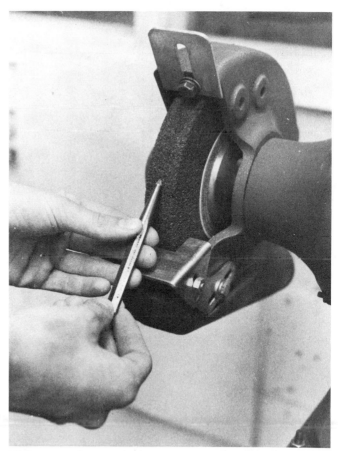

Figure 4–8 Sharpening a center punch

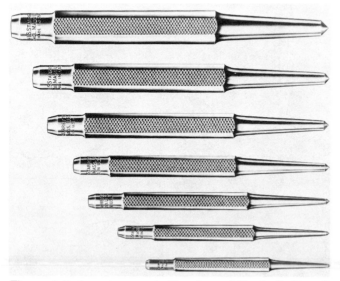

Figure 4–7 A center punch *(Courtesy of The L. S. Starrett Company)*

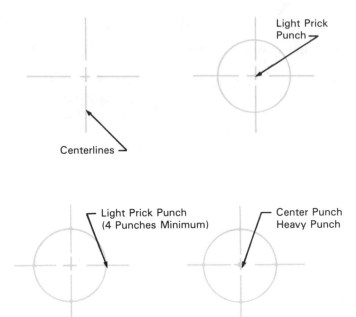

Figure 4–9 An automatic center punch *(Courtesy of The L. S. Starrett Company)*

Figure 4–10 Layout of a hole

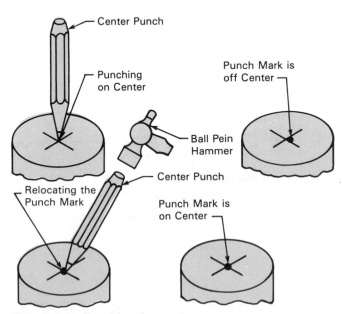

Figure 4–11 Punching the center

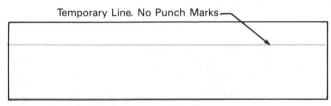

Figure 4–12A Temporary straight line

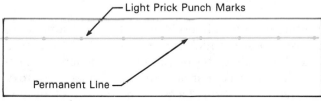

Figure 4–12B Permanent straight line

1. What is a surface plate?
2. What is the accuracy of heavy granite surface plates for flatness?
3. List four characteristics of a master pink granite surface plate by which it is compared to other surface plates.
4. What use is made of hammers in layout work?
5. What is the most common hammer in a machine shop?
6. What is a scriber and what is its primary purpose in layout work?
7. What is the difference between a prick punch and a center punch?
8. Describe the preferred method of sharpening a center punch.
9. Describe the action of an automatic center punch.
10. List three uses for center punches and prick punches.

UNIT 4-3

LAYOUT TOOLS: DIVIDERS AND HERMAPHRODITE CALIPERS

OBJECTIVES

After completing this unit, the student will be able to:

- use dividers properly.
- use hermaphrodite calipers properly.

KEY TERMS

Divider
Hermaphrodite calipers
Locating a center

DIVIDERS

A divider is a tool similar to a compass, Figure 4–13. The extremely sharp points on both legs are hardened and tempered. Dividers are used mainly for scribing circles or stepping off distances, Figure 4–14.

HERMAPHRODITE CALIPERS

Hermaphrodite calipers, Figure 4–15, are sometimes referred to by the nicknames "Jennies" or "odd legs." The single, sharp scriber point is hardened and tempered. One use of hermaphrodite calipers is approximating the location of a center on the end of a round shaft, Figure 4–16. A similar use is locating a center in a hole where no center exists. The hole must first be bridged with a thin, flat piece of wood. The hermaphrodites are set with the bent foot turned out. The center is then located in a similar manner, Figure 4–17, with the four scribed arcs being marked on the wooden bridge. These calipers can be set on a scale, Figure 4–18, and used to locate and scribe shoulder lengths on a lathe, Figure 4–19. Scribing lines parallel to the edge of the work is another typical use, Figure 4–20.

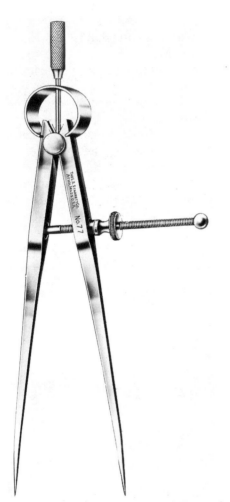

Figure 4–13 Spring dividers *(Courtesy of The L. S. Starrett Company)*

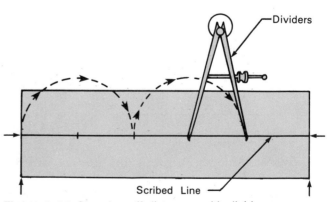

Figure 4–14 Stepping off distances with dividers

Figure 4–15 Hermaphrodite calipers *(Courtesy of The L. S. Starrett Company)*

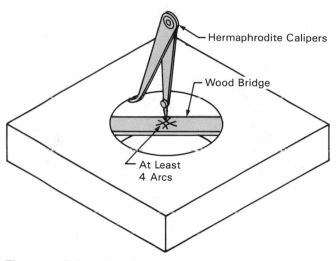

Figure 4–17 Locating the center of a hole

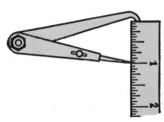

Figure 4–18 Setting the hermaphrodite caliper *(Courtesy of South Bend Lathe Inc.)*

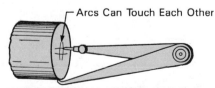

Figure 4–16 Locating the center of a round shaft. The arcs can touch each other *(Courtesy of South Bend Lathe Inc.)*

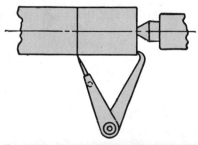

Figure 4–19 Locating the shoulder length on a lathe *(Courtesy of South Bend Lathe Inc.)*

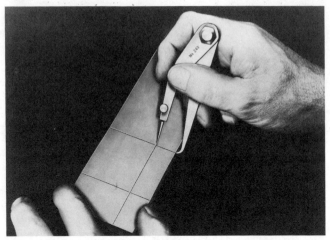

Figure 4–20 Scribing lines parallel to an edge *(Courtesy of The L. S. Starrett Company)*

REVIEW QUESTIONS

1. What is the purpose of a divider?
2. What is a hermaphrodite caliper?
3. Using a hermaphrodite caliper, how would you locate the center of a shaft?
4. Using a hermaphrodite caliper, how would you locate the center of a hole where no center exists?
5. List two other typical uses of the hermaphrodite caliper.

UNIT 4-4

LAYOUT TOOLS: SQUARES, PROTRACTORS AND DEPTH GAGES, SURFACE GAGES, AND VERNIER HEIGHT GAGES

OBJECTIVES

After completing this unit, the student will be able to:

- use squares, protractors and depth gages, surface gages, and vernier height gages properly and accurately.
- use surface gages and vernier height gages in conjunction with surface plates.

KEY TERMS

Square
Combination square
Protractor head
Center head

Protractor and depth gage
Surface gage
Vernier height gage

SQUARES

In layout work, a precision square, Figure 4–21, is normally used to scribe a line perpendicular to a straight edge, Figure 4–22, or to align work, Figure 4–23. The combination square is 90 degrees on one edge and 45 degrees on the other. It can be used to scribe a 90-degree line using one edge, Figure 4–24, and a 45-degree line using the other. It can also be used for checking a length, Figure 4–25. The protractor head, Figure 4–26, can be set to any angle for the scribing of lines. The center head, Figure 4–27, will locate the scale across the center of a round shaft. Usually, at least three lines are drawn to intersect each other at the center of the shaft. The combination square, protractor head, and center head make up the combination square set.

PROTRACTOR AND DEPTH GAGE

The protractor and depth gage, Figure 4–28, is made in different styles. This particular style is very useful and convenient for measuring and scribing angled lines on the work. By adjusting the scale, it can also be used as a depth gage to measure hole or slot depths.

SURFACE GAGE

The surface gage, Figure 4–29, is one of the most useful tools for a machinist to own. It is used to set up work accurately on machine tools and it is also used in layout work. The extremely sharp scriber points are hardened and tempered. In layout work, the surface gage is normally used in conjunction with the surface plate, as in setting a height, Figure 4–30 (page 88). A surface gage can be used to locate the center of a round shaft, Figure 4–31 (page 88). Once a height is set, lines parallel to the surface plate can be scribed on the work both easily and accurately, Figure 4–32 (page 88). Testing work for height, Figure 4–33 (page 88), is also a common use.

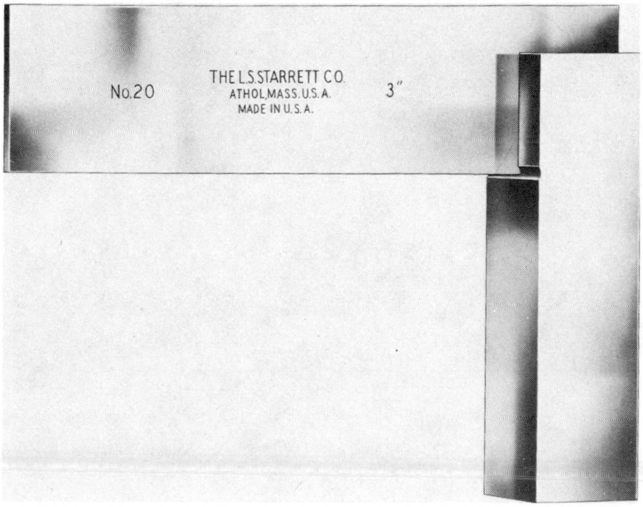

Figure 4–21 A precision square *(Courtesy of The L. S. Starrett Company)*

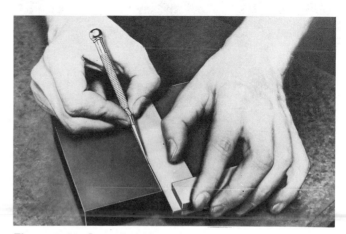

Figure 4–22 Squaring a line with a precision square *(Courtesy of The L. S. Starrett Company)*

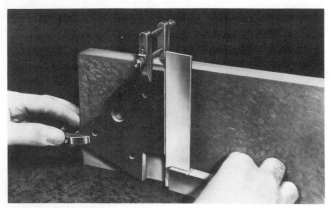

Figure 4–23 Aligning work with a precision square *(Courtesy of The L. S. Starrett Company)*

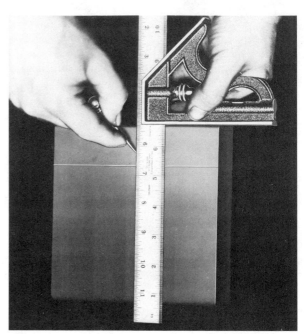

Figure 4–24 Squaring a line with a combination square *(Courtesy of The L. S. Starrett Company)*

Figure 4–25 Checking a length with a combination square *(Courtesy of The L. S. Starrett Company)*

Figure 4–26 Checking an angle with a protractor head *(Courtesy of The L. S. Starrett Company)*

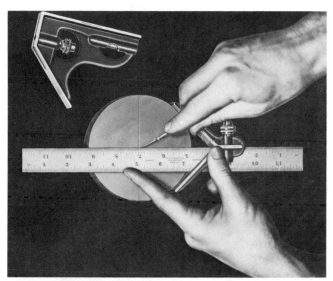

Figure 4-27 Locating center on a round shaft with a center head *(Courtesy of The L. S. Starrett Company)*

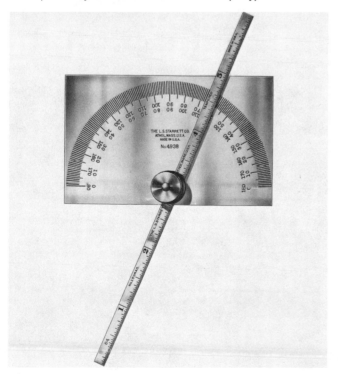

Figure 4-28 The protractor and depth gage *(Courtesy of The L. S. Starrett Company)*

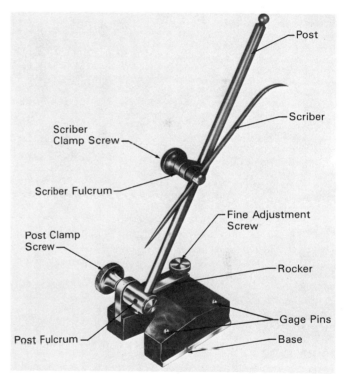

Figure 4-29 A surface gage *(Courtesy of The L. S. Starrett Company)*

Figure 4–30 Setting a surface gage to height *(Courtesy of The L. S. Starrett Company)*

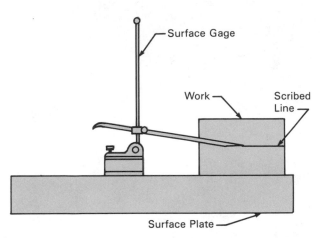

Figure 4–32 Scribing parallel lines with a surface gage

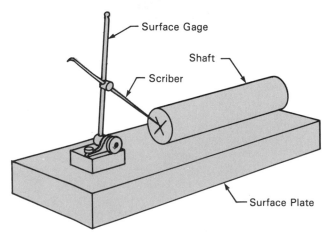

Figure 4–31 Locating the center of a round shaft using a surface gage

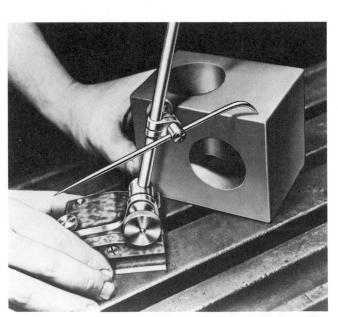

Figure 4–33 Testing work for height *(Courtesy of The L. S. Starrett Company)*

VERNIER HEIGHT GAGE

The vernier height gage, Figure 4–34, can be easily and accurately set to a height within 0.001 in. It is an extremely accurate layout tool. A vernier height gage used in conjunction with a surface plate, a V-block, and an angle plate can be used to scribe lines parallel to a surface plate, Figure 4–35. This illustrates one of the important uses for this tool.

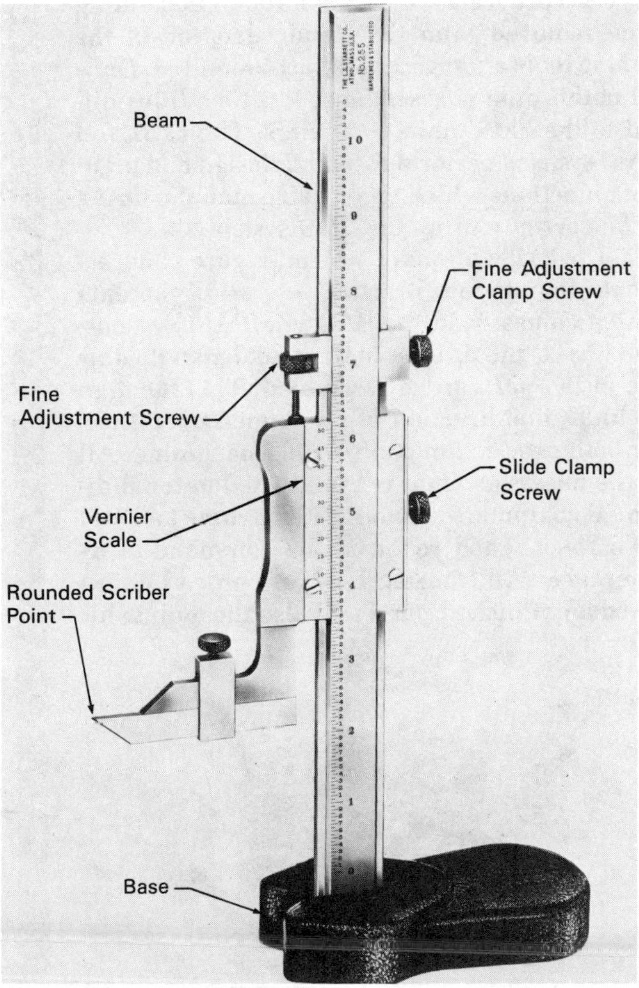

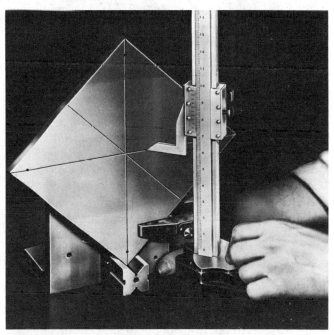

Figure 4–35 Using a vernier height gage to scribe lines parallel to a surface plate *(Courtesy of The L. S. Starrett Company)*

Figure 4–34 A vernier height gage *(Courtesy of The L. S. Starrett Company)*

Beam

Fine Adjustment Clamp Screw

Fine Adjustment Screw

Vernier Scale

Slide Clamp Screw

Rounded Scriber Point

Base

REVIEW QUESTIONS

1. What is a square and for what purpose is it used mainly?
2. List the parts of the combination square.
3. Describe the protractor and depth gage.
4. For what purpose is the protractor and depth gage used?
5. What is a surface gage?
6. List three uses of a surface gage.
7. What is a vernier height gage?

CAD AND CAM IN MANUFACTURING

The modern manufacturing plant is comprised of an integrated arrangement of a Computer Assisted Design system (CAD), a Computer Assisted Manufacturing system (CAM), and a Flexible Manufacturing System (FMS).

The CAD system, Figure 1, is used to design the parts to be manufactured and the machines to be assembled. In addition, it is also necessary to design the jigs and fixtures required to hold the parts on the machine tools during the machining time. Compared to more conventional methods of designing and drafting, this system can significantly shorten the time required. Computers, designing and drawing monitors, and drawing plotters are all tools used by the design engineer to perform tasks such as the structural analysis of the machines being designed and the drawing of all of the parts to be made. Lists of raw materials required to make the parts, drawings of the patterns needed for the molding and casting of the rough castings in the foundry, bills of materials, and drawings of the jigs and fixtures are all part of the work of the design engineer and can all be generated on the CAD system. The work of the design engineer must be coordinated with that of the CAM system.

The CAM system, Figure 2, allows the design engineer to assist in the production engineering tasks that lead to the FMS system where the actual production of the machined parts and the assembled machinery takes place. The work done by the CAM system includes the following: a determination of the different cutting processes required to do the work, selection of the cutting tools required, defining the cutting conditions that will occur during the machining of each of the parts, the making up and compiling of all of the CNC programs, the estimation of machining time required, and the standardization of the cutting tools and jigs and fixtures required. Once all of this data is assembled, it is then transmitted to the FMS main computers. The CAD and CAM systems perform essential design and planning functions which then enable manufacturing to be carried out by the FMS system.

The FMS computer system, Figure 3, implements or puts into practical use all of the data which comes from the CAD and CAM systems, and directs the process of the actual manufacturing of the parts to be assembled. This function includes the direction of unmanned operations for both production control and machining. All of the necessary data is transmitted to the individual computers of each CNC machine tool and each robot. Each robot, at the command of its computer, will transfer, load and unload unfinished and finished parts and also the tooling for

Figure 1 A typical Computer Assisted Design system *(Courtesy of Bausch & Lomb)*

Figure 2 A typical Computer Assisted Manufacturing control *(Courtesy of Computervision Corp.)*

the machine tools. The machine tools, at the command of their computers, will do all of the required machining of each of the required parts. Some shop floor controllers are designed for unmanned stations such as robots which perform sequences of operations that are stable or unchanging over a long period of time. Other more flexible controllers will, in addition, collect and transmit production data from either the shop floor or from a more remote location to a central computer.

The FMS system directs the total machine shop process which includes the transferring of used and new tooling to and from the tool room and each machine tool, the transferring of new stock or castings to the incoming conveyors for each cell, the loading and unloading of machine tools, and the transferring of machined parts from the outgoing conveyors to the storage room or storage area. At the same time the computer system accumulates, updates and reports production data such as the running total of each unmachined part throughout the manufacturing cycle. Each cutting tool — new, worn or broken — is noted. New cutting tools are noted and trans-

ferred as needed, while worn or broken tools are noted and transferred for replacement. Each tool must be cataloged and tracked at all times, as well as the condition of each. Virtually every shop task concerned with production or support systems must be known at all times.

The total system consisting of Computer Assisted Design, with Computer Assisted Manufacturing, and the Flexible Manufacturing Systems, makes up the design capabilities, the production engineering, and manufacturing capabilities of modern manufacturing plants. A typical layout of such a modern factory is illustrated in Figure 4. Machine control is interfaced or coordinated with a master computer control station capable of monitoring more than 200 machine functions. The computer will also perform fault analysis. It can simulate parts processing cycles so that parts scheduling requirements can be determined accurately. If a cutting tool becomes dull or is broken, and a part is not machined correctly, it will be noted immediately. A new cutting tool will be substituted, and machining will be carried on as usual.

Figure 3 A Flexible Manufacturing control unit *(Courtesy of Cincinnati Milacron)*

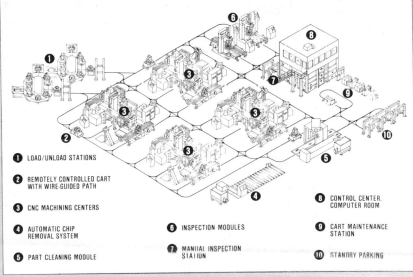

1 LOAD/UNLOAD STATIONS

2 REMOTELY CONTROLLED CART WITH WIRE-GUIDED PATH

3 CNC MACHINING CENTERS

4 AUTOMATIC CHIP REMOVAL SYSTEM

5 PART CLEANING MODULE

6 INSPECTION MODULES

7 MANUAL INSPECTION STATION

8 CONTROL CENTER. COMPUTER ROOM

9 CART MAINTENANCE STATION

10 STANDBY PARKING

Figure 4 Typical layout of a FMS *(Courtesy of Cincinnati Milacron)*

UNIT 4-5

LAYOUT TOOLS: TRAMMELS, KEYSEAT CLAMPS, V-BLOCKS, AND ANGLE PLATES

OBJECTIVES

After completing this unit, the student will be able to:

- use trammels properly.
- use keyseat clamps and V-blocks correctly.
- use angle plates properly.
- use V-blocks and angle plates with surface plates.

KEY TERMS

Trammel	Keyseat clamps
Wooden beam	V-blocks
Beam compass	Angle plates

TRAMMELS

Trammels, Figure 4–36, are very handy for measuring and layout that is normally beyond the capacity of smaller calipers and dividers. The caliper and divider legs are usually clamped to a wooden beam that can be any length. A ball can be placed on the end of one leg to provide a center for a hole where no center exists.

A pencil can be substituted for a scriber leg to convert the trammel into a beam compass.

KEYSEAT CLAMPS

Keyseat clamps, Figure 4–37, will hold a rule parallel to the center of a round shaft. A line can then be scribed lengthwise, parallel to the center of the shaft.

V-BLOCKS

V-blocks are often used on a surface plate, especially when working with round shafting, Figure 4–38. To find the center of a round or square shaft, set the shaft on the V-blocks and set a surface gage or vernier height gage to the center of the shaft, Figure 4–39. Scribe a line across the end and rotate the shaft. Then scribe at least two more lines to intersect the center of the shaft.

ANGLE PLATES

Angle plates, Figure 4–40, are used in conjunction with surface plates. Their main use is to support work in a vertical position. Clamped or bolted to an angle plate, work can be marked and machined. A typical use in layout is illustrated in Figure 4–35 with the vernier height gage in position. Angle plates are either fixed at 90 degrees or are adjustable to different angles. Figure 4–41 illustrates an angle plate used to hold a rule in a vertical position for setting a surface gage to a specific height.

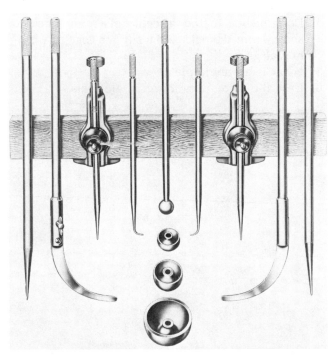

Figure 4–36 Trammels *(Courtesy of The L. S. Starrett Company)*

Figure 4–38 A shaft mounted on V-blocks *(Courtesy of The L. S. Starrett Company)*

Figure 4–37 Keyseat clamps *(Courtesy of The L. S. Starrett Company)*

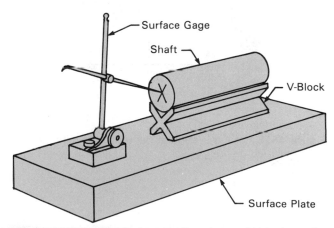

Figure 4–39 Centering a shaft using a V-block and a surface gage on a surface plate

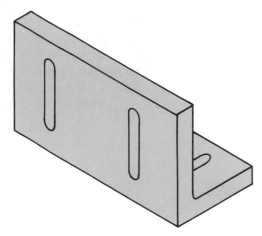

Figure 4-40 An angle plate

1. What is a trammel?
2. What are keyseat clamps?
3. What are V-blocks?
4. Describe how to find the center of a round or square shaft using V-blocks and a surface gage or vernier height gage.
5. What is an angle plate?
6. Describe some uses for an angle plate.

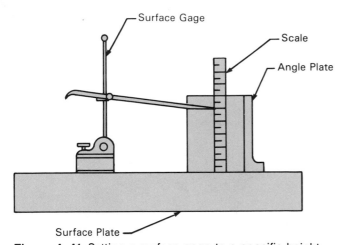

Figure 4-41 Setting a surface gage to a specific height

CHAPTER FIVE

BENCHWORK

Wherever there is machinery there are moving parts. These parts will wear out or they may break down and require replacement. Often it is the responsibility of the machinist to dismantle the defective machine, rebuild and replace the defective part, and reassemble the machine so that it can be returned to service.

The work of rebuilding and replacing defective parts involves the use of hand tools mainly, and possibly a few machine tools as well. Hand tools are equally as important as machine tools and must be looked after with the same dedication and care. A good machinist is just as proficient with hand tools as with machine tools.

HAND TOOL SAFETY

Hand tools may not appear to be dangerous, but any striking tool or cutting tool or sharp-edged tool is potentially dangerous. All that is needed to make them so is a careless or reckless operator.

Screwdrivers

1. Select screwdrivers to fit the screw head being used; e.g., don't use a small screwdriver in a big slot.
2. A screwdriver is not a chisel. Do not use it as such.
3. It is poor practice to strike a screwdriver handle with a hammer unless the handle is built for such treatment.

4. When grinding a screwdriver blade, do it properly. Avoid grinding a chisel edge.

Wrenches

1. Select open-end/box-end wrenches that fit properly. Avoid using wrenches that are loose (see Figure 5–8).
2. It is better to use a solid wrench than an adjustable wrench.
3. When an adjustable wrench is used, it is preferable to position the wrench so that the force acts against the solid jaw rather than against the moveable jaw (see Figure 5–9).

4. It is safer to pull on a wrench than to push it. Striking a wrench with the heel of the hand is good practice because the momentum of the hand becomes a factor in loosening the bolt or nut. Avoid using a hammer on a wrench.

Pliers

1. Use pliers for the purposes for which they are intended. Don't try to substitute pliers for a wrench.

Hammers

1. Never use a hammer with a loose handle.
2. If a handle is split or damaged, replace it.
3. If a hammer face is chipped, discard it and replace it.
4. Never strike two hammer faces together. Both are hardened and tempered. A flying chip from either one could become a lethal projectile.
5. Avoid hammering case-hardened or hardened and tempered surfaces with a hammer. If it is necessary, use protective material between the surface and the hammer.

Chisels

1. Avoid using a chisel with a mushroomed or burred head. Dress the head properly on a bench grinder (see Figure 5–20).
2. Be careful of flying chips. Use a protective screen to protect those around you.
3. Avoid getting oil or grease on chisel heads or hammer faces.

Scrapers

1. Store scrapers in a separate container away from other tools.

2. Keep scraper blades in protective sheaths to preserve sharpness.
3. Always use a handle on a scraper.
4. Keep scrapers sharp. Hone them often with a good oil stone.

Files

1. Always use a file handle (see Figure 5–37). Never file without a handle, except in the case of riffler files. They are not made for handles.
2. Keep files separate from other tools and from each other. Remember that they are cutting tools.
3. On a lathe, it is best to learn to file left-handed. Running a lathe in reverse and filing right-handed at the back of the lathe is poor practice because the controls are at the front of the lathe, out of reach.
4. Use the most suitable file for the work at hand.
5. Grip the file firmly and take a long flat stroke, about one stroke per second.
6. Make sure that your hands and work surfaces are clean and free of oil and grease before filing.
7. Discard broken handles.

Hacksaw

1. Use the right blade for the work at hand; i.e., the correct number of teeth.
2. Do not saw too fast. Use approximately one stroke per second.
3. When the saw starts to break through the work, ease up on the pressure and slowly "feel" the saw through the material.
4. The cutting stroke is forward. Check the teeth to see that they point forward. Apply cutting pressure only on the forward stroke.

UNIT 5-1

BENCHWORK TOOLS

OBJECTIVES

After completing this unit, the student will be able to:
- define the term benchwork.
- provide the proper care and storage of tools used in benchwork.

KEY TERMS

Benchwork
Floorwork
Aversion

INTRODUCTION TO BENCHWORK

A major part of a machinist's work lies in the construction, setup, maintenance, and repair of industrial machinery of all kinds. Machine tools, pumps, compressors, cranes, and hoists are typical examples of machinery a machinist must attend to. A large piece of work might be placed on the floor and be called floorwork. Smaller work might be done on a bench and be referred to as **benchwork**. The same tools, however, are used regardless of the size of the work or the location in which it is carried out. Benchwork tools include wrenches, hammers, screwdrivers, pliers, hacksaws, punches, chisels, pry bars, taps, dies, measuring instruments, and various portable power tools.

Good machinists know how to use and care for all of their tools. They take great pride in their tools and in the condition of their tools. Using the proper tool for the job is important to them. In fact, professionals have an aversion to using the wrong tool for the job, e.g., a sledgehammer on a prick punch, or a screwdriver in place of a cold chisel. They will avoid using tools in poor condition, such as a cold chisel with a mushroomed head.

CARE OF TOOLS

When tools are not in use, they should be stored in a safe and proper place. It is best to have a separate storage area for each tool. Never pile tools, particularly files, on top of each other. All tools should be sharpened and cleaned before storage so they will be in good condition for the next job.

REVIEW QUESTIONS

1. Define the term benchwork.
2. Explain the difference between benchwork and floorwork.
3. Name 12 tools used in benchwork.

4. What is the most important consideration in using tools in benchwork?
5. List three considerations in the care and storage of benchwork tools.

UNIT 5-2

USING BENCHWORK TOOLS: HAMMERS, SCREWDRIVERS, AND WRENCHES AND PLIERS

OBJECTIVES

After completing this unit, the student will be able to:

- use hammers, screwdrivers, wrenches, and pliers properly and safely.
- maintain hammers, screwdrivers, wrenches, and pliers.

KEY TERMS

Hammer
Striking tool
Ball pein hammer
Sledgehammer
Soft-face hammer
Screwdriver
Standard screwdriver blade
Stubby screwdriver

S wrench
T wrench
Tap wrench
Pipe wrench
Adjustable wrench
Turning force
Pliers
Rib joint pliers

Offset screwdriver
Wrench

Long nose pliers
Vise grip pliers

HAMMERS

Hammers are striking tools. There are many different styles, each with a particular type of head mounted on a handle. Each hammerhead has a hole in its center called an eye. The eye is used to receive a handle. A hardwood wedge and a soft steel wedge are driven into the exposed end of any wooden handle to keep the head in place. The handle must be square with the head before the wedges are driven home. Afterwards, the handle must be tight. A loose head is always dangerous, as it may fly off suddenly and ruin work or cause serious injury.

The ball pein hammer is made from a good grade of tool steel, Figure 5–1. The head is hardened and tempered on both of its striking surfaces. The center around the eye is left soft. This hammer is the one most commonly used by machinists. It comes in various sizes from 4 oz to 40 oz (113 gm to 1134 gm).

For larger jobs, a machinist may use a sledge (striking) hammer. Sledgehammers, Figure 5–2, range in size from 4 lb to 20 lb (1814 gm to 9072 gm).

Soft-face hammers, Figure 5–3, are used for forming soft metals such as copper or aluminum. They are also used to move parts and components into position without damaging or bruising their finished surfaces. This feature makes soft-face hammers useful for assembling or dismantling parts.

SCREWDRIVERS

Screwdrivers should be used solely for loosening and tightening screws. They should never be used as pry bars or chisels, or to test electrical circuits. Never hold work in your hand while using a screwdriver. The work and/or the screwdriver may slip and cause you serious injury.

Screwdrivers are made in many varieties and sizes, Figure 5–4. All can be said to consist of a handle, a shank, and a blade. The shank and blade are made of tool steel. The blade is normally hardened and tempered. The handle may be made of wood, plastic, or metal.

Figure 5–1 A ball pein hammer *(Courtesy of Stanley Tools)*

Figure 5–2 A sledgehammer *(Courtesy of Stanley Tools)*

Figure 5–3 A soft-face hammer *(Courtesy of Stanley Tools)*

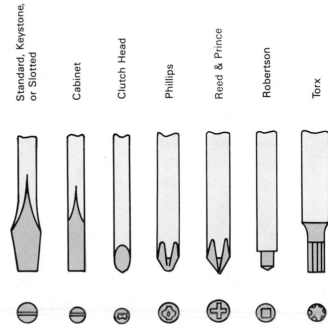

Figure 5–4 Types of screwdriver blades

Standard screwdriver blades are usually ground with tapered sides, Figure 5–5A. Such tapered sides tend to force the blade out of the slot. A correctly ground blade should have parallel sides that fit the slot exactly, Figure 5–5B.

Most screwdrivers are manufactured with round shanks, primarily for reasons of economy, Figure 5–6A. Square shanks, Figure 5–6B, are made from thicker steel. The square shape allows a wrench to be used on the shank to provide greater leverage. Stubby screwdrivers, Figure 5–6C, and offset screwdrivers, Figure 5–6D, are used where long straight shanks or handles will not fit.

WRENCHES AND PLIERS

Wrenches, Figure 5–7, are used primarily to tighten or loosen nuts and bolts. There are many types of wrenches. Some, such as "S" or "T" wrenches, are named for their shape, while others, such as pipe or tap wrenches, are named for the object they are designed to turn. Regardless of type, one or both ends of each wrench is designed to slip over the object to be turned and hold it securely. The shank of the wrench serves as a handle by which the user can then apply the turning force.

The wrench you select must fit snugly over the nut or bolt being turned, Figure 5–8. Adjustable wrenches have a moveable jaw that can be adjusted to fit various sizes of nuts and bolts. When using this type of wrench, it is preferable to position the wrench so that the force acts against the solid jaw rather than against the moveable jaw, Figure 5–9. Wrenches should be pulled, not pushed. A quick blow with the heel of the hand is effective when tightening or loosening

A. Round

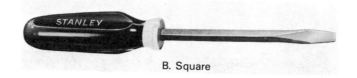
B. Square

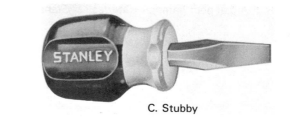
C. Stubby

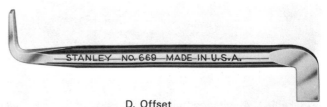
D. Offset

Figure 5–6 Screwdriver shanks *(Courtesy of Stanley Tools)*

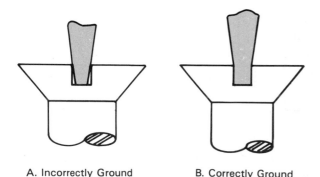

A. Incorrectly Ground   B. Correctly Ground

Figure 5–5 Standard screwdriver blades

nuts and bolts. Wrenches should not be used as hammers, nor should hammers or pipes be used on wrenches. Both practices are dangerous and can result in tool damage and possible injury.

Pliers, Figure 5–10, are tools used mainly to hold, turn, or bend objects. They are usually named for their construction, such as rib joint pliers or long nose pliers, or for their function, such as vise grip pliers. Special pliers are made for getting into hard-to-reach spots and for cutting wire or cable. Pliers should not be used in place of a wrench since slippage can easily result.

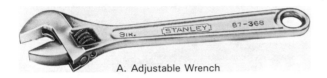

A. Adjustable Wrench

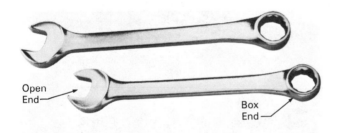

Open End

Box End

B. Combination Wrenchs

Figure 5–7 Wrenches *(Courtesy of Stanley Tools)*

Force

Figure 5–9 Force should act against the solid jaw with all adjustable wrenches

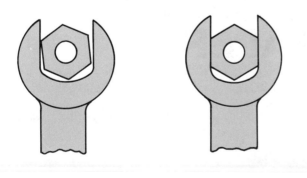

A. Too Loose   B. Correct Fit

Figure 5–8 All wrenches should fit snugly

REVIEW QUESTIONS

1. Name two hammers with tool steel heads commonly used in the machine shop.
2. Describe the uses of soft-face hammers in the machine shop.
3. Name four types of screwdriver blades.
4. Describe the characteristics of a properly ground screwdriver blade.
5. Describe how a wrench is used.
6. When using an adjustable wrench, which jaw should the force act against?
7. What precautions should be observed in the use of wrenches?
8. What is the primary purpose of pliers?

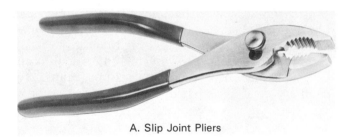

A. Slip Joint Pliers

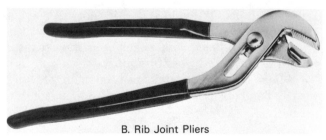

B. Rib Joint Pliers

C. Long Nose Pliers

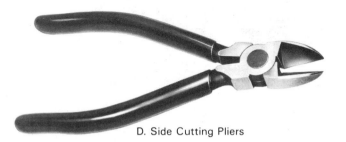

D. Side Cutting Pliers

Figure 5–10 Pliers *(Courtesy of Stanley Tools)*

UNIT 5-3

USING BENCHWORK TOOLS: HACKSAWS, CHISELS, AND SCRAPERS

OBJECTIVES

After completing this unit, the student will be able to:

• use the proper benchwork tools for the work being done.
• work safely with benchwork tools.
• maintain and care for hacksaws, chisels, and scrapers.

KEY TERMS

Hacksaw	Cold chisel
Hacksaw blade	Cape chisel
All hard blade	Flat chisel
Flexible blade	Round nose chisel
Pitch	Curved round nose
Negative rake	chisel
Blade tension	Diamond point chisel
Set of the blade	Parallel facet
Saw kerf	Tapered facet
Chisel	Mushroomed head
Chipping	Scraper
Flat chisel	Hand scraping

HACKSAW

Hacksaws are hand saws designed principally for cutting metal. They come with adjustable frames for holding blades ranging from 10 in (254 mm) to 12 in (305 mm) long. Two types of grips are commonly available: the pistol grip, Figure 5–11A; and the closed grip, Figure 5–11B.

Hacksaw blades are made from high-grade tool steel. This steel may be alloyed with metals such as tungsten or molybdenum. All hard blades are hardened and tempered throughout and stand up well in general use. They are, however, brittle and break if handled improperly. Flexible blades, on the other hand, have flexible, nonhardened backs. Only their teeth are hardened and tempered. As a result, they will withstand more abuse than all hard blades.

Some hacksaw blades have larger teeth than other blades. The number of teeth in a blade is determined by its pitch. Pitch is the number of teeth per inch (or the number of teeth per 25.4 mm). Typical hacksaw blades have pitches ranging from 14 to 32 teeth per inch. Each pitch is designed for use with a particular type of work, as shown in Table 5–1.

Choosing a blade with the correct pitch for a particular piece of work is important. The most common cause of blade breakage is the use of a blade with incorrect pitch for the work being done. Blades with small pitch numbers (large teeth) provide greater chip clearance and will cut through the metal faster. At least two teeth must remain in contact with the work at all times, Figure 5–12, or tooth breakage will result. Thin work requires a higher pitch blade with many small teeth.

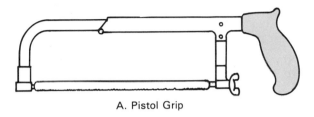

A. Pistol Grip

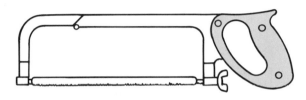

B. Closed Grip

Figure 5–11 Hacksaw grips *(Courtesy of Disston Canada Ltd.)*

Table 5–1 Hacksaw Pitches For Different Uses

PITCH	TYPE OF MATERIAL
14 TPI* (25.4 mm) regular set	Soft solid machinery steel, cast iron, brass, copper, aluminum, etc.
18 TPI (25.4 mm) regular or wavy set	General shop use, machinery steels, tool steels, pipe, angle iron, aluminum, etc.
24 TPI (25.4 mm) wavy set	Hard materials, drill rod, tubing, medium-size sheet metal, etc.
32 TPI (25.4 mm) wavy set	Thin sheet metal, tubing, stock less than 0.085 in (2.16 mm) in thickness.
*TPI = Teeth per inch	

After selecting a blade of the proper pitch, be sure to insert it in the saw frame with its teeth pointing forward, away from the handle, Figure 5–13A. Some blades may have a negative rake at the front end to aid in starting each cutting stroke. Place tension on the blade until it is taut and will not flex in use, Figure 5–13B.

Work to be cut should be mounted in a vise with the cut line as close to the vise jaws as possible, Figure 5–14. This will prevent spring and chatter. Each type

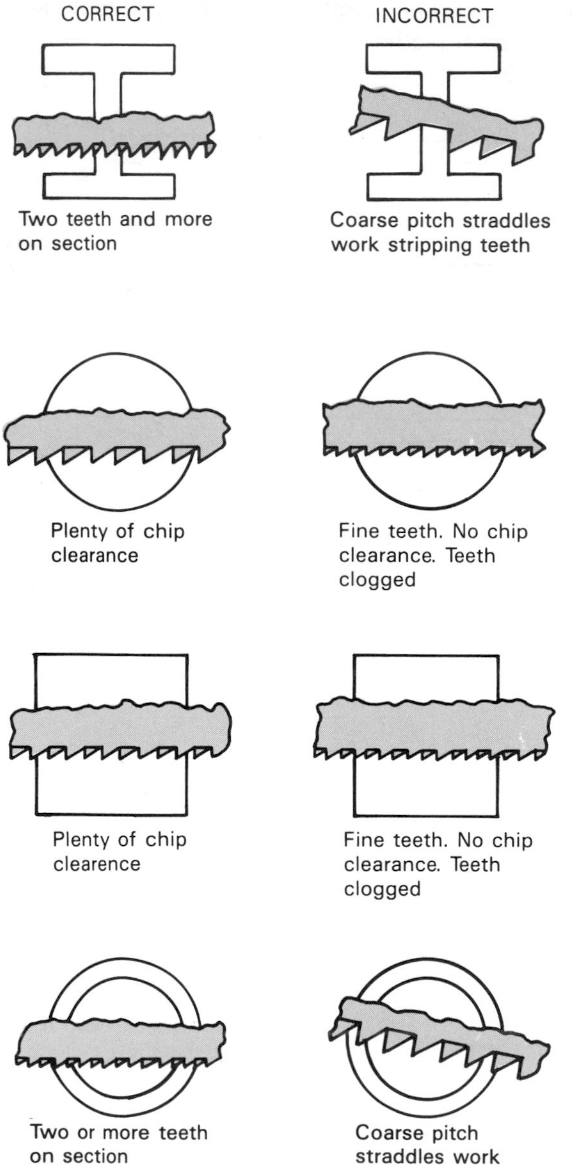

CORRECT   INCORRECT

Two teeth and more on section   Coarse pitch straddles work stripping teeth

Plenty of chip clearance   Fine teeth. No chip clearance. Teeth clogged

Plenty of chip clearence   Fine teeth. No chip clearance. Teeth clogged

Two or more teeth on section   Coarse pitch straddles work

Figure 5–12 Using the correct blade pitch *(Courtesy of Disston Canada Ltd.)*

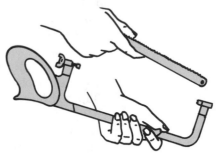

Figure 5–13A Inserting the blade, teeth pointing forward *(Courtesy of Disston Canada Ltd.)*

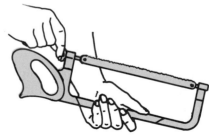

Figure 5–13B Placing tension on the blade *(Courtesy of Disston Canada Ltd.)*

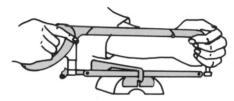

Figure 5–14 Holding a hacksaw properly, with the cut line close to the vise jaws *(Courtesy of Disston Canada Ltd.)*

of work requires a different means of mounting for efficient, accurate, damage-free cutting.

When cutting, hold the hacksaw at a slight angle to the work. If you are right-handed, your left hand should be positioned on the outermost edge of the frame and your right hand should hold the grip firmly. If you are left-handed, these positions should be reversed. Figures 5–15A and 5–15B demonstrate the proper firm grip for both hands while cutting sheet metal and a bolt thread.

As the cutting teeth point forward, force should be applied only on the forward stroke, not on the return stroke. There is a tendency to use too short a stroke. There is also a tendency to cut too fast. A speed of 50–60 strokes per minute is recommended. Near the end of the cut, slow down and use less pressure. "Feel" the saw through the end of the cut. After sawing is finished, clean the blade and slacken off the tension before putting the hacksaw away.

The teeth of any saw blade are set or bent alternately right and left of the plane of the blade. The set makes the **kerf**, or saw cut, wider than the blade thickness. This allows the saw blade to pass freely through the material being cut. With use, however, the blade will lose its set and the kerf will become narrower. For this reason, it is never wise to use a new blade in an old cut. The new blade will have a wider set than the kerf left by the old blade and thus will bind. A new cut should be started or the material should be turned over and the partially completed cut finished from the other side.

CHISELS

Chipping is the name given to the operation of cutting with a hand-held cold chisel to produce a flat surface, a curved surface, or a groove. This is done by striking the chisel head with a hammer. It takes great skill to do this operation properly.

Chisels are forged from a good, tough grade of carbon tool steel. The chisel blade, which is hardened and tempered, usually has a slightly convex cutting edge with one or two facets. The head is left blunt to receive hammer blows and is bevelled to prevent mushrooming.

Chisels are usually named according to the shape of their blades, i.e., flat, cape, round nose, curved round nose, diamond point. Flat chisels, Figure 5–16A, are used to shear sheet metal, cut flat surfaces, cut rivets, and split nuts. Cape chisels, Figure 5–16B, are used to cut keyways, grooves, corners, and slots. Round-nose chisels, Figure 5–16C, will cut round grooves such as oil grooves in bearings, and round fillets. Small round-nose chisels can be used in spot

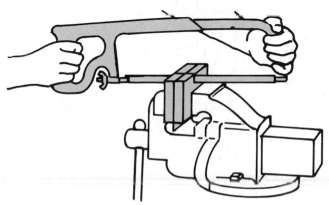

Figure 5–15A Cutting sheet metal held between pieces of scrap wood

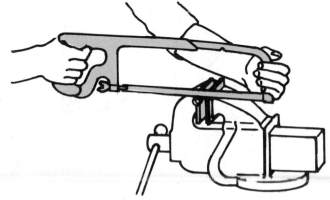

Figure 5–15B Cutting off a bolt thread with nut in position before cutting

drilling to pull drills back on center. Diamond-point chisels, Figure 5–17, are also used for this purpose, as well as for squaring corners in grooves.

Chisels are sharpened on a grinding wheel. They must be supported on the tool rest at the required angle, Figure 5–18A. The angle to be ground will depend on the type of chisel and the hardness of the material to be cut. An angle of about 60–70 degrees is suitable for most metals. A chisel with too large an angle will push metal off instead of shearing it off. It is also important that each facet be parallel rather than tapered. The tendency is to twist the chisel slightly

Figure 5–16A A flat chisel *(Courtesy of Stanley Tools)*

Figure 5–16C A round-nose chisel

Figure 5–16B A cape chisel

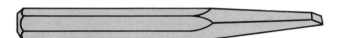

Figure 5–17 A diamond-point chisel

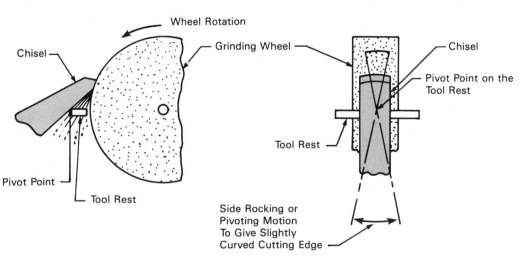

Figure 5–18A Grinding a flat chisel

when grinding it, producing a tapered facet. Efficient chipping cannot be performed with a tapered facet. Both facets should be equal in size and the cutting edge should be slightly convex for general-purpose work, Figure 5–18B.

It is best to mount work to be chipped in a vise. Whenever possible, chip toward the solid vise jaw rather than toward the moveable jaw. It is best to avoid chipping parallel to the jaws, Figure 5–19. Work slowly but steadily. Use sharp, quick blows and watch the cutting edge of the chisel, not the head. Do not try to remove too much material at once. Smaller chips are easier to control and metal removal is quicker.

A mushroomed head is never safe. Dress such heads properly on a grinding wheel before attempting to use them, Figure 5–20. Wear safety goggles when chipping and work behind a protective screen to ensure that others are not hurt by flying chips. Use the proper size hammer, and be certain that there is no grease or oil on the hammer face or chisel head.

SCRAPERS

Scraping is the name given to the operation of pushing or pulling off small metal chips with a very sharp, ground and honed hand tool called a scraper, Figure 5–21. Scraping is carried out for the following reasons: to produce a true flat surface, Figures 5–22 and 5–23, or a true curved surface; to produce surfaces that offer better oil retention; or to decorate an otherwise finished product.

Scrapers have hardened and tempered blades with little or no rake. Handles are usually made of wood or plastic.

Scraping is often carried out on bearing surfaces. A finer, truer surface can be achieved this way. Before starting any scraping operation, it is first necessary to mark the high spots on the work surface. This is done by applying a thin coat of Prussian Blue or Mechanic's Blue to the surface of a surface plate or plug gage. The blue is then transferred to the surface to be scraped by lightly rubbing the treated surface plate or plug gage on the work surface. A few heavy splotches will show

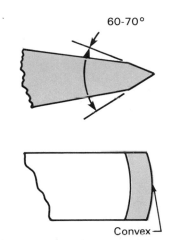

Figure 5–18B A chisel point ground for general-purpose work

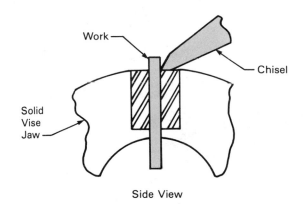

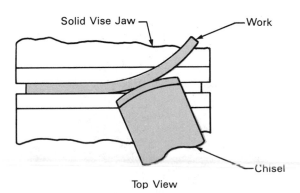

Figure 5–19 Shearing sheet metal with a flat chisel

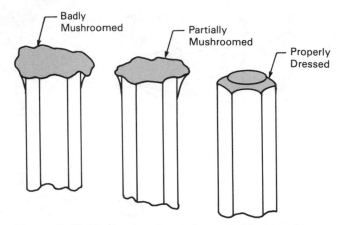

Figure 5-20 Mushroomed and dressed chisel heads

on the work surface at first. These are scraped, Figure 5-24, and the process is repeated. Short, even strokes should be used. Eventually the spots will grow in number until the surface has a great number of evenly distributed small spots, much like the pattern on a speckled trout. As a rule, not more than 0.002 in to 0.003 in (0.05 mm to 0.08 mm) should be removed from a surface by scraping.

When dull, scrapers are honed on a whetstone. The correct method for honing a flat scraper is shown in Figure 5-25.

For safe scraping, always use a handle on a scraper. Be sure to keep scraper edges razor sharp. After use, clean your scrapers and place them in protective sheaths. Store them separately from other tools, preferably in their own container.

Figure 5-23 Using a hook scraper to produce a flat surface

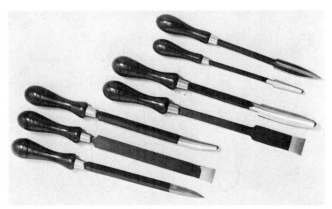

Figure 5-21 Typical hand scrapers *(Courtesy of Neill Tools)*

Figure 5-24 Holding a hook scraper for scraping

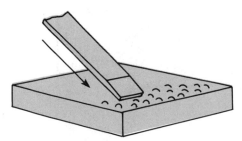

Figure 5-22 Using a flat scraper to produce a flat surface

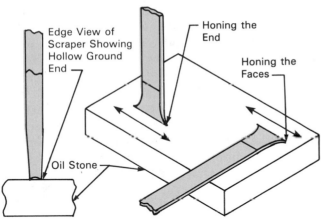

Edge View of Scraper Showing Hollow Ground End

Honing the End

Honing the Faces

Oil Stone

Figure 5–25 Honing a flat scraper

REVIEW QUESTIONS

1. Discuss the difference between all hard hacksaw blades and flexible hacksaw blades.
2. What is meant by the "pitch" of a hacksaw blade?
3. How many teeth are there per inch (25.4 mm) in a general-purpose hacksaw blade?
4. List three considerations in selecting the proper hacksaw blade for the job to be done.
5. Why do some blades have a negative rake at their front ends?
6. Describe the proper method of holding a hacksaw for cutting.
7. When using a hacksaw, which stroke is the cutting stroke?
8. What is meant by the set of the blade?
9. Explain why you should not start a new hacksaw blade in a cut made by an old hacksaw blade.
10. Define the term chipping.
11. Name four types of cold chisels.
12. List four precautions to be followed when chipping.
13. Give three reasons why scraping can be a valuable operation.
14. How are high spots marked on the surface to be scraped?

UNIT 5-4

USING BENCHWORK TOOLS: FILES

OBJECTIVES

After completing this unit, the student will be able to:
- select the proper file for the job to be done.
- mount work properly before filing.
- use all files correctly and safely.
- maintain and store files in the proper manner.

KEY TERMS

Filing	Pillar file
File	Knife file
Tang	Flat file
Heel	Mill file
Safe edge	Three square file
Single-cut file	Triangular saw file
Double-cut file	Swiss-pattern file
Coarse grade	Needle file
Bastard grade	Riffler file
Second-cut grade	Cross filing
Smooth-cut grade	Draw filing
American-pattern file	Lathe filing
Machinist's file	Rotary file
Half-round file	Rotary bur
Round file	Flute

FILES

Filing is the operation of cutting down the surface of a piece of work by hand. Like chipping and scraping, filing is an art. It must be learned and practiced in order to develop real competence.

A **file** is a high-carbon steel hand tool with parallel rows of cutting teeth on its surfaces. Most files have similar parts, Figure 5–26. The length is the distance from the point to the heel. The **tang** is the narrow, tapered section used to seat the file on a handle. To prevent breakage, the tang is not hardened. The face or faces and sometimes one or both edges are

machine cut to have one or two sets of parallel rows of cutting teeth. Edges without teeth are known as safe edges. Teeth are hardened to the maximum for superior cutting.

Most files are classified as either single-cut or double-cut, although rasp-cut and curved tooth files are also available, Figure 5–27. Single-cut files have a single parallel row of teeth running diagonally across their faces. Double-cut files have an additional parallel row of teeth running across their faces at a diagonal to the first row, Figure 5–28.

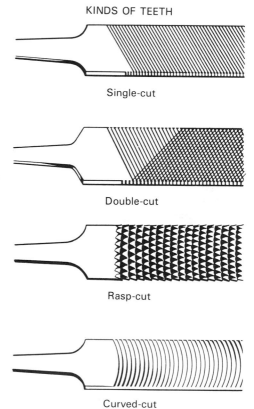

KINDS OF TEETH

Single-cut

Double-cut

Rasp-cut

Curved-cut

Figure 5–27 Kinds of file teeth *(Courtesy of The Cooper Tool Group)*

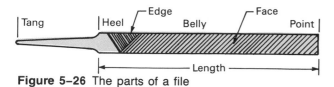

Tang | Heel | Edge | Belly | Face | Point

Length

Figure 5–26 The parts of a file

For any given type of cut, a larger file will produce a coarser cut than a smaller file. A 16-in (405 mm), single-cut file, for instance, will produce a coarser cut than a 4-in (100-mm), single-cut file. Files are made in various degrees of coarseness. Those most commonly used by machinists are the coarse, bastard, second-cut, and smooth-cut grades. Curved tooth and dead-smooth grades are also used occasionally.

Files are made in various shapes, Figure 5–29, styles, Figure 5–30, and patterns, Figure 5–31. Each file has a specific purpose. American-pattern files include machinists' files, mill files, and saw files. Machinists' files are made in shapes designed to accommodate most general machine shop filing needs. For instance, half round files, Figure 5–31, are usually used to file concave surfaces. Round files, Figure 5–32A, are generally used to enlarge or finish holes or fillets. Pillar files, Figure 5–32B, are used to file keyways and slots. Since one or both edges of pillar files are safe, they can also be used when filing against shoulders. Knife files, Figure 5–32C, are used principally by tool and die makers on work having acute angles. Flat files are double cut on either side for fast cutting and are single cut on either edge. Mill files, Figure 5–33, are single cut on all surfaces to produce finer finishes. Mill files are used for lathe work, draw filing, finish work on bronze and brass, and smooth finish filing in general. Three square files, Figure 5–34A, which are double cut, are used to file acute angles, debur square corners, and repair damaged threads. Triangular saw files, Figure 5–34B, are used primarily to sharpen all types of saw blades with teeth of 60 degrees. They are single cut.

Swiss-pattern files are used by tool and die makers and other artisans for precision filing on intricate or delicate work. They are smaller and more slender than American-pattern files. They also have more tapered, finer points and are made in finer cuts to more exacting tolerances. Coarseness grades range from 00 (coarsest) to 6 (finest). Needle files, Figure 5–35 (page 114), are used to true keyways, notches, corners, and grooves, and to finish slots or do any fine precision filing. These files have knurled tangs. Riffler files, Figure 5–36 (page 115), have curved filing surfaces at both ends. These curved surfaces are used to file concave or recessed surfaces such as die cavities.

Before using a file, a handle should be mounted and seated on the tang. Seating is done by striking the end of the mounted file handle against a firm surface, Figure 5–37 (page 116).

Work to be cross or draw filed should be mounted in a vise between protective jaws such as wood, plastic, copper, or aluminum. Use a comfortable stance, and place one foot slightly ahead of the other in line with the direction that the filing strokes will take. **Cross-filing** is done by moving the length of the file across the longitudinal axis of the work, at approximately a 45° angle, alternating in both directions, Figure 5–38A (page 116). **Draw filing** is done by moving the width of the file along the longitudinal axis of the work, Figure 5–38B (page 116). Properly done, draw filing will produce a smoother, flatter surface than cross filing. **Lathe filing** is done by holding the file across the longitudinal axis of the work in a lathe, Figure

Figure 5–28A Differences in double-cut files — flat bastard *(Courtesy of The Cooper Tool Group)*

Figure 5–28B Differences in double-cut files — aluminum type "A" *(Courtesy of The Cooper Tool Group)*

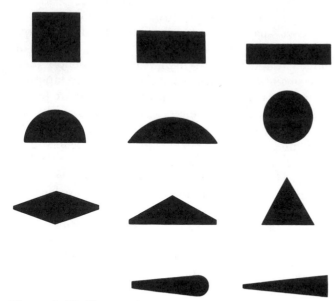

Figure 5–29 Shapes of files *(Courtesy of The Cooper Tool Group)*

5–38C (page 116). Lathe filing should always be done left-handed. This practice keeps the hands and arms away from the revolving parts and within easy reach of the controls. When filing, use light to moderate pressure on the forward stroke, and none on the return stroke. Do not attempt to rush. A speed of about 50–60 strokes per minute is about right.

When ordering a file, be sure to specify its length, pattern, cut, shape, and coarseness grade. This precaution will help to ensure that you get the right file for the job at hand.

After use, a file should be cleaned. Cleaning is done with a file card and brush. The short, stiff wire bristles are used to loosen metal particles embedded in the file cutting surface. The brush side is used to sweep the loose particles away. When your files are clean, store them in a place where they will be both separate from other tools and from each other. This precaution will help keep them sharp and damage free.

Figure 5–30A Styles of file — blunt *(Courtesy of The Cooper Tool Group)*

Figure 5–30B Styles of file — taper *(Courtesy of The Cooper Tool Group)*

Figure 5–31A Patterns of file — American *(Courtesy of The Cooper Tool Group)*

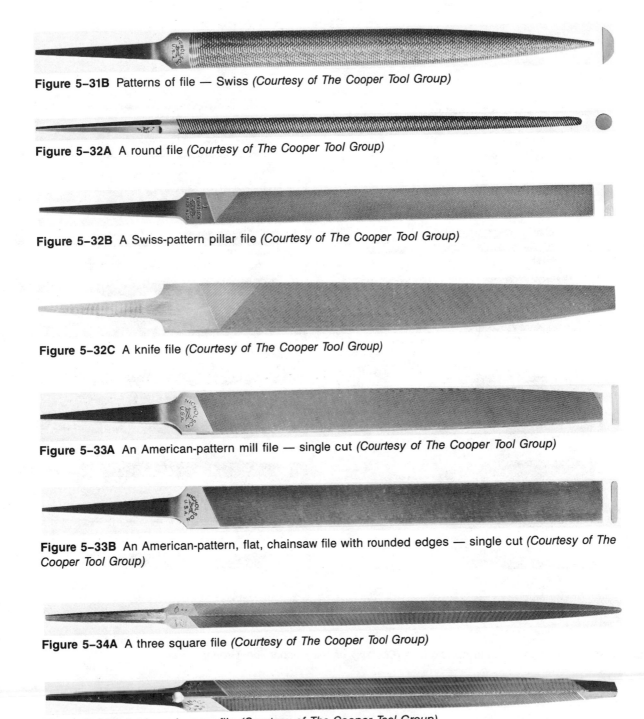

Figure 5–31B Patterns of file — Swiss *(Courtesy of The Cooper Tool Group)*

Figure 5–32A A round file *(Courtesy of The Cooper Tool Group)*

Figure 5–32B A Swiss-pattern pillar file *(Courtesy of The Cooper Tool Group)*

Figure 5–32C A knife file *(Courtesy of The Cooper Tool Group)*

Figure 5–33A An American-pattern mill file — single cut *(Courtesy of The Cooper Tool Group)*

Figure 5–33B An American-pattern, flat, chainsaw file with rounded edges — single cut *(Courtesy of The Cooper Tool Group)*

Figure 5–34A A three square file *(Courtesy of The Cooper Tool Group)*

Figure 5–34B A triangular saw file *(Courtesy of The Cooper Tool Group)*

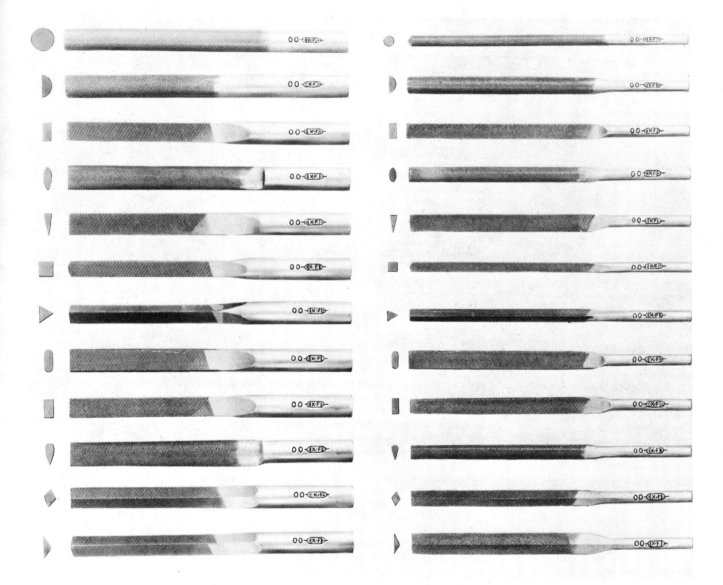

Figure 5–35 A selection of round handle files *(Courtesy of The Cooper Tool Group)*

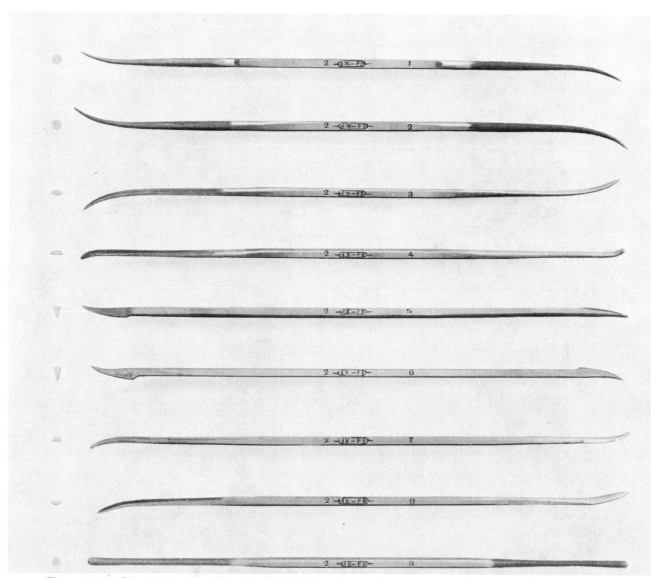

Figure 5–36 Die sinker riffler files *(Courtesy of The Cooper Tool Group)*

Figure 5–37 Seating a file handle

Figure 5–38B Draw filing

Figure 5–38A Cross filing

Figure 5–38C Filing left-handed on a lathe

ROTARY FILES AND BURS

Rotary files and burs were developed essentially for use in portable power tools, Figure 5–39. Each rotary file or bur should be used at a specific speed determined by its head diameter and the type of material being worked, Table 5–2.

Rotary file teeth are hand cut. Together they form broken lines that help to dissipate frictional heat, Figure 5–40. This feature makes rotary files appropriate for use on the toughest steels and forgings. Since the teeth are small, however, they tend to clog easily if used on soft materials such as copper or aluminum.

Flutes are the grooves in a rotary bur or rotary file. Rotary bur flutes are usually machine ground for use on nonferrous metals and alloys such as brass, bronze, magnesium, and aluminum. The comparatively large size of the flutes gives these tools better chip clearance than rotary file teeth. Both high-speed steel and carbide burs are available, Figure 5–41. Carbide burs are usually the preferred choice. Although they are expensive, they can be used on both hard and soft materials and will last much longer than high-speed steel burs.

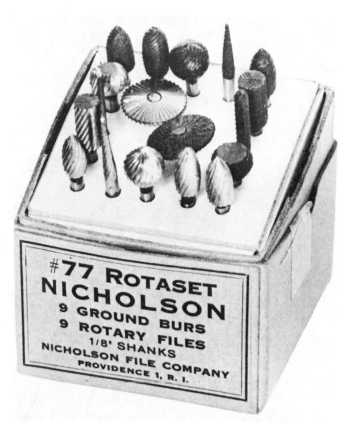

Figure 5–39 A variety of rotary files and burs *(Courtesy of The Cooper Tool Group)*

Table 5–2 Recommended Speeds in Medium Cut

APPROXIMATE SPEEDS (RPM)					
HEAD DIA	MILD STEEL	CAST IRON	BRONZE	ALUMI-NUM	MAGNE-SIUM
1/8″	4600	7000	15000	20000	30000
1/4″	3450	5250	11250	15000	22500
3/8″	2750	4200	9000	12000	18000
1/2″	2300	3500	7500	10000	15000
5/8″	2000	3100	6650	8900	13350
3/4″	1900	2900	6200	8300	12400
7/8″	1700	2600	5600	7500	11250
1″	1600	2400	5150	6850	10300
1 1/8″	1500	2300	4850	6500	9750
1 1/4″	1400	2100	4500	6000	9000

In Fine Cut, decrease speeds about one-third.

A. Suggested speeds for high-speed steel rotary files and burs, provided the chuck grips as close to the bur head as possible, and the bur runs true

RECOMMENDED SPEEDS			
HEAD DIA	RPM	HEAD DIA	RPM
1/8″	45000	5/8″	18000
1/4″	30000	3/4″	16000
3/8″	24000	7/8″	14500
1/2″	20000	1″	13000

In Fine Cut, decrease speeds about one-third.

B. Recommended speeds for carbide burs

Figure 5–40 Tree-shaped radius end rotary files *(Courtesy of The Cooper Tool Group)*

Figure 5–41 Ground bur, tree shape with pointed end *(Courtesy of The Cooper Tool Group)*

REVIEW QUESTIONS

1. Define the operation of filing.
2. List the part of a file.
3. Name four file classifications.
4. Explain what is meant by the "coarseness range" of a file.
5. List five types of machinists' files and state a use for each.
6. Describe the differences between American-pattern files and Swiss-pattern files.
7. What is the difference between draw filing and cross filing?
8. When using a file, which stroke is the cutting stroke?
9. What speed is most appropriate with files?
10. What is the only type of file that can be used without a handle?
11. What is the difference between a rotary file and a rotary bur?
12. What is the purpose of flutes on a rotary file or rotary bur?

CHAPTER SIX

SAWING

Contour band machining includes the cutting of material with a bandsaw blade, the filing of material with a band filing blade, and the band polishing of material with a coated abrasive belt. Sawing is done with either a horizontal machine or a vertical machine. Also used is a combination machine that will cut in either the horizontal position or the vertical position.

The biggest single difference between a reciprocating-type power hacksaw machine and a band machine is that the power hacksaw uses only the center portion of the blade as a rule. It is also extremely slow since it cuts only on the forward or cutting stroke. The band machine has a continuous welded band and all of the teeth are used. The cut is continuous and it is much faster. The blade is much thinner, producing fewer waste chips.

SAFE PRACTICES IN USING SAWS

In power sawing, the most danger is presented to the worker by the moving blade and the method of moving the work into the blade. The hands must always be protected from the moving blade and from ragged edges on the work.

1. Permission should be obtained to operate the machine, at the discretion of the instructor.
2. Always roll up loose sleeves and remove ties, rings, watches, etc., before operating the machine.
3. Always wear specified eye protection.
4. Make all adjustments with the machine off.
5. Always close the saw wheel doors before tensioning the blade or starting the blade in motion.
6. Make sure that the saw band guard on the post is locked in place.
7. Check the saw blade for sharpness and broken teeth; check the width, pitch, and speed of the blade.

121

8. In the case of a horizontal saw, make sure that the work is clamped securely and that the protruding end is well-supported.
9. In the case of a vertical band saw, keep your hands away from the moving blade. Use the proper pushing equipment.
10. Friction sawing produces a sharp bur on the work. Handle it carefully.
11. Never make adjustments when the saw blade is moving.
12. If the blade breaks or jams, shut off the machine, wait for the machine to stop, then make your repairs.
13. It is best to step to one side, away from the welding unit before welding a saw band.
14. Before releasing the blade welding lever, release the right-hand blade clamp to allow for a slight return movement of the blade clamp. This will prevent damage to the blade or clamp jaws.
15. Never walk away from the machine when it is operating.
16. When the work is completed, shut off the power and clean the machine.

UNIT 6-1

BAND MACHINING

OBJECTIVES

After completing this unit, the student will be able to:
- work safely with sawing machines and attachments.
- describe the differences in efficiency between different types of power saws.
- list six characteristics of the contour-band machine.
- explain the advantages of band machining.

KEY TERMS

Cutting off stock
Power hacksaw
Power bandsaw
Abrasive cutting-off wheel

Continuous cut
Throat width
Feed system
Hand screw feed

Contour band machine
Band filing
Band polishing
Filing band
Coated abrasive belt
Reciprocating

Ratchet feed
Power table
Hydraulic feed system
Feathered chip
Uniform chip

CUTTING OFF STOCK IN A MACHINE SHOP

One of the first power sawing operations that is likely to be performed by a machinist apprentice or a student is cutting off stock to be used on another machine tool or on the bench for some form of benchwork.

There are many ways of cutting off stock in a machine shop. Older shops used power hacksaws almost exclusively, and in some cases a power circular saw was added. The power hacksaw was a slow and steady workhorse, but in fact, only a part of the hacksaw blade was used and the method was slow. Just before World War II, many shops started using power bandsaws for cutting off material. Modern shops now use bandsaws, again almost exclusively. They are much faster and much more efficient in that the entire length of the blade is used rather than just a part of the blade.

Abrasive cutting-off wheels are being used now in many shops in addition to bandsaws. They are fast, efficient, and will cut hardened steel as well as annealed steel. Abrasive cutting-off wheels are also being used in many foundries to trim unwanted material from castings before shipment to the machine shop for machining.

CONTOUR BAND MACHINING

The metal cutting contour band machine is an extremely versatile machine tool. It is capable of cutting off metal, cutting intricate shapes in metal, band filing when using a filing band, and band polishing when using coated abrasive belts. A much thinner blade can be used than those required by circular saws and reciprocating power hacksaws because the bandsaw blade can be supported close to the work and depends on the proper tension rather than blade thickness and rigidity for cutting ability and accuracy. Economically, the initial machine is low in cost compared to some other machine tools. Because of the continuous cut, the entire saw blade is used in the cutting operation. A block of metal is cut from the stock using a very thin cut with very little waste

material. The leftover block can be used for other work. For making internal cuts, the blade can be cut, passed through a hole in the work, and the cut ends welded together in order that the internal cut can be completed. Cuts are made rapidly, operation is safe and easy, the savings in material is significant, and less horsepower is required to run the machine. Because all pressure is downward, no elaborate setups are required to hold the work. The work is moved into the blade either by hand feed or by power feed.

BAND MACHINES

There are many different band machines on the industrial tool market today, Figure 6–1. They are sized according to the throat width (which is the distance from the blade to the column) and by the thickness of work that can be cut. One model, for example, can cut work 16 in (406 mm) wide by 12 in (305 mm) thick. Figure 6–2 illustrates the parts of the band machine.

FEED SYSTEMS

A variety of feed systems is available for band machines. These systems range from the very simple hand screw feed and ratchet feed, which clamp to the front of the work table, to the more sophisticated power tables, which move back and forth and which will also tilt to the right or left. The hydraulic feed system, Figure 6–3, provides an accurate power feed for all types of contour sawing. A servovalve keeps the feed rate and force from exceeding the preset amounts. A weight-type power feed system is a mechanical, weight-assisted, power feed system. Both of these systems can be controlled by foot pedals, which leave both hands free to guide the work. It is like having a third hand.

PRINCIPLES OF BAND MACHINING

Figure 6–4 illustrates some of the advantages of band machining. Machining time is sharply reduced because excess material is removed in a solid section rather than by being reduced to waste chips.

A comparison is made in Figure 6–5 between the cutting by a band machine and by a milling machine. The type of chip produced by a milling cutter is a feathered chip rather than the uniform chip produced by the band machine.

Various sawing functions are performed on a band machine. Many band machines use differing terms to describe their cutting operations. The terms used in Figure 6–6 are those that are in general use throughout the sawing industry.

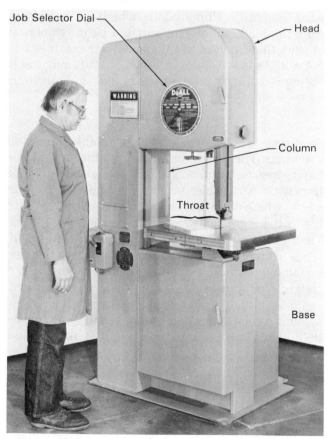

Figure 6–1 A band machine *(Courtesy of DoALL Company)*

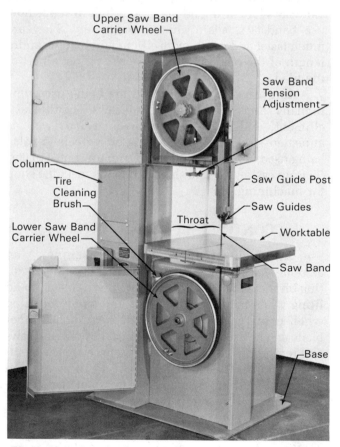

Figure 6–2 Parts of a band machine *(Courtesy of DoALL Company)*

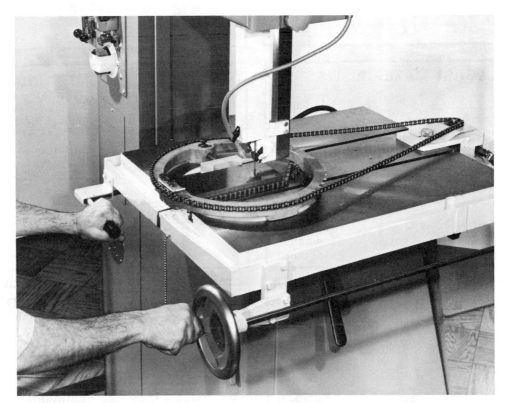

Figure 6-3 Hydraulic contour feed control *(Courtesy of DoALL Company)*

LESS HORSEPOWER

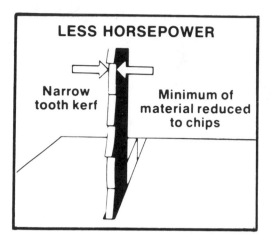

Narrow tooth kerf

Minimum of material reduced to chips

HOLDS SHARPNESS

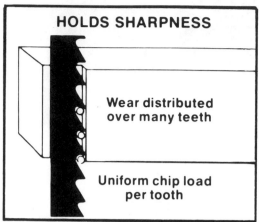

Wear distributed over many teeth

Uniform chip load per tooth

LEAST MATERIAL WASTE

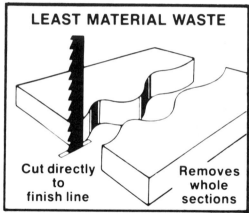

Cut directly to finish line

Removes whole sections

CONTINUOUS CUTTING

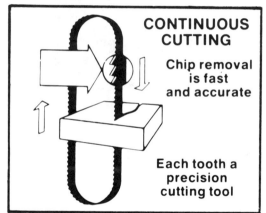

Chip removal is fast and accurate

Each tooth a precision cutting tool

SIMPLE FIXTURING

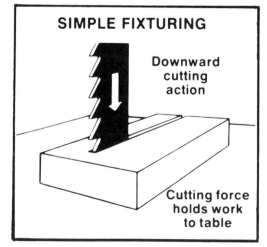

Downward cutting action

Cutting force holds work to table

UNRESTRICTED MACHINING GEOMETRY

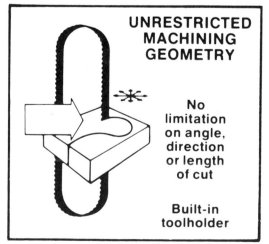

No limitation on angle, direction or length of cut

Built-in toolholder

Figure 6–4 Advantages of band machining *(Courtesy of DoALL Company)*

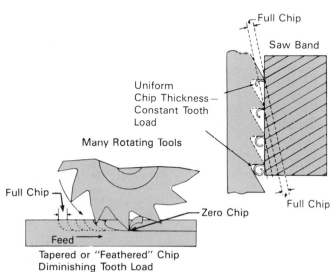

Figure 6-5 Contrast between a feathered milling cutter chip and a uniform band saw chip *(Courtesy of DoALL Company)*

REVIEW QUESTIONS

1. What are five precautions to observe in the use of a contour band machine?
2. Why is it easier to use blades on contour band machines as compared to circular saw and power hacksaw blades?
3. List five applications of the contour band machine.
4. What are six advantages of band machining over other types of machining?

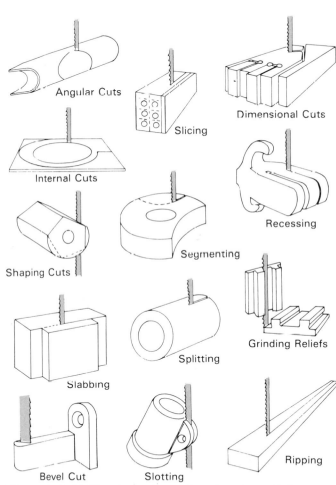

Figure 6-6 Cutting terms for various band machining operations *(Courtesy of DoALL Company)*

UNIT 6-2

THE SAW BLADE

OBJECTIVES

After completing this unit, the student will be able to:
- select the proper saw blade for the job at hand.
- define the terms pitch, width, gage, set, and saw kerf.
- identify three tooth forms available on high-speed steel blades.
- explain how the proper tooth form is selected for a job.
- use the job selector to gain the information necessary for operating the band machine successfully.

KEY TERMS

Saw blade characteristics
Tooth form
Pitch
Width
Gage

Set
Buttress tooth form
Claw tooth form
Precision tooth form
Saw kerf
Job selector

SELECTING SAW BLADES

Terms used to designate the parts of a saw blade are shown in Figure 6–7.

There are five saw blade characteristics that must be considered each time a saw blade is selected. They are tooth form, pitch, width, gage, and set.

Carbon tool steel and high-speed steel blades are available in three tooth forms, Figure 6–8. The forms are buttress, claw tooth, and precision, the last of which is the most generally used form.

In band machining, **pitch** is the number of teeth per inch. At the time of this writing, metric pitch is not given. The thickness of the material to be cut governs the selection of pitch, Figure 6–9. It is recommended that at least two teeth should always be on the work when cutting any material. Claw tooth and buttress forms are available in 2, 3, 4, and 6 pitch. Precision tooth forms are available from 6 to 32 pitch. Figure 6–10 is a guide showing when to use the different tooth forms. Table 6–1 gives blade specifications for Dart and standard carbon blades.

The **width** of the blade is the distance from the tip of the tooth to the back of the blade. The radius guide, Figure 6–11, shows the smallest radius that each width of blade can saw. Measurements for both width and radius are given in inch sizes.

The **gage** of the saw blade, Figure 6–12, is the thickness of the band itself. Some companies give the gage in inch sizes and some in both inch and millimeter sizes.

Set is the amount the teeth are offset from side to side to give clearance for the band in the saw kerf or cut, Figure 6–13A. Set patterns are shown in Figure 6–13B.

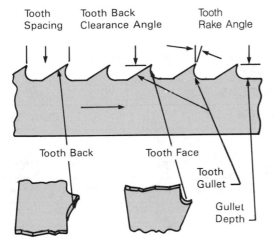

Tooth Spacing   Tooth Back Clearance Angle   Tooth Rake Angle

Tooth Back   Tooth Face

Tooth Gullet

Gullet Depth

Figure 6–7 Parts of a saw blade *(Courtesy of DoALL Company)*

THE JOB SELECTOR

The job selector, Figure 6–14, gives a great deal of useful information very quickly. To find the sawing recommendations on the job selector, turn the dial until the name of the material that you are sawing appears below the window on the dial. The dial will indicate such information as the material to be cut, the recommended pitch and blade type according to work thickness, the blade-width radius chart, the band speed according to the work thickness and blade type, the feed force to be used according to the work thickness, and the recommended method of coolant application.

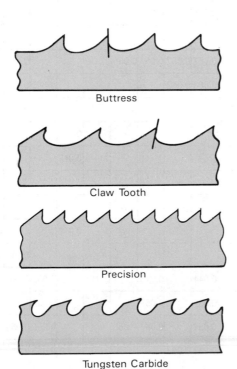

Buttress

Claw Tooth

Precision

Tungsten Carbide

Figure 6–8 Saw tooth forms *(Courtesy of DoALL Company)*

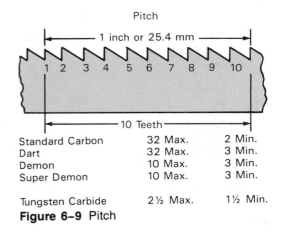

Pitch

1 inch or 25.4 mm

1 2 3 4 5 6 7 8 9 10

10 Teeth

Standard Carbon	32 Max.	2 Min.
Dart	32 Max.	3 Min.
Demon	10 Max.	3 Min.
Super Demon	10 Max.	3 Min.
Tungsten Carbide	2½ Max.	1½ Min.

Figure 6–9 Pitch

TOOTH FORM SELECTION

1. For finer than 6 pitch, *Precision* is only choice.

2. For 6 pitch and coarser, *Claw Tooth* usually gives the best tool life and fastest cutting rate.

3. For best finish, *Precision* and *Buttress* are usually preferred.

Figure 6–10 Tooth form selection *(Courtesy of DoALL Company)*

Table 6–1 Blade Specifications (*Courtesy of DoAll Company*)

SAW BLADE SPECIFICATIONS

		BUTTRESS RAKER SET				PRECISION RAKER SET								PRECISION WAVE SET							CLAW TOOTH RAKER SET			
WIDTH	GAGE	2	3	4	6	4	6	8	10	12	14	18	24	8	10	12	14	18	24	32	2	3	4	6
DART																								
3/16	0.025			0.043				0.044			0.043	0.042												
1/4	0.025			0.043	0.042			0.044			0.043	0.042	0.042						0.042				0.044	0.042
3/8	0.025		0.045	0.043			0.045	0.044			0.430	0.042											0.046	0.044	0.042
1/2	0.025		0.045	0.043		0.045		0.044			0.043	0.042			0.044		0.043	0.043					0.048	0.046	0.042
5/8	0.032							0.055	0.055		0.054	0.052												
3/4	0.032		0.056				0.055	0.055	0.055	0.054	0.054			0.057	0.057	0.057	0.057	0.055				0.056		0.054
1	0.035		0.061				0.058	0.058	0.058	0.058	0.057				0.063	0.063						0.060		0.058
STANDARD CARBON																								
1/16	0.025												0.038						0.038					
3/32	0.025											0.042							0.042					
1/8	0.025										0.043	0.042	0.042											
3/16	0.025			0.043				0.044			0.043	0.042												
1/4	0.025			0.043	0.042			0.044			0.043	0.042	0.042						0.042				0.044	0.042
3/8	0.025		0.045	0.043			0.045	0.044			0.043	0.042											0.046	0.044	0.042
1/2	0.025		0.045	0.043		0.045		0.044			0.043	0.042			0.044		0.043	0.043			0.050	0.048	0.046	0.042
5/8	0.032							0.055	0.055		0.054	0.052			0.057	0.057								
3/4	0.032		0.056				0.055	0.055	0.044	0.054	0.054			0.057	0.057		0.057	0.055			0.058	0.056		0.054
1	0.033	0.065	0.061				0.058	0.058	0.058	0.058	0.057				0.063	0.063					0.065	0.060		0.058

The PITCH headings above each group are labeled NOMINAL SET.

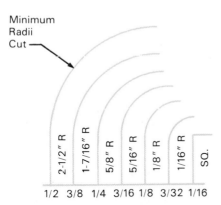

Minimum Radii Cut

2-1/2" R 1-7/16" R 5/8" R 5/16" R 1/8" R 1/16" R SQ.

1/2 3/8 1/4 3/16 1/8 3/32 1/16

Width of Saw Band in Inches

Figure 6-11 Blade width and radius guide *(Courtesy of DoALL Company)*

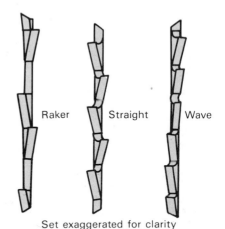

Raker   Straight   Wave

Set exaggerated for clarity

Figure 6-13B Set patterns *(Courtesy of DoALL Company)*

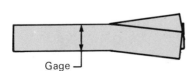

Gage

Figure 6-12 Gage *(Courtesy of DoALL Company)*

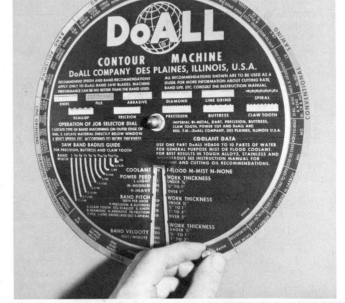

Figure 6-14 The job selector with a good view of the job selector window *(Courtesy of DoALL Company)*

Set

Figure 6-13A Set *(Courtesy of DoALL Company)*

Table 6–2 Selecting Cutting Fluids (*Courtesy of DoALL Company*)

COMPARISON OF CHARACTERISTICS OF TYPES OF CUTTING FLUIDS			
CHARAC-TERISTIC	CUTTING OILS	EMULSIFI-ABLE OILS	CHEMICAL FLUIDS
Heat removal	Fair	Good	Best
Lubricity	Best	Good	Fair
Extreme pressure or anti-weld	Best	Good	Moderate
Water quality tolerance	N/A*	Moderate	Best
Oil content	Almost all	Moderate as mixed	None
Compatibility with welding	Must be cleaned from weld area	Much like cutting oils	Good
Band speeds	Slowest — less than 175 sfpm (52.5 m/min)	Midrange — 150 to 350 sfpm (45–105 m/min)	Fastest — Over 250 sfpm (75 m/min)
NOTE: There is some overlap in application by band speed. This is an advantage in that it allows more flexibility in use.			
Finish	Best	Good	Fair
Materials to use on: Hardest	Best	Intermediate	Free machining only
Toughest	Best	Intermediate	Free machining only
Draggy chips	Best	Intermediate	Not best choice
Scale present	Best	Intermediate	Not best choice
Free machining	Not unless very fine finish required	Good; also similar to cutting oils	Best
Copper or copper alloys	Will stain if active sulfur present; choose right product	Will stain if active sulfur present; choose right product	Will not stain on machine; water rinse after machining
Can be misted	Mist collector mandatory to satisfy OSHA**	Mist collector recommended; may exceed OSHA limits	Yes; collectors usually not required

* N/A — Not Applicable
** OSHA — Occupational Safety And Health Administration

CUTTING FLUIDS

In most metal cutting operations, cutting fluids improve the cutting action for the following reasons.

- They help to dissipate heat caused by the cutting.
- They provide lubrication between the saw tooth and the chip.
- They prevent the formation of a built-up edge.
- They wash away and clear the chips from the cutting area.
- They provide lubrication for the band to keep it from being scored.

Sometimes cutting fluids are not used at all in sawing. When they are used, they must be applied sparingly or as a very light mist. Sometimes a cutting fluid in stick form is applied directly to the blade. A problem occurs if oil accumulates on the rubber surface of the band wheel tires. It may cause the saw band to slip and slide off the wheels. General recommendations for selecting cutting fluids are outlined in Table 6–2.

REVIEW QUESTIONS

1. What five saw-blade characteristics must be considered each time a saw blade is selected?
2. Describe three tooth forms used on high-speed steel blades.
3. Describe the form of a tungsten carbide blade.
4. What is pitch in a saw blade?
5. How is the width of a blade decided upon?
6. What is gage in saw blades?
7. Explain set in saw blades.
8. What is the purpose of cutting fluids in sawing?

UNIT 6-3

WELDING SAW BLADES

OBJECTIVES

After completing this unit, the student will be able to:

- identify a poor weld on a saw band.
- prepare the band and set up the welder to weld a saw band.
- safely use the electric resistance-type butt welder.
- produce a proper weld on a saw band.
- troubleshoot a poor weld.

KEY TERMS

Electric resistance-type butt welder
Blade shear
Blade ends
Welder jaws
Scale
Jaw upset force control
Saw blade alignment
Weld lever
Stationary jaw clamp
Movable jaw clamp
Weld
Flash
Weld thickness gage
Annealing the weld
Anneal heat selector switch
Anneal color
Troubleshooting poor welds
Misaligned weld
Incomplete weld
Burned-out weld
Upset metal

WELDING SAW BLADES

One of the handiest and most important fixtures that can be fixed permanently to the band machine is the electric resistance-type butt welder, Figure 6–15. Often these are built right into the machine. Knowing how

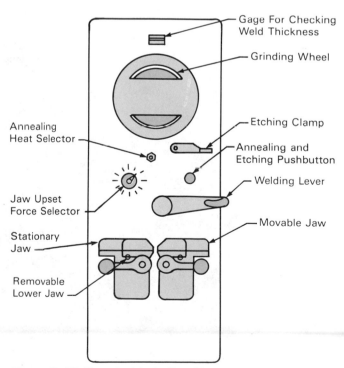

Figure 6–15 The electric butt welder

to weld a blade or band properly is every bit as important as operating the band machine itself. The weld is the most important quarter inch in a saw band. An improperly made weld becomes a weak link in the blade. If a break occurs, it will almost inevitably happen on or beside the weld. Rewelding blades can become a costly, time-consuming operation. On the other hand, if you learn to make good welds every time, the butt welder will very quickly pay for itself over and over again. For more than one reason, then, this section on blade welding is very important and very detailed.

Preparing the Blade

Preparing both the blade and the welder is just as important as the actual welding. A step-by-step procedure follows:

1. Cut the blade to the exact length. The use of a blade shear, Figure 6–16, will insure that the blade ends are flat, square, and smooth.
2. If saw snips or tin snips have been used to cut the blade, it will be necessary to grind the blade ends flat, square, and smooth, Figure 6–17. If the blade ends are held as

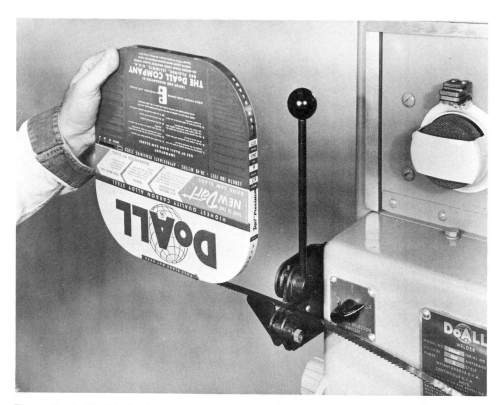

Figure 6–16 The blade shear *(Courtesy of DoALL Company)*

shown, the two ends will match when they are turned over.

3. Clean the blade ends carefully. The part of each blade that comes into contact with the welder jaws must be free of all dirt, scale, grease, and oil. Anything less than a clean blade will prevent a good electrical contact. This is very important.

Preparing the Welder

1. The welder jaws are removable. They must be thoroughly clean and free of scale.

2. The jaw upset force control should be set to the correct position for the width of saw blade being welded. The wider the blade, the more welding heat is required to make the weld.

Saw Blade Alignment Before Welding

1. If internal sawing is to be done, the blade is inserted through the starting hole in the work before welding takes place, Figure 6–18.

2. The blade ends are placed in the welder jaws with the back of the blade against the back of the jaws as shown, Figure 6–19.

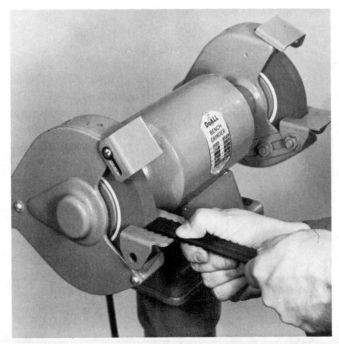

Figure 6–17 How to grind blade ends square (*Courtesy of DoALL Company*)

If the blade ends are not in perfect alignment, remove one blade end and recut or regrind it until it is a perfect fit. A misaligned joint will cause an incomplete weld. Clamp the blade ends securely but not so tightly as to injure the set of the teeth.

MAKING THE WELD

There is more to making a good weld than a simple pressing down of a lever. Once the blade and welder are set and ready, use the following procedure:

1. To weld the band, press and hold down the weld lever. Do not release the weld lever until the weld has cooled to black and the stationary jaw clamp has been released. This will prevent scoring the welder jaw surface when the mov-

able jaw moves back slightly once the weld lever is released.
2. Release the weld lever.
3. Remove the blade and inspect the weld. If the weld is poor, check the Troubleshooting Section, then cut out the old weld completely and make a new weld.

Cleanup After Welding

It is important that the welder jaws be kept clean at all times. The jaws and inserts must be wiped and/or scraped after every weld. This precaution will maintain the strength and fatigue characteristics of the band by holding proper alignment, preventing flash from becoming embedded in the band, and preventing shorts or poor electrical contact.

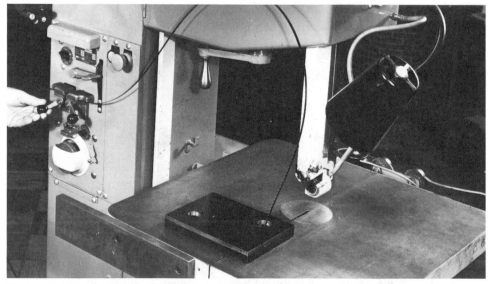

Figure 6–18 Inserting the blade for internal sawing *(Courtesy of DoALL Company)*

Inspection of the Weld

The color of the upset material around the weld should be blue-gray and evenly colored. The teeth spacing should be uniform and the weld should be located at the center of the gullet.

Grinding the Weld (or Flash)

The welded area on both sides of the band should be ground to the same thickness as the rest of the band, Figure 6–20. The blade teeth should face outward at this time and care must be taken not to grind the teeth. The weld area should not be thinner than the band, and the weld area should not be burned or overheated. To test the thickness of the ground blade, insert the blade into the thickness gage above the grinding wheel. Handle the band carefully. Remember that it has not been annealed yet and is very brittle.

Annealing the Weld

Once the blade is welded, it has become air hardened and is very brittle. The weld must be carefully annealed, or softened. Use the following procedure:

1. The welder jaws must be thoroughly clean.
2. Position the blade at the front of the welder jaws with the weld between the jaws, Figure 6–21.
3. Set the "anneal heat" selector switch according to the width of the blade being annealed.

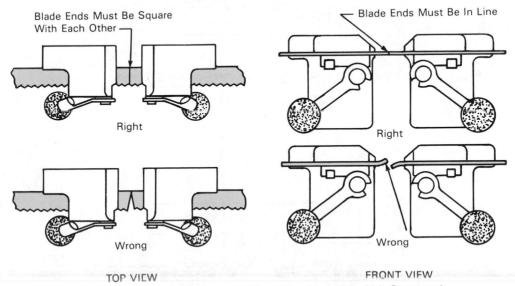

Figure 6-19 Aligning the saw blade for welding *(Courtesy of DoALL Company)*

4. In annealing a carbon tool steel blade, Figure 6–22, joggle the annealing button repeatedly until the weld is a dull cherry to cherry red color. Decrease the joggling action of the anneal button slowly to allow the blade to cool slowly.

To anneal a high-speed steel Dart blade, heat the blade slowly until the blade around the weld is a deep blue color. Allow the blue color to spread to half the length of the blade exposed between the jaws. Do not overheat or the band temper beside the weld will be damaged. Cool quickly by releasing the anneal button.

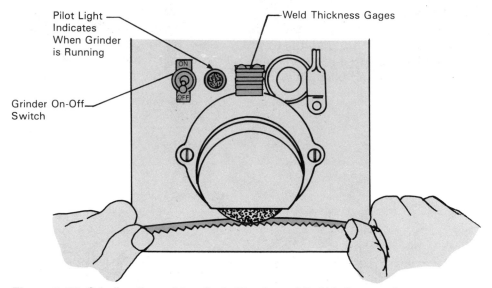

Figure 6–20 Grinding the weld or flash *(Courtesy of DoALL Company)*

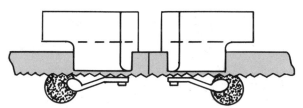

Figure 6–21 Setting the blade for annealing *(Courtesy of DoALL Company)*

Figure 6–22 Annealing the weld *(Courtesy of DoALL Company)*

TROUBLESHOOTING POOR WELDS

No matter how well you think you have the welding operation set up, a poor weld can sometimes be produced. They are unacceptable and dangerous. You must start over again. Cut out the old weld and make a new one. There is no other answer. This section outlines the common errors that are made and how to correct them.

1. The weld is misaligned, Figure 6–23.
   (a.) Dirt or flash scale is on the jaws. Wipe or scrape the jaw faces clean.
   (b.) The blade ends are not square.
   (c.) The blade ends are not correctly aligned when clamped in the jaws.
   (d.) The jaw faces are worn. Replace or repair them.
   (e.) The jaws are not aligned correctly. Adjust them.

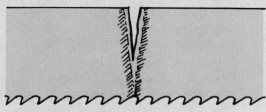

Figure 6–23 Misaligned weld

2. The weld is misaligned with the blade ends overlapped, Figure 6–24.
   (a.) The jaw upset force control is set for a wider blade than the one being welded. Adjust correctly.
   (b.) The jaw ends or jaws are not aligned correctly.
   (c.) The jaw upset spring was replaced by too strong a spring.
   (d.) The jaw-upset-spring adjusting stud is turned in too far, which inceases the upsetting force of the movable jaw.

Figure 6–24 Misaligned weld — overlapped ends

3. The weld is incomplete; the weld is not completely joined, Figure 6–25.
   (a.) The cut-off switch is not adjusted correctly.

(b.) The jaw upsetting force may be set too low. This can be caused by the upsetting force selector being set for a narrower blade than the one being welded. A weak or wrong type of upset spring may have been used. The upset spring adjusting stud may not be adjusted correctly. If the adjusting stud is backed out too far, the upsetting force will be reduced. The slide rod in the adjusting mechanism may be sticking or may be obstructed.

(c.) The voltage may be too low. The welding voltage across the jaws should be 3-1/2 volts. Check for low incoming voltage to the machine.

(d.) Jaw spacing may be too great. The jaw gap should be 0.187 in (4.7 mm) before welding.

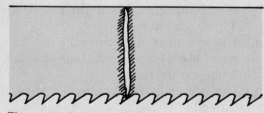

Figure 6–25 Incomplete weld

4. The weld "burns out." The joint is not complete; there is excess metal around the joint, Figure 6–26.

(a.) A defective cut-off switch may not break the circuit at the end of the welding operation. Replace the cut-off switch.

(b.) The cut-off switch is not adjusted correctly.

(c.) The points of the cut-off switch may be welded together. Replace the switch.

(d.) The slide rod in the adjusting mechanism may be sticking because of rust or dirt. Clean and oil the rod.

(e.) Slide rod movement may be obstructed because the stop screw is too tight on the rod. The stop screw should be turned in tight, then backed off 1/2 to 3/4 of a turn.

(f.) Jaw movement may be obstructed by a kinked jaw cable or by tangled wires. Straighten the cable or untangle the wires.

(g.) The movable jaw may be binding on the jaw bearings because the tilt adjustment screw may be turned in too far.

(h.) The voltage may be too high. Check for too high an incoming voltage to the machine. Welding voltage across the jaws should be 3-1/2 volts.

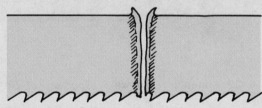

Figure 6–26 Burned-out weld

5. There is excess weld or upset metal around the welds, especially in 1/16 in to 1/8 in blades, Figure 6–27.

(a.) Jaw travel may be too great for narrow blades. Try leaving a 1/64-in gap between the blade ends. This will reduce jaw travel.

Figure 6–27 Excess flash

6. The weld is weak and brittle.
   (a.) The weld has not been annealed sufficiently. Poor annealing can be caused by incorrect annealing heat. Bring the weld up to the correct annealing color as described under "Annealing." There may be scale or oil in the weld, which can cause a slow annealing heat.
   (b.) The voltage may be too low. Check incoming voltage to the machine, which should be 3-1/2 volts.

7. The weld breaks when it is used.
   (a.) The weld has not annealed correctly. Annealing heat may be too low or too high.
   (b.) The weld has been ground too thin.
   (c.) The weld may be "incomplete" or "burned out."

REVIEW QUESTIONS

1. List the steps in the preparation of the saw blade for welding.
2. When preparing to weld a bandsaw blade, what should be done first with the welder jaws?
3. What are the important criteria for blade alignment before welding takes place?
4. How is a bandsaw blade set up for sawing internally?
5. List three characteristics of an acceptable weld.
6. Why is it necessary to anneal a welded joint?
7. What are the differences in annealing practice for carbon tool steel blades and high-speed steel blades?
8. List six common errors made in welding saw blades.
9. What must be done when a welded joint is not quite complete?
10. What causes a welded joint to break?

DIGITAL READOUTS AND COMPUTER CONTROLS

Digital readouts provide an immediate and accurate indication of measurements. One use of digital readouts is to measure a distance travelled, either mechanically or electronically. For example, the odometer in an automobile or truck measures miles travelled on a highway. The reading is accurate to tenths of a mile. This information is shown in easily readable numbers on the odometer. Similarly, the digital readout on the side of a micrometer caliper indicates the distance travelled by the micrometer spindle in thousandths of an inch. Metric micrometers are also available. A measurement of ten-thousandths can be read from a vernier on the sleeve (barrel) of the micrometer. These instruments are mechanical counters, or one type of digital readout.

An electronic digital readout does the same thing as the mechanical digital readout, except that the numbers are expressed electronically and are shown in a large, easily read form on a screen or monitor (CRT). A typical electronic readout may express the distance travelled by one axis on a machine tool, such as a milling machine. If the

Figure 1 A digital readout length gaging system in use. The accuracy of the system is ±0.000 004 inch. *(Courtesy of Heidenhain Corporation)*

table travels longitudinally for a distance of 1.500 inches, this distance registers on the monitor or screen in large numbers. A flick of a switch changes the expression of the same distance to a metric value, or 38.1 mm.

Figure 1 shows a digital length gage used to test lengths of parts. The length is expressed in either millimeters or inches. In this case, the digital readout measures the distance travelled by the vertical spindle of the length gage as it measures the length of the part. Again, this is the distance travelled on only one axis. These digital gages can also be used as linear encoders. In other words, they can convert the linear data expressed on the monitor into data which can be stored in a computer. On some digital gages, the accuracy is expressed in millionths of an inch or in microns (metric system).

Other electronic counters, such as a universal three-axis counter, measure the distance travelled by each of the machine axes — X, Y, and Z. Referring to the vertical milling machine, the X axis is the longitudinal distance travelled by the table, the Y axis is the crosswise distance travelled by the table, and the Z axis is the distance travelled by the vertical spindle or by the vertical raising and lowering of the worktable. Each of the three axes is measured separately, but all three axes are shown together on the monitor. The distance travelled for each is expressed in either millimeters or inches, depending upon which measuring system is being used at the time. Absolute or incremental measuring can be done and linear and rotary encoders can be directly connected so that data can be transmitted to a computer.

Newer digital readouts can be used as either a conventional digital readout for three-axis measurement, or as a numerical control for simple automatic positioning. After the entry of a nominal position value in either the absolute or incremental measuring system (that is, if you want the workpiece to be in a certain position, that position is entered into the numerical control), the machine traverses to the intended position automatically. Automatic positioning is much faster and more accurate than manual positioning.

Point-to-point and straight cut controls, shown in Figure 2, are designed for use right on the machine by the machine tool operator. Such a control can be used on milling machines, drilling machines, and boring mills. The operator is guided by means of questions and hints shown on the monitor in plain language. The control asks for such information as tool data (what type of cutting tool is to be used), nominal positions for the cutting tool (where is the cutting tool to be located), feed rate in inches per minute or meters per minute for table movement, spindle rpm, and so on. X, Y and Z-axis distances can also be expressed on separate screens. Codes used are standard so that programs that have been made up at a separate programming station are readily usable.

Contouring controls, Figure 3, have even more functions and more flexibility than point-to-point and straight cut controls. Contouring controls can be used on the manufacturing lines using a host computer which can transmit programs to the control. Again, X-, Y-, and Z-axis distances can be read directly on the control monitor, if desired. Electronic workpiece alignment is used with this control. This means the control electronically probes the workpiece and sets itself according to the workpiece. When machining to a program, the job can be probed and the control compensates for any deviation in the setting of the job. Contouring controls are used on milling machines, boring mills, and machining centers.

Digital readout controls and computer controls have developed at a phenomenal rate in just a few years. With increasing automation and accuracy in positioning, machine movements (such as automatic positioning) can be speeded up, resulting in time and cost savings.

The new generation of readouts integrated with computer control is playing an increasing role in machining technology. They are almost indispensable in flexible manufacturing system machining. The controls shown here are just a few of the many that are available for modern machining applications.

Figure 2 A point-to-point and straight cut computer numerical control with digital readout *(Courtesy of Heidenhain Corporation)*

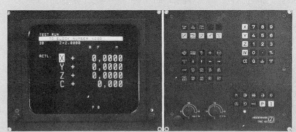

Figure 3 A contouring control with four axes plus an optional fifth axis dedicated to spindle orientation for use on machine tools with toolchangers. Controls feature a conversational prompted programming format. *(Courtesy of Heidenhain Corporation)*

UNIT 6-4

OTHER BAND TOOLS

OBJECTIVES

After completing this unit, the student will be able to:

- use other band machine tools such as filing and polishing bands.
- explain the differences between friction sawing and high-speed sawing.
- state advantages of cutting off material using a band machine.
- safely set up different structural shapes for sawing.

KEY TERMS

Knife blade
Spiral-edge band
Diamond-edge band
Insert-type saw guide
File band
Polishing band
Waste material

File band guide
Polishing band guide
Saw band guide
Friction sawing
High-speed sawing
Cutting off

OTHER BAND TOOLS AVAILABLE

Knife blades, Figure 6–28, are used for cutting soft, fibrous materials that would tear or fray if they were cut with tooth-type bands.

Spiral-edge bands can cut in any direction and are used especially for sawing intricate patterns in plastics, woods, and light gage metals. These bands should be used only on center crowned, rubber-tired band wheels.

Diamond-edge bands, Figure 6–29, are used in

cutting superhard materials. These bands are specially designed to cut abrasive plastics, ceramics, epoxy,

Figure 6–28 Knife blades *(Courtesy of DoALL Company)*

fiberglass, granite, hard carbons, glass, and silicon, just to name a few materials.

Figure 6–30A illustrates the assembled insert-type saw guides. Adjusting the saw guide inserts is shown in Figure 6–30B.

FILE BANDS AND POLISHING BANDS

File bands, Figure 6–31, and polishing bands, Figure 6–32, can be used to quickly convert the band machine into a finishing tool. The work can be roughly cut to shape with saw blades by leaving a small amount of waste material to remove. The work can then be rough and finish filed using the file band, and finally the required surface finish can be applied by using the polishing band. File band guides, Figure 6–33, and polishing band guides, Figure 6–34, must be substituted for saw band guides.

FRICTION SAWING

Friction sawing is cutting by friction. Ferrous metal is fed with heavy pressure into a saw band travelling between 6000 and 15000 surface feet per minute (1800–4500 m/min). This causes tremendous friction, which is so intense that the heat generated instantly softens the metal by raising its temperature to the forging point.

The teeth of the saw band do not actually cut, rather they help produce friction. They then scoop out the softened metal. The molten metal and sparks produced by friction sawing are spectacular to see. Friction sawing is the fastest machining method used for cutting ferrous metals under 1 in (25.4 mm) thick.

HIGH-SPEED SAWING

The name **high-speed sawing** refers to the speed of the saw band as it cuts the work at speeds of 2000 to 6000 surface feet per minute (600–1800 m/min). This should not be confused with the term **high-speed steel**, which refers to the metallurgical characteristics of a particular saw band. It is not friction sawing.

Materials that lend themselves to this type of high-speed sawing include wood, paper, plastics, rubber, aluminum, brass, bronze, zinc, and magnesium. These materials may be in bar, sheet, plate, extruded, cast, or tube form.

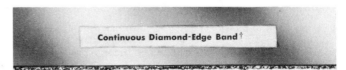

Figure 6–29 Diamond-edge band *(Courtesy of DoALL Company)*

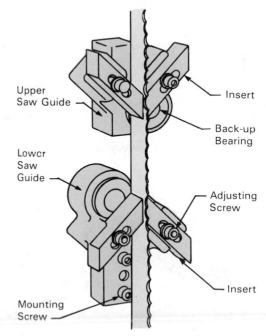

Figure 6–30A Insert-type saw guides *(Courtesy of DoALL Company)*

Insert Gage

Select the correct size gage and inserts to match the saw band to be used.

Insert Gage

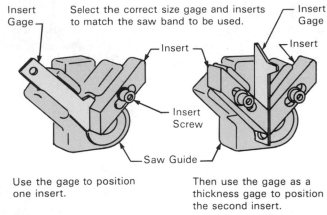

Insert

Insert

Insert Screw

Insert

Saw Guide

Use the gage to position one insert.

Then use the gage as a thickness gage to position the second insert.

Figure 6–30B Adjusting saw guide inserts *(Courtesy of DoALL Company)*

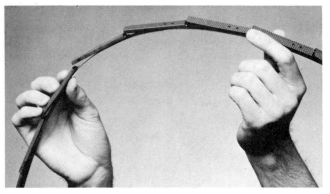

Figure 6–31 A file band *(Courtesy of DoALL Company)*

Figure 6–32 Polishing bands *(Courtesy of DoALL Company)*

CUTTING-OFF WITH BAND MACHINES

There are many different methods and many different machines used to cut off material from stock. Two of the advantages of band machines are the total use of the cutting blade, which is caused by all of the teeth coming into contact with the stock, Figure 6–35, and the small amount of waste material, which results from the thinner cut.

Recommendations for Cutting Off Materials

Tables 6–3, 6–4, and 6–5 show the recommended blade type and pitch for different types of work. Table 6–5 also shows the preferred method of gripping structural shapes for safe cutting.

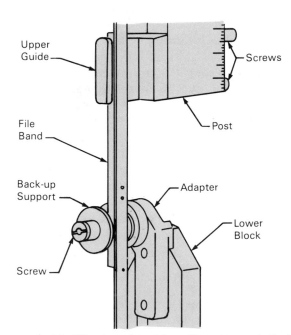

Upper Guide

Screws

File Band

Post

Back-up Support

Adapter

Screw

Lower Block

Figure 6–33 File band guides *(Courtesy of DoALL Company)*

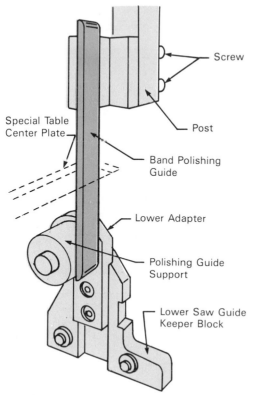

Screw

Special Table
Center Plate

Post

Band Polishing
Guide

Lower Adapter

Polishing Guide
Support

Lower Saw Guide
Keeper Block

Figure 6–34 Polishing band guides *(Courtesy of DoALL Company)*

Figure 6–35 Cutting off stock on a horizontal band machine *(Courtesy of DoALL Company)*

Table 6–3 Blade Selection for Cutting Off Tubing *(Courtesy of DoALL Company)*

TUBING, ALL WALL THICKNESSES			
CROSS SECTION	BEST BLADE PITCH	BLADE TYPE	NEXT BEST BLADE
To 3/8″	1″, 10 pitch	Precision	1″, 8 pitch, Precision
3/8″ to 3/4″	1″, 8 pitch	Precision	1″, 10 pitch, Precision
Over 3/4″	1″, 6 pitch	Precision	1″, 8 pitch, Precision

Table 6–4 Blade Selection for Cutting Off Solids *(Courtesy of DoALL Company)*

SOLIDS AND SCALE-FREE STRAIGHT STACKED BARS			
CROSS SECTION	BEST BLADE PITCH	BLADE TYPE	NEXT BEST BLADE
To 5/8″	1″, 10 pitch	Precision	1″, 8 pitch, Precision
5/8″ to 1″	1″, 8 pitch	Precision	1″, 10 pitch, Precision
1″ to 4″	1″, 6 pitch	Precision	1″, 8 pitch, Precision
4″ to 10″	1″, 4 pitch	Precision	1″, 3 pitch, Claw Tooth
10″ to 18″	1″, 3 pitch	Claw Tooth	1″, 4 pitch, Claw Tooth

Table 6–5 Blade Selection for Cutting Off Structural Shapes and the Preferred Method of Holding Structural Shapes *(Courtesy of DoALL Company)*

STRUCTURAL MATERIALS			
WALL THICKNESS	BEST BLADE PITCH	BLADE TYPE	NEXT BEST BLADE
To 3/8″	1″, 10 pitch	Precision	1″, 8 pitch, Precision
3/8″ to 3/4″	1″, 8 pitch	Precision	1″, 10 pitch, Precision
Over 3/4″	1″, 6 pitch	Precision	1″, 8 pitch, Precision

Angles   Channels   "H" Beams

REVIEW QUESTIONS

1. Describe band filing.
2. List a use for knife blades, spiral-edge bands, and diamond-edge bands.
3. Describe band polishing.
4. What is friction sawing?
5. How is metal removed in friction sawing?
6. What is the danger in friction sawing?
7. What is high-speed sawing?
8. In the range of 3/8″ to 1″ thickness, what type of blade is recommended for cutting off material using a band machine?

UNIT 6-5

SAWING RECOMMENDATIONS AND TROUBLESHOOTING

OBJECTIVES

After completing this unit, the student will be able to:

- plan and organize the sawing operation to ensure that it is done efficiently.
- select the proper saw band or band tool for the work being done.
- saw different materials properly.
- decide when cutting fluids are to be used.
- troubleshoot sawing problems and correct them when they occur.

KEY TERMS

Planning the job
Sequence of cutting
   operations
Feed assist

Blade selection
Operating
   recommendations
Troubleshooting

ORGANIZATION

In any machine tool operation, it is essential that you work from a drawing and/or a layout made directly on the work. From the drawing it is necessary to develop a proper time-saving sequence of cutting operations to use as you proceed with the work. It is easier and cheaper to make mistakes on paper rather than on the stock being cut, so it is preferable that this planning be done on paper first. The following checklist should be followed:

1. Plan a sequence of cuts and lay out the work.
2. Select the correct saw blade type, pitch, width, etc., for the material to be cut (use the job selector on the saw).
3. Install the proper saw guides for the blade being used.
4. Install and tension the saw band. Check the band for tracking and adjust the guide post.
5. Select and install accessories like fences, feed mechanisms, etc.
6. Consult the job selector and select the correct band speed, feed force, etc.

GENERAL SAWING TIPS

1. Use a feed assist whenever possible.
2. If it is needed, use a magnifier for following layout lines. Keep the overhead work light clean.

3. Use a coolant if available. Mist will not obstruct layout lines.
4. Use heavy work slides or heavy-duty clamps (accessories) whenever the work is large or heavy.
5. Use the rip fence or miter bar accessories, when needed, for ripping or making angle cuts.
6. Use a wide gage blade for heavy cuts on contour work.
7. Wear safety glasses when sawing. Brush chips away frequently from the work table. Use a push stick or work-holding jaws on small work pieces.

BAND MACHINING OPERATION RECOMMENDATIONS

Tables 6–6A and 6–6B contain various operation recommendations for all types of band sawing, band filing, and band polishing operations.

Table 6–6A Operation Recommendations (*Courtesy DoALL Company*)

TYPE OF WORK	MATERIAL (AISI CODE NUMBERS)	BAND SPEED RANGE (fpm) (m/min)		RECOMMENDED SAW BAND OR BAND TOOL
		work thickness under ½″	over 3″	
Light-duty sawing Free-machining metals	1112, 1212, & 1213	175 (52.5)	150 (46)	DoALL Standard Carbon or Dart (performance will be improved if coolant is used with Dart) See saw blade specification chart for selection of width, gage, type, set, and pitch
Light-duty sawing Low carbon steel	1015 to 1030	175 (52.5)	145 (44)	
Light-duty sawing Medium carbon steel	1035 to 1050	150 (89)	100 (30)	
Light-duty sawing High carbon steel	1060 to 1095	125 (108)	80 (24)	
Light-duty sawing Alloy steels	Ni, Cr, Mo, Mn, Ni-Cr, Ni-Mo	125 (108)	70 (21)	
High-speed sawing	Al, Mg	4500 (1365)	3000 (910)	DoALL Standard Carbon or Dart
High-speed sawing	Cu, bronze, brass	2750 (834)	800 (243)	
High-speed sawing	Zinc, lead	2750 (834)	1000 (303)	

Table 6–6A Continued

TYPE OF WORK	MATERIAL (AISI CODE NUMBERS)	BAND SPEED RANGE (fpm) (m/min)		RECOMMENDED SAW BAND OR BAND TOOL
		work thickness under ½″	over 3″	
High-speed sawing	Wood	3000 (910)	5200 (1577)	Woodworking or carbon
High-speed sawing	Thermo-plastic	2300 (698)	1200 (364)	Carbon or Dart
High-speed sawing	Thermo-setting plastic	5500 (1668)	3000 (910)	Carbon or Dart
Light-duty sawing	Cast iron	150 (46)	100 (30)	Carbon or Dart
Band filing	Brass, Al, Mg, Cu, Zn	230 (70)	150 (46)	10 teeth, flat or oval short angle, coarse cut
Band filing	Mild, Carbon steel	145 (44)	80 (24)	Flat or oval bastard, medium cut, 14 or 16 teeth
Band filing	Tool, Alloy steel	90 (27)	50 (15)	Flat or oval bastard, medium cut, 16, 20, 24 teeth
Band filing	Cast iron	150 (46)	80 (24)	Flat or oval short angle, coarse cut, 10 teeth
Band polishing	Al, Mg, Cu	1900 (576)	1500 (455)	DoALL Abrasive bands alum. oxide or silicon carbide, 3 grit sizes, all bands are 1 in wide
Band polishing	Carbon steel, stainless	3300 (1000)	2000 (607)	
Contour sawing	All materials	Low speeds as required		Carbon or Dart, find band width in radius chart
Contour sawing intricate shapes, small radius	Metals, woods, and plastics	Low or medium speeds as required		Spiral saw band, four diameters available
Slicing cloth, rubber, paper, etc.	Cloth, rubber, paper, cork, leather, felt, etc.	High speeds		Special DoALL Knife-edge band
Friction sawing	Thin, ferrous materials (16 ga)	Highest speed possible		Special DoALL Friction saw band

Table 6–6B Operation Recommendations (*Courtesy DoALL Company*)

RECOMMENDED SAW GUIDE AND INSERTS	RECOMMENDED COOLANT	COOLANT MIXING INSTRUCTIONS	METHOD OF COOLANT APPLICATION
Insert type — use correct size insert and guide block for width of saw band.	DoALL coolants such as: Kleen-Kool concentrate or #470 soluble oil concentrate if your machine has drip or mist coolant applicators. If not, use Saw-Eze or saw dry.	Follow mixing instructions on container for Kleen-Kool. Mix 30 parts water with 1 part #470 soluble oil concentrate.	Apply drip or very light mist to saw band where it enters work. Apply Saw-Eez through upper saw guide fitting. Too much coolant will cause band to slip off wheels.
Roller-type recommended, insert type can be used.			
	None	None	None
Insert or roller type	None	None	None
Insert type	None	None	None
Special file guides for each size file band. Use special table center disk also.	Kleen-Kool or none	Mix per instructions on container.	Light mist or dry
	None	None	None
Special polishing guides. Use special table center disk also.	None	None	None
Insert type	Kleen-Kool or #470	See above for mixing instructions.	See above.
Special DoALL spiral band guides	None	None	None

RECOMMENDED SAW GUIDE AND INSERTS	RECOMMENDED COOLANT	COOLANT MIXING INSTRUCTIONS	METHOD OF COOLANT APPLICATION
Insert-type saw guides	None	None	None
Roller-type saw guides	None	None	None

TROUBLESHOOTING

When trouble occurs because a mechanical or electrical part fails, the following notes will help to locate the trouble and suggest a remedy.

1. The machine will not start.
   - (a.) Check the main fuses and control circuit fuse.
   - (b.) Check the reset button on the band drive motor starter.
   - (c.) Check the transformer.
2. There is severe machine vibration.
   - (a.) The band wheels may not be balanced.
   - (b.) The variable-speed pulley components may not be balanced.
   - (c.) The variable-speed drive belts may be unbalanced.
3. There is saw band vibration (while sawing).
   - (a.) The band speed may be incorrect.
   - (b.) The choice of saw band pitch may be incorrect.
   - (c.) The choice of coolant may be incorrect.
   - (d.) The feed pressure may be incorrect.
   - (e.) The work piece may not be firmly held.
   - (f.) The saw guide inserts may be worn or improperly adjusted.

(g.) The saw guide backup bearing may be worn.

(h.) A special support has not been used under the work when using a heavy-duty work slide or heavy-duty work clamps.

4. There is no coolant flow.
(a.) The coolant applicator nozzle may be jammed.
(b.) The coolant hose may be clogged or kinked.
(c.) The coolant reservoir may be empty.
(d.) The coolant reservoir may need cleaning out.

5. The transmission will not shift into gear.
(a.) Check the shift linkage for loosened set screws or broken roll pins.
(b.) The shift mechanism in the transmission may be jammed.
(c.) The sliding clutch jaws in the transmission may be jammed or damaged.

6. The transmission will not stay in gear.
(a.) Worn gears may need to be replaced.

7. The saw band is cutting inaccurately.
(a.) Check for worn blade teeth.
(b.) Scale on the work was not removed.
(c.) The work may be work-hardened by grinding to remove scale.
(d.) The blade may be too wide for the radius being cut.
(e.) The saw band or inserts may be incorrect.
(f.) The post may not be square to the work table.
(g.) The feed force may be incorrect.
(h.) The band speed may be incorrect.
(i.) The coolant may not be applied evenly to both sides of the saw band.
(j.) The saw guide on the upper post may not be located close enough to the work.

(k.) The saw band tension may be incorrect. More tension is needed for a Dart blade.

8. The surface finish on the work is too rough.
(a.) The saw guide inserts may be worn.
(b.) The saw band speed may be too low.
(c.) The saw band pitch may be too coarse.
(d.) The feed may be too heavy.
(e.) There may be too much vibration in the saw band, or in the work.

9. The saw band teeth are stripping (usually when chips weld themselves to the teeth).
(a.) The saw band pitch may be too coarse for a thin work section.
(b.) The work may not be held firmly enough to stop vibration.
(c.) The feed pressure may be too high.
(d.) The band speed may be too low.

10. There is premature saw band breakage (usually caused by teeth stripping).
(a.) The saw band speed may be too low.
(b.) The feeding force may be too high.
(c.) The pitch of the saw band may be too coarse.
(d.) The saw guide inserts and backup bearings may not be guiding the band properly.
(e.) You may be using the wrong type of coolant.
(f.) The band tension may be too high.
(g.) There may be a defective blade weld. Check the welder manual.

11. There is premature dulling of the saw band teeth.
(a.) Reduce feed pressure and speed on the first few cuts.
(b.) The band speed may be too high, causing abrasion.
(c.) The saw band pitch may be too coarse.

(d.) The wrong type of coolant or no coolant is used.

(e.) The feed pressure may be too light.

(f.) The coolant may not be covering the saw band properly.

(g.) The material being cut may be faulty (scale, inclusions, hard spots, etc.).

(h.) Material analysis may be incorrect.

(i.) There may be saw band vibration.

(j.) A chipped tooth may be lodged in the cut.

(k.) Chip welding may be taking place.

(l.) Inserts may be too large for the blade width, allowing inserts to hit set teeth.

12. The motor runs, but the band does not move.

(a.) A drive belt may be broken or off the pulley.

(b.) The variable-speed pulley may have been over oiled. There may be oil on the pulley and belts.

(c.) The drive belt tension may be too low.

(d.) The drive belts may be the wrong size.

(e.) The band tension may be incorrect.

(f.) The transmission may be in neutral.

13. The band slips off the wheels.

(a.) The upper wheel may not be aligned correctly. The band does not track on the center of the wheel tire.

(b.) Too much coolant or the wrong type of coolant may have been used, causing the band to slip off the wheel tires.

(c.) The initial machine alignment may have been wrong.

14. There is excessive insert and blade wear.

(a.) The inserts may have been adjusted too tightly on the blade.

(b.) High band speed causes friction. Use Saw-Eez or coolant. Use roller guides, if possible.

(c.) Rollers on the roller guides may be adjusted too tightly on the blade.

(d.) The chip brush may be worn or may not be adjusted properly. Chips stay on the wheel.

15. The file band breaks.

(a.) The feed force may be too high.

(b.) The wrong type of file band may have been used. See the operator's manual.

16. The finish from band filing is poor.

(a.) The feed force may be too high.

(b.) The file band may not be assembled correctly.

(c.) The band tension may be too high. Set the tension the same as that used for a 1/8-in (3.175-mm) wide carbon band.

17. The file band teeth become loaded.

(a.) Clean the teeth immediately.

(b.) The feed force may be too high.

(c.) The wrong shape of file is being used, or the wrong number of teeth for the material being filed.

REVIEW QUESTIONS

1. List the sequence of steps recommended to set up a band machining job.
2. Why is proper planning important before starting a job?
3. List three general safety tips for sawing.
4. What information does the job selector provide for setting up the work?
5. What is the purpose of troubleshooting in band machining?
6. If the sawband is cutting inaccurately, what are six likely causes of the problem?
7. If the surface finish on the work is too rough after sawing, list three likely causes of the problem.
8. List three likely causes of saw band breakage.

Cutting off pipe using a hand-held cutting-off wheel. *Courtesy of Cominco Ltd.*

Feeding the work by hand. This illustrates the simplicity of holding the work. The moving band-saw blade forces the work into the work table, exerting a downward pressure. The operator's hands are mostly on the push bar with little or no pressure needed to hold the work down. *Courtesy of DoALL Company*

Some of the many cuts made by a vertical band machine. The finished product held by the operator is shown beside the excess material, much of it still usable for other jobs. This work is sawed to rough shape, filed to size, and finish polished on one machine. *Courtesy of DoALL Company*

Cutting off large stock. Thin steel discs are cut off on a horizontal cut-off bandsaw. *Courtesy of DoALL Company*

Cutting off stock. A high-speed steel bimetal band-saw blade is being used to cut off stock. *Courtesy of DoALL Company*

Band filing. A curved surface is being band filed on a vertical bandsaw. *Courtesy of DoALL Company*

Band polishing. A curved surface is being polished on a vertical bandsaw. *Courtesy of DoALL Company*

Contour cutting. An internal contour cut is being made with a thin blade on a vertical bandsaw. A power assist feed is being used. *Courtesy of DoALL Company*

Band sawing. Plastic material is being cut on a vertical bandsaw using a fence. *Courtesy of DoALL Company*

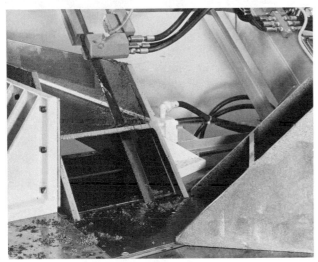

Band sawing. An angular cut is being made on a structural steel beam. *Courtesy of DoALL Company*

Band sawing. An angular cut is being made on steel tubing. *Courtesy of DoALL Company*

Heavy sections are being cut off on a heavy-duty, twin post, cut-off band saw. *Courtesy of DoALL Company*

CHAPTER SEVEN

DRILLS AND DRILLING OPERATIONS

After cutting off stock, the next machine tool operation commonly performed by the student or apprentice machinist is drilling holes in solid material. In fact, an apprentice will probably spend a portion of the apprenticeship learning about drills and drilling machines.

The most common drills used are twist drills and spade drills. The most common drilling machines are electric or air hand drills, drill presses, and radial drilling machines. A type of drill press was probably one of the earliest machine tools to be developed. If a drill press is examined closely, it is really a very simplified version of a lathe, except that it is standing vertically and the cutting tool does the revolving rather than the work. But there is no doubt that the similarity is there.

With the right combination of cutting tools and attachments, the basic machine operations that can be performed on drilling machines are drilling, reaming, boring, countersinking, counterboring, spot-facing, and tapping. Many vertical milling cuts are also possible using drilling machines, as are external threading cuts making use of threading dies and holders.

SAFE PRACTICES IN DRILLING

The moving drill and spindle are the chief dangers in drilling. The possibility of the work breaking loose and rotating and the chips being ejected are also potential dangers.

1. Permission should be obtained to operate drilling equipment, at the discretion of the instructor.
2. Always roll up loose sleeves and remove ties, rings, watches, and so on before operating the machine.
3. Always wear specified eye protection.
4. A drill should always be properly sharpened. Set the proper speed according to the drill size.
5. Chuck keys should never at any time be left in the chuck. If the machine were turned on, the key could become a lethal projectile.
6. Never hold the work by hand. The work should always be clamped properly, Figures 7–1 and 7–2.

Figure 7–1 A tool and die maker demonstrates the unsafe practice of drilling unclamped work *(Courtesy of Cominco Ltd.)*

Figure 7–2 A tool and die maker demonstrates the safe practice of drilling clamped work *(Courtesy of Cominco Ltd.)*

7. Clear the chips from the drill flutes often, particularly when using the smaller drill sizes. Use the proper cutting fluids.
8. When the drill starts to break through the work, ease up on the feed pressure and gently "feel" the drill through the work.
9. When you are making any adjustments, shut the machine off.
10. Never use your hand to stop the chuck from revolving.
11. If the work slips from the clamps, shut off the machine and step back clear of the machine. Never attempt to stop the work with your hands.
12. Never walk away from a machine when it is operating.
13. When the work is complete, shut off the power and clean the machine.

UNIT 7-1

DRILLING MACHINES AND TOOL HOLDING METHODS

OBJECTIVES

After completing this unit, the student will be able to:

- work safely with drills and drilling machines.
- list differences between the sensitive drill press, the radial arm drill, gang drilling machine, and multi-spindle drilling machine.
- select and install the proper tool for the type of tool-holding device on the machine.
- hold the work safely on the machine.

KEY TERMS

Drill
Spindle
Self-holding taper
Drill drift
Sleeve
Work table
Drill press
Post drill press
Upright drill press
Drift slot
Tang
Retang sleeve
Drill socket or extension socket

Sensitive drill press
Radial arm drilling
   machine
Gang drilling
   machine
Multi-spindle
   drilling machine
Morse taper
Straight shank drill
Taper shank drill
Taper shank
Drill chuck
Keyless chuck
Drill sleeve

Drill press vise
Angle vise
Float-lock vise
Universal table
Rotary table
V-block
V-block clamp
Strap clamp
Support block
Filler block
Step block
Universal adjustable
   clamp

DRILLING MACHINES

One of the most common machine shop operations is the drilling of holes in solid material. The cutting tool used is called a **drill**. There are many different types of drilling machines, all of which perform essentially the same types of drilling operations. All drilling machines have one thing in common: they have one or more spindles rotating within individual sleeves that can move the spindles vertically up or down. The work is clamped to the drilling-machine work table. A drill is placed in the spindle which rotates and is moved downward to force or press the drill into the material. When using any drilling machine, it should be remembered that all work, no matter how small, should be clamped to the work table to prevent it from twisting or rotating as the drill is pressed into it.

The drill press is one of the most common types of drilling machines. It is also sometimes called a post drill press or upright drill press. A sensitive drill press, Figure 7–3, has hand feeds only and is made in both floor and bench models. A larger drill press, such as the 24-in (600-mm) machine, Figure 7–4, is equipped with power feeds. The size or capacity of a drill press is determined by the maximum work diameter that can be centrally drilled. Another measure of the size is the

distance from the center of the spindle to the front of the column.

The radial arm drilling machine, or more simply, radial drill, Figure 7–5, is also a common drilling machine. The radial arm can be rotated through 360 degrees around a vertical column. It can be raised or lowered and can be clamped in any position. The drilling head can be moved to any position on the radial

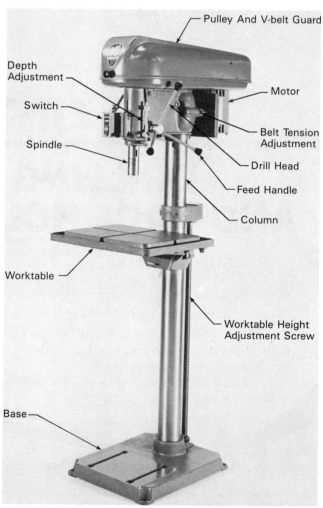

Figure 7–3 Parts of a standard, floor-model, sensitive drill press *(Courtesy of Gate Machinery Company Ltd.)*

arm and can also be clamped in any position. This machine is used for large work where one or more holes must be drilled. The work can be clamped to the work table or to the floor, and the drill head can be moved to wherever it is needed. The size of a radial drilling machine is usually equal to the length of the radial arm or the distance from the column to the center of the spindle when the drill head is set to its maximum position on the radial arm. Any number of holes within the range of the radial drill head can be drilled without moving the work.

Other drilling machines that are usually adaptations of a standard drill press are used. A gang drilling machine, for example, is used to perform several successive operations on one piece of work. There is more than one drill head set over a common work table. Each drill head can be set to perform a separate operation such as drilling, reaming, tapping, etc. Once each operation is performed, the work is moved to the next drill head.

A multi-spindle drilling machine has more than one spindle connected to and driven by the main spindle using universal joints. Such a machine can drill many holes at once.

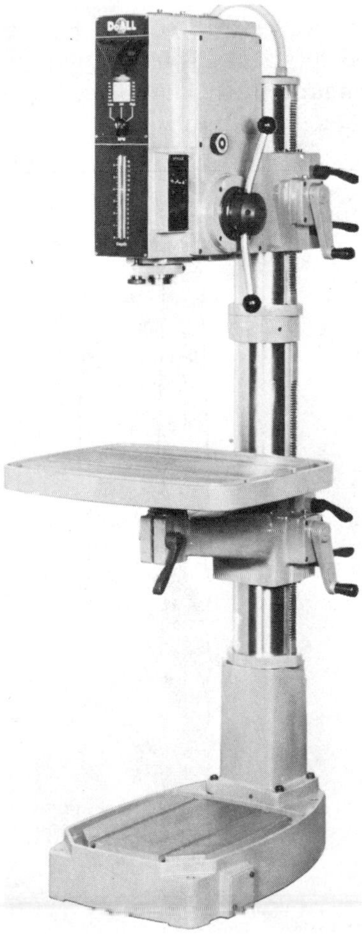

Figure 7–4 A large drill press equipped with power feeds *(Courtesy of DoALL Company)*

Figure 7–5 A radial arm drill *(Courtesy of DoALL Company)*

TOOL HOLDING METHODS

Most drilling machine spindles have a hole bored in the lower end that is tapered with a standard Morse taper, Figure 7–6. The spindle will accept taper shank drills, Figure 7–7, and reamers, or taper shank drill chucks, Figure 7–8A, which are used to hold and drive straight shank drills, Figure 7–8B. Keyless chucks, Figure 7–9, for light and medium duty, do not require chuck keys. They are tightened or loosened by hand pressure only. A drill chuck should never be used to hold and drive a taper shank drill.

There are eight tapers in the Morse taper series, which are numbered from 0 (the smallest) to 7 (the largest). Each has a slightly different taper, varying from 0.598 in per ft (1:20, a metric ratio) to 0.631 in per ft (1:19). These tapers are usually referred to as slow or self-holding tapers, or sometimes self-driving tapers, meaning that the angle of the taper is shallow. A taper shank drill is jammed into the drill spindle which, when revolving, causes the drill to revolve. Driving power to the drill comes from a combination of the jamming ability of the slow taper and also the tang that fits into the drift slot in the spindle.

Some machinists dispute the fact that the tang is part of the driving force. However, both *Machinery's Handbook* and the *American Machinist Handbook* state that most taper shank drills have a driving tang. If a tang is twisted off, it is usually due to a poor fit between the shank and the socket caused by dirt, bruises or burs on the sleeve, or a worn spindle bore.

A standard drill sleeve, Figure 7–10, is used to adapt a small, taper shank drill to a larger taper bore of a spindle, Figure 7–11. If the tang on a taper shank drill breaks, it can be reground and fitted into a retang sleeve, Figure 7–12. Two annular rings on the retang sleeve indicate the normal position of the drift slot on a standard sleeve. A drill socket, Figure 7–13 (page 170), is used to adapt a small tapered spindle to a larger taper shank drill, Figure 7–14 (page 170).

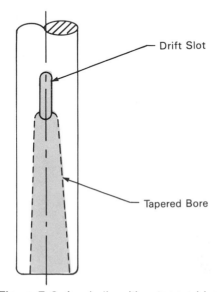

Figure 7–6 A spindle with a tapered bore

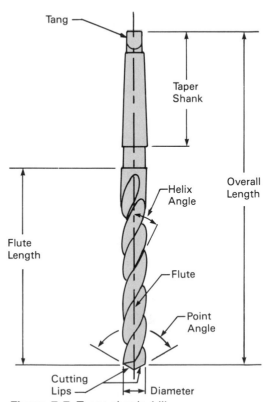

Figure 7–7 Taper shank drill

A drill drift, Figure 7–15 (page 170), is used to remove taper shank drills from spindles or sleeves. The sloping flat edge of the drift fits against the drill tang and the rounded edge fits the upper rounded edge of the drift slot. The end of the drift is struck lightly with a hammer to remove the drill. A wooden block will protect the work table from being damaged by a falling drill.

WORK HOLDING METHODS

All work that requires drilling should be clamped securely to the work table. Work can be held in various types of vises that can be clamped or bolted to the work table, or the work can be clamped and bolted directly to the work table. If, for any reason, the work breaks free and starts to spin, immediately shut off the machine and step well back clear of the machine. Never try to stop the spinning work by hand.

A standard drill press vise, Figure 7–16A (page 171), is one of the most common devices for holding work. The work is placed in the vise and the vise is clamped to the table. A variation of the drill press vise is the angle vise, which can be adjusted to hold the work at any angle to the drill press spindle. A simple frame, Figure 7–16B (page 171), can be made to prevent a drill press vise from rotating. The frame also allows easy lateral adjustment.

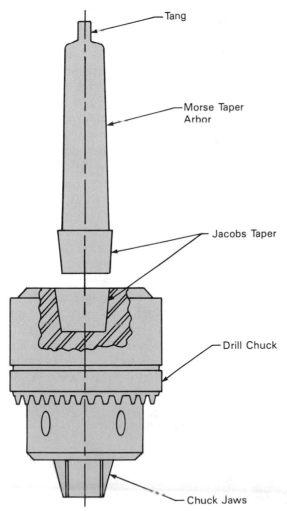

Figure 7–8A A taper shank drill chuck for straight shank drills

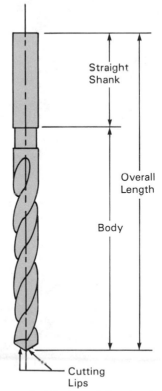

Figure 7–8B A straight shank drill

There is a universal table to which work can be clamped. It has an upper and a lower slide, both of which have horizontal graduated swivels. The slides can be positioned at any angle in a horizontal plane, and can be moved individually or together. Both slides are fitted with hand cranks and graduated collars. Work can be moved horizontally and accurately in any direction within the range of the slides.

The float-lock vise, Figure 7–17, is one of the better vises for clamping work. The jaws can be set in either a vertical or horizontal position and will hold a variety of work. The T-anchor clamp locks the vise securely at any position on the table.

A universal rotary table, Figure 7–18, is used for all circular and rotary machining operations or when holes are to be drilled on a specified hole circle.

V-blocks can be used to hold round stock. The work can be secured either with a V-block clamp or a strap clamp or both.

There are many patterns of strap clamps used to hold work securely to a work table. Figure 7–19 (page 172) illustrates some of the different patterns available, which are used with T-slot nut studs, T-slot nuts, flange nuts, coupling nuts, and step blocks or plain blocks.

Strap clamps must be used correctly so that the work is held securely. The bolt or stud should be closer to the work than the support block so that most of the pressure is exerted on the work, not the support block. A step block or a filler block can be used to support a strap clamp, Figure 7–20 (page 172). A bolt or stud must be placed correctly when clamping work, Figure 7–21 (page 172). The right and wrong ways to clamp work are illustrated in Figure 7–22 (page 173).

Figure 7–9 A keyless drill chuck *(Courtesy of DoALL Company)*

Figure 7–10 A standard drill sleeve *(Courtesy of The Cleveland Twist Drill Company)*

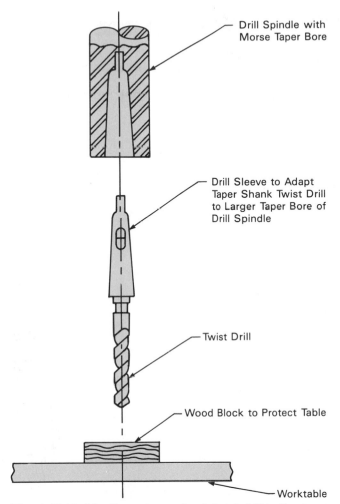

Drill Spindle with Morse Taper Bore

Drill Sleeve to Adapt Taper Shank Twist Drill to Larger Taper Bore of Drill Spindle

Twist Drill

Wood Block to Protect Table

Worktable

Figure 7–11 Adapting a taper shank twist drill to the taper bore of a drill spindle

Figure 7–12 A retang sleeve *(Courtesy of SKF & Dormer Tools Ltd.)*

Figure 7–13 A drill socket *(Courtesy of SKF & Dormer Tools Ltd.)*

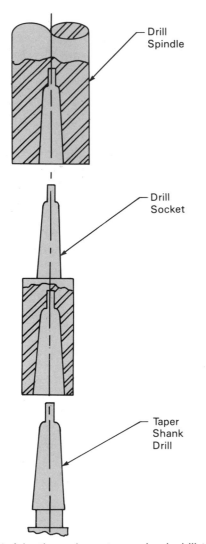

Drill Spindle

Drill Socket

Taper Shank Drill

Figure 7–14 Adapting a large taper shank drill to a small taper shank drill spindle

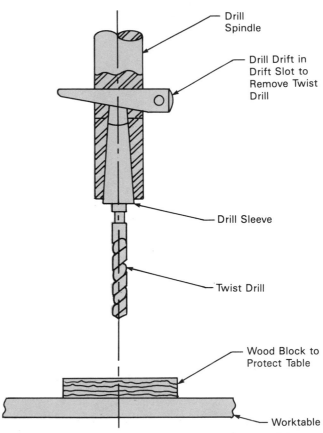

Drill Spindle

Drill Drift in Drift Slot to Remove Twist Drill

Drill Sleeve

Twist Drill

Wood Block to Protect Table

Worktable

Figure 7–15 Using a drill drift to remove a twist drill from a drill spindle

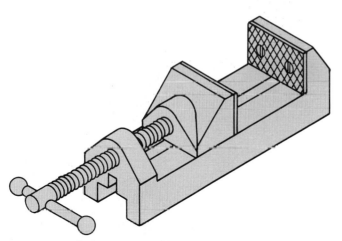

Figure 7–16A A standard drill press vise

Figure 7–17 A float-lock vise *(Courtesy of DoALL Company)*

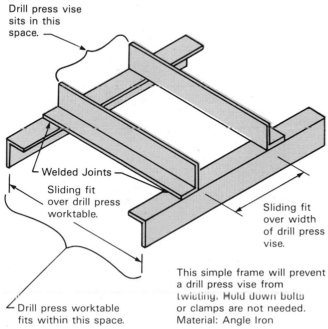

Drill press vise sits in this space.

Welded Joints

Sliding fit over drill press worktable.

Sliding fit over width of drill press vise.

Drill press worktable fits within this space.

This simple frame will prevent a drill press vise from twisting. Hold down bolts or clamps are not needed. Material: Angle Iron

Figure 7–16B Safety frame for a drill press vise

Figure 7–18 A universal rotary table *(Courtesy of DoALL Company)*

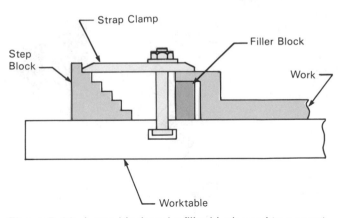

Figure 7-20 A step block and a filler block used to support a strap clamp

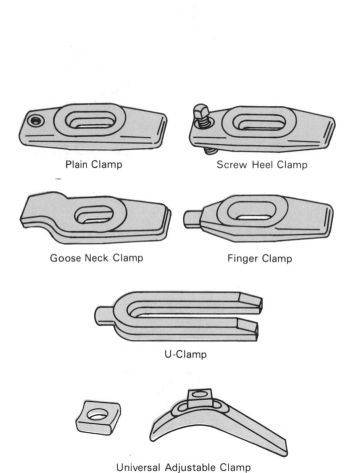

Plain Clamp

Screw Heel Clamp

Goose Neck Clamp

Finger Clamp

U-Clamp

Universal Adjustable Clamp

Figure 7-19 Strap clamps *(Courtesy of Armstrong Brothers Tool Company)*

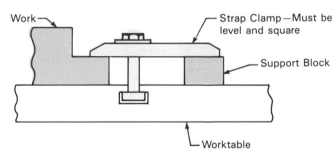

Correct—Most of the clamping pressure is on the work.

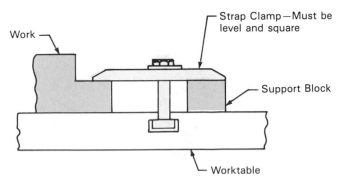

Wrong—Most of the clamping pressure is on the support block.

Figure 7-21 Bolt should be closer to the work than to the support block

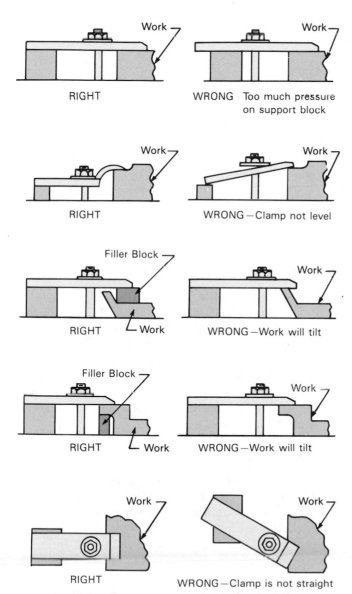

Figure 7–22 Right and wrong ways to clamp work

REVIEW QUESTIONS

1. List five precautions to be observed when setting up for work on a drilling machine.
2. What is one common element in drilling machines?
3. What is a sensitive drill?
4. What is the primary advantage of a radial drill?
5. What determines the size or capacity of a drill press?
6. What is the purpose of a gang drilling machine? Of a multi-spindle drilling machine?
7. Describe the standard method of holding a tool in a drilling machine spindle.
8. What shanks are found on twist drills?
9. What is a slow taper?
10. Should a taper shank drill be used in a drill chuck? Why?
11. What provides the driving power in a taper shank drill?
12. What provides the driving force in a straight shank drill?
13. What is the purpose of a standard drill sleeve?
14. What is the purpose of a drill socket?
15. What is a drill drift?
16. How is the standard drill press vise used?
17. What is a float-lock vise?
18. Name six strap-type clamps.
19. What additional devices can be used with strap clamps to hold the work correctly?
20. List three considerations in the proper clamping of work to the work table.

UNIT 7-2

DRILLS AND THE USE OF DRILLS

OBJECTIVES

After completing this unit, the student will be able to:

- state the purpose of drills.
- define the terms flute, margin, land, body clearance, web, point, chisel edge, and cutting lips.
- describe the differences among the six major types of drills.
- list the twist drill series and specify the size range in each series.
- explain the difference between drill point angle and clearance angle.
- sharpen drills properly for drilling different materials.
- list the characteristics of a properly ground drill.

KEY TERMS

Drill	Low helix twist drill
End cutting tool	High helix twist drill
Chips	Straight fluted drill
Twist drill	Coolant feeding drill
Flutes	High-speed coolant
Margin	inducer drill
Neck	Spade drill
Body clearance	Pilot hole
Cutting lip	Chip splitter groove
Helix angle	Spade drill holder
Axis of the drill	Spade drill point
Land	Lip clearance
Point angle	Twist drill series
Web	Rake angle
Chisel edge	Drill sharpening
General-purpose	Drill point gage
twist drill	Chatter

DRILLS

Drills are end cutting tools with one or more cutting lip. They perform several operations at the same time. They penetrate directly into solid material and eject the chips up through the drill flutes. They maintain a straight direction into the material and control the size of the drilled hole. Most drills today are made of high-speed steels. Some special carbide-tipped drills wear longer and can be operated at three times the speed of high-speed steel drills.

Twist Drills

The type of drill used most often is called a twist drill, Figure 7–23. The helical grooves called **flutes** provide space for chips to be ejected from the cutting lips on the end of the drill. The **margin** is a narrow raised surface on the leading edge of each flute. The diameter formed by the margins is the body size of the drill. Because of the margin, friction between the body of the drill and the sides of the hole is reduced. The

land, which is the area of the body between each flute, contains both the margin and the **body clearance** (that portion of the body cut below the margin). The **helix angle** is the angle between the leading edge of the flute and the axis of the drill. The **web** is the solid tapered core between the flutes, Figure 7–24.

The sharpened lands at the end of the drill form the **point**, the **chisel edge** at the dead centre of the drill, and the **cutting lips**, Figure 7–25. After repeated sharpenings, the web and chisel edge become thicker, Figure 7–26.

General-purpose twist drills are made with either straight or taper shanks, Figure 7–27A and 7–27B. These drills can be used in many different applications and for a wide variety of materials by varying the point angles, rakes, speeds, and feeds.

Low helix twist drills, Figure 7–28, have wide polished flutes designed to eject chips rapidly and prevent clogging. The slow helix reduces the rake angle at the cutting lips, which makes these drills especially useful for drilling brasses, phosphor-bronze, hard plastics such as bakelite, hard rubber, and materials made of compressed fibers.

High helix twist drills, Figure 7–29, are designed for drilling aluminum alloys, copper, and other soft metals. The fast helix increases the rake angle at the cutting lips. The flutes are wider than standard and are polished to prevent them from becoming choked with cuttings.

Straight fluted drills, Figure 7–30, produce short chips and are best suited for brass and nonferrous materials like certain brittle plastics. The straight flutes eliminate pull-in or the tendency of a drill to grab. Drills tipped with carbide are recommended for drilling steels harder than Rc 48, with a cutting speed of 80–100 sfpm (24–30 meters per min).

Coolant feeding drills are heavy-duty twist drills used largely in vertical drilling applications for tough jobs. They are also used effectively for drilling deep holes in soft materials. The straight shank drill, Figure 7–31, is tapped at the shank end with a pipe thread for attaching a coolant feed fitting. The two coolant holes run through the lands to the cutting lips. The taper shank high-speed coolant inducer drills, Figure 7–32, are used for drilling both tough materials and soft materials. The notch thinned point and ground taper shank ensure accurate hole location. A high helix angle ejects chips rapidly even under varying liquid- and mist-coolant pressure settings. The coolant is fed into a coolant inducer socket, then into the side of the taper shank into the coolant holes, and down through the drill to the cutting lips.

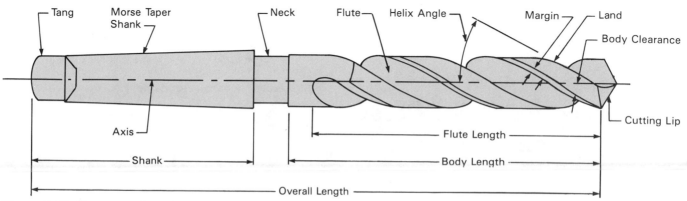

Figure 7–23 The parts of a twist drill

Spade Drills

Twist drills are manufactured in sizes ranging from as small as 0.006 in (0.15 mm) to as large as 3-1/2 in (90 mm) or even larger, but in sizes over 1 in (25.4 mm) they start to become expensive. Spade drills are much more economical, especially when drilling large holes. They are used regularly in N/C and CNC drilling applications. A spade drill is a flat, replaceable blade with two sharp cutting lips mounted in a spade drill holder. Some spade drill holders will hold as many as 21 different blade sizes. Chip splitter grooves on the clearance faces of each blade curl and break the chips into small pieces that are ejected easily. Chip disposal is safe and convenient.

Spade drills tend to drill rougher holes than those drilled by twist drills. Subsequent boring or reaming may be required for precision work. Blades are made in various styles, Figure 7–33. Each style is available in either high-speed steel or carbide tips. They can be sharpened many times before needing replacement. A pilot hole is not recommended when starting spade drills. The spade drill point of 130 degrees is much greater than the 118-degree point on a general-purpose twist drill. The larger angle may cause the spade drill to be thrown off center. The best practice is to start with a short holder using the same size spade drill as the required hole diameter. The hole should be drilled at least 1/8 in (3 mm) deep using the full diameter of the spade drill. This can be followed with a long series holder, if necessary, to complete the hole.

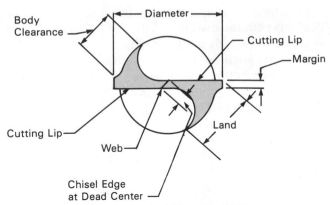

Figure 7–25 Viewing the end of a twist drill

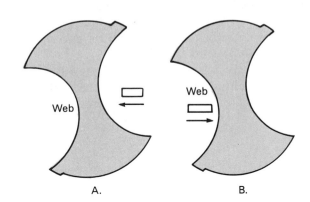

The drill section at "A" represents the web thickness at the point of the drill. The drill section at "B" represents the web thickness at the shank end of the drill.

Figure 7–26 After repeated sharpenings, the web becomes thicker

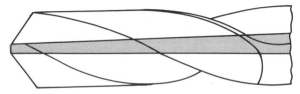

The blue section represents the web of the drill. Its thickness increases as the drill gets shorter.

Figure 7–24 The web of the twist drill

Figure 7–27A A general-purpose, straight-shank, twist drill *(Courtesy of The Cleveland Twist Drill Company)*

Figure 7–27B A taper-shank twist drill *(Courtesy of The Cleveland Twist Drill Company)*

Figure 7–28 A low-helix twist drlll *(Courtesy of The Cleveland Twist Drill Company)*

Figure 7–29 A high-helix twist drill *(Courtesy of The Cleveland Twist Drill Company)*

Figure 7–30 A straight fluted drill *(Courtesy of The Cleveland Twist Drill Company)*

Figure 7–31 A straight-shank, coolant feeding drill *(Courtesy of The Cleveland Twist Drill Company)*

Spade drill holders are made in three series: short, long, and extra long. All are available in straight shank, Figure 7–34A, and taper shank, Figure 7–34B. Sizes of blades and holders can range from microholes with diameters of 0.020 in (0.5 mm) to 8 in (203 mm), or even up to 14 in (55 mm). These larger sizes are usually used in a few specialized industries.

TWIST DRILL SERIES AND SIZES

Twist drill designations are determined by both their diameter and the series to which they belong, that is, fractional drills, wire or number drills, letter drills, or metric drills. Typical twist drill series and sizes are illustrated in Table 7–1 (page 180). Most manufacturers regularly stock drills in fractional sizes from 1/64 in to 2 in, wire or number sizes from #80 (0.013 in) to #1 (0.228 in), letter sizes from A (0.234 in) to Z (0.413 in), and metric sizes from 0.3 mm to 50.5 mm. Other sizes are available on request.

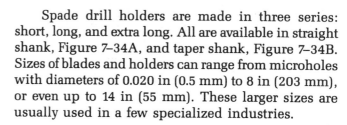

Figure 7–32 A taper-shank, coolant inducer drill *(Courtesy of The Cleveland Twist Drill Company)*

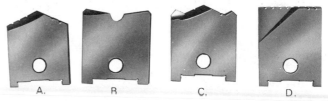

Figure 7–33 Spade drill blades *A.* Standard spade *B.* Coring *C.* Special-purpose *D.* Flat bottom *(Courtesy of DoALL Company)*

DRILL POINT ANGLES AND CLEARANCES

Different point angles and clearances are needed for different materials, depending on the nature of the material being cut and the operation being done, Figure 7–35 (page 182). Contrast the general-purpose point angle of 118 degrees and the lip clearance of 8 degrees to 12 degrees, Figure 7–36 (page 183), with the 60-degree point angle and 12-degree to 15-degree lip clearance used for aluminum, which deforms or bulges easily.

To summarize, the most common general-purpose clearance angles are 8–12 degrees. For high strength, tough steels, clearance angles of 7–10 degrees are better, with the lower angles used for larger diameter drills and the higher angles for smaller diameter drills. In certain cases, clearances can sometimes be increased beyond the listed maximum. For cutting soft, free machining materials like aluminum, which deforms easily, clearances can be 20–26 degrees when using drills less than 1/4 in (6 mm) in diameter. Larger drills can have a clearance as much as 12–18 degrees

The helix angle, Figure 7–37 (page 183), of a twist drill is commonly 24–32 degrees, but can vary from 0 degrees in the case of a straight flute drill to almost

Figure 7–34A A straight-shank, spade drill holder *(Courtesy of DoALL Company)*

Figure 7–34B A taper-shank spade drill holder *(Courtesy of DoALL Company)*

40 degrees in some high helix twist drills. It is common practice to say that the helix angle and the rake angle are the same. This would be true if the point angle were 180 degrees or flat. However, as the point angle decreases, the rake angle increases slightly since rake is usually measured at right angles to the cutting lip of the drill.

Use a protractor to measure the helix angle by holding the protractor perpendicular to the center line of the drill. Using the same setting, hold the protractor at right angles to the cutting lip at the outer periphery of the cutting lip. It will be seen that the rake angle is slightly more than the helix angle. It will also be seen that the rake angle varies from almost 0 degrees at the center of the drill to a maximum at the outer periphery of the drill.

DRILL SHARPENING

Sharpening a drill is a very necessary operation, but it is not a difficult one. The easiest and most comfortable grip is to hold the main body of the drill in your right hand with the forefinger resting against the front of the guard. The left hand should hold the drill shank, and the angle should be half the point angle. The preferred method of holding the drill is shown in the top views of Figure 7–38 (page 183).

The side view, Figure 7–39 (page 184), shows the drill being held horizontally. At the start of the procedure, the cutting edge of the drill is on the center of the wheel or slightly above.

To obtain the proper clearance, drop the left hand straight down an amount equal to the intended clearance angle. Do not rotate the drill during this movement. The drill is rotated only to bring the next cutting edge into position for sharpening, Figure 7–40 (page 184).

The appearance of the point of the drill will indicate the correct clearance. A general-purpose point should be approximately 45 degrees to the cutting edges, Figure 7–41A (page 184). Too much clearance is indicated in Figure 7–41B, and Figure 7–41C shows no clearance.

A drill point gage, Figure 7–42 (page 185), is used to measure the point angle and the length of the cutting lips. The point angle, clearance angle, and lip length should be the same on both sides of the drill.

DRILLING PROBLEMS

A properly ground drill, Figure 7–43 (page 185), should have the following features:

A. The drill point angle should have equal angles on both sides of the center line of the drill.
B. The cutting lips should be the same length.
C. The cutting lip clearance should be the same on both sides.
D. The chisel point should be on center.

A properly ground drill will cut evenly. The chips will be ejected equally and smoothly from both cutting lips, Figure 7–44 (page 185).

A drill that is improperly ground will cause problems. The chisel point will be off center if the cutting lips are of unequal lengths, Figure 7–45A (page 185). The resulting hole will be oversize in diameter, Figure 7–45B (page 185).

If a drill has unequal lip angles, Figure 7–46A (page 186), all of the cutting will be done by one cutting lip. The resulting hole will be oversize and egg shaped, Figure 7–46B (page 186).

If the cutting lips of a drill have insufficient clearance, Figure 7–47A (page 186), too much pressure will be exerted on the drill point, which can result in a split drill, Figure 7–47B (page 186). Excessive feed can also cause a drill to split, even with adequate clearance.

With too much lip clearance, Figure 7–48A (page 186), there is a lack of support behind the cutting edges. This results in a vibrating cutting edge called **chatter** and a quick dulling and breakdown of the cutting edges, Figure 7–48B (page 186).

Excessive speed will cause wear of the outer corners of the cutting lips. Burning and corner breakdown can be the result. When the point is reground, much of the drill length is lost.

Table 7–1 Drill Sizes: Fractional, Wire or Number, Letter and Metric Sizes *(Courtesy of The Cleveland Twist Drill Company)*

Decimal	Fract. Wire Letter	mm.	Decimal	Fract. Wire Letter	mm.	Decimal	Fract. Wire Letter	mm.	Decimal	Fract. Wire Letter	mm.
.0059	97		.0413		1.05	.1065	36		.1960	9	
.0063	96		.0420	58		.1083		2.75	.1969		5.
.0067	95		.0430	57		.1094	**7/64"**		.1990	8	
.0071	94		.0433		1.1	.1100	35		.2008		5.1
.0075	93		.0453		1.15	.1102		2.8	.2010	7	
.0079	92	.2	.0465	56		.1110	34		.2031	**13/64"**	
.0083	91		.0469	**3/64"**		.1130	33		.2040	6	
.0087	90	.22	.0472		1.2	.1142		2.9	.2047		5.2
.0091	89		.0492		1.25	.1160	32		.2055	5	
.0095	88		.0512		1.3	.1181		3.	.2067		5.25
.0098		.25	.0520	55		.1200	31		.2087		5.3
.0100	87		.0531		1.35	.1220		3.1	.2090	4	
.0105	86		.0550	54		.1250	**1/8"**		.2126		5.4
.0110	85	.28	.0551		1.4	.1260		3.2	.2130	3	
.0115	84		.0571		1.45	.1280		3.25	.2165		5.5
.0118		.3	.0591		1.5	.1285	30		.2188	**7/32"**	
.0120	83		.0595	53		.1299		3.3	.2205		5.6
.0125	82		.0610		1.55	.1339		3.4	.2210	2	
.0126		.32	.0625	**1/16"**		.1360	29		.2244		5.7
.0130	81		.0630		1.6	.1378		3.5	.2264		5.75
.0135	80		.0635	52		.1405	28		.2280	1	
.0138		.35	.0650		1.65	.1406	**9/64"**		.2283		5.8
.0145	79		.0669		1.7	.1417		3.6	.2323		5.9
.0156	**1/64"**		.0670	51		.1440	27		.2340	A	
.0157		.4	.0689		1.75	.1457		3.7	.2344	**15/64"**	
.0160	78		.0700	50		.1470	26		.2362		6.
.0177		.45	.0709		1.8	.1476		3.75	.2380	B	
.0180	77		.0728		1.85	.1495	25		.2402		6.1
.0197		.5	.0730	49		.1496		3.8	.2420	C	
.0200	76		.0748		1.9	.1520	24		.2441		6.2
.0210	75		.0760	48		.1535		3.9	.2460	D	
.0217		.55	.0768		1.95	.1540	23		.2461		6.25
.0225	74		.0781	**5/64"**		.1562	**5/32"**		.2480		6.3
.0236		.6	.0785	47		.1570	22		.2500	**1/4"** E	
.0240	73		.0787		2.	.1575		4.	.2520		6.4
.0250	72		.0807		2.05	.1590	21		.2559		6.5
.0256		.65	.0810	46		.1610	20		.2570	F	
.0260	71		.0820	45		.1614		4.1	.2598		6.6
.0276		.7	.0827		2.1	.1654		4.2	.2610	G	
.0280	70		.0846		2.15	.1660	19		.2638		6.7
.0292	69		.0860	44		.1673		4.25	.2656	**17/64"**	
.0295		.75	.0866		2.2	.1693		4.3	.2657		6.75
.0310	68		.0886		2.25	.1695	18		.2660	H	
.0312	**1/32"**		.0890	43		.1719	**11/64"**		.2677		6.8
.0315		.8	.0906		2.3	.1730	17		.2717		6.9
.0320	67		.0925		2.35	.1732		4.4	.2720	I	
.0330	66		.0935	42		.1770	16		.2756		7.
.0335		.85	.0938	**3/32"**		.1772		4.5	.2770	J	
.0350	65		.0945		2.4	.1800	15		.2795		7.1
.0354		.9	.0960	41		.1811		4.6	.2810	K	
.0360	64		.0965		2.45	.1820	14		.2812	**9/32"**	
.0370	63		.0980	40		.1850	13	4.7	.2835		7.2
.0374		.95	.0984		2.5	.1870		4.75	.2854		7.25
.0380	62		.0995	39		.1875	**3/16"**		.2874		7.3
.0390	61		.1015	38		.1890	12	4.8	.2900	L	
.0394		1.	.1024		2.6	.1910	11		.2913		7.4
.0400	60		.1040	37		.1929		4.9			
.0410	59		.1063		2.7	.1935	10				

Table 7–1 Continued

Decimal	Fract. Wire Letter	mm.	Decimal	Fract. Wire Letter	mm.	Decimal	Fract. Wire Letter	mm.	Decimal	Fract. Wire Letter	mm.
.2950	M		.4688	15/32"		.9843		25.	1.4961		38.
.2953		7.5	.4724		12.	.9844	63/64"		1.5000	1-1/2"	
.2969	19/64"		.4844	31/64"		1.0000	1"		1.5156	1-33/64"	
.2992		7.6	.4921		12.5	1.0039		25.5	1.5157		38.5
.3020	N		.5000	1/2"		1.0156	1-1/64"		1.5312	1-17/32"	
.3031		7.7	.5118		13.	1.0236		26.	1.5354		39.
.3051		7.75	.5156	33/64"		1.0312	1-1/32"		1.5469	1-35/64"	
.3071		7.8	.5312	17/32"		1.0433		26.5	1.5551		39.5
.3110		7.9	.5315		13.5	1.0469	1-3/64"		1.5625	1-9/16"	
.3125	5/16"		.5469	35/64"		1.0625	1-1/16"		1.5748		40.
.3150		8.	.5512		14.	1.0630		27.	1.5781	1-37/64"	
.3160	O		.5625	9/16"		1.0781	1-5/64"		1.5938	1-19/32"	
.3189		8.1	.5709		14.5	1.0827		27.5	1.5945		40.5
.3228		8.2	.5781	37/64"		1.0938	1-3/32"		1.6094	1-39/64"	
.3230	P		.5906		15.	1.1024		28.	1.6142		41.
.3248		8.25	.5938	19/32"		1.1094	1-7/64"		1.6250	1-5/8"	
.3268		8.3	.6094	39/64"		1.1220		28.5	1.6339		41.5
.3281	21/64"		.6102		15.5	1.1250	1-1/8"		1.6406	1-41/64"	
.3307		8.4	.6250	5/8"		1.1406	1-9/64"		1.6535		42.
.3320	Q		.6299		16	1.1417		29.	1.6562	1-21/32"	
.3346		8.5	.6406	41/64"		1.1562	1-5/32"		1.6719	1-43/64"	
.3386		8.6	.6496		16.5	1.1614		29.5	1.6732		42.5
.3390	R		.6562	21/32"		1.1719	1-11/64"		1.6875	1-11/16"	
.3425		8.7	.6693		17.	1.1811		30.	1.6929		43.
.3438	11/32"		.6719	43/64"		1.1875	1-3/16"		1.7031	1-45/64"	
.3445		8.75	.6875	11/16"		1.2008		30.5	1.7126		43.5
.3465		8.8	.6890		17.5	1.2031	1 13/64"		1.7188	1 23/32"	
.3480	S		.7031	45/64"		1.2188	1-7/32"		1.7323		44.
.3504		8.9	.7087		18.	1.2205		31.	1.7344	1-47/64"	
.3543		9.	.7188	23/32"		1.2344	1-15/64"		1.7500	1-3/4"	
.3580	T		.7283		18.5	1.2402		31.5	1.7520		44.5
.3583		9.1	.7344	47/64"		1.2500	1-1/4"		1.7656	1-49/64"	
.3594	23/64"		.7480		19.	1.2598		32.	1.7717		45.
.3622		9.2	.7500	3/4"		1.2656	1-17/64"		1.7812	1-25/32"	
.3642		9.25	.7656	49/64"		1.2795		32.5	1.7913		45.5
.3661		9.3	.7677		19.5	1.2812	1-9/32"		1.7969	1-51/64"	
.3680	U		.7812	25/32"		1.2969	1-19/64"		1.8110		46.
.3701		9.4	.7874		20.	1.2992		33.	1.8125	1-13/16"	
.3740		9.5	.7969	51/64"		1.3125	1-5/16"		1.8281	1-53/64"	
.3750	3/8"		.8071		20.5	1.3189		33.5	1.8307		46.5
.3770	V		.8125	13/16"		1.3281	1-21/64"		1.8438	1-27/32"	
.3780		9.6	.8268		21.	1.3386		34.	1.8504		47.
.3819		9.7	.8281	53/64"		1.3438	1-11/32"		1.8594	1-55/64"	
.3839		9.75	.8438	27/32"		1.3583		34.5	1.8701		47.5
.3858		9.8	.8465		21.5	1.3594	1-23/64"		1.8750	1-7/8"	
.3860	W		.8594	55/64"		1.3750	1-3/8"		1.8898		48.
.3898		9.9	.8661		22.	1.3780		35.	1.8906	1-57/64"	
.3906	25/64"		.8750	7/8"		1.3906	1-25/64"		1.9062	1-29/32"	
.3937		10	.8858		22.5	1.3976		35.5	1.9094		48.5
.3970	X		.8906	57/64"		1.4062	1-13/32"		1.9219	1-59/64"	
.4040	Y		.9055		23.	1.4173		36.	1.9291		49.
.4062	13/32"		.9062	29/32"		1.4219	1-27/64"		1.9375	1-15/16"	
.4130	Z		.9219	59/64"		1.4370		36.5	1.9488		49.5
.4134		10.5	.9252		23.5	1.4375	1-7/16"		1.9531	1-61/64"	
.4219	27/64"		.9375	15/16"		1.4531	1-29/64"		1.9685		50.
.4331		11.	.9449		24.	1.4567		37.	1.9688	1-31/32"	
.4375	7/16"		.9531	61/64"		1.4688	1-15/32"		1.9844	1-63/64"	
.4528		11.5	.9646		24.5	1.4764		37.5	1.9882		50.5
.4531	29/64"		.9688	31/32"		1.4844	1-31/64"		2.0000	2"	

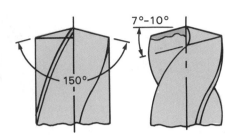

A. Manganese steel and hard materials. Note the flattened cutting lips to reduce rake.

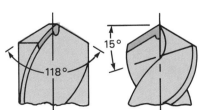

B. Wood, fiber, hard rubber.

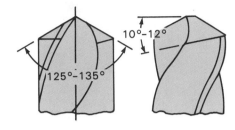

C. Heat-treated steels and drop forgings.

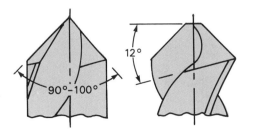

D. Copper and some of the stringier, tougher copper alloys.

E. Brass and soft bronze. Note the flattened cutting lips to reduce rake.

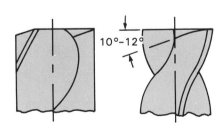

F. A flat point (180°) is used for counterboring an existing hole to a flat bottom.

Figure 7–35 Point angles and clearances for different materials

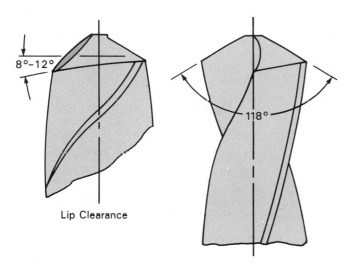

Lip Clearance

Point Angle

Figure 7–36 Lip clearance and point angle for general-purpose work

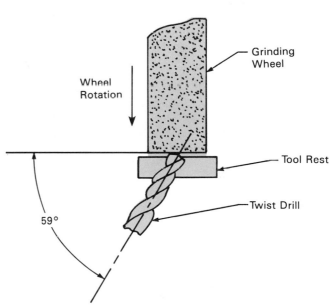

Grinding Wheel

Wheel Rotation

Tool Rest

Twist Drill

59°

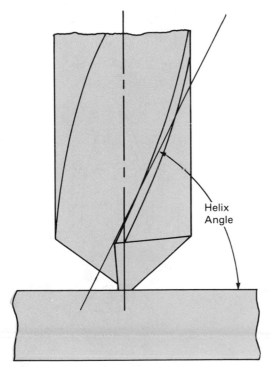

Helix Angle

Figure 7–37 Helix angle

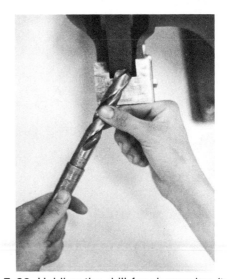

Figure 7–38 Holding the drill for sharpening (top view)

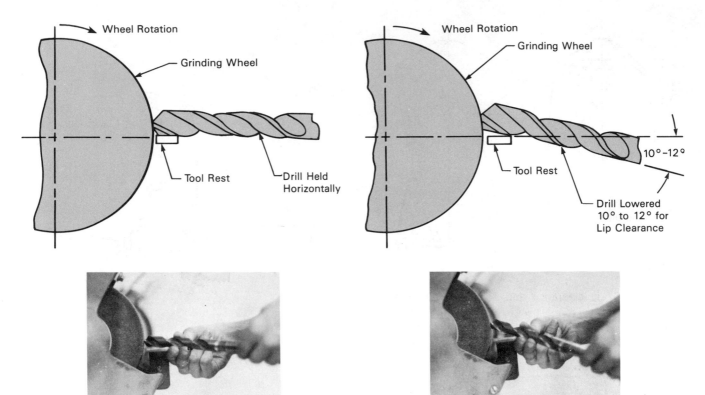

Wheel Rotation
Grinding Wheel
Tool Rest
Drill Held Horizontally

Wheel Rotation
Grinding Wheel
Tool Rest
Drill Lowered 10° to 12° for Lip Clearance
10°–12°

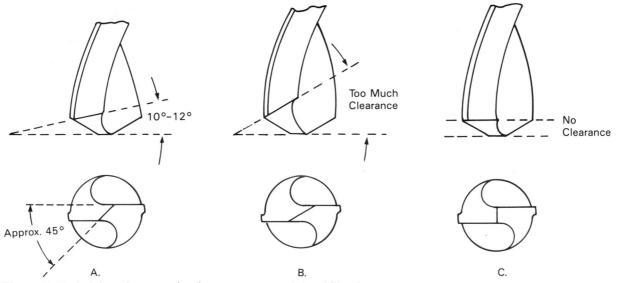

Figure 7–39 Starting position for drill grinding

Figure 7–40 Dropping the left hand to produce lip clearance

10°–12°
Approx. 45°
A.

Too Much Clearance
B.

No Clearance
C.

Figure 7–41 Judging clearance by the appearance of the drill point

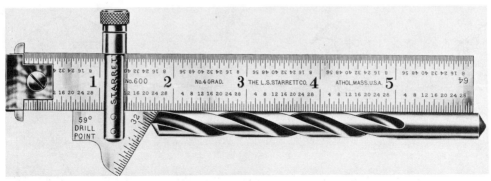

Figure 7–42 Using a drill point gage *(Courtesy of The L. S. Starrett Company)*

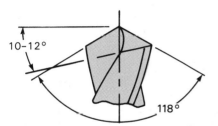 

Figure 7–43 A properly ground drill

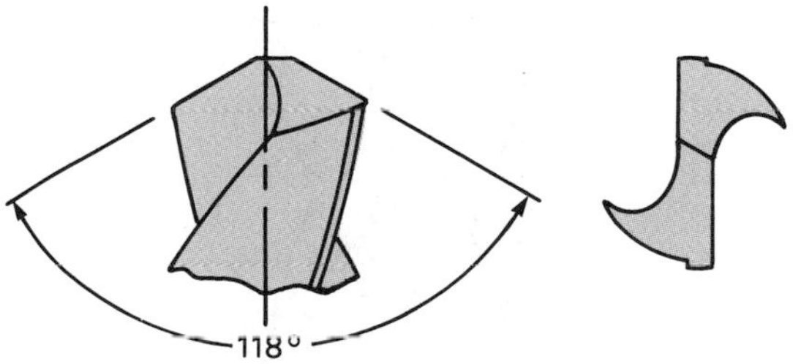

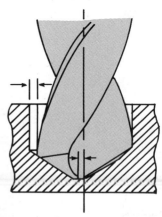

Figure 7–45A Unequal length cutting lips

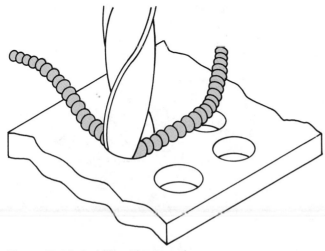

Figure 7–44 A drill cutting evenly

Figure 7–45B Unequal length cutting lips producing an oversize hole

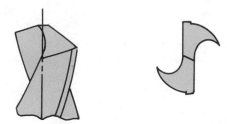

Figure 7–46A A drill with unequal lip angles

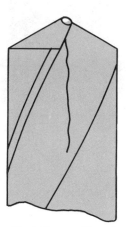

Figure 7–47B A split drill resulting from insufficient lip clearance

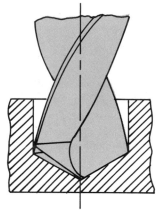

Figure 7–46B Unequal lip angles producing an oversize, egg-shaped hole

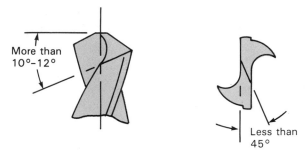

Figure 7–48A Excessive lip clearance

More than 10°–12°

Less than 45°

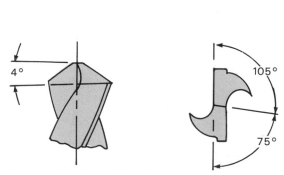

4°

105°

75°

Figure 7–47A Insufficient lip clearance

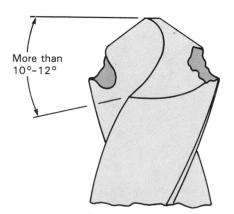

More than 10°–12°

Figure 7–48B The result of excessive lip clearance

REVIEW QUESTIONS

1. Sketch and name the parts of a twist drill.

2. Why does the point of a twist drill get larger as the drill gets shorter?

3. Name six different drill types and give a characteristic of each type.

4. What is the difference between the helix angle and the rake angle of a twist drill?

5. When are spade drills used? List the steps in the use of a spade drill to drill a hole.

6. List the series of twist drills and give the size range of each.

7. What factors govern the choice of drill clearance angles?

8. What are the general-purpose values for the drill point angle and lip clearance?

9. How is the drill to be held for sharpening?

10. How does the appearance of the point of a twist drill indicate the correct clearance?

11. What are the characteristics of a properly ground drill?

12. Give the cause of the following problems.
    a. The drilled hole is oversize and egg-shaped.
    b. The drilled hole is oversize.

13. What is the result of too much lip clearance?

14. If a drill point is off center, how is the problem corrected?

15. A drill has no clearance. What would the result be?

16. What would happen if the drill ran too fast?

UNIT 7-3

DRILLING OPERATIONS

OBJECTIVES

After completing this unit, the student will be able to:

- lay out a hole for spot drilling to locate it exactly.
- perform the spot drilling operation to locate a hole exactly.
- list the methods of producing chips in drilling.
- describe the methods of drill point thinning to relieve pressure on the chisel point.
- list the factors to be considered in selecting the cutting speed and rate of feed for a job.
- select the proper cutting speed and rate of feed for a job.

KEY TERMS

Spot drilling
Chip formation
Chip breaker
Chisel edge
Gash-type thinning
Conventional-type
  thinning
Flattening the lips
Pilot hole
Lead hole
Drill point
  thinning
Web thinning
Split point thinning
Positive rake cutting
  edge
Cutting speed
Feed

LAYING OUT A HOLE FOR SPOT DRILLING

Most drills tend to wander off center, partly because of the rotating action of the chisel point. Another cause is the extreme length of the drill in relation to its diameter. The drill will bend as it is forced into the work at the start of the drilling action. To ensure that a hole is drilled on center, it must be properly laid out, Figure 7–49. If more than one hole is to be drilled on the one center, each hole must be properly laid out from the same center, Figure 7–50.

SPOT DRILLING

Figure 7–51A indicates that the drill has run off center. To correct this situation and cause the drill to move back to center, a small groove is cut with either a diamond-shaped chisel or a small, round nose chisel, Figure 7–51B. The groove is cut on the side toward which you want the drill to move. The drill will move toward the groove. As the drill revolves, each cutting lip catches on the groove, causing the drill to vibrate its way back to center. The cutting of the groove may have to be repeated more than once. The drill should run exactly on center leaving only half of the center punch marks showing, Figure 7–51C.

CHIP FORMATION IN DRILLING

The formation of long, coiling chips when drilling may be spectacular to watch, but such chips are dangerous to the operator and should be avoided whenever possible. Disposal of this type of chip is a problem. Drill flutes offer only a limited amount of room for chips, so it is best to break the chips into smaller pieces so they can be ejected more easily.

There is more than one method of creating small chips. Decreasing the rake angle of the cutting lips causes chips to bend sharply and break. Increasing the feed will increase the thickness of the chip, which makes it less flexible. As it bends back on itself it will tend to fracture rather than coil. **Chip breakers**, Figure 7–52, which are notches ground into the lips or webs

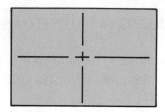

A. Lay out centerlines

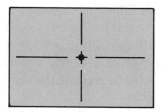

B. A light punch mark at the center

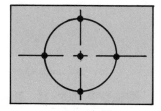

C. At least four light punch marks on the circumference. A deeper punch mark at the center

Figure 7–49 Proper layout procedure for spot drilling

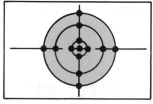

Figure 7–50 Laying out more than one hole on the same center. Each hole to be drilled must be laid out properly.

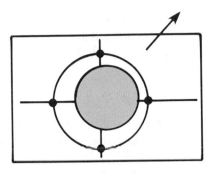

A. Drilled hole has moved off center in the direction of the arrow.

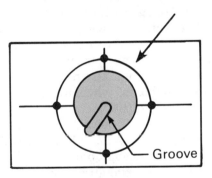

B. A groove has been cut on the side toward which the drill must move.

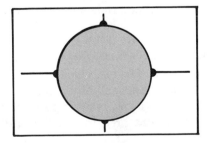

C. Drilled hole is properly centered with half of each punch mark showing.

Figure 7–51 Spot drilling a hole

of the drill, will also effectively break up chips. It takes extreme skill to grind chip breakers. Care must be taken to keep the depth of the chip breaker to a minimum, so as to avoid forming a pocket into which the chip will wedge. The flow of the chip should not be hindered.

PRESSURE ON DRILL POINTS

The chisel edge of the drill point is the least efficient part of a drill. Actually, it is a nuisance since it does not cut but squeezes or extrudes (forces up) the work material. With its rotating motion, it can also cause the drill to walk sideways from the intended position of the hole to be drilled. To compound this effect, the chisel edge gets wider as the drill gets shorter with repeated sharpening. To improve this condition, the machinist can either drill a pilot hole or thin the point of the drill. Either of these moves will relieve the intense pressure that is normally exerted on the drill point.

Pilot Holes

The main purpose of a pilot drill is to remove the material normally encountered by the point of a drill, Figure 7–53. The **pilot hole**, sometimes called a lead hole, should be the size of the drill point so that the chisel edge does not bear on the work at all. This means that the cutting lips start cutting immediately.

It is important that the pilot hole be laid out properly and accurately on the same center as the larger hole. It is equally important that the pilot hole be spot drilled exactly on center. The larger drill tends to follow the pilot hole closely.

If the pilot hole is larger than the drill point, it is possible for the larger drill to wander off center because the drill tends to vibrate, or chatter, as it rotates at the start of the cutting. If a series of drills is to be used to drill out a larger hole, each drill size used should be laid out accurately on the same center, and

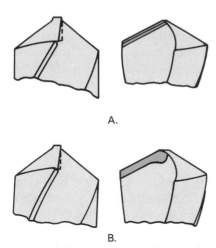

A.

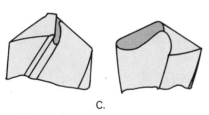

B.

"A" and "B" cause the chip to bend sharply and fracture.

C.

"C" causes the chip to curl tightly and break up into smaller units.

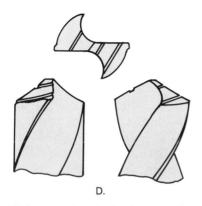

D.

"D" causes lengthwise fractures that curl back and break up into smaller units.

Figure 7–52 Types of chip breakers *(Courtesy of The Cleveland Twist Drill Company)*

a close watch should be kept to make sure that each drill stays exactly on center. If any of the drills wanders off center, it should be pulled back to the center by grooving. See the section on spot drilling.

Drill Point Thinning

Drill point thinning or **web thinning** is used to reduce the width of the chisel point. There are several ways of thinning a drill point. Whatever method is used, it is very important that an equal amount of metal be removed from each side of the drill. Drill point thinning must be done very carefully.

Gash-type thinning, Figure 7–54, is a very common practice. Metal is removed from the back side of the drill flute, taking as little as possible from the cutting lip. In **conventional-type thinning**, Figure 7–55, metal is removed from the drill so as to blend in with the flute contour. **Flattening the lips**, Figure 7–56, thins the drill point but rake is also reduced. This prevents the drill from "grabbing" when drilling materials such as brass and some of the more brittle plastics.

Another method of point thinning is called **split point thinning**, Figure 7–57. This is also called an offset, notched, or crankshaft point. The chisel edge is actually eliminated, extending positive rake cutting edges to the center of the drill and reducing thrust in drilling. This type of thinning is recommended for drilling deep holes and for drilling stainless steels, titanium alloys, high-temperature alloys, nickel alloys, very high-strength materials, and some tool steels. It is important that all traces of the original chisel edge be removed. Failure to do this results in a dull chisel edge having excessive negative rake and extremely poor cutting action. The two notching cuts should also just meet. If they pass each other, the exact center of the drill is ground away, resulting in a weak spot at which chipping and breakage of the cutting edges could start.

To summarize, then, whatever type of point thinning is done, it is important that it be done evenly, with the same amount of stock removed from each cutting edge. Always keep in mind that chips are formed on the cutting edges and flow into the flutes. The shape of any thinning should be such that it does not interfere in any way with the chip flow.

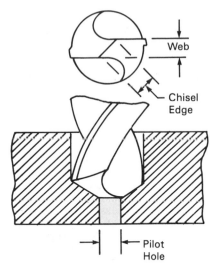

Figure 7–53 Using a pilot hole

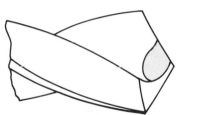

Figure 7–54 Gash-type point thinning

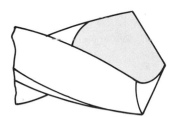

Figure 7–55 Conventional-type point thinning

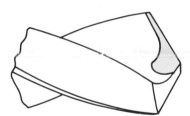

Figure 7–56 Flattening the lips to thin the point

DRILLING SPEEDS AND FEEDS

Many factors must be considered before a machinist decides on what speed or feed should be used on a particular piece of work. Some of these factors are the condition of the machine tool, the rigidity of the work setup, the rigidity and strength of the cutter, the type of cutter, the material to be cut, and whether or not a roughing or finishing cut is being made. There are no hard and fast rules for either speed or feed, but there are some guidelines that can be used.

Cutting **speed** is the rate at which a point on the cutter moves past the work, or it is the rate at which a point on the work moves past the cutter. This will depend on which machine tool is being used. For example, a drill on a drilling machine does the moving, whereas revolving work on a lathe does the moving. Cutting speed is stated either in feet per minute in the English (U.S. Customary) system or in meters per minute in the metric system. Cutting speed is converted into revolutions per minute to suit each machine tool.

Each material that is cut has its own recommended cutting speed. This speed depends upon what material is used for the cutter. A carbide cutter, for example, cuts material three times faster than a high-speed steel cutter.

The recommended relationship between the cutting speed and revolutions per minute, according to diameter, is given in Table 7–2 for the English (U.S. Customary) system, and in Table 7–3 for the metric system. Each chart is followed by the formula for calculating rpm.

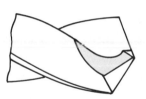

Figure 7–57 Split point thinning

Table 7–2 Cutting Speeds and RPM (inch)

Feet Per Min	FEET PER MINUTE										
	30	40	50	60	70	80	90	100	150	200	250
Dia (in)	REVOLUTIONS PER MINUTE										
1/16	1833	2445	3056	3667	4278	4889	5500	6112	9167	12223	15279
1/8	917	1222	1528	1833	2139	2445	2750	3056	4584	6112	7639
3/16	611	815	1019	1222	1426	1630	1833	2037	3056	4074	5093
1/4	458	611	764	917	1070	1222	1375	1528	2292	3056	3820
5/16	367	489	611	733	856	978	1100	1222	1833	2445	3056
3/8	306	407	509	611	713	815	917	1019	1528	2037	2546
7/16	262	349	437	524	611	698	786	873	1310	1746	2183
1/2	229	306	382	458	535	611	688	764	1146	1528	1910
5/8	183	244	306	367	428	489	550	611	917	1222	1528
3/4	153	204	255	306	357	407	458	509	764	1019	1273
7/8	131	175	218	262	306	349	393	473	655	873	1091
1	115	153	191	229	267	306	344	382	573	764	955
1 1/8	102	136	170	204	238	272	306	340	509	679	849
1 1/4	92	122	153	183	214	244	275	306	458	611	764
1 3/8	83	111	139	167	194	222	250	278	417	556	694
1 1/2	76	102	127	153	178	204	229	255	382	509	637
1 5/8	71	94	118	141	165	188	212	235	353	470	588
1 3/4	66	87	109	131	153	175	196	218	327	437	546
1 7/8	61	82	102	122	143	163	183	204	306	407	509
2	57	76	96	115	134	153	172	191	287	382	477
2 1/4	51	68	85	102	119	136	153	170	255	340	424
2 1/2	46	61	76	92	107	122	138	153	229	306	382
2 3/4	42	56	70	83	97	111	125	139	208	278	347
3	38	51	64	76	89	102	115	127	191	255	318
3 1/4	35	47	59	71	82	94	106	118	176	235	294
3 1/2	33	44	55	66	76	87	98	109	164	218	273
3 3/4	31	41	51	61	71	81	92	102	153	204	255
4	29	38	48	57	67	76	86	96	143	191	239
4 1/2	26	34	42	51	59	68	76	85	127	170	212
5	23	31	38	46	54	61	69	76	115	153	191
5 1/2	21	28	35	42	49	56	63	70	104	139	174
6	19	26	32	38	45	51	57	64	96	127	159
6 1/2	18	24	29	35	41	47	53	59	88	118	147
7	16	22	27	33	38	44	49	55	82	109	136
7 1/2	15	20	26	31	36	41	46	51	76	102	127
8	14	19	24	29	33	38	43	48	72	96	119

$$RPM = \frac{4\ CS\ (\text{Cutting speed})}{D\ (\text{inches})}$$

Table 7-3 Cutting Speeds and RPM (metric)

METERS PER MINUTE										
Meters Per Min	9	12	15	18	21	24	27	30	45	60	75
Dia (mm)	REVOLUTIONS PER MINUTE									
1.6	1800	2400	3000	3600	4200	4800	5400	6000	9000	12000	15000
3	955	1274	1592	1911	2229	2548	2866	3185	4777	6667	7962
5	573	764	955	1146	1338	1529	1720	1911	2866	3822	4777
6	500	638	798	957	1117	1277	1436	1596	2394	3191	3989
8	360	480	600	720	840	960	1080	1200	1800	2400	3000
10	286	382	478	573	669	764	860	955	1433	1911	2389
11	261	348	435	522	609	696	783	870	1304	1740	2174
12	239	318	398	477	557	637	716	796	1194	1592	1989
16	180	240	300	360	420	480	540	600	900	1200	1500
19	150	200	250	300	350	400	450	500	750	1000	1250
22	130	174	217	261	304	348	391	435	652	870	1087
25	114	152	190	228	266	304	342	380	570	759	949
30	96	128	160	191	223	255	287	319	479	638	798
32	90	120	150	180	210	240	270	300	450	600	750
35	82	109	136	164	191	218	245	273	409	545	682
38	76	101	126	151	176	202	227	252	378	504	630
42	68	91	114	136	159	182	205	227	341	455	568
45	64	85	106	128	149	170	191	213	319	426	532
48	59	79	99	118	138	158	178	197	296	395	493
50	57	76	96	115	134	153	172	191	287	382	478
56	51	68	85	102	119	136	153	170	256	341	426
64	45	60	75	90	104	119	134	149	224	298	373
70	41	55	68	82	95	109	123	136	205	273	341
76	38	50	63	75	88	100	113	126	188	251	314
82	35	47	58	70	82	93	105	117	175	233	292
90	32	42	53	64	74	85	95	106	159	212	265
95	30	40	50	60	70	81	91	101	151	201	252
100	29	38	48	57	67	76	86	96	143	191	239
115	25	33	42	50	58	66	75	83	125	166	208
125	23	31	38	46	53	61	69	76	115	153	191
140	20	27	34	41	48	55	61	68	102	136	170
150	19	25	32	38	45	51	57	64	96	127	159
165	17	23	29	35	41	46	52	58	87	116	145
180	16	21	27	32	37	42	48	53	80	106	133
190	15	20	25	30	35	40	45	50	75	101	126
200	14	19	24	29	33	38	43	48	72	96	119

$$RPM = \frac{CS\ (m/min)}{\pi D\ (meters)} \quad \text{or} \quad RPM = \frac{CS\ (m/min) \times 1000}{D\ (mm) \times \pi}$$

Feed is the amount that the cutter will move into a piece of work in one revolution, in inches or millimeters. Again, there are no set rules to follow. Suggestions can be made, such as the Dormer chart for using high-speed steel twist drills for general-purpose work, Table 7–4A. It should be stated, however, that it is wise to start with the lightest feed and gradually increase until the maximum feed that is satisfactory for a particular piece of work is found. Table 7–4B contains suggested cutting speeds for drilling.

Table 7–4A Suggested Feeds for High-Speed Twist Drills for General-Purpose Work

DRILL DIAMETER		FEED PER REVOLUTION	
METRIC	ENGLISH (U. S. CUSTOMARY)	METRIC	ENGLISH (U. S. CUSTOMARY)
Up to 3 mm	Up to 1/8″	0.025 – 0.050 mm	0.001″ – 0.002″
3.0 mm to 6.3 mm	1/8″ to 1/4″	0.05 – 0.10 mm	0.002″ – 0.004″
6.3 mm to 12.5 mm	1/4″ to 1/2″	0.10 – 0.18 mm	0.004″ – 0.007″
12.5 mm to 25.0 mm	1/2″ to 1″	0.18 – 0.38 mm	0.007″ – 0.015″
Over 25.0 mm	Over 1″	0.38 – 0.63 mm	0.015″ – 0.025″

This chart should be used as a guide only.

Table 7–4B Suggested Cutting Speeds for Drilling

MATERIAL	CARBON TOOL STEEL		HIGH-SPEED STEEL	
	FT/MIN	m/MIN	FT/MIN	m/MIN
Steel: SAE 1020	40–50	12–15	80–100	24–30
1050	30–40	9–11	60–80	18–24
1095	20–40	6–12	40–60	12–18
Cast Iron (Gray)	40–60	12–18	80–100	24–30
Stainless Steel	25–40	7–12	50–80	15–24
Aluminum	50–125	15–37	100–250	30–76
Brass, Bronze (free cutting)	50–75	15–21	100–200	30–61

This chart should be used as a guide only.

REVIEW QUESTIONS

1. What are the steps in laying out the center for spot drilling?

2. What is the reason for laying out a hole for drilling?

3. If a drill wanders off center, what can be done about it?

4. Why is it necessary to break chips formed during machining?

5. List three methods of breaking up chips.

6. For what reasons is it necessary to relieve the pressure exerted on the drill point?

7. State the purpose of a pilot hole. List three factors to consider in making a pilot hole.

8. What is the purpose of drill point thinning? List four methods of drill point thinning.

9. What is the most important precaution to be followed in drill point thinning?

10. In split point thinning, what happens if the two notching cuts do not meet exactly, but pass each other?

11. List five factors to be considered in selecting the cutting speed and feed for a job.

12. State the relationships between rpm and cutting speed for both the English (U. S. Customary) and metric systems.

AUTOMATED GUIDED VEHICLES – ROBOTS IN MOTION

One manufacturing problem solved by the use of robots is the loading and unloading of parts within a machine tool group or cell. Another problem is the transfer of tools from a toolroom to the cell, or the transfer of "green parts" from a stock room to the cell, or the transfer of finished parts to a store room for shipping. A solution to these manufacturing needs is the use of moving robots, or automated guided vehicles (AGVs). These robots can be made to move in set patterns about a machine shop floor. At present, this type of movement is accomplished with the robot guided on tracks or along a smooth floor by means of special induction wires or electronic sensors embedded in the floor. Automated guided vehicles can move virtually anywhere in the machine shop complex and can be made to do many different transporting jobs.

In a completely automated system, the main computer controls the selection of the kind and type of tools needed for a particular set of machining operations. The computer also stores data on the following: tool dimensions, tool breakage detection, tool life, and spare tools. The toolroom contains all of the tools necessary for manufacturing operations. The operator in the toolroom, Figure 1, is responsible for presetting all of the necessary tools required for a particular set of operations. This means that the operator sets the tools in a tool storage drum and also in spare storage drums if required. A drum loader robot, Figure 2, is directed by the central computer to pick up the drum and transfer the tooling to a machine tool in the manufacturing cell. Since cutting tools have only a limited cutting life before they become dull and must be replaced, a means must be provided to substitute a new tool quickly. If a tool storage drum is empty of spare tools, a computer message will notify the toolroom operator to change the drum. The operator will send the proper command to the drum loader robot which will then pick up the used storage drum from the machine tool and transfer it to the toolroom to be refilled. The robot then automatically picks up a new preset tool storage drum and transfers it to the machine tool. Some tool storage drums can carry 40 or 80 preset,

Figure 1 Toolroom operator *(Courtesy of Mazak Corporation)*

Figure 2 Drum loader robot *(Courtesy of Mazak Corporation)*

Figure 3 Transfer pallet loader robot *(Courtesy of Mazak Corporation)*

sharp cutting tools. In this way, tool setup time is reduced to a fraction of that required for a conventional manufacturing system.

Another type of moving robot is a transfer pallet loader robot, Figure 3. Loads in excess of 30 tons (27 216 kg) can be transported by these machines. Two loader sizes commonly used are the three-ton and the 16-ton loader (2727 and 14 545 kg). One of the functions of the three-ton loader is to restock an incoming conveyor with green (unmachined) parts. Another function is to unload finished parts from an outgoing conveyor. Speeds of 195 ft/min (60 m/min) are common. The 16-ton loader, with a capacity of two eight-ton workpieces, moves at speeds of 130 ft/min (40 m/min). These pallet loaders commonly run on a track. This system is driven by a servo motor with an encoder feedback system. (An encoder is an instrument which transfers signals into a code for the computer.) The system provides extremely reliable automatic work transfer and positioning accuracy is very high.

For unmanned machining operations, the pallet loader robot automatically carries unmachined and machined workpieces to and from the machining centers under the command of the main computer. Positioning or docking of the pallet loader robot must be precise. One type of locating system uses inverted cones in the floor at each work station to position the robot exactly to ensure X-, Y-, and Z-axis orientation of the vehicle with the pallet dock. Using compressed air, an airlift pallet on the vehicle will raise the pallet to the proper height. The pallet is removed and set on the worktable, and the vehicle moves away.

A pallet stocker is used to store the workpieces in the required quantity prior to the unmanned machining operations which are done after midnight. When unmanned machining is being performed during the third work shift, the pallet loader robot automatically carries unmachined and machined workpieces to and from the machining centers under the command of the main computer. Robots can be programmed to work alone (without supervision) under the command of the main computer for eight or ten hours.

Automated Guided Vehicles (AGVs) can be controlled to transfer materials or tools anywhere in a machine shop, from the tool room to the machining center or cell, from the stock room to the machining center or cell, or from the machining center or cell to the storage room. The control mechanisms can be designed to meet any set of requirements.

UNIT 7-4

OTHER DRILLING MACHINE OPERATIONS

OBJECTIVES

After completing this unit, the student will be able to:

- select the proper size of tool based on the job requirements.
- perform the following basic operations on drilling machines: reaming, boring, spot-facing, counterboring, countersinking, and tapping.
- identify problems in tapping screw threads and correct them.
- calculate tap drill sizes given the thread size required.
- remove broken taps.
- use threading dies to produce an external screw thread.
- care for cutting tools properly.

KEY TERMS

Reaming
Boring
Spot-facing
Counterboring
Countersinking
Tapping
Reamer
Hand reamer
Machine reamer
Helical flute reamer
Straight flute reamer
Rose-type chucking
   reamer
Taper reamer

Emergency reamer
Reamer action
Undersize
Oversize
Precision boring head
Boring bar
Center drill (combined
   drill and countersink)
Threading tap
Hand tap
Taper tap
Plug tap
Bottoming tap
Blind hole

Taper pin reamer
Shell reamer
Expansion reamer
Adjustable reamer
Bell mouth
Eccentric
Tap drill size
Tap extractor
Threading die
Die stock

Through tap
Spiral point tap
Cold forming tap
Spiral fluted tap
Collet
Dies
Solid-adjustable die
Solid die
Hexagon rethreading
   die

INTRODUCTION

Other basic operations performed on drilling machines include reaming, boring, spot facing, counterboring, countersinking, and tapping. **Reaming**,

Figure 7–58, produces a round, smooth, accurately sized hole from a slightly smaller drilled hole. **Boring**, Figure 7–59, is the operation of trueing and enlarging a hole by means of a single-point cutting tool held in a boring bar. **Spot-facing**, Figure 7–60, produces a flat surface at right angles to an existing hole to provide a seal for the head of a bolt or for a nut. **Counterboring**, Figure 7–61, enlarges the end of an existing hole to produce a straight hole with a square shoulder that will act as a recess for a bolt head. **Countersinking**, Figure 7–62, enlarges the end of an existing hole to a conical shape for a flat head or oval head screw (82 degrees). Other countersinking drills are made with angles of 60, 90, 100, and 120 degrees. **Tapping**, Figure 7–63, produces an internal screw thread in an existing hole with a cutting tool called a tap. It is best to use a tapping attachment for this operation.

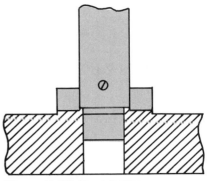

Figure 7–60 Spot facing

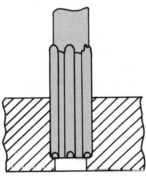

Figure 7–58 Reaming

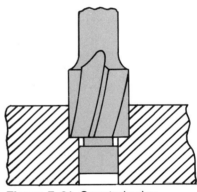

Figure 7–61 Counterboring

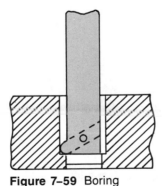

Figure 7–59 Boring

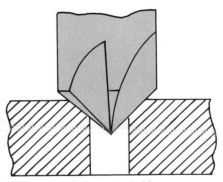

Figure 7–62 Countersinking

REAMERS

As a rule, drills will cut a hole slightly oversize in diameter. Two reasons for this are the greater length of the drill relative to its smaller diameter and the rotating action of the chisel point, which tends to force the drill off center. In order to obtain an accurate diameter and a round smooth hole, a machinist will drill a hole slightly undersize and finish it to size with a reamer. As a rule of thumb, it is common practice to drill 1/64 in (0.39 mm) undersize for holes up to 1/2 in (13 mm) in diameter, and 1/32 in (0.79 mm) undersize for holes over 1/2 in (13 mm) in diameter.

Reamers are used under the same general conditions as in drilling. However, it is usually best to ream at slower speeds compared to those used in drilling. In actual practice, it has been found that speeds should be 25% to 50% slower. Current thinking by some manufacturing companies tends to reduce speeds even further, to around 2/3 less than drilling speeds. Feeds are medium to heavy, two to three times faster than for drilling. Too coarse a feed produces rough walls. A good rule of thumb is to feed 0.002 in (0.05 mm) per tooth per revolution, and then to adjust the feed either up or down as you proceed.

Reamers are made in hundreds of sizes and are of many types in both hand-held and machine-held styles. Reamers are available with straight and tapered bodies, straight and tapered shanks, and with either helical cut or straight cut flutes.

Reamer Nomenclature

The names of the main parts of a reamer are shown in Figure 7–64.

The square end indicates a hand reamer. The taper shank indicates a machine reamer. A hand reamer should not be used in machines. Helical flutes have a better shearing cut than straight flute reamers and are particularly good where there is a keyway in the hole.

A rose-type chucking reamer is held in a chuck and is also a machine reamer. Rose-type reamers cut only on the end and are particularly adapted for reaming cored holes. Flutes are cylindrical and are cleared on the chamfer only.

Taper reamers are made in all standard tapers in both roughing and finishing styles. They are used for finishing or refinishing taper holes.

Taper pin reamers, Figure 7–65, are made with both straight and helical flutes. Square shanks are for hand use and straight shanks are for machine use.

Shell reamers are fluted almost their entire length and have a tapered hole with a 1/8-in taper per foot or a metric taper ratio of 1:96. Different reamer sizes can be used on the same tapered arbor. Slots in the reamer engage driving lugs on the arbor, which eliminate any slippage of the reamer on the arbor. Shell reamers are more economical to use than solid-type reamers.

Expansion reamers can be adjusted by a central setscrew so that the reamer can be repeatedly sharpened and adjusted to its original size.

Adjustable reamers can be adjusted over a considerable range of sizes, much more so than an expansion-type reamer. Blades can be repeatedly sharpened until they are discarded. One reamer body will outwear several sets of blades.

The chucking reamers illustrated are a straight flute, right-hand cut reamer, Figure 7–66, and a right-hand helical flute with right-hand cut, Figure 7–67.

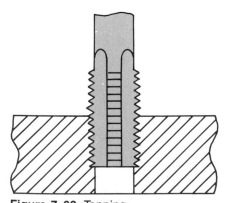

Figure 7–63 Tapping

Emergency Reamers

Twist drills can be used quite satisfactorily as reamers. It is best to hone the corners to a slight round. The initial hole is drilled undersize just as when using a regular reamer. Speeds and feeds are the same as in reaming.

Reamer Action

Chatter is the most troublesome factor in reaming. Chattering is the vibration of the cutting edge, which produces a rough surface. Causes of chattering include the lack of rigidity in the machine, in the fixture that holds the work, in the work itself, or improper design of the reamer.

Some reamers are fluted unevenly, that is, the teeth are unevenly spaced. This design lessens the tendency of the reamer teeth to set up a synchronized vibration, thereby producing a rough cut. Helical fluted reamers cut extremely well, with a better shearing cut than straight fluted reamers. The helix is cut

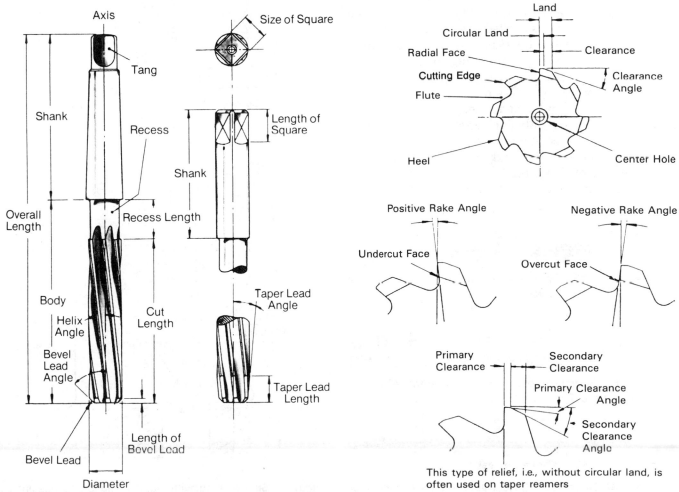

Figure 7–64 Parts of a reamer *(Courtesy of SKF & Dormer Tools Ltd.)*

opposite that of a drill to eliminate the tendency of the reamer to pull into the work. Such reamers cut extremely well in holes with keyways and oil grooves. The majority of reamed holes are through holes. A left-hand helix is preferable to a right-hand helix, in this case, because chips are pushed forward through the hole. A right-hand helix cuts more freely than a left-hand helix because of its chip clearing ability, giving it an advantage in reaming blind holes.

Reamer Care

As with other cutting tools, reamers should be kept clean and sharp. Burs should be carefully honed away. If resharpening is needed, it should be taken care of immediately, before the reamer is used again. Nothing is more frustrating than trying to use a dull or chipped cutting tool. Using the proper cutting fluid is always good practice. Pay close attention to using the proper feeds and speeds. Keep in mind that too heavy a feed can not only damage a reamer, it can produce a rough surface. A reamer should never be turned backward. This would quickly ruin the cutting edges. One method of using reamers is to use guides and guide bushings, Figure 7–68. After using a reamer, it

Figure 7–65 Taper pin straight flute *(Courtesy of The Cleveland Twist Drill Company)*

Figure 7–66 Straight flute, right-hand cut reamer *(Courtesy of The Cleveland Twist Drill Company)*

Figure 7–67 Right-hand, helical flute reamer *(Courtesy of The Cleveland Twist Drill Company)*

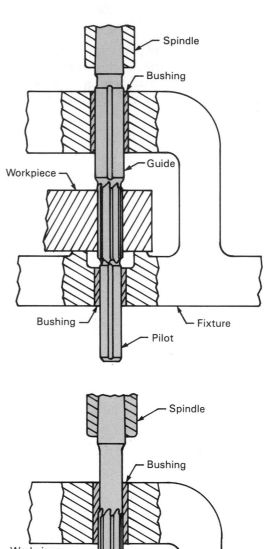

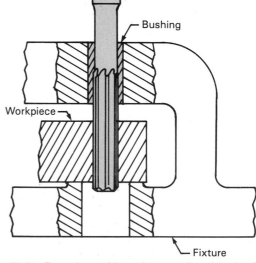

Figure 7–68 Reaming with guides and guide bushings. These are also used with boring bars and counterbores.

should be thoroughly cleaned and inspected, lightly oiled, and stored in its own separate container.

There are no hard and fast rules for reaming, but Table 7–5 lists suggested feeds and speeds for some of the more common materials.

BORING

An early method of boring employed the use of a simple boring bar and a single-point tool bit. A system of pilots and pilot bushings kept the boring bar running true.

In modern practice, it is easier and more accurate to use a precision boring head. This high-precision tool utilizes a sliding radial arm, a boring bar, and a single-point cutter to do the job of boring, Figure 7–69, and recessing, Figure 7–70. The radial arm has a micrometer adjustment with accuracies on some heads to 0.0001 in (0.002 mm). The smaller heads are capable of a radial adjustment from 0 to 4-1/2 in (0 to 115 mm).

SPOT-FACING

Spot-facing is sometimes done with a boring bar, a pilot and bushing, and a double-edged cutting tool. More modern practice makes use of appropriate-sized counterbores to do the same operation. End milling cutters adapted to a drill press or used on a vertical milling machine also do an excellent job. Speeds are calculated according to the diameter of the cutter.

COUNTERBORING

Counterbores, Figure 7–71, enlarge previously drilled holes to provide recesses for bolt heads or cap screws. Counterbores come with interchangeable pilots, Figure 7–72, that keep the cutter concentric with the hole being counterbored. Speeds are calculated according to the diameter of the counterbore.

COUNTERSINKING

A countersink, Figure 7–73, is used to drill a cone-shaped recess in one end of a previously drilled hole. The recess may just be a light chamfer or it may be made wider and deeper to allow the head of a flat-headed or oval-headed screw to be mounted flush or just below the work surface. Countersinks with included angles of 60 degrees and 82 degrees are the most common, but 90, 100, and 120 degrees are also available. The combined drill and countersink, Figure 7–74, or more commonly, the center drill, has an

Table 7–5 Suggested Speeds and Feeds for Reaming Various Materials

MATERIAL	SPEEDS				FEEDS
	HIGH-SPEED STEEL		CARBIDE		
	ft/min	m/min	ft/min	m/min	
Aluminum	100–150	30–46	300–450	90–150	Medium-hard
Brass, Bronze — soft	75–150	23–46	225–450	70–150	Medium
Brass, Bronze — tough	35–75	12–23	100–225	36–70	Medium
Cast Iron — gray	50–75	15–23	150–225	45–70	Hard
Cast Iron — hard	15–25	4.5–8	45–75	14–24	Low
Nickel Steel	15–40	4.5–12	45–120	14–36	Low
Stainless Steel — free machining	15–40	4.5–12	45–120	14–45	Medium
Steel — SAE 1020	40–55	15–20	120–165	45–60	Medium-hard
SAE 1050	35–40	12–12	100–120	36–45	Low-medium
SAE 1080	25–30	8–9	75–90	24–30	Low-medium

included angle of 60 degrees. This specialized drill is used to make center holes in workpieces. They are covered in Chapter 8, Lathework. Speeds of center drills are calculated on the largest diameter that will contact the work.

CUTTING SCREW THREADS

Cutting internal screw threads is a basic operation performed on a drill press. This is done with a threading tap, which cuts the internal thread in a hole of designated size.

Taps are made in many different styles and types. Hand taps, Figure 7–75, are usually made in sets of three as shown. The taper tap, Figure 7–75A, is normally used to start a threaded hole, followed by the plug tap, Figure 7–75B, which cuts the thread a little deeper. Very often on through holes, (holes that go all the way through the work), these two taps are all that are needed to complete the thread. The bottoming tap, Figure 7–75C, is used to cut a thread to the bottom of a blind hole (a hole that does not go all the way through the work). Hand taps should not be used in

machine tools unless they are driven by a tapping attachment. A tapping attachment has a special clutch that is designed to slip when too much torque is used on the tap or if the tap suddenly jams. Tap wrenches, Figure 7–76, normally supply the rotating power for hand taps. Tap nomenclature is given in Figure 7–77.

Spiral point taps, Figure 7–78, are designed primarily for machine tapping of through holes. The main difference between this machine tap and a hand tap is the spiral point and the shallower, narrower flutes that make the machine tap much stronger. The spiral point is ground in such a way that the chips tend to flow through the hole ahead of the tap. This action explains the nickname "gun tap." The tap has the ability to "shoot" the chips ahead. Chips do not accumulate in the flutes as they do in hand taps. The flutes, being free, act mainly as grooves for the cutting fluid. The improved cutting action reduces the torque required to drive the tap, and higher peripheral speeds are possible.

Cold forming taps, Figure 7–79, do not have cutting edges or conventional flutes. They are designed to cold form internal threads in aluminum, brass,

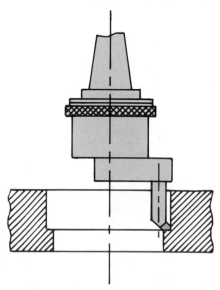

Figure 7–69 Boring with a precision boring head

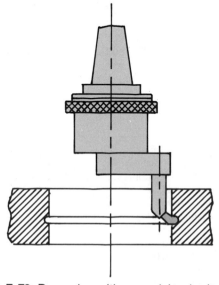

Figure 7–70 Recessing with a precision boring head

copper, zinc, magnesium, and other ductile metals. The initial hole is drilled larger than the tap size for conventional taps. Some of the metal is squeezed up to form a part of the thread, rather than being cut. Metal is displaced or shifted rather than being cut. Cold forming produces a much stronger thread because of this.

Another type of tap is the spiral fluted tap, Figure 7–80, which has helix angles between 15 and 52 degrees. Helix angles can be right-hand or left-hand in a right-hand thread. When the helix angle and thread are the same hand, chips are directed back out of the hole. If they are not the same hand, chips flow ahead of the tap. The same applies to left-hand threads. A right-hand spiral fluted tap is shown being used on a drill press in Figure 7–81. These taps are recommended especially for tapping holes that have a keyway or some other form of groove or slot.

Figure 7–73 A countersink *(Courtesy of The Cleveland Twist Drill Company)*

Figure 7–74 Combined drill and countersink *(Courtesy of The Cleveland Twist Drill Company)*

Figure 7–71 A counterbore *(Courtesy of The Cleveland Twist Drill Company)*

Figure 7–72 A pilot for a counterbore *(Courtesy of The Cleveland Twist Drill Company)*

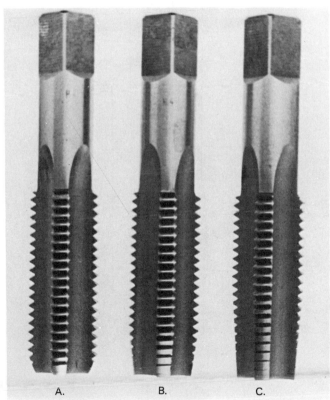

Figure 7–75 Hand taps   *A.* Bottoming tap   *B.* Plug tap   *C.* Taper or starting tap *(Courtesy of SKF & Dormer Tools Ltd.)*

Figure 7–76A Straight-style tap wrench *(Courtesy of DoALL Company)*

Figure 7–76B T-handle-style tap wrench *(Courtesy of DoALL Company)*

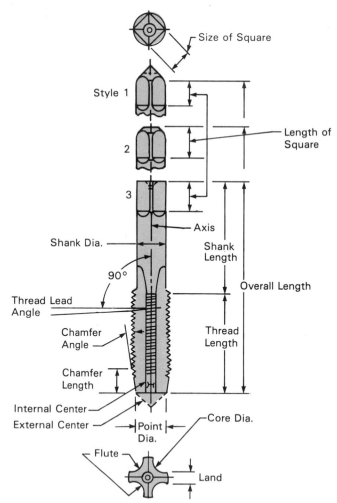

Figure 7–77 Tap nomenclature *(Courtesy of SKF & Dormer Tools Ltd.)*

Figure 7–78 A spiral-point tap *(Courtesy of SKF & Dormer Tools Ltd.)*

Figure 7–79 A cold forming tap *(Courtesy of SKF & Dormer Tools Ltd.)*

Figure 7–80 A spiral fluted tap *(Courtesy of SKF & Dormer Tools Ltd.)*

Figure 7–81 A spiral fluted tap being used on a drill press
(Courtesy of SKF & Dormer Tools Ltd.)

TAPPING FAULTS

1. Oversize and bell mouth hole.
   (a.) The tap is misaligned in the drilled hole, Figure 7–82, causing the tap to cut more heavily on one side than the other. Be sure the tap is aligned correctly in the drilled hole.

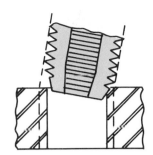

Figure 7–82 Tap misalignment

   (b.) The taper lead of the tap has been reground, probably by hand, and is eccentric to the tap diameter. Regrind the tap properly by machine.
   (c.) An incorrect feed rate, when the tap is fed too heavily, will cause the cut threads to be too thin, Figure 7–83. It is best to use a floating type of tapping attachment.

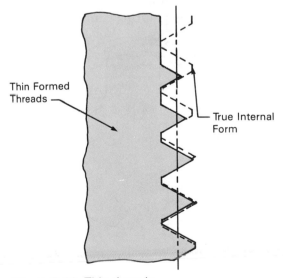

Thin Formed Threads

True Internal Form

Figure 7–83 Thin threads

2. Poor finish on the threads.
   (a.) The tap is dull, Figure 7–84. Rough, torn threads are the result. Either regrind or replace the tap.

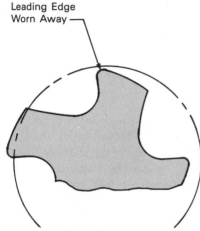

Leading Edge
Worn Away

Figure 7–84 Dull tap

   (b.) If a tap has been reground with insufficient clearance, Figure 7–85, the result will be similar.
   (c.) The wrong cutting fluid or not enough

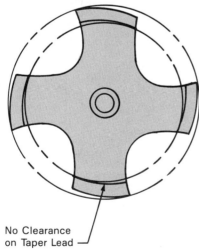

No Clearance
on Taper Lead

Figure 7–85 Insufficient clearance

cutting fluid has been used. Use the correct cutting fluid for the work at hand.

3. Tap teeth are chipping.
   (a.) Too much clearance has been ground, Figure 7–86, or too much rake. This weakens the cutting edges. Regrind the tap according to the manufacturer's recommendations.

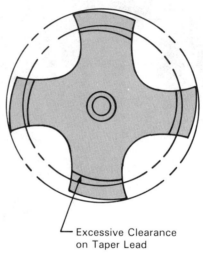

Excessive Clearance
on Taper Lead

Figure 7–86 Excessive clearance

   (b.) The tap is allowed to hit the bottom of a blind hole, Figure 7–87. Allow for clearance at the bottom of the hole.
   (c.) Check work for hard spots or work hardening.

4. Excessive rate of wear..
   (a.) The speed of the tap is too high. Select and use the correct speed.
   (b.) Not enough cutting fluid or incorrect cutting fluid is used. Use the correct cutting fluid.
   (c.) Drilling may have caused work-hardening in the hole. Use correct drilling procedure for the work at hand.

5. Tap breakage.
   (a.) Choked flutes can cause a torsional

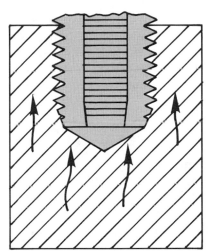

Figure 7–87 Tap strikes the bottom of the hole

overload. Remove the tap and clean the flutes as often as necessary.

(b.) Misalignment of the tap in the hole, Figure 7–88, can cause breakage. Ensure correct alignment.

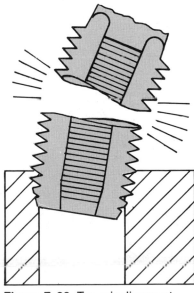

Figure 7–88 Tap misalignment

(c.) A dull tap can cause breakage. Sharpen the tap or use a new one.

(d.) The tap drill is too small. Use the correct tap drill.

CALCULATING TAP DRILL SIZES

Figure 7–89 shows the comparative diameters for a coarse and a fine thread. The major diameters are the same in both cases, but the minor diameters are different. The finer the thread is, the larger the tap drill hole must be.

The following formulas should be used to calculate tap drill size:

$$\text{Tap drill size (inch)} = D - \frac{1}{N}$$

where D is the diameter of the tap (inches) and N is threads per inch.

$$\text{Tap drill size (millimeters)} = D - p$$

where D is the diameter of the tap (millimeters) and p is the pitch of the tap (millimeters).

EXAMPLES:

1. Calculate the tap drill size for a 1/2 in – 20 UNF thread.

$$\begin{aligned} TDS &= 1/2 - 1/20 \\ &= 0.5 - 0.05 \\ &= 0.450 \text{ in} \end{aligned}$$

A 29/64 drill is 0.4531 in. Use a 29/64-in drill.

2. Calculate the tap drill size for a M6 × 1 thread.

$$\begin{aligned} TDS &= 6 - 1 \\ &= 5 \text{ mm} \end{aligned}$$

Use a 5-mm drill.

BROKEN TAP REMOVAL

A broken tap is annoying and is sometimes difficult to remove. If a tap extractor is available, clean out the flutes thoroughly, insert the tap extractor legs into the flutes, and carefully try to remove the broken end of the tap. Another method is to attempt to jar the tap loose with a pin punch. You can also try to break up a tap with a pin punch since the tap is very brittle. Taps can also be eroded away with an electrical discharge machine (EDM). Whichever method you use, it will take some valuable time to remove the broken tap.

THREADING DIES

The main purpose of a threading die is to cut an external screw thread on a round shaft. This can be done by hand with a hand die or by machine, with the machine supplying the rotating power. Following is a selection of hand dies as used in the machine shop.

The four-part hand die, Figure 7–90, is made up of the cap, the dies or cutters, the guide, and die stocks. This is an adjustable die that can be adjusted to cut larger or smaller than the nominal diameter of the screw thread. The cap positions the dies or cutters. The guide keeps the dies or cutters from moving and

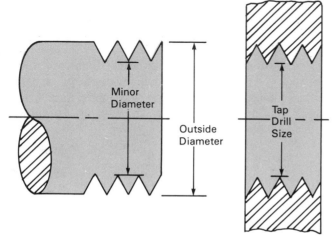

Coarse Thread

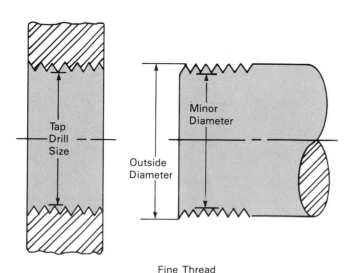

Fine Thread

Figure 7–89 A comparison of fine and coarse threads. The major diameter is the same in both cases.

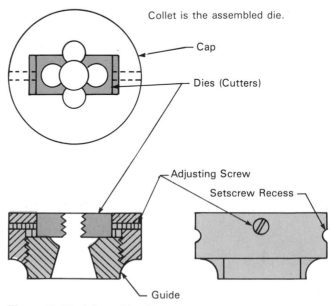

Figure 7–90 Adjustable dies

keeps them square on the shaft. The stocks provide the rotating power. The cutters must never be loose in the cap.

There are two cutters in the adjustable die, Figure 7–91. The size of thread, the number of threads per inch, and the form of thread are stamped on one side of one cutter. Metric sizes and form can also apply here. Using the example 3/8 in – 16 UNC, 3/8 in is the nominal major diameter with 16 threads per inch. The thread is a unified coarse thread. In the example 8 mm × 1.25 mm ISO, 8 mm is the nominal major diameter, 1.25 mm is the pitch, and ISO is the thread form. On one side of the other cutter is the maker's trademark. Either a number or a letter or a combination of both a number and a letter is stamped on the opposite side of both cutters to identify the cutters as a matched set. This notation must always be on the same side of the cutters when they are assembled and contained within the cap. It also indicates the starting side of the dies, and, as a rule, is placed on the inside of the collet out of sight. The **collet** is the assembled threading die.

The collet must be placed in the stocks so that the guide is toward the shoulder within the stocks. Only then will the set screw be centered on the cone-shaped thumb screw hole in the cap. Finger tightness of the set screw is all that is necessary to keep the dies in place.

Another style of hand die is the round split or round adjustable die, Figure 7–92, sometimes called a solid-adjustable die. The adjustment is very slight, sometimes being done by adjusting a tapered screw located in the split or, as in this case, by tightening

the die stocks. These dies will fit either way within the die stocks.

Still another style of hand die is the solid die with no adjustment. A hexagon rethreading die, or die nut, Figure 7–93, is normally used on a thread that has already been cut. The usual purpose is to repair a bruised or battered screw thread. Rotation is usually supplied by a wrench.

RESOURCE TABLES FOR THREADING

The following tables contain useful data for the machinist or student in the production of screw threads: Table 7–6, Tap Drill Sizes (inches); Table 7–7, Tap Drill Sizes (millimeters) (page 214); and Table 7–8, Cutting Speeds for Taps (page 215).

EMERGENCY DRILLS

If there is no other option, it is quite possible to make an emergency spade-type drill fairly quickly and easily. Two possibilities are illustrated in Figures 7–94A and 7–94B (page 216). To make either drill, use a piece of high-carbon drill rod with a carbon content of approximately 0.9%, or 90-point carbon.

The wider drill needs forging to flatten it out. This must be followed immediately by normalizing. Refer to the heat treatment unit in Chapter 14.

The narrower drill does not need forging, but in both cases, the work must be done carefully since the point of the drill must rotate exactly on center. Grind a general-purpose point angle of 118 degrees. Clear-

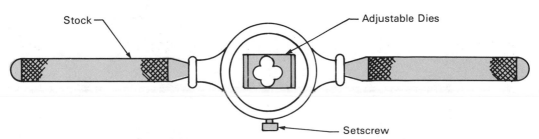

Figure 7–91 Die stocks with adjustable dies in position

ances, rakes, and the drill point angle can be shaped by hand file or bench grinder. A chip curler groove can be used for ductile metals and a zero rake for those materials that require it. Rake and clearance are explained in Chapter 9, Cutting Tools.

The drill points require hardening and tempering. For this procedure, refer to the heat treatment unit in Chapter 14. If your drill rod is a water hardening steel, both the hardening and tempering heats should be quenched in water for best results.

Figure 7–92 Round adjustable or round split die *(Courtesy of SKF & Dormer Tools Ltd.)*

Figure 7–93 Hexagon rethreading die or die nut *(Courtesy of SKF & Dormer Tools Ltd.)*

Table 7–6 Tap Drill Sizes (inches) (*Courtesy of SKF & Dormer Tools, Limited*)

TAP SIZE	PITCH	TAP DRILL	ALT. TAP DRILL mm	TAP SIZE	PITCH	TAP DRILL	ALT. TAP DRILL mm	TAP SIZE	PITCH	TAP DRILL	ALT. TAP DRILL mm
0	80 UNF	$3/64$	1.25	$5/16$	18 UNC	F	6.50	$1\,1/4$	7 UNC	$1\,7/64$	28.00
1	64 UNC	53	1.55		24 UNF	1	6.90		12 UNF	$1\,11/64$	29.50
	72 UNF	53	1.55	$3/8$	16 UNC	$5/16$	8.00	$1\,3/8$	6 UNC	$1\,7/32$	30.75
2	56 UNC	50	1.85		24 UNF	Q	8.50		12 UNF	$1\,19/64$	32.75
	64 UNF	50	1.90	$7/16$	14 UNC	U	9.40	$1\,1/2$	6 UNC	$1\,11/32$	34.00
3	48 UNC	47	2.10		20 UNF	$25/64$	9.90		12 UNF	$1\,27/64$	36.00
	56 UNF	45	2.15	$1/2$	13 UNC	$27/64$	10.80	$1\,3/4$	5 UNC	$1\,9/16$	39.50
4	40 UNC	43	2.35		20 UNF	$29/64$	11.50	2	4 UNC	$1\,25/32$	45.00
	48 UNF	42	2.40	$9/16$	12 UNC	$31/64$	12.20	TAPER PIPE TAPS NPT			
5	40 UNC	38	2.65		18 UNF	$33/64$	12.90	$1/16$	27 NPT	D	6.30
	44 UNF	37	2.70	$5/8$	11 UNC	$17/32$	13.50	$1/8$	27 NPT	R	8.70
6	32 UNC	36	2.85		18 UNF	$37/64$	14.50	$1/4$	18 NPT	$7/16$	11.10
	40 UNF	33	2.95	$3/4$	10 UNC	$21/32$	16.50	$3/8$	18 NPT	$37/64$	14.50
8	32 UNC	29	3.50		16 UNF	$11/16$	17.50	$1/2$	14 NPT	$23/32$	18.00
	36 UNF	29	3.50	$7/8$	9 UNC	$49/64$	19.50	$3/4$	14 NPT	$59/64$	23.25
10	24 UNC	25	3.90		14 UNF	$13/16$	20.40	1	$11\,1/2$ NPT	$1\,5/32$	29.00
	32 UNF	21	4.10	1	8 UNC	$7/8$	22.25	$1\,1/4$	$11\,1/2$ NPT	$1\,1/2$	38.00
12	24 UNC	16	4.50		12 UNF	$59/64$	23.25	$1\,1/2$	$11\,1/2$ NPT	$1\,47/64$	44.00
	28 UNF	14	4.70		14 UNS	$15/16$	23.50	2	$11\,1/2$ NPT	$2\,7/32$	56.00
$1/4$	20 UNC	7	5.10	$1\,1/8$	7 UNC	$63/64$	25.00	$2\,1/2$	8 NPT	$2\,5/8$	67.00
	28 UNF	3	5.50		12 UNF	$1\,3/64$	26.50	3	8 NPT	$3\,1/4$	82.50

Drill sizes based on approximately 72%–77% of full thread.

Table 7–7 Tap Drill Sizes (millimeters) (*Courtesy of SKF & Dormer Tools, Limited*)

TAP SIZE mm	PITCH mm	TAP DRILL mm	ALT. TAP DRILL
ISO METRIC COARSE			
1.6	0.35	1.25	3/64
1.7	0.35	1.35	55
1.8	0.35	1.45	54
2	0.40	1.60	1/16
2.2	0.45	1.75	50
2.3	0.40	1.90	49
2.5	0.45	2.05	46
2.6	0.45	2.15	44
3	0.50	2.50	40
3.5	0.60	2.90	33
4	0.70	3.30	30
4.5	0.75	3.70	27
5	0.80	4.20	19
5.5	0.90	4.60	15
6	1.00	5.00	9
7	1.00	6.00	15/64
8	1.25	6.80	H
9	1.25	7.80	5/16
10	1.50	8.50	O
11	1.50	9.50	3/8
12	1.75	10.20	Y
14	2.00	12.00	15/32
16	2.00	14.00	35/64
18	2.50	15.50	39/64
20	2.50	17.50	11/16
22	2.50	19.50	49/64

TAP SIZE mm	PITCH mm	TAP DRILL mm	ALT. TAP DRILL
24	3.00	21.00	53/64
27	3.00	24.00	61/64
30	3.50	26.50	1 3/64
33	3.50	29.50	1 5/32
36	4.00	32.00	1 1/4
36	4.00	35.00	1 3/8
ISO METRIC FINE			
3	0.35	2.65	37
4	0.35	3.65	27
4	0.50	3.50	29
4.5	0.45	4.05	21
5	0.50	4.50	16
5	0.70	4.30	18
5	0.75	4.25	18
5.5	0.50	5.00	9
6	0.50	5.50	7/32
6	0.75	5.25	5
7	0.75	6.25	D
8	0.50	7.50	M
8	1.00	7.00	J
9	0.50	8.50	Q
9	1.00	8.00	O
10	0.50	9.50	3/8
10	0.75	9.25	U
10	1.00	9.00	T
10	1.25	8.75	11/32

TAP SIZE mm	PITCH mm	TAP DRILL mm	ALT. TAP DRILL
11	1.00	10.00	X
12	1.00	11.00	7/16
12	1.25	10.75	27/64
12	1.50	10.50	Z
13	1.50	11.50	29/64
13	1.75	11.25	7/16
14	1.25	12.75	1/2
14	1.50	12.50	31/64
15	1.50	13.50	17/32
16	1.00	15.00	19/32
16	1.25	14.75	37/64
16	1.50	14.50	9/16
18	1.00	17.00	43/64
18	1.25	16.75	21/32
18	1.50	16.50	41/64
18	2.00	16.00	5/8
20	1.00	19.00	3/4
20	1.50	18.50	47/64
20	2.00	18.00	45/64
22	1.00	21.00	53/64
22	1.50	20.50	13/16
22	2.00	20.00	25/32
24	1.00	23.00	29/32
24	1.50	22.50	7/8
24	2.00	22.00	55/64
24	2.50	21.50	27/32

Table 7–8 Cutting Speeds for Taps *(Courtesy of SKF & Dormer Tools, Limited)*

FRACTIONAL SIZE TAPS

nom. tap dia.	decimal equivalent	CUTTING SPEED ft./min. 20 / m./min. 6	30 / 9	70 / 21	100 / 30
		REVOLUTIONS PER MINUTE			
1/32"	0·0312"	2,449	3,673	8,570	12,244
3/64"	0·0469"	1,629	2,443	5,701	8,144
1/16"	0·0625"	1,222	1,833	4,278	6,112
5/64"	0·0781"	978	1,467	3,424	4,891
3/32"	0·0938"	814	1,222	2,851	4,072
7/64"	0·1094"	698	1,047	2,444	3,492
1/8"	0·1250"	611	917	2,139	3,056
9/64"	0·1406"	543	815	1,902	2,717
5/32"	0·1562"	489	734	1,712	2,445
11/64"	0·1719"	444	667	1,555	2,222
3/16"	0·1875"	407	611	1,426	2,037
7/32"	0·2188"	349	524	1,222	1,746
1/4"	0·2500"	306	458	1,070	1,528
9/32"	0·2812"	272	408	951	1,358
5/16"	0·3125"	244	367	856	1,222
11/32"	0·3438"	222	333	778	1,111
3/8"	0·3750"	204	306	713	1,019
7/16"	0·4375"	175	262	611	873
1/2"	0·5000"	153	229	535	764
9/16"	0·5625"	136	204	475	679
5/8"	0·6250"	122	183	428	611
11/16"	0·6875"	111	167	389	556
3/4"	0·7500"	102	153	356	509
13/16"	0·8125"	94	141	329	470
7/8"	0·8750"	87	131	306	437
15/16"	0·9375"	81	122	285	407
1"	1·0000"	76	115	267	382
1 1/8"	1·1250"	68	102	238	340
1 1/4"	1·2500"	61	92	214	306
1 3/8"	1·3750"	56	83	194	276
1 1/2"	1·5000"	51	76	178	255
1 5/8"	1·6250"	47	71	165	235
1 3/4"	1·7500"	44	65	153	218
1 7/8"	1·8750"	41	61	143	204
2"	2·0000"	38	57	134	191
2 1/4"	2·2500"	34	51	119	170
2 1/2"	2·5000"	31	46	107	153
2 3/4"	2·7500"	28	42	97	139
3"	3·0000"	25	38	89	127
3 1/4"	3·2500"	24	35	82	118
3 1/2"	3·5000"	22	33	76	109
3 3/4"	3·7500"	20	31	71	102
4"	4·0000"	19	29	67	95

R.P.M. for cutting speeds not listed, can be obtained by simple multiplication or division e.g.
50 ft./min.＝100÷2＝611 r.p.m. (for 1/16" dia.)
90 ft./min.＝30×3＝393 r.p.m. (for 3/8" dia.)

METRIC SIZE TAPS

nom. tap dia. m/m	decimal equivalent	CUTTING SPEED ft./min. 20 / m./min. 6	30 / 9	70 / 21	100 / 30
		REVOLUTIONS PER MINUTE			
1·0	0·0394"	1,939	2,908	6,786	9,695
1·2	0·0472"	1,619	2,428	5,665	8,093
1·4	0·0551"	1,386	2,080	4,853	6,932
1·5	0·0591"	1,293	1,939	4,524	6,463
1·6	0·0630"	1,213	1,819	4,244	6,063
1·7	0·0669"	1,142	1,713	3,997	5,710
1·8	0·0709"	1,077	1,616	3,771	5,387
2·0	0·0787"	971	1,456	3,397	4,854
2·2	0·0866"	882	1,323	3,088	4,411
2·3	0·0906"	843	1,265	2,951	4,216
2·5	0·0984"	776	1,165	2,717	3,882
2·6	0·1024"	746	1,119	2,611	3,730
3·0	0·1181"	647	970	2,264	3,234
3·5	0·1378"	554	832	1,940	2,772
4·0	0·1575"	485	728	1,698	2,425
4·5	0·1772"	431	647	1,509	2,156
5·0	0·1969"	388	582	1,359	1,941
5·5	0·2165"	353	529	1,235	1,764
6·0	0·2362"	323	485	1,132	1,617
7·0	0·2756"	277	416	970	1,386
8·0	0·3150"	243	364	849	1,213
9·0	0·3543"	216	323	755	1,078
10·0	0·3937"	194	291	679	970
11·0	0·4331"	176	265	617	882
12·0	0·4724"	162	243	566	809
13·0	0·5118"	149	224	522	746
14·0	0·5512"	139	208	485	693
16·0	0·6299"	121	182	424	606
18·0	0·7087"	108	162	377	539
20·0	0·7874"	97	146	340	485
22·0	0·8661"	88	132	309	441
24·0	0·9449"	81	121	283	404
27·0	1·0630"	72	108	252	359
30·0	1·1811"	65	97	226	323
33·0	1·2992"	59	88	206	294
36·0	1·4173"	54	81	189	270
39·0	1·5354"	50	75	174	249
42·0	1·6535"	46	69	162	231
45·0	1·7717"	43	65	151	216
48·0	1·8898"	40	61	141	202
52·0	2·0472"	37	56	131	187
56·0	2·2047"	35	52	121	173
60·0	2·3622"	32	49	113	162

R.P.M. for cutting speeds not listed, can be obtained by simple multiplication or division, e.g.
10 ft./min.＝100÷10＝571 r.p.m. (for 1·7 m/m dia.)
80 ft./min.＝20×4＝484 r.p.m. (for 16·0 m/m dia.)

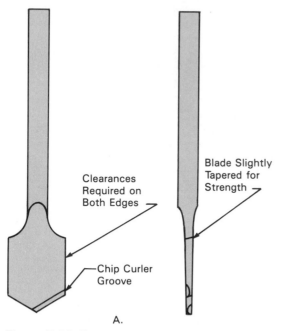

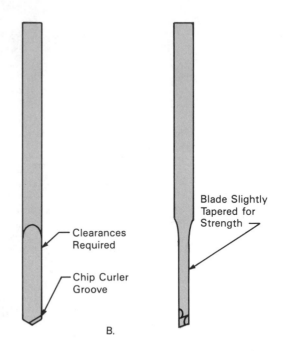

Figure 7–94 Emergency spade-type drill

REVIEW QUESTIONS

1. Provide a brief definition of each of the following operations: reaming, boring, spot-facing, counterboring, countersinking, and tapping.
2. What is the purpose of a reamer?
3. Sketch and name the parts of a reamer.
4. How does a reamer differ from a twist drill?
5. As a rule of thumb, what size hole should be drilled for reamers under ½-inch diameter?
6. How should the speed of a reamer compare to that of a twist drill?
7. How do feeds of a reamer compare to those of a twist drill?
8. What is a good rule of thumb to use for reamer feeds?
9. How does a machine reamer differ from a hand reamer?
10. Should a hand reamer be used in a machine?
11. List six types of reamers and give a characteristic of each.
12. What is the most troublesome factor in reaming? Give two possible causes.
13. Why are reamers fluted unevenly?
14. How does the cutting action of a helical fluted reamer differ from the action of a straight fluted reamer?
15. List five recommendations for proper reamer care.
16. What is the preferred method of boring in current practice?
17. What is the difference between countersinking and counterboring?
18. What are the basic steps in cutting an internal screw thread?
19. Should a hand tap be used in a machine? Explain.
20. How does the cutting action of a spiral point tap differ from that of a hand tap?

21. How does a machine tap differ from a hand tap?
22. Describe how a cold forming tap differs from a conventional tap.
23. What is a spiral fluted tap especially suited for?
24. What are two causes of an oversize, bell mouth tapped hole?
25. What is a principal cause for a poorly finished thread?
26. What are two causes of excessive wear on a tap?
27. Give three causes of tap breakage.
28. State the formula for calculating the tap drill size in English (U. S. Customary) units.
29. Calculate the tap drill size for 5/8 in-11 UNC thread.
30. Calculate the tap drill size for 5/8 in-18 UNF thread.
31. State the formula for calculating tap drill size in metric units.
32. Calculate the tap drill size for 8 mm × 1.25 pitch.
33. Calculate the tap drill size for 10 mm × 1.5 pitch.
34. What is the purpose of a threading die?
35. Name three styles of threading dies.

Chip formation. Note that smaller chips are being formed, rather than longer coils. These chips present less danger to the operator and chip removal is much less of a problem. *Courtesy of Cleveland Twist Drill Company*

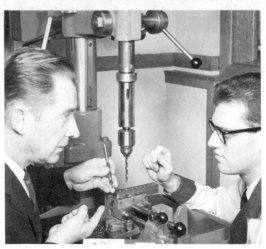

An apprentice pays close attention to gain the benefits of years of experience. A machinist explains the intricacies of the twist drill. *Courtesy of DoALL Company*

A side view showing a hand method of sharpening a twist drill. This view shows clearance being formed by raising the twist drill. The left hand is still lowered to form the clearance behind the cutting edge. The forefinger is prevented from moving closer to the grinding wheel because the knuckle of the adjacent finger rests against the front of the tool rest. The potential danger is that the forefinger will be hit by the wheel. *Courtesy of Cominco Ltd.*

Slot milling. *Courtesy of The Cleveland Twist Drill Company*

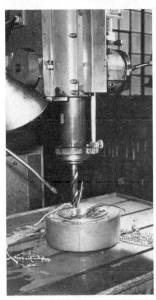

Drilling a large piece of steel using a high-speed steel twist drill with a tapered shank. *Courtesy of DoALL Company*

Drilling half-hard free-cutting brass using a good grade of insoluble cutting fluid. The drill is a high-speed steel straight shank twist drill. *Courtesy of DoALL Company*

Drilling large holes in a steel block using a large taper shank high-speed steel twist drill. The drill is ground to break the chips into small pieces rather than forming long, coiling chips. *Courtesy of DoALL Company*

Using a pin chuck to grip and drive a small drill. The pin chuck is gripped in a standard drill chuck. *Courtesy of DoALL Company*

CHAPTER EIGHT

THE LATHE AND LATHE OPERATIONS

The lathe is one of the most versatile machine tools in a machine shop. Any cut made by other manual machine tools is possible on a lathe. It is really the backbone of the machining industry. The lathe was one of the earliest machine tools developed. Advances in its design ultimately led to the development of other machine tools, such as turret lathes and boring mills.

The basic machine operations performed on a lathe are facing, turning, boring, tapering, and threading. These basic operations lead to other operations such as center drilling, drilling, reaming, and knurling. Cylindrical grinding and internal grinding are also possible. With the addition of a milling arm, many milling operations can also be performed on the lathe. It is essential that students learn all they can about the lathe and become proficient in its use.

SAFE PRACTICES IN LATHEWORK

Rotating chucks and work are the major dangers in lathework. Chips and the tool bit are also potential dangers if the operator does not pay attention to what is happening.

1. Permission should be obtained to operate the machine, at the discretion of the instructor.

2. Always roll up loose sleeves and remove ties, rings, watches, and so on before operating the machine.

3. Always wear specified eye protection.

4. Make all adjustments with the machine off.

5. When installing or removing chucks, place a

safety board on the ways in case the chuck falls. This will prevent damage to both the chuck and ways.

6. When loosening a chuck on a threaded spindle, do not use a pry bar between the chuck jaws. It is too hard on the jaws. Use a wrench on one of the jaws, with the jaw fully supported by the chuck.

7. Never leave a chuck wrench in a chuck. If the machine were turned on, the wrench could become a lethal projectile.

8. To check the clearance of the chuck and the work, rotate by hand before switching on the machine.

9. Always keep your hands away from moving parts.

10. Always keep your hands away from chips. They are hot, sharp, and dangerous.

11. When filing, always use a file handle. It is also best to learn to file left-handed rather than by placing your left arm over the revolving chuck. Running the lathe in reverse and filing right-handed at the back of the lathe is poor practice because the controls are at the front of the lathe, out of reach. It is also best to remove the tool post assembly.

12. Do not measure work with the lathe running. It is poor practice.

13. Do not adjust the cutting tool with the machine running.

14. In general, it is poor practice to change gears while the lathe is running.

15. As a general rule of thumb, no more than 3 times a diameter of the work should be out of a chuck without being supported by the tailstock or a steady rest.

16. Never walk away from a lathe when it is operating.

17. When the work is complete, shut off the power and clean the machine.

UNIT 8-1

THE LATHE

OBJECTIVES

After completing this unit, the student will be able to:

- explain the basic function of a lathe.
- state the primary function of these lathe components: bed, headstock, tailstock, carriage.
- describe the two types of feed systems (feed and threading) normally found on metal lathes.
- select and use the spindle nose that will do the best job.

KEY TERMS

High speed turning operation	Gear train
Lathe	Feed rod
Geared head lathe	Sliding gib key
Bed	Lead screw gear (or, lead gear)
Reverse feed mechanism	Lead screw
Spindle speed	Feed clutch
Headstock	Power screw thread
Quick change gear box	Threading lever
Compound tool rest (or, compound)	Friction clutch
Cross slide	Swing
Saddle	Between centers
Apron	Spindle nose
Carriage	Threaded spindle nose
Tailstock	Long taper keydrive spindle nose
Feed mechanism	
Thread mechanism	Cam lock spindle nose

INTRODUCTION TO THE LATHE

Figure 8–1 illustrates a high-speed turning operation performed on a modern, heavy-duty lathe using a carbide cutter. A deep cut and a moderate feed cause the chips to form a classic number nine shape, indicating good high-speed cutting.

A **lathe**, Figure 8–2, is the basic machine tool in a machine shop. It is the one machine that can duplicate any cut made on other manual machine tools. All lathes, no matter what style, make, or size, make the same basic cuts. The main function of a lathe is to remove excess material from a revolving piece of work.

The parts of a geared head lathe are illustrated in detail in Figure 8–3. The major parts of a lathe consist of the bed, the headstock, the carriage, the tailstock, the feed mechanism, and the threading mechanism.

The bed is the backbone of any lathe. It supports the headstock, the tailstock, the carriage, and both power feed systems. The headstock contains the main spindle, all speed controls controlling spindle speeds, the spindle gear that starts the feed systems, and sometimes the reverse feed mechanism. The reverse feed mechanism can also be located on the quick change gear box below the headstock. The carriage is composed essentially of the compound tool rest (more commonly called the compound), the cross slide, the saddle, and the apron. The main function of the carriage is to carry the tool bit into the revolving work. It is also possible for the work to be set up on the carriage and to be carried into a revolving cutter. The tailstock does such jobs as drilling and reaming, and acts as a support for work that is too long to support

Figure 8–1 A modern turning operation using a carbide cutter *(Courtesy of DoALL Company)*

itself. A lateral adjustment on the tailstock allows taper turning to be done with the work suspended between centers.

THE LATHE FEED SYSTEMS

There are normally two feed systems on metal lathes. The feed mechanism is used for all general cutting and the thread mechanism is used for cutting screw threads.

Both mechanisms start at the spindle gear, which is rotated from the outside end of the main spindle on a cone pulley drive lathe or from the main spindle within the headstock of a geared head lathe. Motion is transmitted from the spindle gear to a gear train located at the outside end of the headstock, then to a quick change gear box on the bed below the headstock. Reverse feed gears sometimes are located following the spindle gear or are located on the quick change gear box. The purpose of the quick change gear box is to vary the amount of feed per revolution of the spindle or to vary the number of threads per inch that can be cut or the thread pitch in millimeters. The last gear in the gear train is called the lead screw gear, or more commonly, the lead gear.

The quick change gear box transmits motion to the apron in one of two ways. Some lathes have two shafts, Figure 8–4A, to connect the gear box to the apron. In this case, one shaft called the feed rod is plain except for one or more keyways, or splines, running along most of its length. The second shaft

Figure 8–2 A modern geared head lathe showing a heavy duty turning operation with the work held in a three-jaw chuck and supported by a revolving center *(Courtesy of DoALL Company)*

is called the lead screw and is threaded with a power screw thread, usually an Acme thread, for most of its length.

Some lathes have only one shaft, Figure 8–4B, to connect the gear box to the apron. In this case, a single keyway is cut along the threaded portion of the lead screw. The keyway converts the lead screw so that it becomes both a feed rod and a lead screw.

The feed rod passes through a worm located behind the apron. A gib key connects the feed rod to the worm. The worm is in mesh with a worm gear, on which there is a friction clutch. As the feed rod rotates, so does the worm and worm gear. If the friction clutch is engaged against the face of the worm gear, Figure 8–5A, it also rotates. A small gear on the clutch shaft will then rotate and transmit motion to

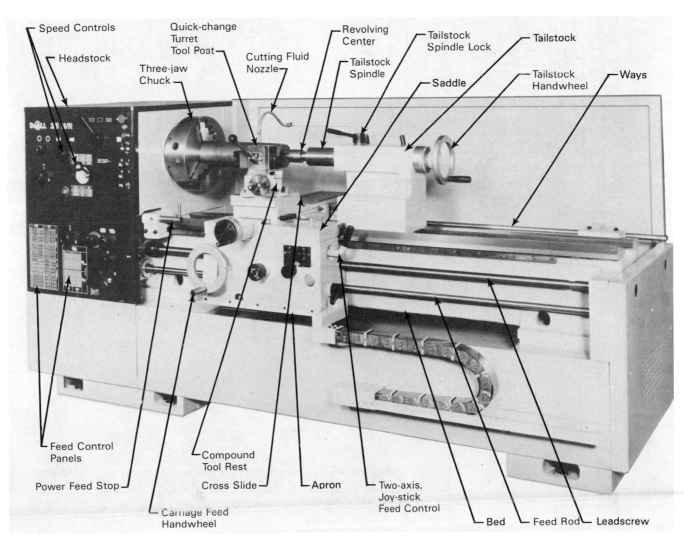

Figure 8–3 Parts of a geared head lathe *(Courtesy of DoALL Company)*

either the longitudinal hand wheel or to the cross slide. If the friction clutch is disengaged, Figure 8–5B, it stops rotating and the small gear on the clutch shaft stops transmitting motion to the hand wheel or the cross slide. The friction clutch located on the front of the apron either starts or stops the feed movement.

If the friction clutch is tight and the carriage moves along the ways of the lathe, the worm and worm gear connected to the back of the apron move with it.

The connecting gib key slides along the keyway, hence the name sliding gib key.

The Thread Mechanism

The lead screw passes through a split nut. One half of the split nut is located below the lead screw and the other half is located above it. The split nut is threaded internally to match the Acme screw thread

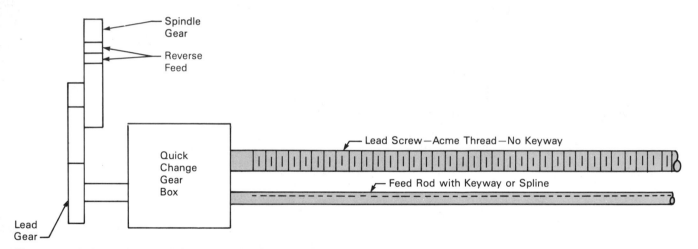

Figure 8–4A Lathe feed and thread mechanisms

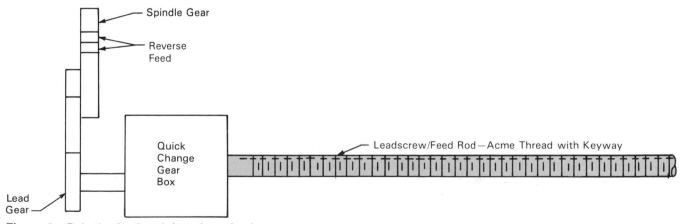

Figure 8–4B Lathe feed and thread mechanisms

of the lead screw. It is also connected to the back of the apron. As the lead screw rotates, the threading lever located on the front of the apron is moved to close the split nut around the Acme screw thread, Figure 8–6A. This causes the carriage to move along the ways of the lathe. Moving the threading lever to open the split nut, Figure 8–6B, causes the movement to stop.

In the case of a lathe with a single feed rod/lead screw, it will pass through both the worm and the split nut. Otherwise, the individual actions of the feed rod and the lead screw are the same as for any other lathe.

It is important to remember that for both the feed mechanism and the thread mechanism, the lead screw and/or the feed rod must be revolving for any power feed to occur.

Figures 8–4, 8–5, and 8–6 are simple diagrams illustrating how a feed mechanism and a thread mechanism will work. Modern lathes have a system of gears, which are often spring loaded, on the feed mechanism clutch. A simple movement of a lever

quickly engages or disengages the feed. Too much pressure on the tool bit or a depth stop at the end of the ways can cause the feed to disengage, quickly stopping all feed movement.

TYPES OF HEADSTOCKS

There are two styles of headstocks made for lathes. They are the cone pulley headstock and the geared headstock. Today, the geared headstock is by far the most common style used in industry. The cone pulley headstock, which lacks power, is rapidly disappearing from industrial shops.

SIZE OF A LATHE

Figure 8–7 shows where three important lathe sizes are measured. In America, the size at *A* is called **swing**, which is the maximum diameter that can be revolved over the ways of the lathe. The center-to-center distance

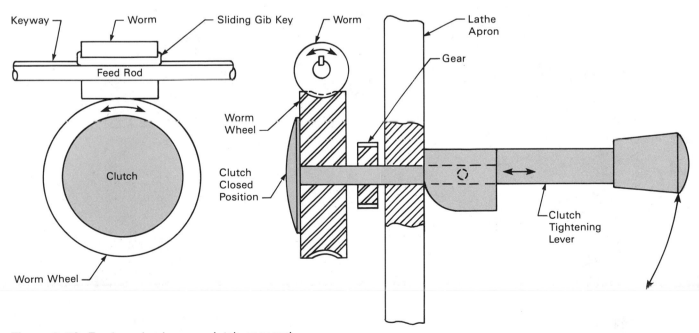

Figure 8–5A Feed mechanisms — clutch engaged

at B is the maximum length of the work that can be supported between centers. Manufacturers once preferred to give the bed length, but it is seldom considered now. The distance C is the radius or half of the swing and is the preferred size given by most British and European manufacturers.

SPINDLE NOSES

Spindle noses hold and locate chucks and face plates on center. Three types of spindle noses are the threaded spindle, the cam lock spindle, and the long taper spindle.

The threaded spindle, Figure 8–8, was once a common type of spindle nose. It had a precision thread that allowed interchangeability of chucks and face plates. It is seldom used now in industry. It was slow to set up and a chuck would often unscrew if the rotation was reversed, as it often was.

A type D–1 cam lock spindle, Figure 8–9, is by far the most popular type of spindle used today. It is the easiest to clean and the easiest to operate. In CNC work, it is sometimes referred to as the American standard spindle nose. One important point about the cam lock is that each cam must be rotated clockwise to lock it. The cams will not lock if they are rotated counter-

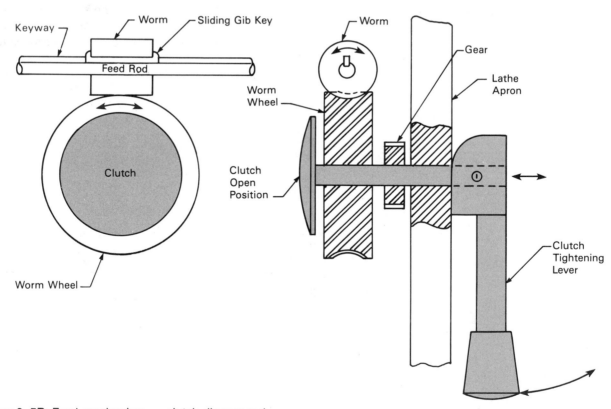

Figure 8–5B Feed mechanism — clutch disengaged

clockwise. This can cause the chuck to come off the spindle nose.

A long taper keydrive spindle is another type of spindle nose. It is commonly called the American long taper spindle nose. It also is not as widely used as it was once. The main disadvantage of this spindle nose involves the threaded ring used to tighten the chuck in position. This type of spindle nose is slower to use, and cleaning is not easy.

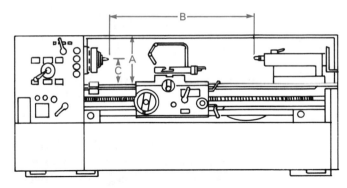

A = Swing
B = Distance between centers
C = Radius or swing/2

Figure 8-7 The size of a lathe

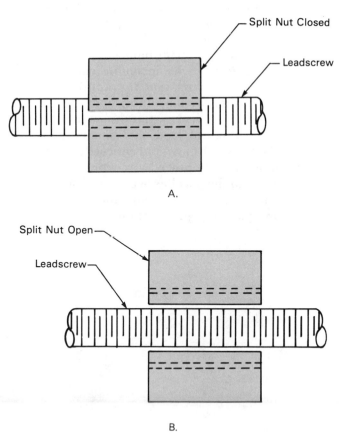

A.

B.

Figure 8-6 Thread mechanism

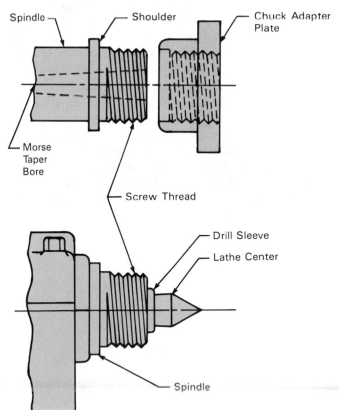

Figure 8-8 The threaded spindle

ROBOTS – LOADING AND UNLOADING MACHINE TOOL GROUPS OR CELLS

Certain jobs in manufacturing are performed in less than optimum conditions. The jobs may require very hard physical labor and/or they may be boring, repetitive, dirty, noisy, or dangerous. In many cases, robots can be used in place of human operators. In modern manufacturing, another important reason exists for the increasing use of robots on the shop floor. Production machines are meant to generate a specific amount of work during a standard shift. If a human operator loads and unloads the work, the machine's production goal may not be met because the operator may get tired and lose concentration. The robot does neither. It can be set to operate at a certain speed and it will maintain that speed throughout the work shift. Operator delays for any reason can idle a production machine tool by as much as 70% to 80% in a work shift.

Industrial robots today can do most of the jobs traditionally done by human operators. The decision to buy and use a robot is usually an economic one. Will the robot pay for itself in a reasonable period of time? Will the operator be more cost effective? Productivity in modern factories must be improved if they are to survive. In many cases, robots are the answer. However, when a robot is placed on the shop floor, one or more human operators may be replaced. Many of these operators are being retrained for other manufacturing positions in the factory. Jobs relating to the operation, programming or maintenance of the robots represent opportunities for retrained operators.

Modern factories utilize robots more efficiently by setting up machine tools in groups. With a group of tools, the robot can be placed in a central position within the group so that it can service all of the tools. This group of machine tools is often called a cell. In Figure 1, the robot is programmed to pick up the unmachined rough castings or parts from an incoming conveyor. Next, it loads the parts into the appropriate machine tool. After the machining operation, it then removes the finished part, transfers the part to the next machine tool, loads that machine tool, unloads the finished part, and proceeds to the next machine tool. The procedure continues until the machining of the part has been completed on all of the machine tools. The robot then unloads the finished part and places it on an outgoing

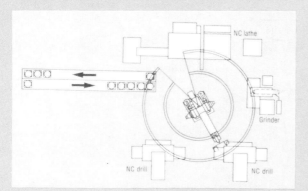

Figure 1 Robot loading and unloading a group of production machines *(Courtesy of ASEA Robotics Inc.)*

Figure 2 Robot picking up "green parts" from an incoming conveyor *(Courtesy of ASEA Robotics Inc.)*

conveyor. Figure 2 shows a robot picking up parts from an incoming conveyor which is kept full to prevent the robot from waiting. The outgoing conveyor, on the right, shows finished parts passing through an automatic inspection machine. Another robot machine tools conveyor, Figure 3, shows a machine-serving robot application in a high-production operation.

A group of machine tools (cell) usually can be operated by one operator. The job includes overseeing the total operation of the cell and ensuring that everything progresses as programmed. The operator may be required to change tools and set up new jobs. If the operator is also a trained robot programmer, then he or she can easily and quickly reprogram the robot for new jobs.

The major problem with a group of machine tools serviced by a robot is that of downtime — when a machine tool stops for some reason. If one machine tool stops, the whole group stops. The most common reason for downtime is tool adjustment because of tool wear. If an automatic size control system is included in the cell, the dimensions of finished parts are automatically checked. If variations from the programmed dimensional information exist, the control automatically signals a device that then adjusts the tool position,

or activates the automatic tool changers to change the tools. Downtime would be kept to a minimum.

In addition to the single-arm robot fixed to one position on the factory floor, there is a similar type that can be suspended above a machine tool. By adding a travel unit such as a rotating track, the robot is given the ability to move along the floor, resulting in greatly expanded capability. The ability of the robot for extremely accurate positioning adds to its flexibility. For example, Figure 4 demonstrates the flexibility of the robot in moving along an assembly line to assemble parts at an automotive or aircraft assembly plant.

Robots that are normally used for loading and unloading are not considered to be sophisticated. A sophisticated robot is one that may be equipped with artificial vision by means of special cameras, or it may have a sense of touch by means of special electronic sensors placed under a thin plastic skin on the grippers. Sophisticated robots such as these are primarily used for inspection or assembly applications. They can recognize parts and pick them out of a group of mixed parts. They can also gently pick up fragile parts without damaging them. At the moment, for all their versatility, robots, like computers, cannot think. They must still be programmed and told what to do.

Figure 3 A conveyor installation in a high-production operation *(Courtesy of ASEA Robotics Inc.)*

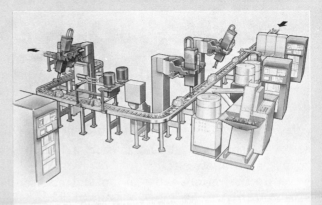

Figure 4 A travelling robot moves along a track assembling parts *(Courtesy of ASEA Robotics Inc.)*

Figure 8–9 A cam lock spindle nose *(Courtesy of Monarch Sidney)*

REVIEW QUESTIONS

1. What is the function of a lathe?
2. Name the six essential parts of the lathe.
3. Explain the primary function of each of the parts listed in item 2.
4. How is motion transferred from spindle to tool bit in the lathe feed system?
5. How is motion transferred from spindle to tool bit in the lathe threading system?
6. What is the purpose of the quick change gear box?
7. Name two types of lathe headstock.
8. Define the three important lathe sizes: swing, center-to-center distance, and radius.
9. What is the function of a spindle nose?
10. Describe three types of lathe spindle noses.
11. Why is the cam lock spindle nose the most popular of the three types?
12. What is an important point to remember about the cam lock spindle nose?

UNIT 8-2

THE TOOLS OF THE LATHE

OBJECTIVES

After completing this unit, the student will be able to:

- state the primary function of each of the following: tool holder, tool post, chuck, face plate, steady rest, and follower rest.
- set up the various types of tool holders on a lathe.
- set up the various types of lathe chuck.
- set up and use steady and follower rests.

KEY TERMS	
Tool holder	Three-jaw universal
Cut-off tool	scroll chuck
Standard tool post	Combination chuck
Rocker ring	(three jaw or four jaw)
Rocker	Concentric circles
Step ring	Counterweight
Turret tool holder	Face plate
Chuck	Driving plate
Four-jaw independent	Center or steady rest
chuck	Follower rest

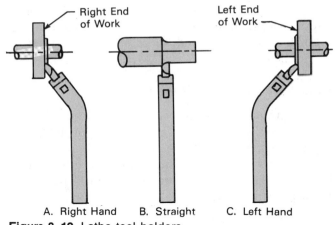

Figure 8–10 Lathe tool holders

THE TOOLS OF THE LATHE

TOOL HOLDERS

Lathe Tool Holders

Lathe tool holders hold the lathe tool bits in position. They are made in three styles: straight, right hand, or left hand, as shown in Figure 8–10.

Commercial Tool Holders

Commercial tool holders are used almost universally when the standard tool post is used. Tool bits in this type of tool holder, Figure 8–11, are usually held at a 15-degree angle.

The tool holders in Figure 8–12 are usually used with carbide-tipped tools, which are held at a zero-degree angle.

Special tool holders are used with cast alloy tools that are held at a 0-degree angle. A soft pad is used between the setscrew and the cutting tool to prevent chipping since the tools are quite brittle.

For standard tool bits, many machinists prefer the carbide cutter tool holder with the 0-degree angle rather than the standard tool holder with the 15-1/2 degree angle. Often a machinist will make a set of tool holders with a 0-degree angle. The resulting tool bit

Figure 8–11 Standard lathe tool holders *(Courtesy of DoALL Company)*

is heavier, carries away more heat, and gives more support to the cutting edge. On the other hand, when a positive rake is needed, more of a rake angle must be ground.

Commercial Cut-off Tool Holders

Commercial cut-off tools are held in a variety of tool holders. Figure 8–13 illustrates a rigid style, with the blade held at a 0-degree angle. Figure 8–14 illustrates a spring style, which tends to "give" a little bit if the blade sticks or jams in the cut. This action often prevents jamming.

LATHE TOOL POST SYSTEMS

Tool posts are set on the compound tool rest and hold the tool holders. A standard tool post, Figure 8–15A, is shown with a retainer ring, tool post with clamp screw, rocker ring, and rocker. The standard tool post assembly is shown in Figure 8–15B. The position of the rocker in the rocker ring sets the height of the tool bit.

The difference between the standard tool post shown in Figure 8–16A and the one in Figure 8–15 is the step ring. The various steps set the height of the tool bit. For a fine set of height, in both cases, the tool

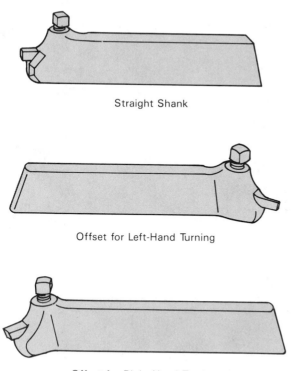

Straight Shank

Offset for Left-Hand Turning

Offset for Right-Hand Turning

Figure 8–12 Lathe tool holders for carbide cutters *(Courtesy of Armstrong Brothers Tool Company)*

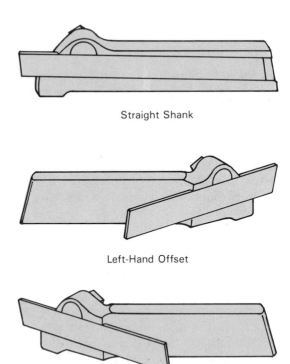

Straight Shank

Left-Hand Offset

Right-Hand Offset

Figure 8–13 Cut-off tool holders, rigid style *(Courtesy of Armstrong Brothers Tool Company)*

bit can be moved out of the tool holder to raise it or into the tool holder to lower it. The tool post assembly is shown in Figure 8–16B. Keep in mind that the tool bit should not stick out more than three times its thickness for maximum efficiency.

TURRET TOOL HOLDERS

As cutting has become faster, the need for a more rigid style of tool holder has increased. Although they are still considered to be an optional item in engine lathework, turret tool holders are being used more and more. Two major styles of turret tool holders, the rigid style and the quick-change style, Figure 8–17, are available for use on the compound of an engine lathe. Both styles are similar in the way in which cutting tools are held. Figure 8–18 illustrates commonly used attachments for the quick change turret.

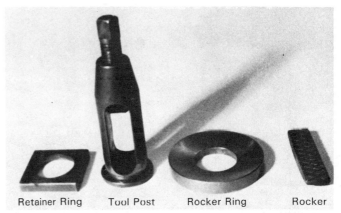

Retainer Ring   Tool Post   Rocker Ring   Rocker

Figure 8–15A Standard tool post with rocker, disassembled

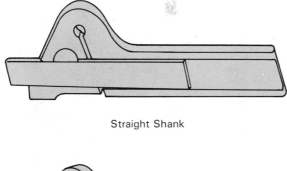

Straight Shank

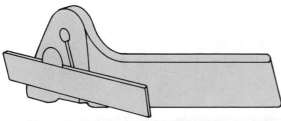

Right-Hand Offset

Figure 8–14 Spring-style cut-off tool holders *(Courtesy of Armstrong Brothers Tool Company)*

Figure 8–15B Standard tool post, assembled

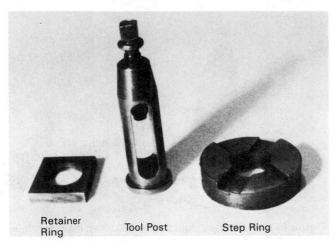

Figure 8–16A Standard tool post with step ring, disassembled

In CNC work, large, heavy-duty, revolving-type turret tool holders are used exclusively. Special tooling is made for each operation and is adapted for holding on the turrets. Some of these holders are shown in Chapter 12.

TYPES OF LATHE CHUCK

There are many styles of chuck available for lathes. Three of the more common chucks in use are the four-jaw independent chuck, the three-jaw universal scroll chuck, and the three- or four-jaw combination chuck. The four-jaw independent chuck requires only one set of jaws, with each jaw being individually adjustable and reversible. Two sets of jaws are normally supplied with the three-jaw universal scroll chuck — one inside clamping set and one outside clamping set. In the three-jaw chuck, all jaws move at the same time. The three- or four-jaw combination chuck has two sets of adjusting screws. Each jaw can be moved indepen-

Figure 8–16B Standard tool post with step ring, assembled

Figure 8–17 A quick-change, turret-style, tool holder *(Courtesy of KAR Industrial Ltd.)*

dently on its own screw, or all jaws can be moved at the same time. Only one set of jaws is necessary for the combination chuck. Other common chucks include magnetic chucks and collet chucks. The principles of setting up work in these chucks are similar.

Four-jaw Chuck Work

The four-jaw chuck, Figure 8–19, is the most versatile chuck for general machine shop work. It exerts the strongest hold on a piece of work and the work can be set exactly on center by the use of surface gages and dial indicators. Each jaw can be set individually and each jaw can be used in either the outside position or the inside position, Figure 8–20, depending upon the requirements. A four-jaw chuck will usually grip any shape.

SETTING UP A FOUR-JAW CHUCK

Concentric circles (from the same center) are cut into the faces of four-jaw chucks. These circles help when used to set a piece of work approximately on center. The work shown in Figure 8–21 is typical. The jaws are just tight enough to hold the work in position, but not so tight as to prevent the work from moving as it is being set on center. To set the work on center, proceed as follows:

1. Revolve the chuck slowly and mark the high spot on the work with chalk.
2. Place the high spot at the bottom by rotating the chuck by hand, slightly loosen the jaw opposite A, and slightly tighten the jaw at A. This will move the work slightly away from A.
3. Repeat this procedure until the work is on center. Use a surface gage, Figure 8–22, for more accuracy and a dial indicator, Figure 8–23 (page 240), to center the work exactly.
4. Once the work is on center, tighten each jaw in rotation a little at a time until the work is secure.
5. Check the setting to be sure that the work is still on center.

It should be noted that if the major portion of the weight of the work is off center, a counterweight will be necessary for balance. The counterweight should be bolted to the opposite side of the four-jaw chuck to eliminate vibration and bearing wear on the lathe.

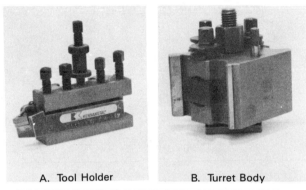

A. Tool Holder

B. Turret Body

C. Boring Tool Holder

D. Morse Cone Tool Holder

Figure 8–18 Attachments for quick-change, turret-style, tool holder *(Courtesy of KAR Industrial Ltd.)*

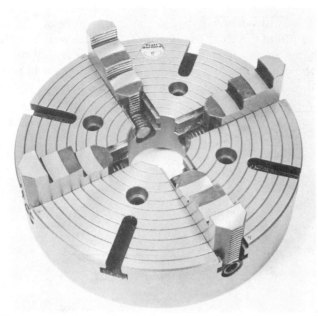

Figure 8-19 A four-jaw independent chuck *(Courtesy of Pratt Burnerd International)*

Using the Three-jaw Chuck

The three-jaw universal scroll chuck, Figure 8–24A, is a self-centering chuck, that is, all three jaws move at the same time. An extra set of outside jaws, Figure 8–24B, is used for gripping larger work, as shown in Figure 8–25A (page 242). Figure 8–25B shows the jaws in the normal or more common position for gripping a shaft. Figure 8–25C illustrates the gripping of work on an inside diameter so that, in this case, turning and facing may be done in the same setting.

Using a Combination Chuck

In a three-jaw combination chuck, Figure 8–26A (page 242), one adjusting screw operates a scroll plate to move all jaws at the same time, exactly as is done in a three-jaw universal chuck. In addition, each jaw has its own adjusting screw so that each jaw can be moved individually. A combination chuck has the centering ability of the three-jaw universal chuck, plus

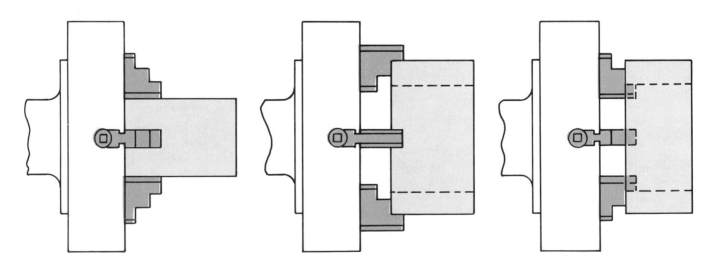

A. Jaws in normal position   B. Chuck jaws reversed   C. Work chucked on the inside

Figure 8-20 Holding work in a four-jaw chuck

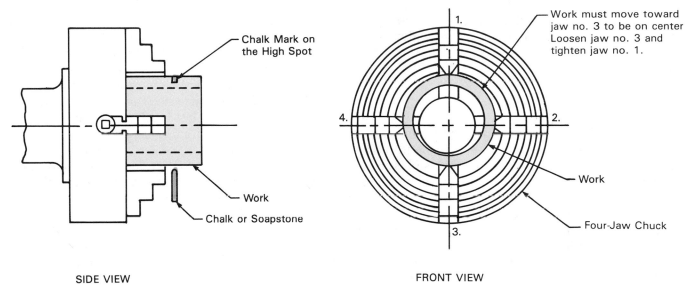

SIDE VIEW

FRONT VIEW

Figure 8–21 Rough centering with chalk or soapstone

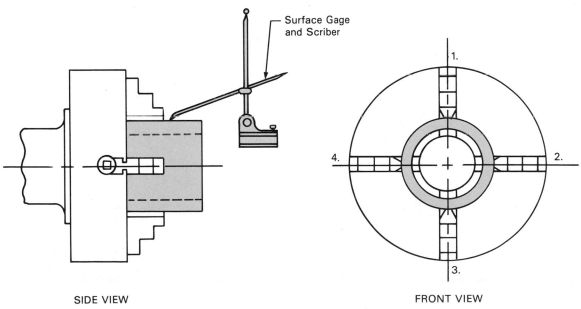

SIDE VIEW

FRONT VIEW

Figure 8–22 Centering with a surface gage

the individual adjustment of each jaw necessary for the fine, accurate adjustment that can be done with a four-jaw chuck. A four-jaw combination chuck is shown in Figure 8–26B.

FACE PLATE

A face plate, Figure 8–27A, is needed when work is too large or awkward for a chuck to handle. The smaller type of face plate with the open slot is called a driving plate, Figure 8–27B. The open slot receives the tail of a lathe dog in order to rotate the lathe dog. The lathe dog, clamped to a mandrel, causes it to rotate. As a result the work held by the mandrel is rotated for turning.

When using a face plate, the work is clamped and centered on the face plate, using bolts, strap clamps, or a combination of both.

A typical operation using a face plate would be the boring of a number of bearings with flat bases. An angle plate would be bolted to the face plate in such a way that the bore of the bearing is centered. The setup is relatively simple and each bearing can be bored and, if necessary, faced. The angle plate remains in position until the last bearing is done.

Recall that whenever the major portion of the total weight of the work is off center in a lathe, a counterweight must be bolted to the opposite side of the face plate or chuck to eliminate any vibration and unnecessary bearing wear on the lathe.

CENTER OR STEADY RESTS

A center rest, more commonly called a steady rest, has one main purpose. It is an outer support on a lathe for round stock that is too long to support itself. If,

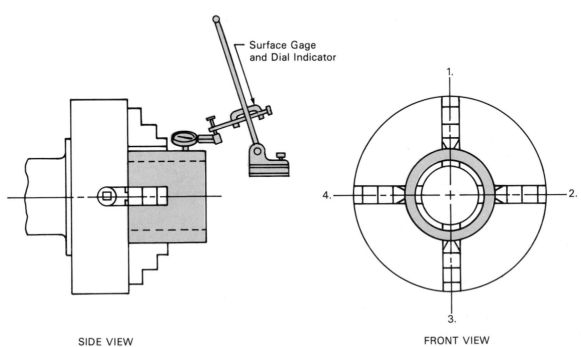

SIDE VIEW

FRONT VIEW

Figure 8–23 Centering with a dial indicator

for example, the work protrudes from the chuck more than three times its diameter, it must be supported either by the tailstock or by a steady rest. The steady rest is clamped firmly in position to the inner ways of the lathe. This allows the saddle to pass on either side of the steady rest so that the cross slide can move in and actually touch the steady rest. This allows turning to be done as close as possible to the jaws. If the end of the shaft is projecting just beyond the jaws, it can be faced, center drilled, drilled, bored, or threaded without problems.

All steady rests have three adjustable jaws, two on the bottom and one on the top. The jaws are adjusted to fit the shaft that is being worked on to hold the shaft on center. The jaws should be tight enough to support the shaft but loose enough to allow the shaft to rotate easily. Jaws are made of cast iron, or bronze-tipped steel, or they can have roller bearing tips. They act as

bearings to support the shaft. A hinge top model, Figure 8–28, is the most common style. Figure 8–29 (page 244) shows an open-sided model. Figure 8–30 (page 245) illustrates typical facing and center drilling work using a steady rest for support.

THE FOLLOWER REST

The follower rest, Figure 8–31A (page 246), often called a follower, prevents a long shaft from springing away from the cutting tool. As a cutter exerts

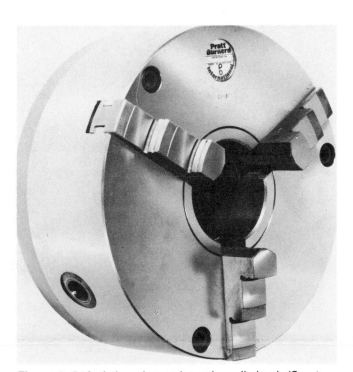

Figure 8–24A A three-jaw, universal scroll chuck *(Courtesy of Pratt Burnerd International)*

Figure 8–24B A three-jaw chuck with two sets of jaws *(Courtesy of Pratt Burnerd International)*

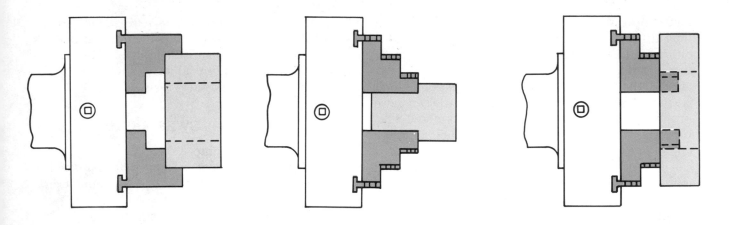

A. Outside jaws gripping outside   B. Jaws in normal position   C. Work chucked on the inside

Figure 8–25 Holding work in a three-jaw chuck

Figure 8–26A A three-jaw combination chuck *(Courtesy of Pratt Burnerd International)*

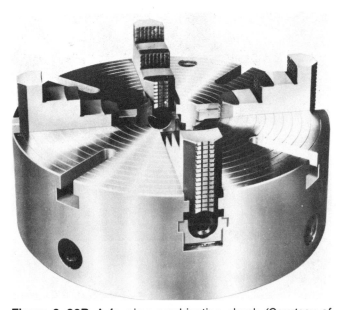

Figure 8–26B A four-jaw combination chuck *(Courtesy of Pratt Burnerd International)*

pressure on a long shaft, the shaft tends to spring back and lift up from the pressure of the cutting tool. The follower rest is bolted solidly to one side or the other of the cross slide, or on the saddle, depending on the manufacturer. This allows the follower to move with the cutting tool.

Two adjustable jaws, one located behind and one located above the shaft, bear on the machined portion of the shaft and follow along slightly behind the cutting tool. This setting prevents the shaft from springing back or from lifting. Again, the cutting tool must be set slightly ahead of the jaws, and the jaws must be adjusted to just touch the machined portion of the shaft for each succeeding cut. The jaws are made of cast iron or bronze-tipped steel, or they can have roller bearing tips.

A steady rest and a follower rest can be used together. The steady rest can be set to support the center of a long shaft, and the follower rest can be set to follow the cutting tool so that machining can be done between the steady rest and the tailstock. Depen-

ding on requirements, the shaft can also be turned end for end, and machining can be done on the other end.

Figure 8–31B (page 247) illustrates a typical follower rest setup for long shaft turning. Long shaft threading also requires this setup.

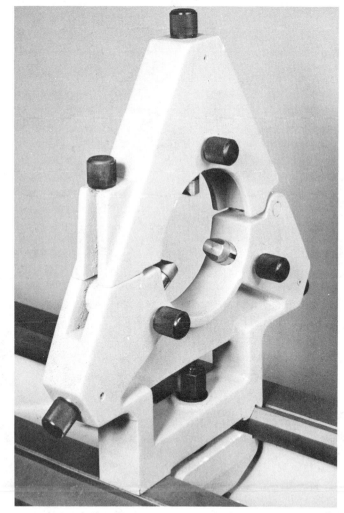

Figure 8–28 A center or steady rest, hinge-top model *(Courtesy of DoALL Company)*

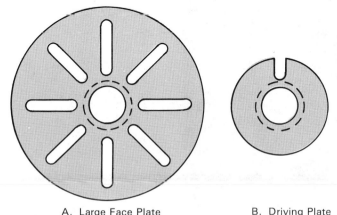

A. Large Face Plate   B. Driving Plate

Figure 8–27 A face plate and a driving plate

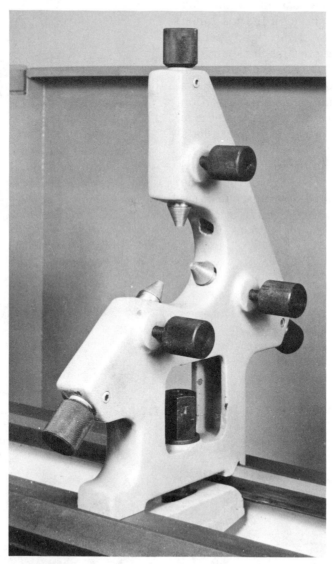

Figure 8–29 A steady rest, open-sided model *(Courtesy of DoALL Company)*

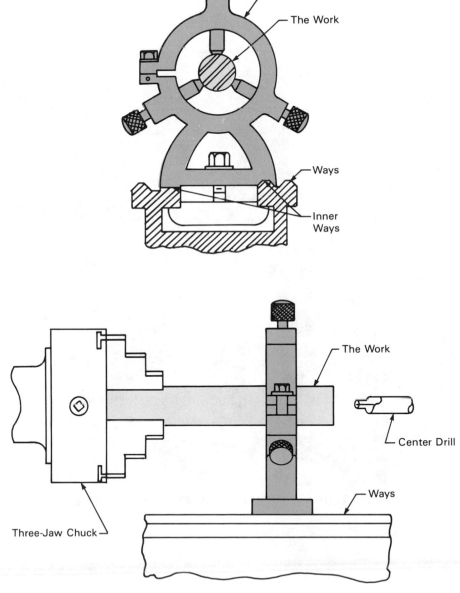

Figure 8–30 A typical, steady-rest setup for facing and centering

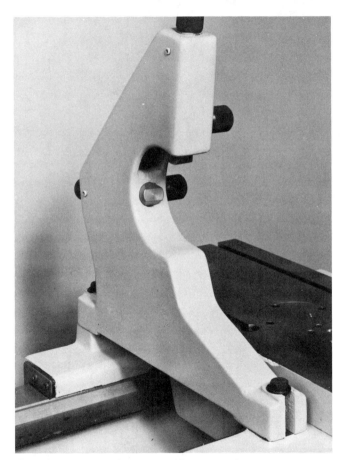

Figure 8–31A A follower rest *(Courtesy of DoALL Company)*

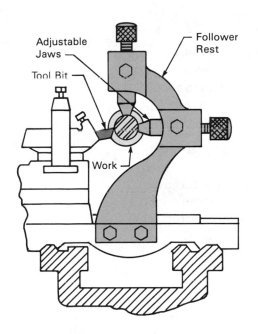

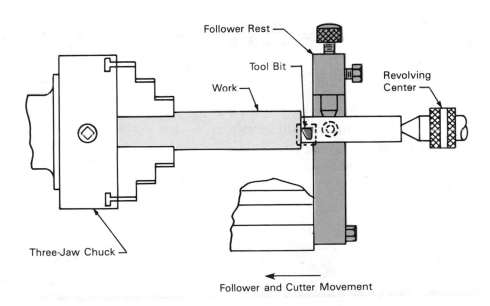

Figure 8–31B A typical follower rest setup

REVIEW QUESTIONS

1. What is the function of a lathe tool holder?
2. At what angle are carbide-tipped tools held in a tool holder? At what angle are tools held in a standard tool holder?
3. For maximum efficiency, how far should a tool bit project from the tool holder?
4. Describe a standard tool post and state its function.
5. What is one advantage of a turret tool holder over standard tool posts? What type of work requires the use of a turret tool holder at all times?
6. Name three types of chuck for lathe spindles and state the differences between them.
7. When the four-jaw chuck is used, list the steps in centering the work.
8. What is the purpose of a counterweight in setting up work in a chuck?
9. What is the primary function of a face plate? How does a driving plate differ from a face plate?
10. What is the purpose of a steady rest?
11. What is the purpose of a follower rest?
12. Describe how a steady rest and a follower rest can be used together.

UNIT 8-3

LATHE OPERATIONS – FACING

OBJECTIVES

After completing this unit, the student will be able to:

- explain the facing operation on a lathe.
- describe the five different methods for facing on a lathe.
- set the cutting tool properly for the different facing methods.

KEY TERMS

Facing or squaring	Turning cut
Rough facing	Finishing cut
Undercut	Dragging the tool bit

LATHE OPERATIONS

There are five basic lathe operations: facing, turning, boring, tapering, and threading. With the proper attachments, other operations such as cylindrical grinding and milling are possible.

FACING (SQUARING)

Facing (or **squaring**) is a basic lathe operation. It is the machining operation where the work is revolved and excess material is removed from the face or end of the work. The cutting tool moves at right angles to the centerline of the work. In facing, the cutting edge should be set on center.

Methods of Facing

In Figure 8–32A, the main cutting edge is on the nose of the tool bit. This is a common method of rough facing. The danger here is that pressure could force the tool holder to revolve, undercutting the surface of the face.

In Figure 8–32B, the main cutting edge is on the side of the tool bit, but some cutting can be done with the nose, and it often is. This tool bit can also be forced into undercutting the surface of the face.

Some machinists prefer to rough face by taking a series of turning cuts, Figure 8–32C. This method is generally followed when quite a bit of material must be removed from the face.

In Figure 8–32D, the main cutting edge is on the side of the tool bit. This is a good method of rough facing, especially when quite a bit of material must be removed from the face. There is no danger of undercutting the surface of the face.

A finishing cut, Figure 8–32E, should be light. The cutting edge should cut from the center outward. This is called **dragging** the tool bit, which is pulled rather than pushed. The final finish is smoother and the surface is much flatter.

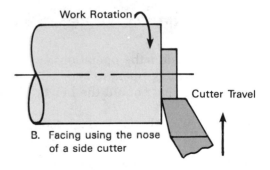

B. Facing using the nose of a side cutter

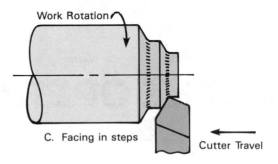

C. Facing in steps

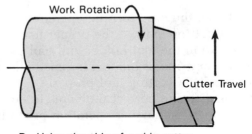

D. Using the side of a side cutter

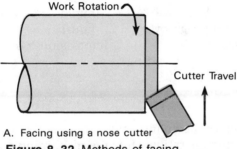

A. Facing using a nose cutter

Figure 8–32 Methods of facing

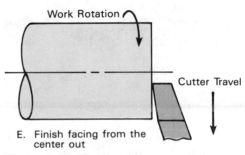

E. Finish facing from the center out

Figure 8–32 Continued

REVIEW QUESTIONS

1. List the five basic lathe operations.
2. Describe the facing operation.
3. When facing, what should the height of the cutting edge be?
4. Describe three methods of rough facing.
5. For which two methods is there a possibility of undercutting the surface of the face?
6. When facing, what does "dragging" the tool bit mean?
7. When would "dragging" the tool bit take place when facing?

UNIT 8-4

LATHE OPERATIONS – TURNING

OBJECTIVES

After completing this unit, the student will be able to:

- define the term turning operation.
- set cutting tools at the proper height for turning.
- position the tool holder and tool post properly for maximum tool efficiency.
- use any one of the turning methods for turning a diameter or shoulder.
- grind the parting tool properly.
- identify and correct problems in cutting off.
- perform a knurling operation properly.

KEY TERMS

Turning
Tool position
Cutting position
Turning a shoulder
Grooving
Straight groove

Squaring a cut-off tool
Necking tool
Knurling
Diamond pattern knurls

Double V groove
Half-round groove
Plunge cutting
Cutting off or parting

Straight line pattern knurls
Impression tool

TURNING ON A LATHE

Turning, as shown in Figure 8–1, is another of the five basic lathe operations. It is the machining operation where the work is revolved and excess material is removed from the outside diameter by a cutting tool called a tool bit. As in each of the basic operations, the height of the tool bit can be critical to the cutting action. Theoretically, all lathe cutting tools should be set on center, but most machinists will set tools either on center, Figure 8–33A, or approximately 5 degrees above center, Figure 8–33B. Both settings are correct under different circumstances. Rarely is a setting of more than 1/32 inch above center used for any reason.

The work tends to lift when it is being cut on a lathe. One of the reasons for setting the tool bit above center is to stop chatter or vibration, particularly of small diameter shafts. A tool bit set above center tends to rub or heel slightly on the work. This helps to steady the work and hold it down.

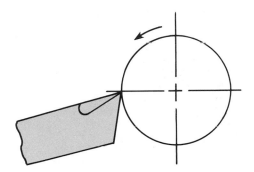

Figure 8–33A Cutter set on center

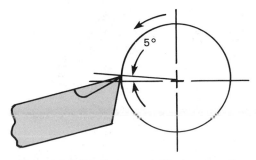

Figure 8–33B Cutter set 5° above center

Grooving tools are usually set on center, especially in cutting off. However, many machinists will set grooving tools slightly above center to stop chatter. Then they will make adjustments in height as the cutting tool approaches center.

Play or looseness in the compound, cross slide, or carriage will also cause problems. The tool bit could actually fall below center, effectively changing the rake and clearance angles and increasing the tendency of the work to lift. Using a setting slightly above center helps to offset this tendency.

In the machining of tough metals, the great pressures encountered could force the tool bit below center. A setting slightly above center would also help in this situation.

Summary

In most cases, especially in facing, tapering, or cutting-off, the tool bit should be set on center and the setup should be as rigid as possible. There are times, however, when a setting slightly above center is preferred.

TOOL POSITION

Tool position on a lathe during a machining operation is a very important consideration if you are to expect the maximum efficiency of the tool bit.

In Figure 8–34A, the tool holder is held short with room only for the tool post wrench. This placement is best since the tool holder is rigid.

In Figure 8–34B, the tool holder protrudes too far, permitting spring and vibration in the holder, which result in chatter or vibration of the cutting edge and inaccuracy in the cut.

The tool post, Figure 8–34C, is set to the left side of the compound to permit the tool bit to cut much closer to the end of the shaft before the lathe dog strikes the compound. A nose-type turning tool, Figure 8–34D, will cut even closer. Care must always be taken to prevent anything that is revolving from striking the compound or cross slide.

The tool bit being used, Figure 8–35, has the cutting edge on the left-hand side so that the movement will be to the left. If pressure on the tool bit is great enough to rotate the tool holder, Figure 8–35A, the work will be undercut, reducing the diameter. If the tool holder rotates as in Figure 8–35B, the diameter of the work will increase and no harm will be done. The setup in Figure 8–35 may be necessary at times. In such cases, cutting must be done carefully.

METHODS OF TURNING

There is more than one way of turning material from one diameter down to another diameter.

Figure 8–36A shows one method of rough turning used by many machinists. The cutting edge is on the nose of the tool bit, which can get in close to the chuck without the compound or cross slide hitting the chuck. The danger is that if pressure causes the tool holder to rotate, the diameter of the work could be cut undersize. Figure 8–36B shows another version of the same type of cutter. Again, there is a danger of undercutting the work if the tool holder rotates.

The cutting edge in Figure 8–36C is on the side of the tool bit. In this case, there is no lead angle. If pressure causes the tool holder to rotate, the diameter is not cut undersize.

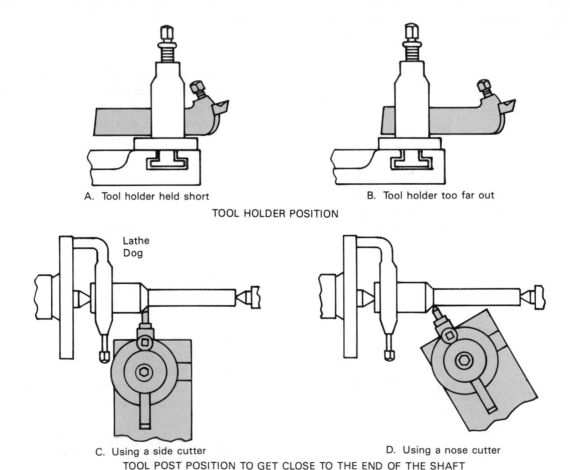

A. Tool holder held short

B. Tool holder too far out

TOOL HOLDER POSITION

Lathe Dog

C. Using a side cutter

D. Using a nose cutter

TOOL POST POSITION TO GET CLOSE TO THE END OF THE SHAFT

Figure 8–34 Tool position

THE LATHE AND LATHE OPERATIONS

In Figure 8–36D, the cutting edge is again on the side of the tool bit, but in this case there is a lead angle. The advantages of a lead angle are explained in Chapter 9.

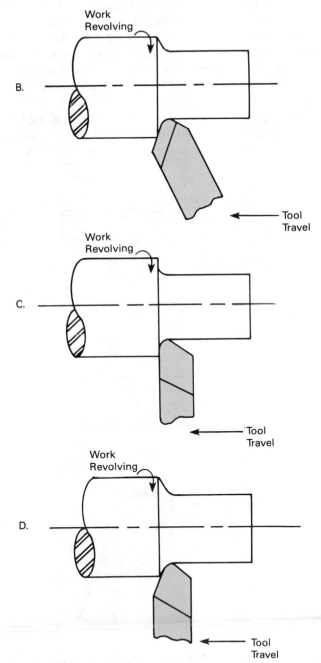

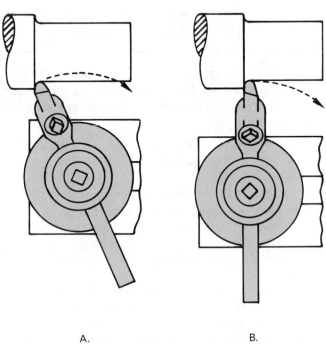

A.   B.

Figure 8–35 Tool holder position

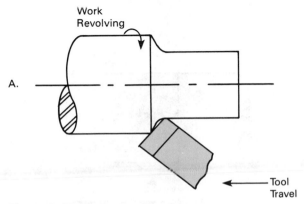

Figure 8–36 Methods of turning

Figure 8–36 Continued

CUTTING POSITION

With the tool bit in position, Figure 8–37A, and with a lead angle presented to the job, pressure actually tends to push the tool bit away from the work, rather than pull it into the work. The center of the round nose is doing the cutting. The result is a smooth, accurate surface. Pressure against the cutting edge is less because the chip is thinner. The cut is more of a shearing action because of the angle of the leading edge.

The tool bit shown in Figure 8–37B has a tendency to dig in and bounce slightly. The surface is actually cut by the back side of the round nose, not by the center of the round nose. The result is a very poor, inaccurate, bumpy surface.

TURNING A SHOULDER

One of the most common turning operations is the turning of a shoulder. The following sequence of steps is an accepted method of accomplishing this operation.

1. Face the end of the shaft and lay out the shoulder length using hermaphrodite calipers, Figure 8–38A.

2. Establish a rough diameter (approximately 1/16 inch oversize) using outside calipers, Figure 8–38B.

3. Rough turn the diameter (approximately 1/16 inch oversize) in one or more roughing cuts, leaving the shoulder length about 1/16 inch short, Figure 8–38C.

4. Rough square the shoulder, Figure 8–38D.

5. Finish turn the required diameter.

6. Finish facing the shoulder, taking the final cut from the center outward. Use a rounded point if a rounded corner is required, Figure 8–38E.

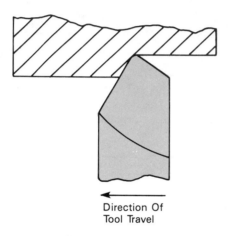

Direction Of
Tool Travel

A. Cutter position with lead angle

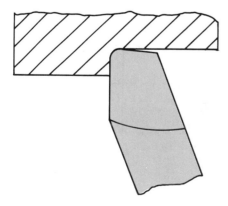

B. Cutter position with lead angle at right angles to the work

Figure 8–37 Cutter positions

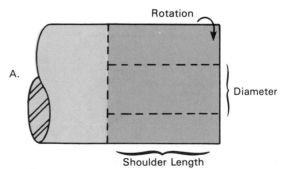

Figure 8–38 Procedure for turning a shoulder

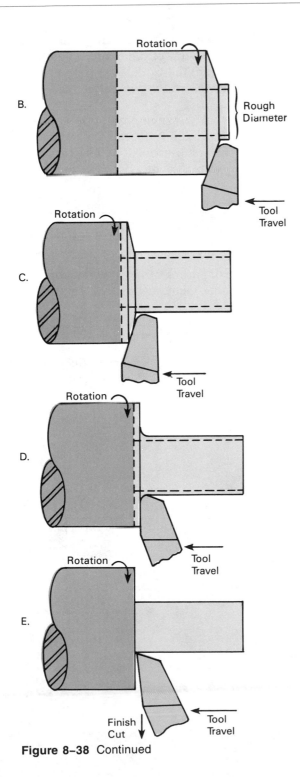

Figure 8–38 Continued

GROOVING

Grooving, Figure 8–39, is a turning operation where the blade, with the cutting edge on the end, moves straight into the work at right angles to the centerline of the work. The blade is referred to as a cut-off tool or parting tool, or sometimes as a necking tool. This type of cutting is called **plunge cutting**.

Types of Grooves

Figure 8–40 illustrates three types of grooves commonly cut on a lathe. They are the straight groove, the single or double V-groove, and the half-round groove.

CUTTING OFF OR PARTING

In cutting off material, it is good practice to grind a leading point on the right side of the cutter, Figure 8–41A. The leading point breaks through first and the work drops off cleanly leaving the bur on the stock. The bur is easily removed by moving the cutter in further. Another version of the cutter is shown in Figure 8–41B.

Preparing the work for threading into a neck is shown in Figure 8–42.

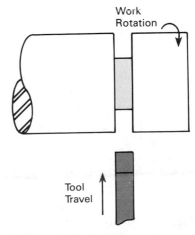

Figure 8–39 Grooving

Squaring a Cut-off Tool

One of the most important aspects of cutting-off material is the setting of the cutter perpendicular to the centerline of the work, Figure 8–43. First, make sure that the end of the shaft is square. Then, set a scale across the square end and check the cut-off tool on both sides.

Cut-off Tools

Cutting-off is often one of the hardest machine operations to master and one of the least discussed. The correct grinding of the cutter is most important. This operation is often referred to as **parting** and the blade used is often called a parting tool.

The height of the cutting edge should always be at least the same height as the top of the cut-off tool, Figure 8–44A. A commercial cut-off tool is tapered on both sides so that it has three clearances — one side clearance on each side and one front clearance.

If the cutting edge is ground lower than the top of the cut-off tool, Figure 8–44B, as it often is with repeated sharpening, the cutting edge becomes narrower than the top of the blade. The blade soon jams as it cuts into the work. This is one of the most common problems and is one of the major causes of broken blades.

A cut-off tool (or necking tool) may also be ground from a tool bit, Figure 8–45. Such a tool can have as many as five clearances: one front clearance, two side clearances, and two lengthwise clearances.

PROBLEMS IN CUTTING-OFF (PARTING)

Cutting-off or parting can be a difficult operation. The following recommendations can make the work easier.

1. If the blade is too far out of the holder, the cutting edge will vibrate (chatter). The cutting edge may break down or the blade may be forced to drop below center, causing the

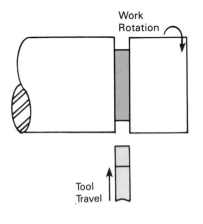

A. Straight

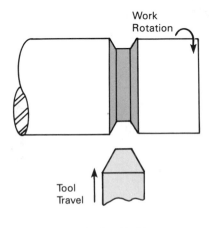

B. Double V

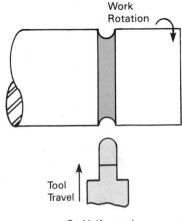

C. Half-round

Figure 8–40 Types of grooves

work to ride up on the cutting edge. This is the chief cause of breakage in cut-off tools.

2. If the blade is not square with the work, the cutting edge will bind (rub), cutting action will be poor, and breakage will be the probable result. Cut-off tools should always be carefully checked on both sides to ensure that they are square with the work.

3. If the cutting edge has been ground lower than the top of the blade, the blade will bind in the narrower cut. Breakage will result.

4. Speed should be much lower than in the usual turning operations. Remember that you are using a wide, flat nose, plunge cutting tool. If the speed is such that the cutting edge vibrates (chatters), the work is rotating too fast and the cutting edge could break down.

5. Cutting fluids should be used for those materials that require them.

6. Use the correct rakes for various materials. Front clearance should be at a minimum.

7. Do not be timid with a cut-off tool. Feed it in firmly and steadily. Do not hesitate. Do not give the cutting edge a chance to rub, as this dulls the cutting edge.

8. Do not allow the work to protrude too far from the chuck. The work and the blade should be rigid.

9. Using a spring tool holder will help. This usually gives the tool a resiliency that takes up any chatter and keeps the work from climbing up on the tool.

10. Using a cut-off tool upside down is always good practice. As in any upside down cutting, pressure exerted by the blade is downward. The main lathe spindle is actually forced to ride on the bottom of the main

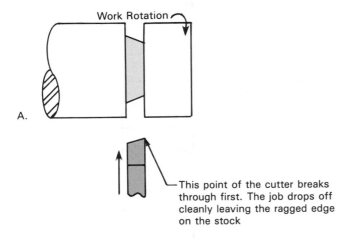

This point of the cutter breaks through first. The job drops off cleanly leaving the ragged edge on the stock

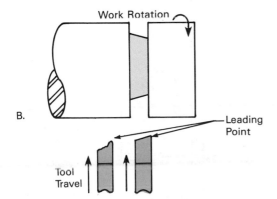

Figure 8–41 Cutting off material with a leading point

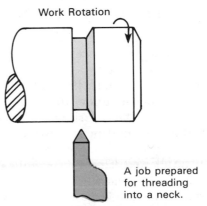

A job prepared for threading into a neck.

Figure 8–42 Preparing work for threading into a neck

bearings. Cutting fluid can be applied easily, and chips will drop away from the cutting edge.

11. Setting the cutting edge very slightly above center will cause a slight heeling action that will prevent the work from riding up on the blade. This will also prevent chatter. However, the cutting edge must be readjusted to center as the smaller diameters are approached, especially when cutting-off.

KNURLING

Knurling is not a cutting operation in the sense that material is removed from the job. A knurling tool, Figure 8–46, is an impression tool. Serrated rollers are pressed into the revolving work causing a pattern to form on the outside diameter. The metal is actually squeezed up, increasing the diameter. Two common styles of knurling tools are a knurling tool with a single set of knurls, and a revolving head knurling tool with coarse, medium, and fine knurls, Figure 8–47. Figure 8–48 illustrates the rollers available and the patterns that are formed. The main reasons for knurling a piece of work are: (1.) to provide a positive, non-slip hand grip, (2.) to improve the appearance of a surface, and (3.) to increase the diameter of the work for a press fit.

THE KNURLING OPERATION

To knurl a piece of work successfully, the following points should be observed:

1. The knurling tool should be absolutely clean. No chips should be allowed under the knurling tool or tool post ring, etc. The knurling tool must sit perfectly vertical and should not be upside down.

2. The work must be rigid and no bending of the work should be allowed. The knurling rolls should run flat on the surface of the work at all times, and the rolls should be clean. Some machinists use a slight 3- to 5-degree lead to achieve a correct pattern.

3. The knurling rolls should be located equally above and below the centerline of the work. The revolving head should be free to move.

4. Set a slow spindle speed with a feed of about 0.030 in (0.75 mm) per revolution of the chuck.

5. Start the lathe and press the rolls into the surface. Both rolls must bear on the surface.

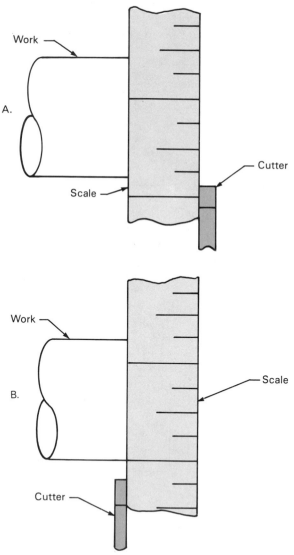

Figure 8–43 Squaring a cutting-off tool using a scale

6. As the impression begins to form, stop the lathe, leaving the rolls in contact with the surface. The correct impression should be a full diamond pattern, not a double impression, Figure 8–49. If the impression is doubled, start again in a new spot. Check the setting of the knurling tool. Check the work. It should be rigid with no bending taking place.

7. Once the diamond pattern forms, force the rolls into the work, engage the feed, and use cutting fluid as required. The lubricating qualities of a cutting fluid help to prevent binding of the knurls on their shafts.

8. Knurl the full required length, then reverse the feed and run back over the knurl. Repeat until the diamond patterns are raised to a point with no flat spots showing. Run the full length of the knurled surface without disengaging the feed. This will prevent the formation of rings on the knurled surface. As long as the lathe speed is slow, there should be no grinding of gears when the direction of the feed is reversed.

9. Any wiping of the knurled surface or the knurling rolls is best done with a brush. Do not use rags, or your fingers. Keep your hands out of the way as long as the work is revolving.

10. All of these recommendations also apply to straight line rolls.

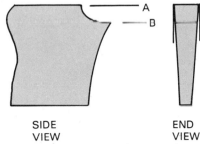

SIDE VIEW

END VIEW

Figure 8–44B An improperly ground cut-off tool

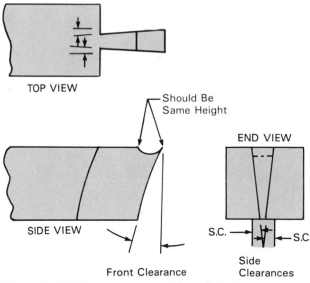

TOP VIEW

Should Be Same Height

END VIEW

SIDE VIEW

Front Clearance

S.C. — — S.C.

Side Clearances

Figure 8–45 Clearances in cutting-off tools

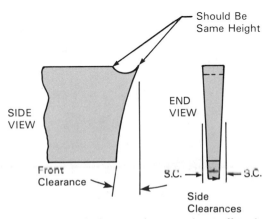

Should Be Same Height

SIDE VIEW

END VIEW

Front Clearance

S.C. — — S.C.

Side Clearances

Figure 8–44A A properly ground cut-off tool

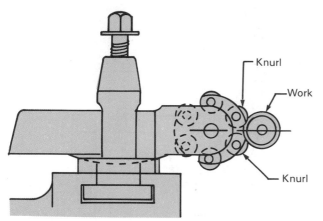

Figure 8-46 A knurling tool setup *(Courtesy of Armstrong Brothers Tool Company)*

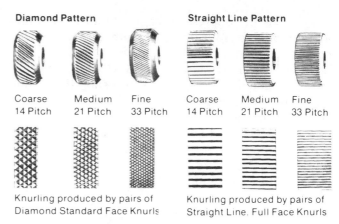

Figure 8-48 Knurling rollers and patterns *(Courtesy of Armstrong Brothers Tool Company)*

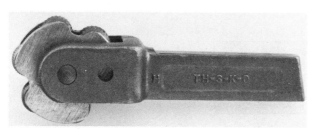

Figure 8-47 A revolving-head knurling tool with coarse, medium, and fine knurls *(Courtesy of DoALL Company)*

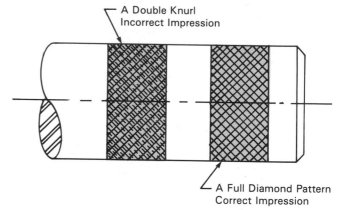

Figure 8-49 Knurling impressions

REVIEW QUESTIONS

1. Describe the turning operation on a lathe.
2. A tool bit can be set on center or above center depending on conditions. Explain this statement.
3. Why is tool position so important? List three considerations when setting the position of the tool holder and tool post.
4. What is one danger in turning that may result in cutting undersize?
5. Describe four methods of turning material from one diameter down to another diameter.
6. With a lead angle presented to the job, describe the type of cut produced by the tool bit.
7. List the steps in turning a shoulder.
8. Describe the operation of grooving.
9. When grinding a cut-off tool, what happens when the cutting edge is ground lower than the top of the blade?
10. What are some methods of preventing chatter in a cut-off tool?
11. How does the speed of a cutting-off operation compare to that of other turning operations?
12. What is the purpose of using a cut-off tool upside down?
13. What is the purpose of a knurl?
14. List five conditions that are necessary to knurl a piece of work successfully.

UNIT 8-5

LATHE OPERATIONS – BORING

OBJECTIVES

After completing this unit, the student will be able to:

- select the proper boring tool holder and boring bar for the work to be done.
- set up the work and boring tool for conventional boring and for upside-down boring.
- perform drilling and reaming operations on a lathe.
- state the characteristics of an accurately made center drill.
- perform the center drilling operation properly.
- remove a broken center drill from the work.
- align the centers of a lathe.
- turn work accurately between lathe centers.
- use lathe centers and spigots properly.
- use spiders, catheads, and mandrels properly.

KEY TERMS

Boring	Dead center
Boring tool bit	Turned center
Boring tool holder	Revolving center
Boring bar	Automatic driving
End cap pattern	center
boring bar	Lathe dog
Conventional boring	Spigot
Upside-down boring	Half center
Center drilling	Cathead
Center drill	Standard lathe
Aligning lathe centers	mandrel
Lathe center	Gang mandrel
Live center	Expansion mandrel

GENERAL

Boring is another basic lathe operation. It is the machining operation where the work is revolved and excess material is removed from the inside diameter of the work. The tool bit is usually held in one of three ways. It can be held at 90 degrees to the boring bar, with the cutting edge on the side, Figure 8–50A, or it can be held at 45 degrees or 30 degrees to the boring bar, with the cutting edge on the nose, Figure 8–50B. The cutting edge should be on center. Figure 8–51 illustrates a high-speed boring operation using a carbide insert tool and boring bar. The metal chips curled into a number 9 shape are easy to see. When a boring operation is being done, always use the largest and steadiest boring setup that is practical.

BORING TOOL HOLDERS, COMMERCIAL TYPES

There are many styles of commercial boring tool holders. A 3-bar tool holder, Figure 8–52, shows three different sizes of boring bar. Boring bars should always be as large as possible. A yoke boring tool holder, Figure 8–53, can be used with forged-type boring bars. This type of holder can also be used to hold a tool bit which can be ground for boring, or can be used with other boring bars, Figure 8–54. A heavy-duty boring tool holder with end cap pattern bars is shown with tool bits that can be held at 90, 45, and 30 degrees, Figure 8–55.

Many machinists prefer to make their own V-block and boring bar to fit in the tool post. The boring bar sits in the V-block and is clamped directly in the tool post, Figure 8–56.

Many styles of boring bars are available on the market. Plain bars, Figure 8–57, can be bought with or without flats, or can be made from good, tough bar steel. Forged types of boring bars, Figure 8–58, can be bought or made from high-carbon tool steel, stellite,

or carbide. End cap pattern bars, Figure 8–59, have interchangeable heads which hold the tool bits at 90, 45, or 30 degrees. A clamp-type boring bar is shown in Figure 8–60, and a boring bar for indexable carbide inserts is shown in Figure 8–61. In many cases, especially in smaller work, a machinist will grind a small boring tool from a tool bit, Figure 8–62.

Figure 8–51 A boring operation using a carbide cutter and boring bar *(Courtesy of Kennametal Inc.)*

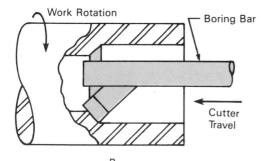

Figure 8–50 Holding the boring cutter

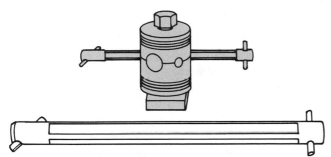

Figure 8–52 A three-bar boring tool *(Courtesy of Armstrong Brothers Tool Company)*

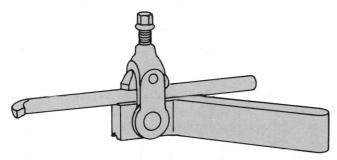

Figure 8–53 A yoke boring tool holder with solid forged boring bar *(Courtesy of Armstrong Brothers Tool Company)*

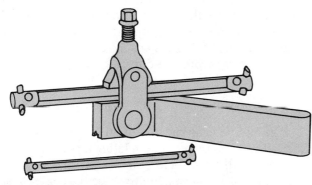

Figure 8–54 A yoke boring tool holder with clamp-type boring bars *(Courtesy of Armstrong Brothers Tool Company)*

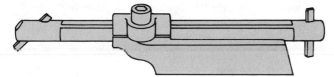

Figure 8–55 A heavy-duty boring tool holder with end cap boring bars *(Courtesy of Armstrong Brothers Tool Company)*

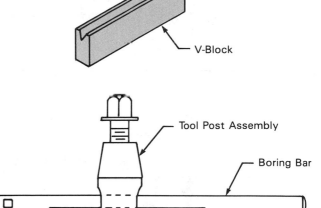

Figure 8–56 V-block to support the boring bar in the tool post

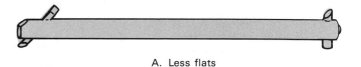

A. Less flats

B. With flats

Figure 8–57 Plain boring bars *(Courtesy of Armstrong Brothers Tool Company)*

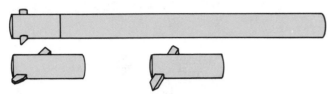

Figure 8–58 Forged boring bar (Courtesy of Armstrong Brothers Tool Company)

Figure 8–59 End cap pattern boring bars (Courtesy of Armstrong Brothers Tool Company)

Figure 8–60 Clamp-type boring bars (Courtesy of Armstrong Brothers Tool Company)

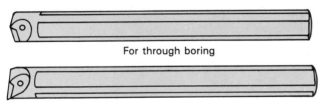

For through boring

For bottoming and facing

Figure 8–61 Boring bars for indexable carbide inserts (Courtesy of Armstrong Brothers Tool Company)

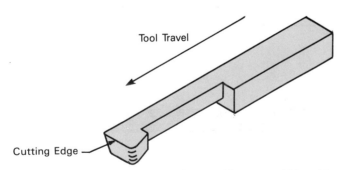

Tool Travel

Cutting Edge

Figure 8–62 A small boring tool ground from a tool bit, with zero rake

BORING METHODS

The method of holding the cutter for conventional boring is shown in Figure 8–63. The cutting edge should be set on center. Pressure exerted by the boring tool is upward, causing the work to lift. However, this does not present too much of a problem in boring. Always remember to keep the boring bar as large and as short as possible. Figure 8–64 shows a different approach, called upside-down boring. The cutting edge is upside down and on center, with the cut taking place on the backside of the work. Pressure exerted by the cutting edge is downward. The boring tool tends to lift, but the work is forced downward. There are several advantages in using the upside-down boring method. In the case of an old lathe with loose main bearings, the cutting edge exerts pressure downward, forcing the lathe spindle to run on the bottom of the main bearings. Vibration from the loose bearings is eliminated and the cutting action is smooth and accurate. In the case of boring a long bushing supported by a steady or center rest, the downward pressure holds the bushing down on the two bottom support jaws of the steady rest. The third jaw on top is not needed. This effectively eliminates the heat generated by the three jaws of the steady rest bearing on the bushing. Chattering is minimal. Cutting oil is easy to apply and chips fall naturally away from the cutting edge.

DRILLING AND REAMING

The following is an outline of an effective procedure for drilling and reaming on a lathe. It is good practice to center drill a hole before drilling with a twist drill, Figure 8–65. This gives the twist drill a center to start on. The hole is drilled undersized, ready for reaming, Figure 8–66. Normally, holes are 1/64 in (0.4 mm) undersize for holes up to 1/2-in (12.7-mm) diameter and 1/32 in (0.8 mm) undersize for holes over 1/2-in (12.7-mm) diameter. The hole is reamed to the correct diameter, Figure 8–67, in this case with a machine reamer (taper shank). A hand reamer (square shank) should not be used under power. It is not strong enough to take the pressure.

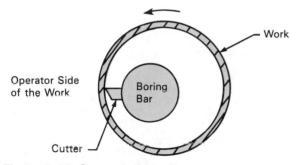

Work

Operator Side
of the Work

Boring
Bar

Cutter

Figure 8–63 Conventional boring

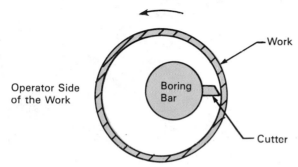

Work

Operator Side
of the Work

Boring
Bar

Cutter

Figure 8–64 Upside-down boring

Figure 8–66 Drilling

Figure 8–65 Center drilling

Figure 8–67 Reaming

Milling Operations on a Lathe

By the use of milling attachments and grinding attachments, a lathe can be converted to perform milling and grinding operations such as milling a dovetail and milling a keyway. A tool post grinder will perform such operations as cylindrical grinding, Figure 8–68. Not all machine shops have a cylindrical grinder. For a job shop that does not have such a machine, a tool post grinder mounted on a lathe is a viable alternative.

The tool post grinder with the internal grinding spindle, Figure 8–69, is a good rigid setup for the work being done. Note the good machining practice of protecting the ways of the lathe from the grinding wheel cuttings. This setup demonstrates the versatility of the lathe.

Center Drilling

Center drilling is a very common and a very important machine shop operation. The center hole, which is usually drilled at the center of the face on the end of a shaft, acts as a support hole for the shaft and also provides a starting hole for drilling either on a lathe or on a drilling machine.

The combined drill and countersink, Figure 8–70, is most commonly called a center drill. It drills a straight hole and also drills a 60-degree countersink.

A properly shaped center hole will form a 60-degree support hole for long work that lacks support. A 60-degree lathe center, either revolving or not, will provide the support. The center drill will also provide a starting hole for another drill, to help prevent that drill from wandering off center.

With no center holes, Figure 8–71A, the work has no support. As soon as the cutter touches it, the work will be forced out from between the lathe centers in the direction of the arrows.

The center hole in Figure 8–71B is too shallow. The result is poor support for the work and possible damage to both the lathe center and the work.

In the example shown in Figure 8–71C, there is a bearing surface to match the 60-degree lathe center, but there is no clearance for the point of the lathe center and no room for lubrication when using a

Figure 8–68 Cylindrical grinding with a tool post grinder *(Courtesy of Dumore Corporation)*

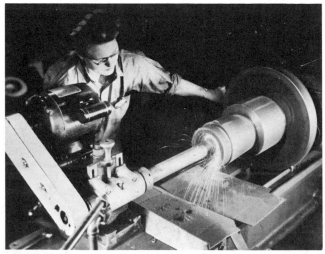

Figure 8–69 Internal grinding with a tool post grinder *(Courtesy of Dumore Corporation)*

"dead" center. The point, the weakest part of the lathe center, will overheat and could burn off.

The center hole in Figure 8–71D is a good combination. It has clearance for the point and also a 60-degree bearing surface to match the 60-degree point of the lathe center.

Care must be taken not to drill too deep, Figure 8–72. The lathe center point and the drilled center hole will not make contact. Support is poor and damage could result to both the lathe center and the work.

Care must be taken to face the end of the shaft, Figure 8–73, at least around the center hole. If this is not done, uneven pressure will be exerted on the lathe center, and especially in the case of a "dead" center, uneven wear will result.

Size of Center Drills. The size of a center drill, Figure 8–74, is the diameter of the drill point C, not the body size.

Table 8–1 lists the recommended size of the center drill for different shaft sizes. The speed calculation is based upon the largest diameter entering the work:

$$Rpm = 4\frac{CS}{D}$$

The dimensions of plain and bell-type center drills are given in Table 8–2.

Methods of Center Drilling. Some machine shops have a center drilling machine that holds work on center while the ends are center drilled. The main disadvantage of this machine is that larger work and irregularly shaped work often cannot be held. In addition, another setup is required on another machine to face the ends of the shafts before center drilling takes place.

By far the easiest, most common, and most accurate method of center drilling is shown in Figure 8–75. It involves the use of the lathe, a 3- or 4-jaw chuck, a steady or center rest, and the tailstock. With this setup, the end of the shaft can be held safely, it can be faced accurately, and the shaft can be center drilled exactly on center. The tailstock spindle must be set exactly on center before starting.

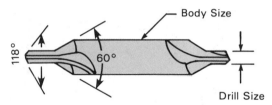

Figure 8–70 Combination countersink and center drill

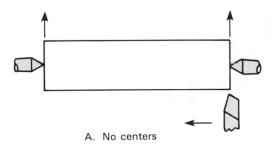

A. No centers

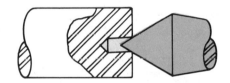

B. Too shallow

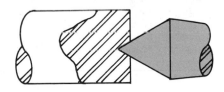

C. No point clearance

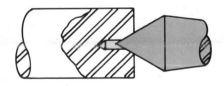

D. Good

Figure 8–71 The center hole

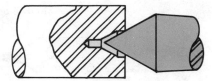

Figure 8–72 Hole is too deep

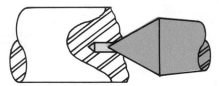

Figure 8–73 Shaft is not square

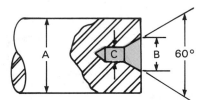

Figure 8–74 Size of the center drill

With the shaft revolving, lightly touch the end of the shaft with the center drill. If the shaft is off center, the center drill point will mark a circle on the end of the shaft. Adjusting the steady rest jaws will move the end of the shaft so that the center drill is exactly on center. The center hole can then be drilled.

Table 8–2 Dimensions of Combined Drills and Countersinks

PLAIN TYPE				
DIMENSIONS IN INCHES				
Size No	Body Dia	Drill Dia	Drill Length	Overall Length
00	3/32	0.025	0.030	1 1/8
0	3/32	1/32	0.038	1 1/8
1	1/8	3/64	3/64	1 1/4
2	3/16	5/64	5/64	1 7/8
3	1/4	7/64	7/64	2
4	5/16	1/8	1/8	2 1/8
5	7/16	3/16	3/16	2 3/4
6	1/2	7/32	7/32	3
7	5/8	1/4	1/4	3 1/4
8	3/4	5/16	5/16	3 1/2

Plain-type combined drills and countersinks have an included angle of 60 degrees.

Table 8–1 Recommended Center Drill Sizes for Different Shaft Diameters

SHAFT DIA A	LARGEST DIA OF CENTER B	DRILL SIZE C		DEPTH OF HOLE D
in (mm)	in (mm)	SIZE NO in (mm)	DRILL SIZE in (mm)	in (mm)
1/4 (6.35)	1/8 (3.18)	1 (M-1)	3/64 (1.00)	5/32 (3.96)
3/8 (9.53)	3/16 (4.76)	2 (M-4)	5/64 (2.00)	7/32 (5.56)
1/2 (12.7)	1/4 (6.35)	3 (M-5)	7/64 (2.50)	5/16 (7.94)
5/8 (15.88)	5/16 (7.94)	4 (M-6)	1/8 (3.15)	3/8 (9.51)
3/4 (19.05)	7/16 (11.11)	5 (M-7)	3/16 (4.00)	7/16 (11.11)
1 (25.4)	1/2 (12.7)	6 (M-8)	7/32 (5.00)	19/32 (15.09)
1 1/4 (31.75)	5/8 (15.88)	7 (M-9)	1/4 (6.3)	21/32 (16.66)
1 1/2 (38.1)	3/4 (19.05)	8 (M-9)	5/16 (6.3)	11/16 (16.66)
2 (50.8)	3/4 (19.05)	8 (M-9)	5/16 (6.3)	3/4 (16.66)

BELL TYPE					
DIMENSIONS IN INCHES					
Size No	Body Dia	Drill Dia	Drill Length	Overall Length	Bell Dia
11	1/8	3/64	3/64	1 1/4	0.100
12	3/16	1/16	1/16	1 7/8	0.150
13	1/4	3/32	3/32	2	0.200
14	5/16	7/64	7/64	2 1/8	0.250
15	7/16	5/32	5/32	2 3/4	0.350
16	1/2	3/16	3/16	3	0.400
17	5/8	7/32	7/32	3 1/4	0.500
18	3/4	1/4	1/4	3 1/2	0.600

Bell-type combined drills and countersinks have an included angle of 60 degrees and also an included angle of either 82 degrees or 120 degrees, which form a protected center.

Table 8-2 Continued

PLAIN-TYPE METRIC			
Stock No	Drill Point mm	Body Dia mm	Length Overall
M-1	1.00	3.15	29.5
M-2	1.25	3.15	29.5
M-3	1.60	4.00	33.5
M-4	2.00	5.00	38.0
M-5	2.50	6.30	43.0
M-6	3.15	8.00	48.0
M-7	4.00	10.00	53.0
M-8	5.00	12.50	60.0
M-9	6.30	16.00	68.0

Once the steady rest is set, any number of shafts of the same diameter can be accurately drilled in succession.

Another less common method of center drilling involves the use of a laid out center hole and a drill press. The work either must be clamped vertically to the drill press table and lined up exactly with the drill press spindle, or it must be held vertically in a drill vise.

CENTER DRILLING ACCURATELY

Accurate center drilling on a lathe is extremely important if your work is to be machined accurately.

1. The overall setup, Figure 8-76A, shows the shaft in a three-jaw chuck and supported by a steady rest. The shaft can be faced and center drilled accurately using this method.

2. The shaft in Figure 8-76B is set off center. The center drill point marks a circle on the end of the shaft.

3. By adjusting the three jaws of the steady rest, the shaft can be set exactly on center, Figure 8-76C.

4. The completed center hole is shown drilled exactly on center, Figure 8-76D.

Tips on Center Drilling

1. A center drill provides a starting hole for a twist drill to prevent the twist drill from wandering off center. Once the hole has been drilled, reaming or boring can follow if they are required.
2. The proper speed should be calculated on the largest size of the drill that enters the work.
3. Be gentle with a center drill. Do not jam it into the shaft. Feed it into the shaft with fingertip pressure only.
4. Clear the center drill of chips often. Do not allow the chips to jam the drill.
5. Use the appropriate cutting fluid when required.

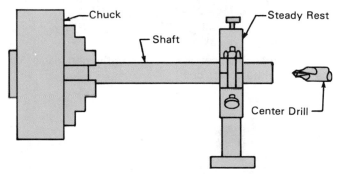

Figure 8-75 Center drilling

Figure 8-76A Setup for accurate center drilling

Figure 8–76B Shaft set off center. Note the ring marked by the point of the center drill

Broken Center Drills

One of the most annoying aspects of center drilling is the broken center drill, Figure 8–77A. When the drill breaks, the small high-speed steel tip is left in the end of the shaft. Since there is no way of softening the tip, it usually must be cut out.

One of the quickest and easiest methods of cutting out the tip is to regrind the broken end of the center drill as shown in Figure 8–77B, and redrill with it. The new point will move down the side of the broken tip to expose it. You can then grip and remove it. Redrill a new center hole using a center drill the next size larger.

Alignment of Centers

Before turning a job between centers, the live center at the headstock end must run dead true. Both centers must be exactly in line with each other, Figure 8–78. If the dead center or revolving center at the tailstock end is off center, a taper will result.

Figure 8–76C Shaft set exactly on center

Figure 8–76D The finished center hole

One method of checking alignment is to use a dial indicator on a test bar. Use the same cross slide setting at each test point, Figure 8–79.

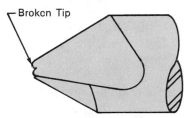

Figure 8–77A Side view

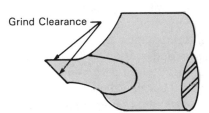

Figure 8–77B Top view

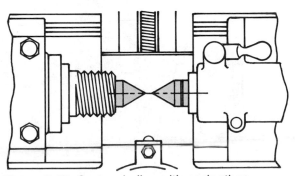

Figure 8–78 Centers in line with each other

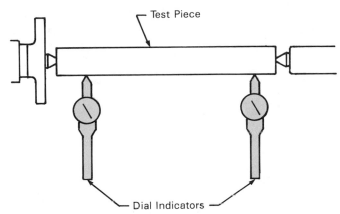

Figure 8–79 Using a dial indicator

Another method of checking alignment is to take a test cut on each end of a test bar, using the same cross slide setting for each cut. Each diameter must measure exactly the same, Figure 8–80. Similarly, a test bar can be made with collars on each end. Test cuts are made on each collar, and they should measure exactly the same diameter.

Figure 8–81 illustrates the result of turning a shaft with the live center running "out of true."

Turning Between Centers

A large proportion of work on a lathe is done between centers, Figure 8–82. For example, if the work has more than one diameter and all of the diameters are to be concentric or on the same centerline, this method of turning is most accurate. As long as both centers are in alignment, the work can be removed and replaced easily and accurately or turned end for end easily and accurately. The sketches show a conventional setup using a driving plate and a live and dead center. A revolving center is more commonly used today than a dead center.

It is not always possible to locate a driving plate or enough lathe centers in most machine shops. A much more common and much more practical method of turning between centers involves the use of a turned center held in a chuck, Figure 8–83. To do this, turn

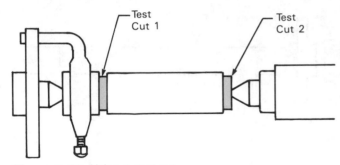

Figure 8-80 Using a test cut

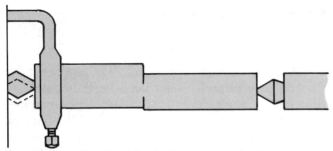

Figure 8-81 Result of live center not running true

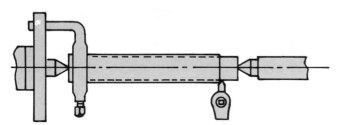

Figure 8-82A Centers are in line. Work is turned parallel

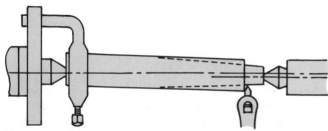

Figure 8-82B "Dead" center is out of line. Work is turned tapered

a shoulder on a piece of round scrap steel. Then place the turned portion in the chuck with the shoulder tight against the chuck jaws. This will prevent the steel center from being forced into the chuck. Then set the compound and turn a 60-degree center. This is your live center.

Whenever this center is removed for any reason and replaced in the chuck, a light cut must be taken on the 60-degree point to make sure it is running true.

The lathe dog is driven by one of the chuck jaws. It is also more common today to use a revolving center in the tailstock rather than a dead center.

Lathe Centers

At one time, standard lathe centers were used almost exclusively for turning between centers. Both soft and hardened and tempered centers were common. The soft center was plain, but the hardened and tempered center had an identifying ring cut around the circumference just behind the point. The soft center was called a **live center**. It was set into the headstock spindle and revolved with the work. The hardened and tempered center was called a **dead center**. It was set into the tailstock spindle and did not revolve with the work.

The lathe center in Figure 8-84 has a hardened and tempered, high-speed steel point. Since the center in the tailstock does not revolve, an extremely high grade of grease must be used in the center hole of the work for lubrication. The center acts as a support bear-

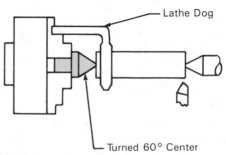

Figure 8-83 Using a turned center

ing for a long shaft. If the center is the least bit too tight, heat will build up very quickly to the point where the center will actually weld itself to the end of the work. At the same time, if the center is too loose, the shaft will vibrate, resulting in poor and inaccurate machining. Constant adjustment and relubrication are necessary.

Heavy-duty, live, ball bearing centers or revolving centers, Figure 8–85, are now preferred for most work. These centers remove most of the worry of heat build-up when turning between centers. Usually, an oil hole is provided for lubricating the bearings of revolving centers. It is essential that lubrication be done, otherwise the bearings could fail. Revolving centers are by far the most commonly used tailstock centers in modern machining practice.

Another problem in turning between centers is the lathe dog that does the driving. It is always a potential hazard to the operator as it rotates. In addition to this, in order to turn the shaft full length, the shaft is turned as far as possible and then must be switched end-for-end to complete the other end.

The automatic driving center, with self-aligning driving discs, is one answer to these problems. It is set into the corresponding Morse taper in the lathe spindle, Figure 8–86. The center is spring loaded, Figure 8–87, which compensates for variations in the depths of center holes. The self-aligning driving discs adjust automatically to work that is not cut to an exact right angle. The automatic driving center also makes it possible for the work to be machined over the whole length without rechucking.

Turning Between Centers with a Spigot

Another method of turning between centers involves the use of a turned spigot held in a chuck, Figure 8–88. Typical work includes taking a light cut from the outside diameter of a bushing or a bearing. This cut must be concentric with the inside diameter

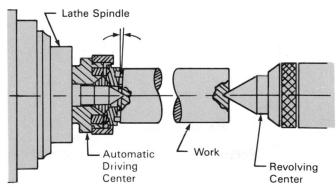

Figure 8–86 Automatic driving center *(Courtesy of KAR Industrial Ltd.)*

Figure 8–84 Lathe center *(Courtesy of The Cleveland Twist Drill Company)*

Figure 8–85 Revolving center *(Courtesy of DoALL Company)*

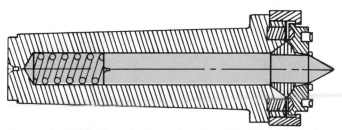

Figure 8–87 Sectional view of automatic driving center *(Courtesy of KAR Industrial Ltd.)*

of the bushing. Similar to the turned center, the spigot needs a shoulder, but instead of a 60-degree point, the compound is set at only 1-1/2 degrees. The resulting 3-degree taper should just enter the bore of the bushing. The taper in this case, like the taper shank of a drill, will drive the bushing or the bearing without having to use a lathe dog. Light cuts can be taken on the outside diameter, concentric to the bore of the bushing or the bearing. It is best to use a revolving center rather than a dead center for this type of work.

Facing the end of a shaft using a half-center allows the tool bit to get right into the center. This is a dead center, not a revolving center, Figure 8–89.

When too large a lathe dog is used in a driving plate, it may bind on the sides or the bottom of the slot, preventing the live center from seating properly, Figure 8–90. The work will not run true because of this.

Spiders and Catheads

Spiders and catheads have a very specific purpose when using center holes. A **spider** acts as a support for pipe or tubing, Figure 8–91. A **cathead** provides a support and bearing for a shaft in a steady rest.

To support pipe or tubing for facing or turning, it is necessary that you make a spider, Figure 8–92. It can have three or four legs, or more if necessary, with the end of each leg bent at a right angle. An alternative is to weld nuts at right angles to the end of the legs. The end of each leg is drilled and tapped as shown for a pointed setscrew. The setscrews are adjusted outward to press against the inside of the pipe or tubing. In this way, the spider is held in place and on center to provide a center-drilled hole for a lathe center.

Similarly, a cathead, Figure 8–93, provides a bearing surface for a steady rest. The cathead is adjusted centrally on the shaft, using the setscrews with pro-

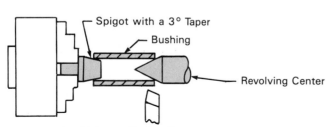

Figure 8–88 Using a "spigot"

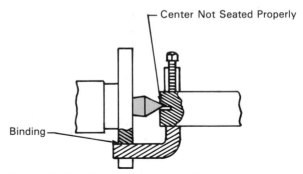

Figure 8–90 Using too large a lathe dog

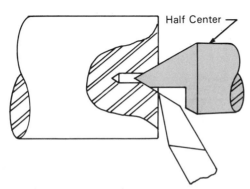

Figure 8–89 Using a half center

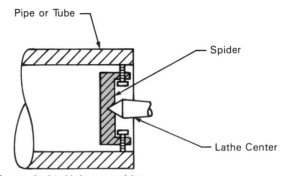

Figure 8–91 Using a spider

tecting pads if necessary. The shaft in question may be rough on the outside and, therefore, may not provide a good bearing surface for the steady rest. The shaft may also be accurately finished and polished. It may not be desirable to use the steady rest legs directly on the shaft because of this.

Mandrel

A **mandrel** is a round tool steel shaft that is hardened and tempered, and center drilled at both ends. Its purpose is to provide a pair of centers for a piece of work with a finished hole or bore that can be supported and revolved so that machining can be done.

Mandrels are made in different styles. One of the more common styles is a standard lathe mandrel as shown in Figure 8–94. This mandrel is slightly tapered along its length. In the figure, for example, a 1.000-in (25.4-mm) diameter is the median diameter at the center of the mandrel. The small end has a diameter of 0.999 in (25.375 mm). The large end has a diameter of 1.002 in (25.45 mm) in a total length of 7 in (178 mm). The size of the mandrel is marked on the large end.

Both ends have a turned shoulder less than the diameter of the mandrel, and both contain a slight flat. The smaller diameter of the shoulders permits free movement of the work on the tapered portion of the mandrel. The flats give maximum holding power to the lathe dog setscrew. It is good practice to use a soft pad of metal, such as copper, between the end of the setscrew and the flat to prevent damage to the mandrel. The small end of the mandrel should just be able to enter the finished hole in the work.

The mandrel is usually pressed into place with an arbor press, or it is sometimes driven into place with a rawhide mallet or a soft-faced hammer. A ball pein hammer should *never* be used for this purpose, since the end of the mandrel and the centers would be damaged. Machining should be done with all pressure exerted toward the large end of the mandrel. Always keep in mind that a mandrel is a precision tool and should be treated carefully.

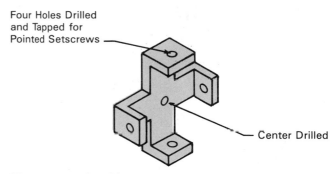

Four Holes Drilled and Tapped for Pointed Setscrews

Center Drilled

Figure 8–92 A spider

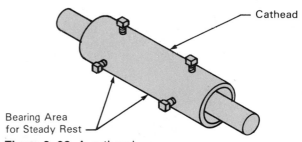

Cathead

Bearing Area for Steady Rest

Figure 8–93 A cathead

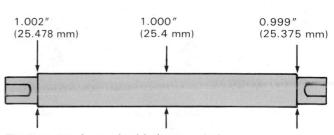

1.002″ (25.478 mm) 1.000″ (25.4 mm) 0.999″ (25.375 mm)

Figure 8–94 A standard lathe mandrel

A standard lathe mandrel is commonly used for turning gear blanks and cutting gear teeth. Figure 8–95 illustrates the setup used.

Another style of mandrel is shown in Figure 8–96. The work is mounted on a parallel shoulder with a spacer collar and washer, all held in position with a finished hexagon nut. The lathe setup is the same as that used with the standard lathe mandrel. More than one piece of work can be mounted on this style of mandrel, converting it into a gang mandrel. This mandrel can easily be made of cold rolled steel and case-hardened or tool steel that is hardened and tempered.

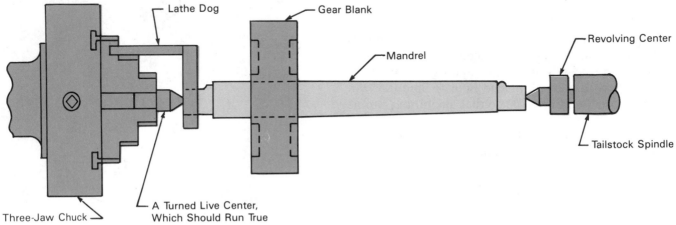

Figure 8–95 A standard lathe mandrel setup for turning a gear blank between centers

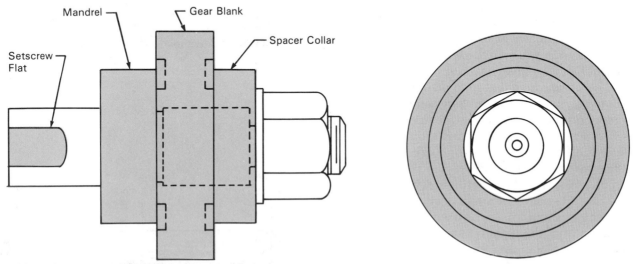

Figure 8–96 A homemade parallel-type mandrel

In both cases, a ground finish is best.

An adaptation of this style of mandrel has a tapered shank with either a Morse or a Brown and Sharpe taper, Figure 8–97.

An expansion mandrel, Figure 8–98, has a tapered shaft on which is mounted a cast-iron expansion sleeve. The work is placed on the sleeve which is then pressed or driven onto the tapered shaft. This expands the sleeve, which tightens within the bore of the work, holding it firmly.

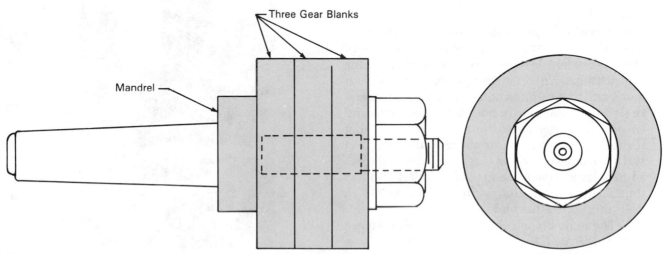

Figure 8–97 A taper-shank mandrel set up as a gang mandrel

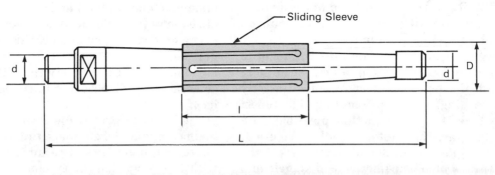

Concentricity 0.0006″

Figure 8–98 An expansion mandrel

ROBOTS – FUTURE TRENDS

The use of robots is expected to grow in manufacturing areas such as assembly, welding, metal processing, material handling, machine tending, applying adhesives and sealants, and thermal spraying. One of the most exciting innovations is the use of vision in robots. That is, robots will be able to see where they are going and what they are doing. Rudimentary vision systems are now being placed in selected machine applications. Vision systems in both two dimensions and three dimensions will play an increasingly important role in both robot guidance and inspection. Two-dimensional vision means that a robot can recognize an object and compare it with information in its memory. Three-dimensional vision means that a robot also has depth perception; that is, it has the ability to determine distance and depth.

The majority of robots now in operation are unsophisticated. In other words, they do not have special qualities such as vision or the sense of touch. They go about their work blindly. If something gets in their way, such as a human operator, the robot goes right on with its programmed movement. If the human is struck by the robot, the human can be hurt or killed. But if the robot has vision, it can be programmed to recognize a human and stop all movement rather than do damage by striking the human.

A robot vision system permits the robot to identify different objects quickly, accurately determine the position of each object (exactly where it is lying) and the orientation of the object (exactly how it is lying in that position), and inspect the parts for uniform quality, Figure 1. Special computer controlled cameras serve as the eyes of the robot. The cameras feed signals into an image processing system which is integrated with the robot's control system. The same teach pendant and method used for programming the

Figure 1 Robot vision — parts inspection *(Courtesy of ASEA Robotics Inc.)*

robot through its movements is also used to program the vision system. This makes all programming functions uniform and simple. The addition of vision reduces the need for specially designed peripheral equipment and provides flexibility for part changes. In some vision systems, cameras are mounted on the arm or wrist of the robot. The cameras may also be mounted on the fixture itself.

The image processing system contains two main components, an image preprocessor and a microcomputer. These two units are integrated with the robot control system. A robot vision system, at this time, is a gray scale system and works with 64 levels of gray. As a result, the system has increased capacity to work in an industrial environment with normal lighting conditions.

An adaptive control can be a part of the system. This is a software function that allows automatic adjustment of programmed robot movements within preprogrammed correction parameters. When sensors inform the system that there is a discrepancy between programmed movements and required movements, the robot's path is adjusted to follow the required movement. This change affects only the tool center point (TCP) position or the positioning speed, not the orientation of the robot. Adaptive control software reacts to sensor-transmitted electrical voltages to perform functions such as searching for a point, controlling the speed of the robot's path, or following a contour within preprogrammed correction parameters.

An example of an adaptive control is the seam finder for arc welding applications, Figures 2 and 3. The seam finder is a sensor system that enables the robot to determine quickly the actual position of a joint or workpiece. At the same time, the robot can select the correct welding parameters. This development increases the number of welding applications that can be automated and reduces the requirement for fixtures and line tolerances in the fabrication of the workpiece. The seam finding system consists of a laser-type sensor and a microcomputer which evaluates the sensor signals. The results are then transmitted to the adaptive control portion of the robot control system. In the search process, the joint is defined in three dimensions and the welding gun is positioned simultaneously. A complete search in three dimensions and location of the gun takes about 1.5 seconds. A search in two dimensions is often sufficient. The robot is programmed with information relating to the material thickness and the search type (edge/overlap joint or fillet joint, and so on). Correction of the welding parameters is performed automatically.

The seam finding system is primarily suited to the welding of thin steel sheets because of its rapidity and high degree of accuracy. Other parameters that can be evaluated by the system include the location of the elevation of a surface, the location of edges, and the location of overlap and fillet joints. The seam finder can also measure the gap in overlap joints. The robot can then be programmed to select the appropriate welding data and movement pattern for the gap measured. It is also possible to check that a workpiece is within predetermined dimensional limits.

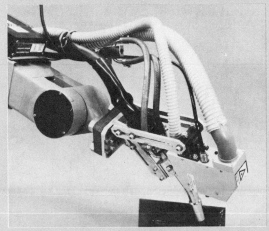

Figure 2 A seam finder *(Courtesy of ASEA Robotics Inc.)*

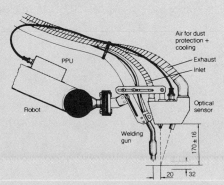

Figure 3 Schematic of a seam finder *(Courtesy of ASEA Robotics Inc.)*

REVIEW QUESTIONS

1. Compared with a turning tool bit, how is a boring tool bit held?
2. Give two important considerations for holding the cutter for conventional boring.
3. What is upside-down boring?
4. List two advantages of upside-down boring.
5. What is the purpose of a center drill?
6. Name two things that a properly shaped center hole will do.
7. What will happen to work with no center holes if you try to set that work between centers?
8. Make a cross-sectional sketch of a properly shaped center hole and say why it is properly shaped.
9. What would be the result if a center hole were drilled too deep?
10. In judging the size of a center drill, what diameter is used?
11. What would happen if a center hole were drilled too shallow?
12. Is it necessary to face the end of a shaft that has been center drilled?
13. On what diameter is the speed of a center drill calculated?
14. Describe four methods of locating centers on stock.
15. When setting up to center drill a shaft, how is it determined if the shaft is off center? How is the shaft centered?
16. What is the procedure for removing a broken center drill from the work?
17. Describe two methods for accurately aligning lathe centers.
18. What is the purpose of a turned "live" center?
19. What is the most commonly used tailstock center? Give one reason why this center is preferred.
20. When turning between centers, what is an advantage of using an automatic driving center?
21. What is the purpose of a spigot?
22. What is the reason for using a half center?
23. What is the possible result of using too large a lathe dog in a driving plate?
24. What is a spider and how is it used?
25. What is the purpose of a cathead?
26. Describe the purpose of a standard lathe mandrel.

UNIT 8-6

LATHE OPERATIONS – TAPERING

OBJECTIVES

After completing this unit, the student will be able to:

- describe the operation of tapering.
- measure tapers properly.
- define the terms "slow taper" and "fast taper."
- explain the differences between the Morse taper, Brown and Sharpe taper, and Jarno taper.
- set up and turn tapers using any one of three methods: compound tool rest method, tailstock set-over method, and using a taper attachment.

KEY TERMS

Tapering

Taper

Flow taper

Fast taper

Self-holding taper

Shallow taper

Steep taper

Morse taper

Brown and Sharpe taper

Jarno taper

Taper pin

Compound tool rest method of taper turning

Tailstock set-over method of taper turning

Offset

Taper attachment

TAPERING

Tapering is a basic lathe operation. It is the machining operation where the work is revolved and excess material is removed either from the outside diameter, Figure 8–99A, or from the inside diameter, Figure 8–99B, by a tool bit moving at an angle to the centerline of the work. In all methods of taper turning, the cutting edge should be set on center. If the cutting edge is not set on center, it is on a different plane. A bellied taper can be the result, and the wrong taper per foot can be cut.

TAPERS

A **taper**, Figure 8–100, is a difference in diameters in a unit of length. For example, if the large diameter is 1 inch, the small diameter is 1/2 inch, and the length is 12 inches, then the taper is 1/2 inch in 12 inches, or 1 foot.

A taper is measured as a ratio in metric measure — 1 mm per unit of length; e.g., 1:20 would mean a taper of 1 mm in 20 mm of length. In inch measure, taper is measured in inches per inch of length or in inches per foot of length; i.e., taper per inch or taper per foot.

A "slow" taper has a shallow angle. The taper shank of a drill is called a slow taper. It is also referred to as a self-holding taper.

A "fast" taper has a steep angle. A milling machine taper is called a fast taper, or sometimes a steep taper, at 3-1/2 in per ft (1:3.5 in metric). It is not a self-holding taper, but it does center the arbor exactly in the center of the milling machine spindle.

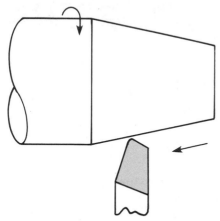

Figure 8–99A An external taper

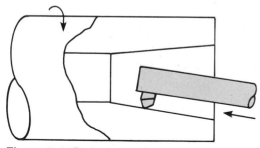

Figure 8–99B An internal taper

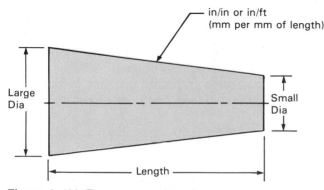

in/in or in/ft
(mm per mm of length)

Large Dia

Small Dia

Length

Figure 8–100 Taper nomenclature

Standard Tapers

Standard tapers are those tapers that are made to specific standard sizes. The following tapers are just a few of those available.

The Morse taper is one of the most common tapers in the machine shop. It is used on twist drills, drill press spindles, lathes, etc. From #0 (the smallest) to #7 (the largest), each of the 8 tapers is slightly different, ranging from 0.598 in per ft to 0.631 in per ft, or approximately 5/8 in per ft. In metric measure, the Morse taper would be between 1:19 and 1:20; i.e., between 1 mm of taper in 19 mm of length and 1 mm of taper in 20 mm of length.

The Brown and Sharpe taper was once commonly used on milling machines, milling machine collets, end mills, reamers, etc. The taper is approximately 1/2 in per ft. In metric measure, this would be a ratio of 1:24.

The Jarno taper is the least used taper system. Each taper, of which there are 19, has a taper of 0.600 in per ft, or approximately 1:19 in metric measure. It is used on such machine tools as profiling machines and die sinking machines. It has also been used for the headstock and tailstock of some lathes.

Taper pins, Figure 8–101, are used to hold parts together. For example, a collar on a shaft can be held by a taper pin that passes through the collar and shaft at right angles to the centerline. There are 17 different sizes of taper, all with a taper of 1/4 in per ft. The metric ratio would be 1:48. Diameters range from 1/16 in (1.6 mm) at the large end for the #7/10 taper pin to 0.706 in (18 mm) at the large end for a #10 taper pin.

American National pipe threads are tapered at 3/4 in per ft (or approximately 1:16 in metric). As a pipe fitting is attached to a pipe, it tightens and seals the joint.

TAPER TURNING METHODS

In lathework, there are three methods used for cutting tapers. They are the compound tool rest method, the tailstock set-over or offset method, and cutting a taper by using a taper attachment.

In the compound tool rest method, the compound is set at half the taper angle, Figure 8–102. The tool bit is moved by hand feed. Only the compound moves, not the carriage. The carriage is locked.

In the tailstock set-over method, when the tailstock is set exactly in line with the headstock, Figure 8–103, the work is set for parallel turning. For taper turning, the tailstock is set off center and the power feed can be used, Figure 8–104.

When the length of the work varies, the amount of taper will be different if the same offset is used, Figure 8–105. In calculating the offset, length is the total length of the stock, not just the length of the taper, Figure 8–106.

To calculate the amount of offset:

$$\text{Offset} = \frac{T\,L}{2}$$

where

T = taper per inch
L = total length of the work in inches

When the taper attachment is used, Figure 8–107, it is set at the required angle and then is clamped solidly to the ways of the lathe so that only the slide will move. As the tool bit moves to the left, as in this case, the slide moves along the taper attachment and forces the cross slide to move outward, causing the tool bit to move at the angle set by the taper attachment. Power feed can be used. The taper attachment is the same slope as the taper being cut.

Morse tapers and Brown and Sharpe tapers are compared in both inch and millimeter sizes in Table 8–3.

Table 8–4 compares taper per foot and metric tapers to angles in degrees.

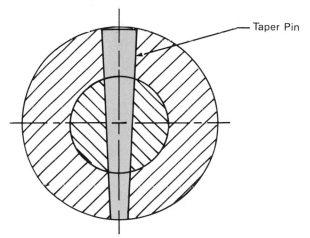

Figure 8–101 A taper pin holding two shafts together

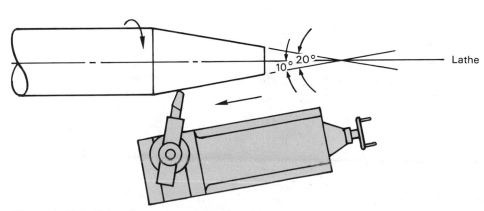

Figure 8–102 Using the compound tool rest

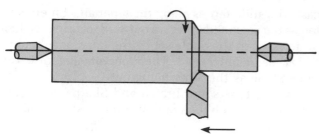

Figure 8–103 Job set for parallel turning

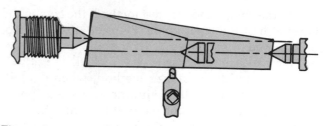

Figure 8–105 Length of stock in taper turning

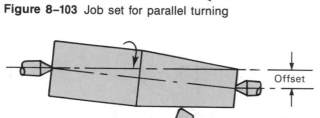

Figure 8–104 Job set for taper turning

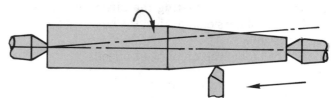

Figure 8–106 Length in taper turning is the total length of stock

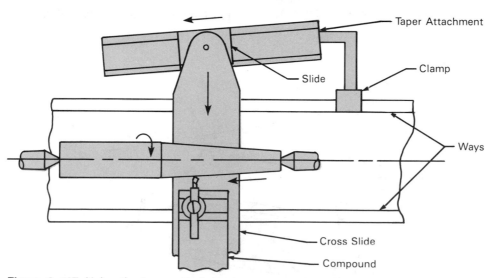

Figure 8–107 Using the taper attachment

Table 8-3   A Comparison of the Morse Taper and the Brown and Sharpe Taper

No. of Taper	TAPER		
	in/ft	mm/mm	mm/100mm
MORSE TAPERS			
0	0.625	1 : 19	5.21
1	0.599	1 : 20	4.99
2	0.599	1 : 20	4.99
3	0.602	1 : 20	5.02
4	0.623	1 : 19	5:19
5	0.631	1 : 19	5.26
6	0.626	1 : 19	5.22
7	0.624	1 : 19	5.20

No. of Taper	TAPER		
	in/ft	mm/mm	mm/100mm
BROWN AND SHARPE TAPERS			
1	0.502	1 : 24	4.18
2	0.502	1 : 24	4.18
3	0.502	1 : 24	4.18
4	0.502	1 : 24	4.18
5	0.502	1 : 24	4.18
6	0.503	1 : 24	4.19
7	0.501	1 : 24	4.18
8	0.501	1 : 24	4.18
9	0.501	1 : 24	4.18
10	0.516	1 : 23	4.30
11	0.501	1 : 24	4.18
12	0.500	1 : 24	4.17
13	0.500	1 : 24	4.17
14	0.500	1 : 24	4.17
15	0.500	1 : 24	4.17
16	0.500	1 : 24	4.17
17	0.500	1 : 24	4.17
18	0.500	1 : 24	4.17

Table 8-4 Comparing Taper per Foot and Millimeters per Millimeter of Length to Angles in Degrees *(Courtesy of South Bend Lathe Inc.)*

		TAPERS AND ANGLES							
mm PER mm OF LENGTH	TAPER PER FOOT	INCLUDED ANGLE			ANGLE WITH CENTERLINE			TAPER PER INCH	TAPER PER INCH FROM CENTERLINE
		Deg	Min	Sec	Deg	Min	Sec		
1 : 96	1/8	0	35	47	0	17	54	0.010416	0.005208
1 : 64	3/16	0	53	44	0	26	52	0.015625	0.007812
1 : 48	1/4	1	11	38	0	35	49	0.020833	0.010416
1 : 38	5/16	1	29	31	0	44	46	0.026042	0.013021
1 : 32	3/8	1	47	25	0	53	42	0.031250	0.015625
1 : 27	7/16	2	5	18	1	2	39	0.036458	0.018229
1 : 24	1/2	2	23	12	1	11	36	0.041667	0.020833
1 : 21	9/15	2	41	7	1	20	34	0.046875	0.023438
1 : 19	5/8	2	59	3	1	29	31	0.052084	0.026042
1 : 17	11/15	3	16	56	1	38	28	0.057292	0.028646
1 : 16	3/4	3	34	48	1	47	24	0.062500	0.031250
1 : 15	13/16	3	52	42	1	56	21	0.067708	0.033854
1 : 14	7/8	4	10	32	2	5	16	0.072917	0.036456
1 : 13	15/16	4	28	26	2	14	13	0.078125	0.039063
1 : 12	1	4	46	19	2	23	10	0.083330	0.041667
1 : 10	1 1/4	5	57	45	2	58	53	0.104166	0.052084
1 : 8	1 1/2	7	9	10	3	34	35	0.125000	0.062500
1 : 7	1 3/4	8	20	28	4	10	14	0.145833	0.072917
1 : 6	2	9	31	37	4	45	49	0.166666	0.083332
1 : 5	2 1/2	11	53	38	5	56	49	0.208333	0.104166
1 : 4	3	14	2	0	7	1	0	0.250000	0.125000
1 : 3.4	3 1/2	16	35	39	8	17	49	0.291666	0.145833
1 : 3	4	18	55	31	9	27	44	0.333333	0.166666
1 : 2.6	4 1/2	21	14	20	10	37	10	0.375000	0.187500
1 : 2.4	5	23	32	12	11	46	6	0.416666	0.208333
1 : 2	6	28	4	20	14	2	10	0.500000	0.250000

REVIEW QUESTIONS

1. Describe the basic lathe operation of tapering.
2. What is a taper?
3. In all methods of taper turning, what should the height of the cutting edge be? Why should it be this height?
4. How is taper measured?
5. What is a slow taper?
6. What is a fast taper?
7. What is a standard taper?
8. List the three most common tapers and give the sizes available for each taper.
9. What is the purpose of a taper pin? What is the taper in inch and metric sizes?
10. In the compound tool rest method of cutting tapers, at what angle is the compound set?
11. In the tailstock set-over method of cutting tapers, how is the amount of offset calculated?
12. What length is used in calculating the amount of offset?
13. When the taper attachment is used, how is it set on the lathe and at what angle?

UNIT 8-7

LATHE OPERATIONS – THREADING

OBJECTIVES

After completing this unit, the student will be able to:

- define the term threading.
- read the different screw thread representations on drawings.
- explain the differences between a single screw thread, a double start thread, and a triple start thread.
- measure screw threads accurately using fixed gages and a thread micrometer.
- demonstrate the three-wire method of measuring pitch diameter.
- identify the most common screw thread forms: Unified form, International metric (SI) thread, Acme thread, buttress thread.

KEY TERMS

Threading
Helical groove
Straight helix
Tapered helix
Screw thread
External thread
Internal thread
Left-hand screw
  thread
Right-hand screw
  thread
Helix angle
Single screw thread
Double start thread
Triple start thread
Pitch

Major diameter
Minor diameter
Pitch diameter
Plug thread gage
Ring thread gage
Thread micrometer
Three-wire method
American National
  thread form
International metric
  (SI) thread form
Power (translation)
  thread
Acme thread
Buttress thread
Unified screw thread

Lead
Threads per inch (tpi)

ISO thread form

THREADING

Cutting a screw thread is a basic lathe operation. It is one of the most important operations in the machine shop. Threaded screws are used to transmit motion and power, and threaded fasteners of all types are used to hold machinery together.

Threading is the machining operation where a groove of specific shape or form is cut helically on the outside of a diameter or on the inside of a hole. The screw thread is in the form of a straight helix (a straight screw thread) or a tapered helix (a tapered screw thread). Figure 8–108 illustrates a modern high-speed threading operation using a carbide threading tool. Figure 8–109 shows the form of a National Form screw thread.

SCREW THREADS

A screw thread cut on the outside of a cylinder is called an external thread. A screw thread cut on the inside of a cylindrical hole is called an internal thread, Figure 8–110.

Screw threads are represented in different ways on a drawing, Figure 8–111. A right-hand screw thread revolves to the right as it advances, Figure 8–112. A left-hand screw thread revolves to the left as it advances, Figure 8–113.

The parts of a screw thread are detailed in Figure 8–114. The difference in minor diameters of a screw and matching nut are compared in Figure 8–115.

Single and Multiple Threads

In each of the three cases in Figure 8–116, note the difference in the angle of the sloped line. The angle formed between this sloped line and a line perpendicular to the axis of the thread is called the **helix angle** or **lead angle**.

A single screw thread has only one starting point at the end of the thread. A double start thread has two separate threads side-by-side with two equally spaced starting points at the end of the thread. A triple start thread has three separate threads side-by-side with three equally spaced starting points at the end of the thread.

Pitch is the distance from a point on a thread to the corresponding point on the adjacent thread. **Lead** is the distance a point on a thread will move or advance in one revolution of the thread.

In each of the cases in Figure 8–117, note the difference in the pitch and the lead. Note also the difference in the helix angle or lead angle.

Measuring Screw Threads

Figures 8–118 and 8–119 show various methods of measuring the number of threads per inch or the pitch in millimeters. When cutting threads on a lathe or when using any other methods for cutting threads, knowing threads per inch or the pitch in millimeters is absolutely necessary.

Figure 8–108 A modern, high-speed, threading operation being performed using a carbide cutter *(Courtesy of Kennametal Inc.)*

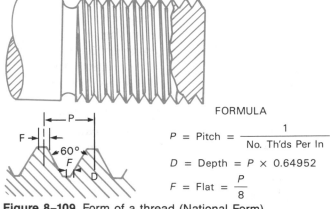

AMERICAN NATIONAL SCREW THREAD
(Formerly U.S. Standard Screw Thread)

FORMULA

$$P = \text{Pitch} = \frac{1}{\text{No. Th'ds Per In}}$$

$$D = \text{Depth} = P \times 0.64952$$

$$F = \text{Flat} = \frac{P}{8}$$

Figure 8–109 Form of a thread (National Form)

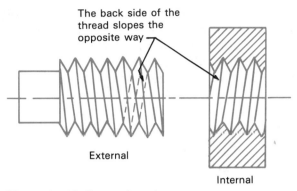

External

Internal

Figure 8–110 Screw threads

As has been stated, pitch is the distance from the center of a thread to the center of the adjacent thread (the thread beside it). Do not confuse pitch with threads per inch (tpi). If a 1-inch UNC thread has 8 threads per inch, the pitch is 1/8 inch. Pitch is the same for a single start, a double start, a triple start, or a quadruple start thread.

$$P = \frac{1}{N}$$

where

P = pitch
N = number of threads per inch

Figure 8–111 Thread representations

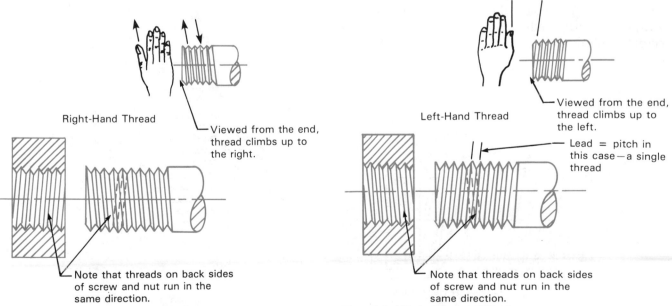

Right-Hand Thread

Viewed from the end, thread climbs up to the right.

Note that threads on back sides of screw and nut run in the same direction.

Figure 8–112 Right-hand thread

Left-Hand Thread

Viewed from the end, thread climbs up to the left.

Lead = pitch in this case—a single thread

Note that threads on back sides of screw and nut run in the same direction.

Figure 8–113 Left-hand thread

Lead is the distance a nut will travel in one revolution. Only in a single start thread does lead equal pitch. In a double start thread, lead = 2 pitches; in a triple start thread, lead = 3 pitches.

Other important measurements in screw threads are the diameters: the outside diameter (more commonly called the **major diameter**); the **minor diameter** (or root diameter); and most important, the **pitch diameter**. The pitch diameter is the point at which the width of the groove equals half the pitch.

One common method for attaining the correct pitch diameter is by making use of fixed gages; a plug thread gage for testing internal threads, and a ring thread gage for testing external threads. For average work, these are usually accurate enough.

A more accurate method for measuring pitch diameters involves the use of a thread micrometer that has a double, 60-degree, V-groove in the anvil and a single, 60-degree point on the spindle. These micrometers are made to measure a range of threads; e.g., 7–9 tpi, 10–13 tpi, 14–18 tpi, etc. Because of the range of threads, a slight error is involved due to the differing lead angles. The best procedure to use, in this case, is to measure the pitch diameter of a plug thread gage, compare it to that of the workpiece, and then add any possible error to the micrometer reading to get the actual pitch diameter.

The most accurate method for measuring the pitch diameter of precision screw threads is the three-wire method, Figure 8–120. A micrometer and three wires or pins of equal diameter are used. Two of the wires are in contact with the thread on one side and the third wire is in contact on the opposite side.

Standard terms for three-wire measurement are given by the Bureau of Standards. Some of these are:

M — measurement over the wires
W — wire or pin diameter
E — pitch diameter
D — major (or OD) of the thread
P — pitch of the thread
N — number of threads per inch.

The degree of accuracy of the three-wire method depends upon three things: the accuracy of the wires (both the diameter and the straightness), the accuracy of the measuring instrument, and the pressure exerted on the wires by the measuring instrument.

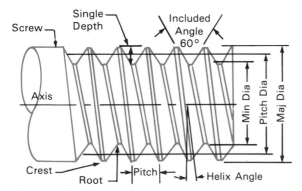

Figure 8–114 Screw thread parts

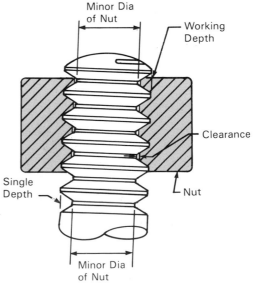

Figure 8–115 Comparison between minor diameters of screw and nut

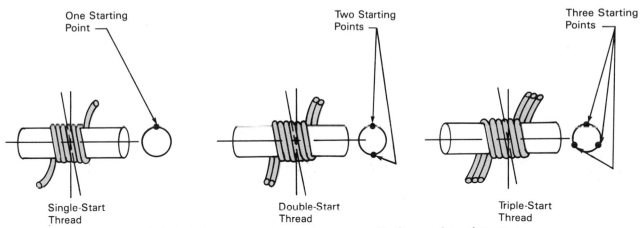

One Starting Point

Two Starting Points

Three Starting Points

Single-Start Thread

Double-Start Thread

Triple-Start Thread

Figure 8–116 A method of showing how the lead angle increases with the number of starts

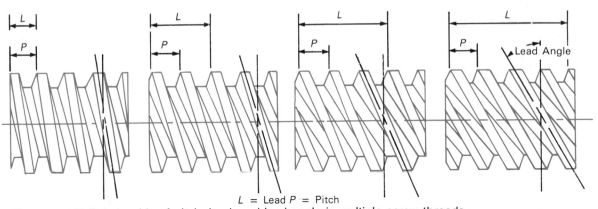

L

P

L

P

L

P

L

P

Lead Angle

L = Lead P = Pitch

Figure 8–117 Relationship of pitch, lead, and lead angle in multiple screw threads

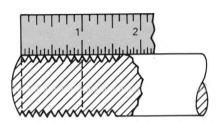

Figure 8–118 Measuring screw threads with a scale *(Courtesy of The L. S. Starrett Company)*

Figure 8–119 Measuring screw threads using a screw pitch gage *(Courtesy of The L. S. Starrett Company)*

If the degree of accuracy is to be within 0.0001 inch, the diameter of the wire must be known to within 0.00002 inch. In addition, the three wires must be the same diameter to within 0.00002 inch. The wires should also be straight to within 0.00002 inch, and they should be finished, polished, and hardened.

The so-called best wire sizes to use in the three-wire method are those that make contact at the pitch line of the thread. Approximate formulas for wire diameters for American Standard Thread Forms follow:

Smallest wire size = 0.56P
Best wire size   = 0.57737P
Largest wire size = 0.90P

The standard approximate formula for calculating the measurement over the three wires is:

$$M = E - 0.86603P + 3W \text{ or } E = M + 0.86603P - 3W$$

This formula does not compensate for differences in the lead angle, but it is accurate enough for most measurements of 60-degree, single-start threads.

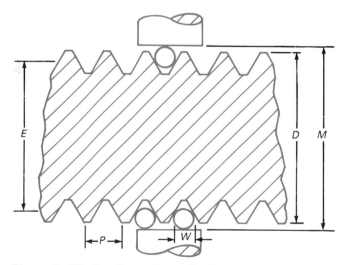

Figure 8–120 The three-wire method

If major diameter D is used in the formula, then the formula is:

$$M = D - (1.5155P) + 3W$$

EXAMPLE: Calculate M and E for a 1-inch, UNC 8, screw thread.

$$
\begin{aligned}
\text{Best size wire} &= 0.57735P \\
&= (0.57735)(0.125) \\
&= 0.072 \text{ in} \\
M &= D - (1.5155P) + 3W \\
&= 1 - 0.189 + 0.216 \\
&= 1.027 \text{ in} \\
E &= M + 0.86603P - 3W \\
&= 1.027 + 0.108 - 0.216 \\
&= 0.919 \text{ in}
\end{aligned}
$$

It should be pointed out at this time that a screw thread may measure correctly but still may be unacceptable. This may be due to the fact that the screw thread form is not correct. There are a number of causes of incorrect thread form. The threading tool may have worn, changing its original shape. The threading tool may have moved during the cutting action. The initial shape of the cutting tool or the setup itself may be inaccurate. In cutting screw threads, each of these points must receive careful attention.

SCREW THREAD FORMS

The threads used most in industry have symmetrical sides that are inclined at equal angles with a vertical centerline through the thread apex. The Sharp-V thread, Figure 8–121, is one of the early forms and is now used only occasionally. The American National Form, Figure 8–122 (external and internal) can have a rounded or flat root. The exact form of rounding is not specified since, in practice, this varies according to tool wear. Whitworth threads, Figure 8–123, are the British coarse thread series. Since the Unified thread is now standard, the Whitworth form is expected to be used only for replacements or spare parts. International metric threads (SI), Figure 8–124, are similar in form to American Standard thread except that the depth is greater. A rounded root profile is recommended.

<h2 style="text-align:center">POWER THREADS
(TRANSLATION THREADS)</h2>

The threads in Figures 8–125 to 8–128 are used specifically for transmitting power or moving machine parts against heavy loads. The square thread, Figure 8–125, is the strongest and most efficient of these threads. It is also the most difficult to cut because of the parallel sides. It has been replaced for the most part by the Acme Thread. The modified square thread, Figure 8–126, with an included angle of 10 degrees, is the practical equivalent of the square thread. It can be produced economically. The Acme thread, Figure 8–127, has none of the disadvantages of the square thread and

is only slightly weaker and less efficient. It is the most common power thread. Buttress threads, Figure 8–128, transmit power in one direction only because of their nonsymmetrical form. This form combines the high efficiency and strength of the square thread with the ease of cutting and adjustment of the Acme thread. Figure 8–128A is a common form with a perpendicular face. The thread in Figure 8–128B, with the face at 1 5 degrees, presents less cutter interference in milling the thread. The thread in Figure 8–128C illustrates a buttress thread that is well-known in Germany and Italy.

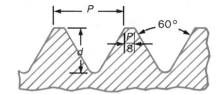

Figure 8–124 International metric (SI) thread

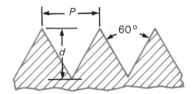

Figure 8–121 Sharp-V thread

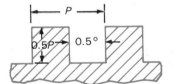

Figure 8–125 Square thread

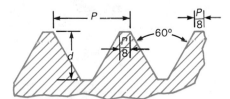

Figure 8–122 National Form thread

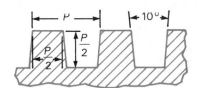

Figure 8–126 Modified square thread

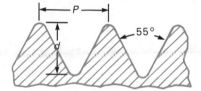

Figure 8–123 Whitworth thread

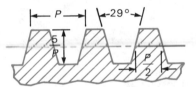

Figure 8–127 Acme thread

UNIFIED SCREW THREADS

The American National Standard for Unified Screw Threads includes certain modifications of the former standard. The basic profile is identical for both UN (Unified) and UNR (Unified Revised) screw threads, Figure 8–129.

In American practice for the UN external thread, Figure 8–130A, a flat root contour is specified with an optional rounded root since there will be threading tool wear. The UNR external thread, Figure 8–130B, has a rounded root, which provides greater fatigue strength, greater root clearance, easier assembly, and longer wear of threading tools. The UN and UNR external threads have flat crests, but in practice, production threads have partially or completely rounded crests as an option. The UN internal thread, Figure 8–130C, has a rounded root, which allows for some threading tool crest wear. There is no internal UNR screw thread.

METRIC THREADING

The Systeme Internationale (SI) thread, Figure 8–131, was adopted at the International Congress for the standardization of screw threads in Zurich in 1898. The proposal for metric threads at the present time is that a single worldwide system for mechanical fasteners should be developed. The International Standards Organization (ISO), which is based in Geneva, is the principal agency for the development of international fastener standards.

On this continent, the American National Standards Institute (ANSI), the Society of Automotive Engineers (SAE), the American Society for Testing Materials (ASTM), and the Canadian Standards Association (CSA) are all working with the Industrial Fasteners Institute (IFI) in the development of stan-

A. $d = \dfrac{P}{75}$

B. $d = 0.69\,P$

C. $d = 0.867P$
$f = 0.263P$

Figure 8–128 Buttress threads

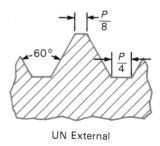

UN External

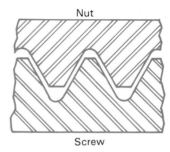

Nut

Screw

Unified

Figure 8–129 Unified threads

dards. Some of the suggestions that have been made by the IFI are:

1. A single series of 25 standard diameter/pitch combinations ranging from 1.6-mm to 100-mm diameter should be used. As a contrast, there are, at present, 34 standard nonmetric thread sizes with 57 diameter/pitch combinations with coarse and fine threads.
2. New identification markings should be used.
3. A new series of hexagon bolts and cap screws that feature reduced material content in the heads of some sizes should be used.
4. A new series of hexagon nuts should be designed to provide a higher degree of resistance to stripping.
5. Standard head types of tapping screws should be reduced from 12 to 5; i.e., flat, oval, pan, hexagon, and hexagon washer heads.
6. Metric slotted and recessed head machine screws should be developed with head dimensions identical to those of tapping screws and with a common 90-degree head angle instead of the present 82 degrees.

The principal metric systems used are the SI, German, British, French, Swiss, and Japanese. The differences are mainly in pitches of various metric diameter threads and differences in tolerances, depending on the use to which the various threads are put; e.g., watch making, instrumentation, general fasteners, etc.

It is generally thought that the American and Canadian governments will eventually use the ISO thread profile, Figure 8–132, which is identical to the Unified thread profile. The ISO thread forms for bolts and nuts are detailed in Figure 8–133. Table 8–5 lists the proposed 25 basic ISO screw threads.

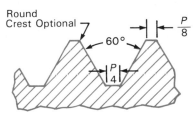

A. UN External

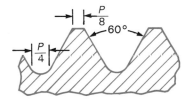

B. UNR External

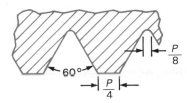

C. UN Internal

Figure 8–130 Unified threads

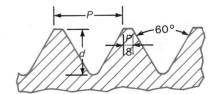

Figure 8–131 International metric (SI) thread

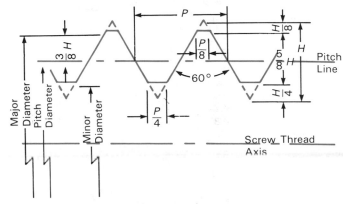

Figure 8–132 Basic ISO thread form

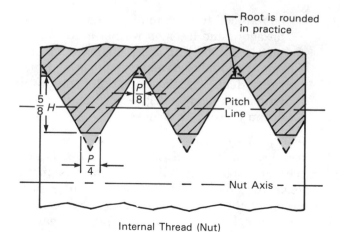

Internal Thread (Nut)

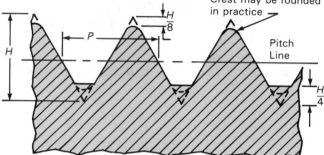

Figure 8–133 ISO thread forms

1. What is a screw thread?
2. What is the difference between a right-hand and a left-hand screw thread?
3. Define the helix angle of a screw thread.
4. What is the pitch of a screw thread?
5. What is the lead of a screw thread?
6. What is the relationship between pitch and lead in single and multiple screw threads?
7. Describe the three-wire method for measuring the pitch diameter.
8. List three factors which determine the accuracy of measurement of the three-wire method.
9. List three conditions that can lead to an incorrect screw thread form.
10. Name three power (translation) threads and state which one is the most commonly used currently.
11. How does the Unified screw thread differ from the American Standard thread?
12. How does the SI metric screw thread differ from the ISO metric screw thread?
13. What metric systems are used most today?
14. What currently used thread profile is identical to the proposed ISO thread profile?

Table 8–5 Proposed 25 Basic ISO Screw Threads

NOMINAL DIA (mm)	THREAD PITCH (mm)	NOMINAL DIA (mm)	THREAD PITCH (mm)
1.6	0.35	20	2.5
2	0.4	24	3
2.5	0.45	30	3.5
3	0.5	36	4
3.5	0.6	42	4.5
4	0.7	48	5
5	0.8	56	5.5
6.0	1.0	64	6
8	1.25	72	6
10	1.5	80	6
12	1.75	90	6
14	2	100	6
16	2		

UNIT 8-8

THREADING TOOLS – COMMERCIAL

OBJECTIVES

After completing this unit, the student will be able to:

- set the threading tool correctly.
- set the compound correctly.

KEY TERMS

rigid threading tool holder
spring-type threading tool holder
setting the compound

One of the commercial tools available for cutting screw threads is shown in Figure 8–134. Two of the more common holders are the standard rigid style and the spring holder style.

SETTING THE THREADING TOOL

If the top of the threading tool is sloped, Figure 8–135B, rather than sitting flat, Figure 8–135A, the wrong included angle is presented to the work. It is less than the intended 60-degree included angle. Some machinists say that this isn't enough of a factor to worry about in general-purpose threading, but in precision threading it is a factor to consider. A center gage, Figure 8–136A, is used for grinding a threading tool to shape and for setting the threading tool square to the work for threading. A 29-degree Acme thread gage, Figure 8–136B, is used in the same way for Acme threading tools. In all cases, threading tools are set perpendicular to the centerline of the work when cutting straight threads or tapered threads, Figure 8–137. The boring tool should be set square with the work and on center for cutting internal threads, Figure 8–138.

SETTING THE COMPOUND: CUTTING A 60-DEGREE THREAD

It is important to set the compound so that the feed moves in the direction of the cut; i.e., in the direction of carriage travel, Figures 8–139 to 8–142.

When the compound is set at 30 degrees, Figure 8–143, the threading tool progresses down the back side of the thread on each successive cut. Cutting is done on the front side only, leaving the back side of the thread rough. The last one or two cuts should be fed straight in. This causes a scraping action on both sides of the thread, which cleans the back side of the thread as well as the front side.

When the compound is set at 29 degrees, Figure 8–144, the threading tool progresses down a 29-degree line. It fills the full 60-degree space, as it does with the 30-degree setting. As the cutter progresses downward on each successive cut, a slight scraping action occurs on the back side of the thread, the shaded portion in the figure, to automatically clean the back side of the thread. Feeding straight in on the final cuts is unnecessary with this setting.

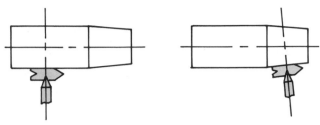

A. Correct   B. Incorrect

Figure 8–137 Setting the threading tool to cut a tapered thread

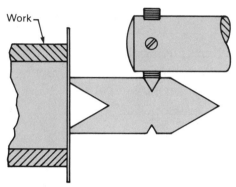

Figure 8–138 Setting the boring tool for internal thread cutting

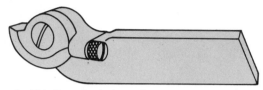

Figure 8–134 Standard threading tool holder, rigid style *(Courtesy of Armstrong Brothers Tool Company)*

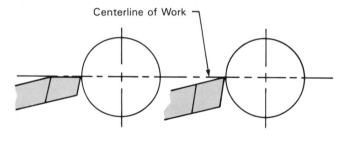

A. Correct   B. Incorrect

Figure 8–135 The slope of the top of the threading tool is correct at *A* and incorrect at *B*

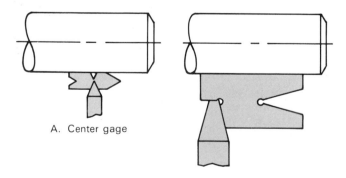

A. Center gage

B. 29° Acme thread gage

Figure 8–136 Setting the threading tool with (A) a center gage and (B) a 29° Acme thread gage

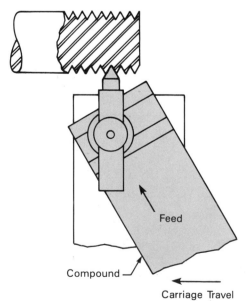

Figure 8–139 Right-hand external thread

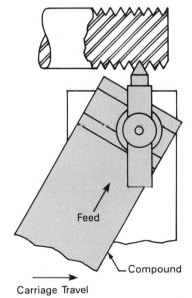

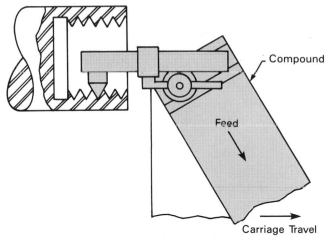

Figure 8–140 Left-hand external thread

Figure 8–142 Left-hand internal thread

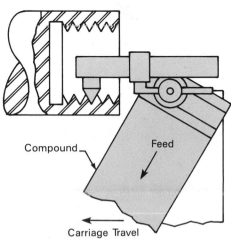

Figure 8–141 Right-hand internal thread

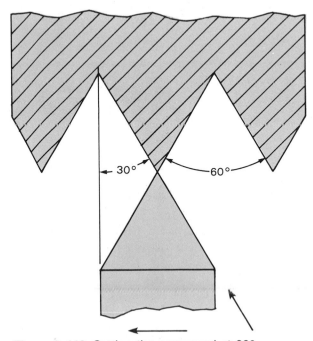

Figure 8–143 Setting the compound at 30°

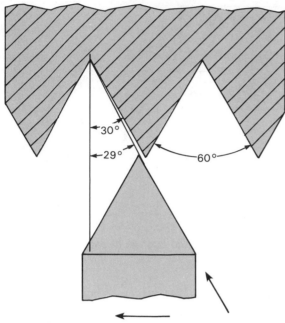

Figure 8–144 Setting the compound at 29°

REVIEW QUESTIONS

1. Name two common holders for threading tools.
2. In cutting straight threads or tapered threads, how is the threading tool set with regard to the centerline of the work?
3. For what purpose is a center gage used?
4. What is an Acme thread gage and for what purpose is it used?
5. Using a simple line diagram (centerlines only), show how the compound is set in relation to the feed and direction of travel for a right-hand external 60-degree screw thread.
6. In setting the compound for cutting a 60-degree thread, what provision is made for machining the back side of the thread?
7. List two advantages of setting the compound at 29° for cutting a 60-degree thread.

UNIT 8-9

THREAD CUTTING

OBJECTIVES

After completing this unit, the student will be able to:
- set the lathe properly for cutting screw threads.
- perform the thread cutting routine for threading to a neck or relief groove and for threading when there is no neck or groove.
- use the threading dial.
- pick up a thread after an interrupted threading cut.
- properly grind a threading tool bit.
- properly grind an Acme threading tool.
- make a coil spring.
- use the proper procedure to cut a metric thread.

KEY TERMS

Thread cutting	Cutting ductile metal
Threading to a neck	Power thread
Threading with no	Acme thread
neck	Negative back rake
Threading dial	Coil spring
Even-geared lathe	Cutting a metric
Picking up a thread	thread
Zero back rake	Cutting a tapered
Positive back rake	thread

THREAD CUTTING

In setting the lathe to cut screw threads, the following is an accepted procedure:

1. Set the lathe to cut the required number of threads per inch; i.e., set the lead of the thread.

2. Set the carriage to move in the right direction. With all right-hand threads, external and internal, the carriage moves toward the headstock, or to the left. With all left-hand threads, external and internal, the carriage moves toward the tailstock, or to the right.

3. Set the rpm. Threading is done at speeds about 1/4 to 1/3 of the speed of turning, or at a speed that the operator can be comfortable with. Some operators are much quicker in reaction time than others. With a coarse lead, for example, the carriage will reach the end of the thread very quickly unless a long thread is being cut. It is best to start with a slow speed, say 100 rpm, and increase the speed as the operator adjusts to the threading routine. Another important factor in setting speed is whether you are threading to a shoulder or a neck. When threading to a shoulder, the operator must move very quickly to stop the carriage and at the same time pull out the threading tool. See the thread cutting routines that follow.

4. Set the compound in the proper direction and at the proper angle before setting the threading cutter.

5. In all cases, set the threading tool perpendicular to the centerline of the work, using the center gage in the case of 60-degree or 55-degree threads or the 29-degree screw thread gage in the case of an Acme thread.

Note that in all cases the height of the cutting edge is on center, Figure 8–145. This is only a suggested sequence of steps for setting the lathe, except for setting the compound. The compound must be set before the threading tool is set in position. In all cases, the lathe must actually be set to cut the lead of a thread. Only in the case of a single start thread does the pitch equal the lead.

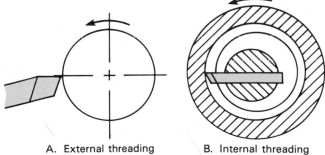

A. External threading   B. Internal threading

Figure 8–145 The height of the cutting edge is on center

THE THREAD CUTTING ROUTINE

Two situations are possible when cutting screw threads: (1) threading to a neck or groove, Figure 8–146A, and (2) not having a neck or groove to thread to, Figure 8–146B.

To thread to a neck or relief groove:

1. Feed the threading tool straight in until it just touches the revolving work.

2. Set the cross slide dial to zero.

3. Feed in, using the compound, about 0.010 in (0.25 mm) for the first trial cut.

4. Close the split nut. When the threading tool reaches the neck, open the split nut. Check the pitch in threads per inch or millimeters.

5. Using the cross slide, pull the threading tool back clear of the work.

6. Return the threading tool to the beginning of the thread.

7. Return the cross slide to the setting of zero.

8. Feed in again, using the compound. The amount of feed for each successive cut is up to the operator, but each succeeding cut should become smaller, with the final 1 to 3 cuts at about 0.0005 in (0.013 mm) to 0.0015 in (0.038 mm) in depth.

9. Take your second cut.

10. Repeat the routine until the thread is cut, fitting the thread to a gage or nut.

If you do not have a neck or groove to thread to, the only difference is that as the threading tool reaches the end of the cut, it must be pulled clear of the work. At the same time, the split nut must be opened to stop the carriage for each cut. The rest of the routine is the same as above.

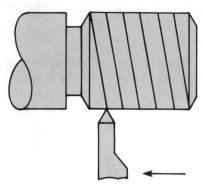

Figure 8–146A Threading to a neck

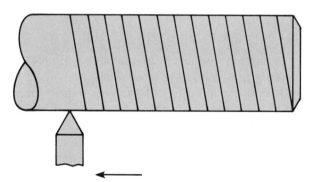

Figure 8–146B Threading without a neck

USING THE THREADING DIAL

In threading, the lead screw, the split nut, the threading lever, and the threading dial, Figure 8–147, are used. The lead screw revolves; the threading lever closes the split nut around the lead screw; the split nut moves along the lead screw taking the carriage with it; and when the threading lever opens the split nut, the carriage stops moving.

When the split nut is open, the carriage does not move. The threading dial, which is driven by the lead screw, revolves past zero. When the split nut is closed, the carriage moves. As long as the carriage moves, the threading dial stops revolving.

If the lathe is even geared, the following rules tell when to close the split nut, by reading the dial:

1. When cutting even-numbered threads, close on numbers and lines both.

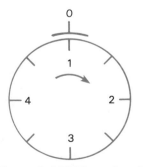

Figure 8–147 Threading dial

2. When cutting odd-numbered threads, close on numbers or lines, but not both.

3. When cutting half threads, close on two opposite points; e.g., 1 and 3.

4. When cutting quarter threads, close on one point only.

A lathe is even geared when the revolutions of the spindle and the revolutions of the reverse spindle stud gear are the same, or if the spindle gear and the inside stud gear of the tumbler gear train are the same size.

If the number of threads you are cutting is a multiple of the number of threads per inch of the lead screw, the split nut can be closed at any time on an even-geared lathe.

PICKING UP A THREAD

During a threading operation, the threading tool may become dull, it may break, or excess pressure may force the point from the center of the cut. It then becomes necessary to resharpen the threading tool and to reset the point on the center of the cut. This is an annoying situation, but it is easily rectified. Use the following procedure:

1. Resharpen the threading tool.

2. Using the center gage, reset the threading tool squarely with the work. Set the cutting lip height on center.

3. Start the lathe, and with the point clear of the work, close the split nut.

4. As the point starts to pass over the cut, shut off the lathe, leaving the split nut closed.

5. Adjust the point to the center of the cut from its probable position, Figure 8–148A, to its proper position, Figure 8–148B. Do this by adjusting both the compound and the cross slide until the point is in the center of the cut. A piece of white paper placed under the threading tool will help in making this adjustment. Any opening between the threading tool and the cut will be seen immediately.

6. Adjust both cross slide and compound dials to zero. Modern machines have adjustable dials, but if the dials are not adjustable, mark them with chalk, soapstone, or pencil, but do not scratch the dials with a tool such as a scriber.

7. Open the split nut and return the threading tool to the start of the cut.

8. Take two or three light trial cuts, feeding straight in with the cross slide. This is to make sure the point is in the center of the cut.

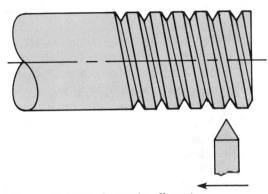

Figure 8–148A Cutter is off center

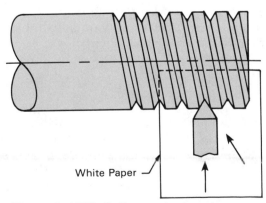

Figure 8–148B Cutter reset on center

9. Readjust or remark the cross slide dial to zero.

10. Proceed with your threading operation, setting the cross slide at zero for each cut, and feeding in with the compound.

GRINDING THREADING TOOL BITS

When grinding a 60-degree threading tool, the top or face should sit flat; i.e., a zero back rake should be presented to the work as in Figure 8–149A. If there is a positive back rake as in Figure 8–149B, a slightly narrower angle will be cut.

A problem occurs in the cutting of ductile metals such as mild steel. Since the top of the threading tool sits flat, Figure 8–150A, it does not cut too well. This situation is corrected by adding 12 to 14 degrees of side rake, Figure 8–150B, and then making any necessary adjustments to the 60-degree included angle so that a true 60-degree angle will be cut.

Another modification for ductile metals is shown in Figure 8–150C. Grind a chip curler type of rake parallel to the cutting edge. This peels off a beautiful chip. Each of the threading tools shown is used for cutting right-hand threads. They cut toward the chuck. There is one caution about using the chip curler rake in Figure 8–150C. There is a very slight hollow on the trailing edge just in back of the point. In precision threading, this might be noticeable. There would be a slight bulge on the back side of the thread with the compound set at 29 degrees. To rectify this, set the compound at 30 degrees and then polish the back of the thread after the thread is cut.

CUTTING POWER THREADS: ACME THREADS

The most common power thread cut today is the Acme thread, Figure 8–151. This thread has an included angle of 29 degrees and the same depth as a square thread. It has largely replaced the square thread, and it is almost as strong, and much easier to cut compared to the square thread.

GRINDING THE THREADING TOOL

Following is an accepted procedure for grinding an Acme threading tool:

1. Grind a negative back rake on top of the threading tool so that the top will sit flat when held in a standard tool holder.

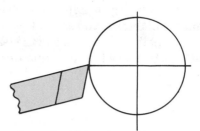

Figure 8–149A Zero back rake

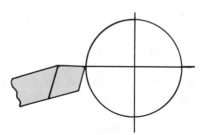

Figure 8–149B Positive back rake

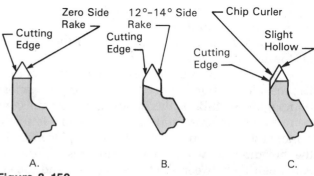

Figure 8–150

2. Shape the threading tool to fit the angle at the end of the Acme thread gage. Grind clearance on the leading edge only. The trailing edge does not need clearance, Figure 8–152, due to the lead angle of the thread.

3. Grind the nose flat and with front clearance, to fit in the required numbered slot on the side of the Acme gage; e.g., for 4 tpi, use the #4 slot.

CUTTING THE ACME THREAD

1. Set the compound to 14-1/2 degrees, or half of the included 29-degree angle.

2. Using the Acme thread gage, set the threading tool on center and at right angles to the work.

3. Set the lathe to cut the correct lead of the thread and proceed as in cutting any other thread.

MAKING COIL SPRINGS

Making coil springs on a lathe is closely linked to cutting screw threads in that the lead screw and split nut are used. A lead is set, but instead of cutting it, the spring steel wire is fed through a special holder, usually shop made, which is held in the tool post, Figure 8–153, and the wire is wrapped around a shaft. Either left-hand or right-hand coiled, tension, or compression springs can easily be made, depending on what lead is used.

Once the spring is completed, you must cut the wire. There will be some "spring back" as the tension is released, and the diameter of the spring will increase in size. The amount of increase depends upon the tension exerted on the wire by holding it back as it is fed through the holder. Always use a shaft smaller than the required diameter of the spring. How much smaller is a matter for experimentation.

CUTTING A METRIC THREAD

Cutting a metric thread on a metric lathe is exactly the same as cutting an inch thread on an inch lathe. Set the lathe to cut the correct lead; set the compound; set the threading tool; and proceed with the thread cutting.

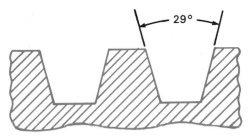

Figure 8–151 Acme thread

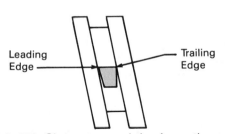

Figure 8–152 Clearances and the Acme thread cutter

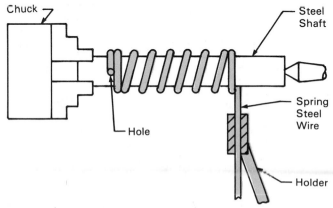

Figure 8–153 A left-hand coil spring being wound

If the lathe has inch/metric capabilities and an inch lead screw, there is one complication. The threading dial cannot be used successfully.

1. Set the lathe to cut the correct lead; set the compound; set the threading tool.

2. Once the initial cut is made, *the split nut must remain closed until the thread is finished.*

3. After each cut, pull the point back clear of the cut. Reverse the work rotation to move the carriage and threading tool to the starting point.

4. Reset the depth plus the additional feed.

5. Take the next cut.

6. Repeat until the thread is completed.

If you are using an inch lathe with an inch lead screw, you need metric tranposing gears to cut metric threads, or English transposing gears in the case of a metric lathe cutting inch threads. Check the chart on the lathe. Substitute the metric transposing gears into the gear train to give the correct lead of the thread you wish to cut, and proceed as though you had an inch/metric lathe with an inch lead screw.

CUTTING TAPERED THREADS

Standard pipe and pipe fittings are an example of the practical use of tapered threads. Pipe threads are tapered 3/4 inch to the foot. As the pipe fitting is turned onto the pipe, it tightens, forming a tight seal. A pipe compound or tape is also used on the thread to further ensure a leak-proof seal.

The most common method of threading pipe is on a pipe threading machine using pipe dies. Hand pipe dies can also be used. In a machine shop, pipe threads can be cut on a lathe by utilizing the taper attachment. Within limitations due to length, some pipe threading between centers could be done using the tailstock set-over method.

RESOURCE TABLES FOR THREADING

The following tables contain useful data for the machinist or student in the production of screw threads: Table 7–6, Tap Drill Sizes (inches); Table 7–7, Tap Drill Sizes (millimeters); Table 8–6, Acme Threads; Table 8–7, Pipe Threads; Table 8–8, Inch to Millimeter Conversions; and Table 3–2, Millimeter to Inch Conversions.

Table 8–6 Acme Threads *(Courtesy of SKF & Dormer Tools Ltd.)*

AMERICAN STANDARD ACME SCREW THREAD DIMENSIONS

$p = \dfrac{1}{n}$

$K = D \text{ minus } p$

$F_c = \dfrac{0.3707}{n}$

$K_r = D \text{ minus } 2h'$

h = Basic depth of thread
h' = Depth of thread with clearance
K = Tap drill. Basic minor diameter of nut
F_c = Width of flat at crest of thread
F_r = Width of flat at bottom of space

n = Number of threads per inch
p = Pitch of thread
K_r = Minor diameter of screw
D = Major diameter of screw
T = Major diameter of tap

For 10 or less threads per inch

$h' = \dfrac{P}{2}$ plus 0.010

$F_r = \dfrac{0.3707}{n}$ minus 0.0052

$T = D$ plus 0.020

For more than 10 threads per inch

$h' = \dfrac{P}{2}$ plus 0.005

$F_r = \dfrac{0.3707}{n}$ minus 0.0026

$T = D$ plus 0.010

Table 8–6 Continued

THREADS PER INCH (n)	DEPTH OF THREAD WITH CLEARANCE (h′)	FLAT AT TOP OF THREAD (F_c)	FLAT AT BOTTOM OF SPACE (F_r)*	SPACE AT TOP OF THREAD	THICKNESS AT ROOT OF THREAD
1	0.5100	0.3707	0.3655	0.6293	0.6345
1⅓	0.3850	0.2780	0.2728	0.4720	0.4772
2	0.2600	0.1854	0.1802	0.3146	0.3198
3	0.1767	0.1236	0.1184	0.2097	0.2149
4	0.1350	0.0927	0.0875	0.1573	0.1625
5	0.1100	0.0741	0.0689	0.1259	0.1311
6	0.0933	0.0618	0.0566	0.1049	0.1101
7	0.0814	0.0530	0.0478	0.0899	0.0951
8	0.0725	0.0463	0.0411	0.0787	0.0839
9	0.0655	0.0412	0.0360	0.0699	0.0751
10	0.0600	0.0371	0.0319	0.0629	0.0681
12	0.0467	0.0309	0.0283	0.0524	0.0550
14	0.0407	0.0265	0.0239	0.0449	0.0475
16	0.0363	0.0232	0.0206	0.0393	0.0419

Table 8–7 Pipe Threads — NPT *(Courtesy SKF & Dormer Tools, Limited)*

AMERICAN STANDARD PIPE THREAD AND TAP DRILL SIZES				
PIPE SIZE INCHES	THREADS PER INCH	ROOT DIAMETER SMALL END OF PIPE AND GAGE	TAP DRILL	
			TAPER NPT	STRAIGHT NPS
1/8	27	0.3339″	Q	11/32″
1/4	18	0.4329″	7/16″	7/16″
3/8	18	0.5676″	9/16″	37/64″
1/2	14	0.7013″	45/64″	23/32″
3/4	14	0.9105″	29/32″	59/64″
1	11 1/2	1.1441″	1 9/64″	1 5/32″
1 1/4	11 1/2	1.4876″	1 31/64″	1 1/2″
1 1/2	11 1/2	1.7265″	1 47/64″	1 3/4″
2	11 1/2	2.1995″	2 13/64″	2 7/32″

Table 8-8 Inch to Millimeter Conversions *(Courtesy of The L.S. Starrett Company)*

Decimals to Millimeters

Decimal	mm	Decimal	mm
0.001	0.0254	0.500	12.7000
0.002	0.0508	0.510	12.9540
0.003	0.0762	0.520	13.2080
0.004	0.1016	0.530	13.4620
0.005	0.1270	0.540	13.7160
0.006	0.1524	0.550	13.9700
0.007	0.1778	0.560	14.2240
0.008	0.2032	0.570	14.4780
0.009	0.2286	0.580	14.7320
0.010	0.2540	0.590	14.9860
0.020	0.5080		
0.030	0.7620		
0.040	1.0160	0.600	15.2400
0.050	1.2700	0.610	15.4940
0.060	1.5240	0.620	15.7480
0.070	1.7780	0.630	16.0020
0.080	2.0320	0.640	16.2560
0.090	2.2860	0.650	16.5100
0.100	2.5400	0.660	16.7640
0.110	2.7940	0.670	17.0180
0.120	3.0480	0.680	17.2720
0.130	3.3020	0.690	17.5260
0.140	3.5560		
0.150	3.8100		
0.160	4.0640	0.700	17.7800
0.170	4.3180	0.710	18.0340
0.180	4.5720	0.720	18.2880
0.190	4.8260	0.730	18.5420
0.200	5.0800	0.740	18.7960
0.210	5.3340	0.750	19.0500
0.220	5.5880	0.760	19.3040
0.230	5.8420	0.770	19.5580
0.240	6.0960	0.780	19.8120
0.250	6.3500	0.790	20.0660
0.260	6.6040		
0.270	6.8580		
0.280	7.1120	0.800	20.3200
0.290	7.3660	0.810	20.5740
		0.820	20.8280
0.300	7.6200	0.830	21.0820
0.310	7.8740	0.840	21.3360
0.320	8.1280	0.850	21.5900
0.330	8.3820	0.860	21.8440
0.340	8.6360	0.870	22.0980
0.350	8.8900	0.880	22.3520
0.360	9.1440	0.890	22.6060
0.370	9.3980		
0.380	9.6520		
0.390	9.9060	0.900	22.8600
0.400	10.1600	0.910	23.1140
0.410	10.4140	0.920	23.3680
0.420	10.6680	0.930	23.6220
0.430	10.9220	0.940	23.8760
0.440	11.1760	0.950	24.1300
0.450	11.4300	0.960	24.3840
0.460	11.6840	0.970	24.6380
0.470	11.9380	0.980	24.8920
0.480	12.1920	0.990	25.1460
0.490	12.4460	1.000	25.4000

Fractions to Decimals to Millimeters

Fraction	Decimal	mm	Fraction	Decimal	mm
1/64	0.0156	0.3969	33/64	0.5156	13.0969
1/32	0.0312	0.7938	17/32	0.5312	13.4938
3/64	0.0469	1.1906	35/64	0.5469	13.8906
1/16	0.0625	1.5875	9/16	0.5625	14.2875
5/64	0.0781	1.9844	37/64	0.5781	14.6844
3/32	0.0938	2.3812	19/32	0.5938	15.0812
7/64	0.1094	2.7781	39/64	0.6094	15.4781
1/8	0.1250	3.1750	5/8	0.6250	15.8750
9/64	0.1406	3.5719	41/64	0.6406	16.2719
5/32	0.1562	3.9688	21/32	0.6562	16.6688
11/64	0.1719	4.3656	43/64	0.6719	17.0656
3/16	0.1875	4.7625	11/16	0.6875	17.4625
13/64	0.2031	5.1594	45/64	0.7031	17.8594
7/32	0.2188	5.5562	23/32	0.7188	18.2562
15/64	0.2344	5.9531	47/64	0.7344	18.6531
1/4	0.2500	6.3500	3/4	0.7500	19.0500
17/64	0.2656	6.7469	49/64	0.7656	19.4469
9/32	0.2812	7.1438	25/32	0.7812	19.8438
19/64	0.2969	7.5406	51/64	0.7969	20.2406
5/16	0.3125	7.9375	13/16	0.8125	20.6375
21/64	0.3281	8.3344	53/64	0.8281	21.0344
11/32	0.3438	8.7312	27/32	0.8438	21.4312
23/64	0.3594	9.1281	55/64	0.8594	21.8281
3/8	0.3750	9.5250	7/8	0.8750	22.2250
25/64	0.3906	9.9219	57/64	0.8906	22.6219
13/32	0.4062	10.3188	29/32	0.9062	23.0188
27/64	0.4219	10.7156	59/64	0.9219	23.4156
7/16	0.4375	11.1125	15/16	0.9375	23.8125
29/64	0.4531	11.5094	61/64	0.9531	24.2094
15/32	0.4688	11.9062	31/32	0.9688	24.6062
31/64	0.4844	12.3031	63/64	0.9844	25.0031
1/2	0.5000	12.7000	1	1.0000	25.4000

REVIEW QUESTIONS

1. List five required steps in setting the lathe to cut screw threads.
2. In which direction does the lathe carriage move for cutting right-hand threads? For cutting left-hand threads?
3. What is the recommended speed for threading as compared to turning?
4. What should be the height of the cutting edge in threading?
5. In setting the lathe to cut a thread, is the lathe set to cut the pitch or the lead of a thread?
6. How does the procedure for threading when there is no neck or groove differ from the procedure when there is a neck or groove?
7. What is an even-geared lathe?
8. On an even-geared lathe, when is the split nut closed for cutting the following threads if the dial is divided into eight equal divisions:
   (a) Even numbered threads?
   (b) Odd numbered threads?
   (c) Half threads?
   (d) Quarter threads?
9. Can you close a split nut at any time?
10. You are cutting a thread and your tool bit breaks. What procedure do you follow to rectify this situation?
11. In grinding a 60-degree threading tool, what provisions can be made regarding rake?
12. Describe two modifications that can be made on cutting tools made from ductile metals to improve the cutting action.
13. What is the most common power thread?
14. Describe the steps in grinding an Acme threading tool.
15. In grinding a tool bit to cut an Acme thread, what provision is made for clearance on the trailing edge?
16. At what angle is the compound set for cutting an Acme thread?
17. How is the making of coil springs on a lathe linked to the cutting of screw threads?
18. When making a coil spring, how does the shaft size compare to the required diameter of the spring? How is this size determined?
19. In cutting metric threads on a lathe that has an inch lead screw, what is done regarding the closing of the split nut?
20. What is the most common method of cutting tapered threads?

Facing stainless steel pipe flanges. *Courtesy of Cominco Ltd.*

Facing stainless steel pipe flanges with an indexable carbide cutting tool. *Courtesy of Cominco Ltd.*

Facing a stainless steel flange with a square indexable insert tool. *Courtesy of Cominco Ltd.*

Facing bronze bushings mounted on a spigot. This spigot is straight with a slight taper at the end. The bushing is placed in position on the spigot and then is forced on to the taper. Cuts are light. A brazed tip carbide cutting tool is being used. *Courtesy of Cominco Ltd.*

Turning steel. Notice the three-bladed cross between the chuck jaws and against the face of the chuck. This prevents the work from being forced into the chuck when using heavy feeds. *Courtesy of Cominco Ltd.*

Turning steel. A closeup of the feeding action using a soluble cutting fluid, showing the cross which prevents the work from being forced into the chuck. *Courtesy of Cominco Ltd.*

Turning steel with a multilayer ceramic-coated, square indexable carbide insert. The insert is a coated carbide grade with a multilayer of titanium nitride, aluminum oxide, and titanium carbide. It is designed with strength to withstand interrupted cuts at ceramic coated insert speeds. Note the classic number "9" shape of the chip. *Courtesy of Kennametal Inc.*

A high-speed turning operation on steel with a square coated insert. Turning steel with a square, indexable insert. This titanium nitride (TiN) coated insert is suitable for finishing to heavy roughing applications. *Courtesy of Kennametal Inc.*

Turning steel with a Kendex square indexable insert. The insert is a general-purpose turning, milling, and threading grade of uncoated carbide. It is particularly suitable for light roughing to finishing of cast irons, nonferrous alloys, and nonmetals at moderate to high speeds and light chip loads. Cutting speeds for steels (up to a 0.4% carbon range) are 180 sfpm to 280 sfpm (60 m/min to 80 m/min). *Courtesy of Kennametal Inc.*

Turning stainless steel using a square cermet (ceramic and metal) insert. A cermet insert combines aluminum oxide, titanium carbide, and titanium nitride. This insert is capable of high cutting speeds and heavy feeds. *Courtesy of Kennametal Inc.*

Parting off steel using a carbide cut-off tool. *Courtesy of Kennametal Inc.*

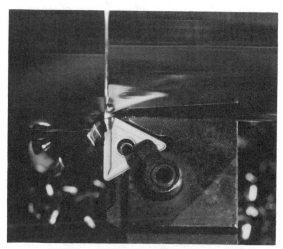

Finish turning steel with a triangular coated insert. *Courtesy of Kennametal Inc.*

Polishing the journals of a ball mill. This large lathe is referred to as the "bull" lathe by the company. It is very suitable for large work such as is shown. *Courtesy of Cominco Ltd.*

Polishing the journals of a ball mill. The closeup shows the polishing block held between the heavy boring bar and the journal surface. Emery cloth does the polishing. *Courtesy of Cominco Ltd.*

Turning using a polycrystalline diamond cutting edge insert. The insert is made from a polished synthetic polycrystalline diamond material with a wear resistant cutting edge chemically bonded to a strong carbide base. *Courtesy of Kennametal Inc.*

Cutting a nickel-base alloy using a round Kyon 2000 insert. This insert is capable of rough machining nickel-base alloys at cutting speeds of 400 sfpm to 800 sfpm (120 m/min to 240 m/min and feed rates of 0.005 to 0.012 ipr (0.13 to 0.30 mm/rev). *Courtesy of Kennametal Inc.*

Turning an aircraft/aerospace alloy using a ceramic insert. Cutting speeds of 400 sfpm to 800 sfpm (114 m/min to 228 m/min) are possible. Compare this to cutting speeds of 180 sfpm to 280 sfpm (55 m/min to 80 m/min) for a standard carbide insert. *Courtesy of Kennametal Inc.*

Turning a bronze journal built up by bronze wire welds. The machine is a vertical boring mill. An indexable carbide cutting tool is used. *Courtesy of Cominco Ltd.*

High-speed turning of cast iron using a square Kyon ceramic insert. The insert will cut cast iron at cutting speeds of 1000 sfpm to 4000 sfpm (305 m/min to 1140 m/min). Feed rates of 0.015 to 0.030 ipr (0.38 – 0.76 mm/rev) are used. Cutting speeds for cast iron using a regular carbide insert are 180 sfpm to 280 sfpm (55 m/min to 80 m/min). *Courtesy of Kennametal Inc.*

Turning cast iron using a square ceramic-coated carbide insert. This insert has a multilayer coat of titanium nitride, aluminum oxide, and titanium carbide. It is suitable for a wide range of machining, from medium roughing to finishing cast irons, high temperature alloys, stainless, low carbon, and alloy steels. This insert is designed to withstand interrupted cuts. Cutting speeds of 1000 sfpm to 4000 sfpm (305 m/min to 1140 m/min) are possible. Compare this to cutting speeds of 180 sfpm to 280 sfpm (55 m/min to 80 m/min) for a regular carbide insert. *Courtesy of Kennametal Inc.*

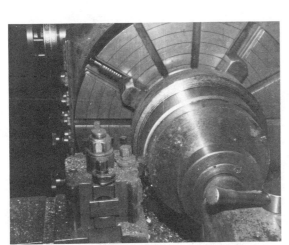

Turning grey cast iron. Notice the small crumbling type of chip and the dust from the casting. *Courtesy of Cominco Ltd.*

Boring steel with a coated carbide insert. This is an interchangeable-head boring bar. The coolant feeds through the tool with a flood to the cutting edge. The objective is to keep the temperature of the cutting edge as constant as possible. Sudden temperature changes can cause cracking of the cutting edges, although this particular insert has superior thermal and mechanical shock resistance. The coating on the insert is titanium nitride. *Courtesy of Kennametal Inc.*

CHAPTER NINE

CUTTING TOOLS

At first glance, this chapter appears to be about tool bits and carbide cutting tools. In fact it is much more than that. All of the principles of clearance, rake, and tool shape apply not only to tool bits but to all cutting tools, regardless of the type of cutting tool material used or the material being cut. A rotary milling cutter is actually a number of tool bits mounted in the form of a wheel. A lathe tool bit is a single-point cutting tool; a rotary milling cutter is a multitooth cutting tool. The same principles apply to carbide cutters, both single-point cutters and multitooth cutters. The only difference in the cutters is in the cutter material, which results in a change in the cutting speed. In essence, we simply use a tool bit to demonstrate the principles of clearance, rake, and tool shape.

The rules regarding speed, feed, and depth of cut apply to all cutting tools. However, there are no absolute rules. In the practical world of the machine shop, the rules are guides only, which are subject to change as the conditions of cutting change. The use of cutting fluids is also determined by actual cutting practices and conditions.

One well-known fact is that in the last forty years cutting speeds have increased several thousand times. Manufacturers are producing more goods in a shorter time than ever before. The development of better, more efficient cutting tools and cutting fluids is an important factor in increased production. Older, slower machine tools are being removed from service and replaced by more efficient machine tools that will do more operations than ever — more quickly and at a lower cost. However, these new improved cutting tools are still based on the principles of clearance, rake, and tool shape.

SAFE USE OF CUTTING TOOLS

Cutting tools of all kinds are always potentially dangerous. They are sharp and the chips are also sharp, ragged, and hot.

1. When making adjustments to a machine or to the

work, always make sure that the cutting tool is moved out of the way. When using a wrench, for example, it can slip, causing the hand to be suddenly jammed into a sharp cutting edge.

2. One of the reasons for not wearing loose clothing or jewelry is that they can be caught by a cutting tool and a hand or an arm can be pulled into the moving tool or moving work.

3. Long hair should be tied back out of the way. A revolving milling cutter or revolving work can easily catch long hair and pull the head into the work or cutter, violently and quickly.

4. Never handle chips with bare hands. When coming from the cutting tool, they are hot, sharp, and ragged. Lying in a chip pan, they are still sharp and ragged.

5. Never put your hands near a moving cutting tool or moving work. If you must adjust a cutting tool or the work, shut off your machine first.

6. It is best to design and grind a tool bit that will produce small, tightly curled chips. Long chips, coiled or flat and tangled, are dangerous to the operator and are difficult to dispose of, Figure 9–1. A small chip shaped like a number "9" can be produced with the right combination of feed and speed, and a properly designed tool bit. When drilling, chips can be caused to break up into smaller pieces in the same way.

7. Both the cutting tool and the work can become extremely hot. To prevent burns, handle both with care.

8. It is permissible to use a brush to sweep away excess chips, but not a rag which can be caught and pulled into the machinery. Never use your hands to brush away chips.

9. Keep your cutting tools sharp. A dull tool is dangerous. It tends to push the work away instead of cutting it. A dull cutting tool can also jam on the work or tear the work from the machine. Figure 9–2 illustrates what can happen if a tool bit suddenly jams and snaps.

Figure 9–1 A demonstration showing a tangle of chips that present a safety hazard to the operator *(Courtesy of Cominco Ltd.)*

Figure 9–2 These safety glasses and the machinist's eyes survived a direct blow from a broken tool bit *(Courtesy of Cominco Ltd.)*

UNIT 9-1

THE PRINCIPLES OF CUTTING

OBJECTIVES

After completing this unit, the student will be able to:

- define the term cutting tool and give examples.
- list four basic criteria by which a cutting tool is judged.
- describe the action of a cutting tool on metal and the pressure forces exerted on the tool.
- explain how the design of the cutting tool affects the pressure exerted on the metal being cut.
- list the basic forms of a cutting edge.
- list four factors affected by the shape and location of the cutting edge.
- state the right-hand rule for turning tools.

KEY TERMS

Cutting tool	Straight edge cutting
Cutting tool geometry	tool
Cutting tool criteria	Shaped or formed
Cutting action	edge cutting tool
Rigidity	Cutting edge
Shape	Cutting efficiency
Area of contact	Surface finish
Pressure forces	Primary cutting edge
Basic form	Plunge cutter
Pointed cutting tool	Right-hand rule

INTRODUCTION

A **cutting tool** removes excess material from a work-piece. Cutting tools include drill bits, lathe tool bits, milling cutters, grinding wheels, and so forth. A machine tool is only as efficient as its cutting tool. For this reason, nothing in machine shop work should be considered as carefully as the principles behind cutting tool design and shape.

These principles are perhaps easiest explained by using lathe tool bits as an example, Figure 9–3. The engine lathe is the fundamental machine tool in a machine shop. All of the principles applicable to lathe tool bits are applicable to other machine cutting tools. Considered closely, there are not many actual variations in cutting tool geometry. A relatively small number of different geometries will do many jobs on many diverse materials.

A cutting tool must meet basic criteria. It must be strong enough to support and maintain its cutting edge; hard enough so that the cutting edge won't wear easily; tough so that the cutting edge won't chip easily; and large enough to carry away the heat generated by the cutting action.

CUTTING TOOL ACTION

The cutting action of a cutting tool is determined primarily by three factors: The rigidity of the workpiece, the rigidity of the cutter, and the shape of the cutting edge. This last factor determines the area of contact between the cutting edge and the work. The greater the area of contact, the greater the amount of force that must be used to drive the cutter into the work at any set feed. The greater the force, the greater the pressure the cutting edge exerts on the work. In the cutting of metal, pressure is the most important factor in the cutting action.

It is largely because of the effect of pressure that a cutting tool is designed in a specific way. This can be appreciated by considering three different lathe tool bits each cutting to the same depth and each set at the same feed.

The round-nose tool bit, Figure 9–4A, has the largest cutting edge in contact with the work. More force is needed to drive it, which results in more pressure, more friction, and more heat. Surface finish on the work, however, is good.

The sharp-nose tool bit, Figure 9–4B, has the least amount of cutting edge in contact with the work. Less force is needed to drive it, resulting in less pressure, less friction, and less heat. However, the sharp point is much weaker than that of the round-nose cutting tool, and the surface finish on the work is poor.

The small round-nose tool bit, Figure 9–4C, is the best compromise between the two tool bits in Figures 9–4A and 9–4B. Less cutting edge is in contact with the work than with the round-nose tool bit, the small round-nose is much stronger than the point of the sharp-nose tool bit, and surface finish is good.

A metal cutting tool actually "pushes" metal apart, Figure 9–5. Pressure exerted against the cutting edge is very high and this pressure increases as the feed and depth of cut increase. Pressure forces that work on a tool bit act against the top or face, against the side or flank, Figure 9–6A, and against the end or nose, Figure 9–6B.

BASIC FORMS OF A CUTTING EDGE

Regardless of the many different shaped tool bits that may be found in a machinist's tool box, there are only three basic forms. These are the pointed, the straight edge, and the shaped or formed edge cutting tools. An example of a pointed cutting tool is the sharp-V threading tool, Figure 9–7A. Some other examples of pointed cutters are a sawtooth, scriber, double angle milling cutter, and diamond point chisel. A right-hand turning tool, Figure 9–7B, is an example of a straight edge cutting tool. The cold chisel, plain milling cutter,

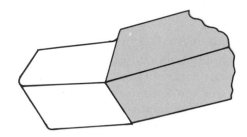

A. Tool bit with standard side and back (compound) rake

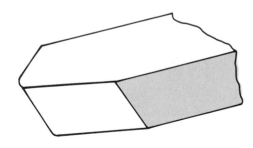

B. Tool bit with zero rake

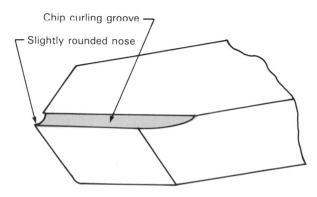

C. Tool bit with chip curler

Figure 9–3 Lathe tool bits

and drill are other examples of straight edge cutting tools. The round-nose tool bit is a *shaped or formed edge cutting tool*, Figure 9–7C. Other examples are the hollow cutter and gear cutter. These basic forms can be combined into one cutting tool in many ways. One example of a combination is shown in Figure 9–8.

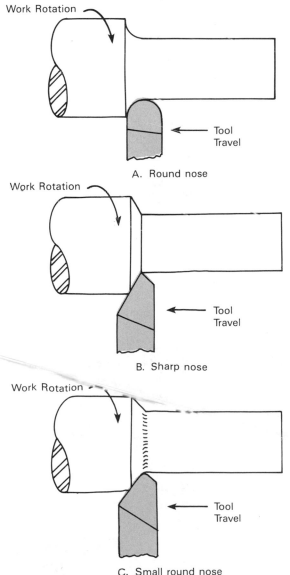

Figure 9–4 Cutting tool design, top view

Importance of the Cutting Edge

Lathe tool bits, which are relatively sturdy and simple in design, contain a considerable amount of metal in proportion to the size of their cutting edge. The **cutting edge** is the only portion of the cutter that actually cuts, but the rest of the tool bit also has a function. It serves as a support for the cutting edge to prevent it from vibrating (chattering). It serves to carry away heat generated by the cutting action. The larger the cutting tool, the more heat is carried away. It aids in the removal of the chip. The chip is formed by the rake or slope of the face and is guided away from the cutting edge.

The cutting efficiency of a cutting tool depends on the proper design and location of its cutting edge or edges. The location and shape of the cutting edge determine how the cutting tool must be applied to the work, the final shape of the work, and the type of surface finish the work will have.

The cutting edge that does most of the cutting is called the **primary cutting edge**. It is located either on the side of the tool bit, as on a right- or left-hand turning tool, or on the end or nose of the tool bit as on cutters used for cutting-off (parting-off) or grooving. These end-type blades are sometimes called plunge cutters.

The Right-hand Rule

Right-hand turning tools begin cutting on the right end of the work and move to the left, Figure 9–9A. Left-hand turning tools begin cutting on the left end of the work and move to the right, Figure 9–9B.

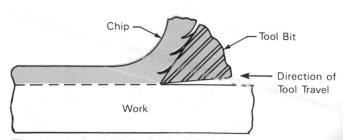

Figure 9–5 A metal cutting tool "pushes" metal apart

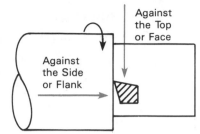

Against the Top or Face

Against the Side or Flank

Figure 9-6A Pressure forces on the top and side

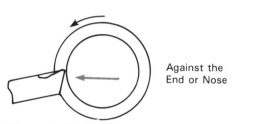

Against the End or Nose

Figure 9-6B Pressure forces against the nose

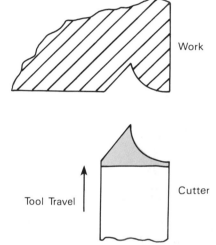

Work

Cutter

Tool Travel

Figure 9-8 Combination of the three basic forms

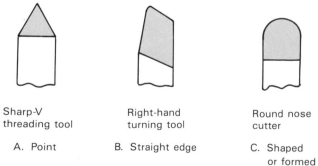

Sharp-V threading tool

Right-hand turning tool

Round nose cutter

A. Point

B. Straight edge

C. Shaped or formed

Figure 9-7 Basic forms of cutting tools

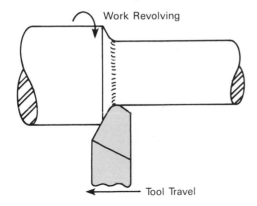

A. Right-hand turning tool

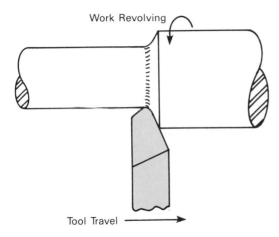

B. Left-hand turning tool

Figure 9-9 The right-hand rule

REVIEW QUESTIONS

1. List six safe practices in the use of cutting tools.
2. What is the function of a cutting tool?
3. What is the relationship between the efficiency of a machine tool and the cutting tool selected for a job?
4. Why do we use a lathe tool bit to explain the principles of cutting?
5. List the four basic criteria a cutting tool must meet.
6. Name three factors upon which the cutting action of a cutting tool depends.
7. What is the most important factor in cutting metal?
8. From question 7, how would this factor affect the design of a cutting tool?
9. State the advantages of a small round-nose tool bit compared to a round-nose tool bit and a sharp-nose tool bit (where the cutting depths and feeds are the same).
10. Where do the pressure forces act on a cutting tool?
11. What are the basic forms of a cutting tool?
12. List four factors affected by the shape and location of the cutting edge of a cutting tool.
13. What is the primary cutting edge of a cutting tool?
14. Where can the primary cutting edge be located on a tool bit?
15. Explain the right-hand rule.

UNIT 9-2

CUTTING TOOLS

OBJECTIVES

After completing this unit, the student will be able to:

* explain the differences between roughing and finishing cuts.
* select the proper tool for either roughing or finishing and for the specific material being cut.
* identify the standard parts and angles of a tool bit.
* care for tool bits properly.
* list the three most widely used cutting tool materials.
* describe the characteristics of cutting tools made from each of the most widely used materials and state applications of the cutting tools.
* define the terms clearance and rake.
* apply the principles of clearance and rake when grinding different cutting tools.
* identify the different forms of clearance and rake and give a benefit of each form.

KEY TERMS

Roughing cut
Finishing cut
Bulk of excess
  material
Surface finish
Right-hand turning
  tool with chip
  breaker
Standard rake
Chip breaker
Shape or contour
Profile
Honing
Carbon tool steel
Fast finishing steel

High-speed steel
Cast material alloy
Cemented carbide
Ceramic (cemented
  oxide) tool
Diamond cutting tool
Clearance or relief
Side clearance
End (front) clearance
End cutting-edge
  angle
Side cutting-edge
  angle (lead angle)
Rake
Back rake

Red hardness
Double or compound
  rake
Standard side rake
Chip curler

Side rake
Clearance (relief) angle
Rake angle
Nose radius

ROUGHING AND FINISHING CUTTING TOOLS

Speed can be made only during the roughing cut when surface finish is not important. The important thing when roughing is to remove the bulk of excess material as quickly as possible. Surface finish is important only when finishing. Unlike roughing, finishing is a slow process.

The right-hand turning tool with chip curler, rather than standard rake, Figure 9–10A, is designed to take a deep cut. For example, a 5/16-in square tool bit will cut 5/16 in deep, and an 8-mm square tool bit will cut 8 mm deep. This tool bit will remove excess metal very quickly. Chip disposal is excellent, with chips being ejected in small tight curls. It is also an excellent finishing tool bit. Do not confuse the chip curler with a chip breaker. A chip breaker actually snaps the chip against a sharp shoulder.

Compare the tool bit with the chip curler groove to the tool bit with the standard rake, Figure 9–10B. This tool bit is commonly used for both roughing and finishing. Although this tool bit is strong enough to make a deep cut, chips come off in a long curved line that pile up around the cutter. This causes a chip disposal problem and is a danger to the operator because the chips are so tangled and sharp. Doubling or tripling the feed will help, but this takes considerably more horsepower and the chips still tend to be long. For light cuts, however, this tool bit will function well and will produce a good surface finish due to the small nose radius. It is excellent for gray cast iron.

Shape or Contour of a Cutting Tool

The **shape** or **contour**, sometimes called the profile is what you see in the plan view or looking down at the top or face. Figure 9–11 illustrates a few suggested shapes that can be ground on a tool bit and used successfully. The National Form threading cutting tool with the slight flat on the end can be used for National Form threads, Unified threads, and ISO threads. The tool is ground to the correct size of flat for a particular pitch of thread.

Holding a Tool Bit

Some tool holders hold a tool bit at an approximate 15-degree angle, while other tool holders hold it flat. These factors must be taken into consideration when grinding clearance and rake as illustrated in Figures 9–12A and 9–12B. The tool bit in Figure 9–12B, which is held at a zero-degree angle, is a heavier cutting tool than that in Figure 9–12A and will carry away more heat because of it. The same amount of front clearance is presented to the work in both cases.

The Parts and Angles of Tool Bits

The names of the parts of a tool bit, Figure 9–13, and the names of the angles, Figure 9–14, are standard in the machining industry.

Grinding Tool Bits

Tool bits should not be ground while they are being held in a tool holder. The reasons for not using this practice follow: It is a clumsy and inefficient method due to the bulk and extra weight of the tool holder. Too much pressure can be exerted on the grinding wheel. This can result in damage to the wheel and/or damage to the tool bit due to overheating and cracking. There is also the possibility of grinding the tool holder.

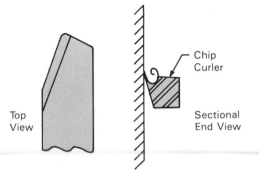

Figure 9–10A A right-hand turning tool with chip curler

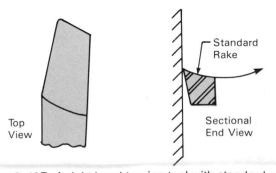

Figure 9–10B A right-hand turning tool with standard compound rake

Honing

A machinist should always have small pocket hones in his tool box. An aluminum-oxide hone is used for carbon tool steel and high-speed steel tools. A silicon-carbide hone is used for carbide tools. Cutting tools should be honed often to keep the cutting edge smooth and keen, resulting in longer tool life and the production of a smoother surface finish.

CUTTING TOOL MATERIALS

Carbon tool steel tool bits usually consist of 0.9% to 1.3% carbon. Cutting tools made from these steels can withstand temperatures up to their tempering temperatures of approximately 400 degrees F (205 degrees C) to 500 degrees F (260 degrees C) depending on the amount of carbon used. Above these temperatures a carbon tool steel cutting tool will soften

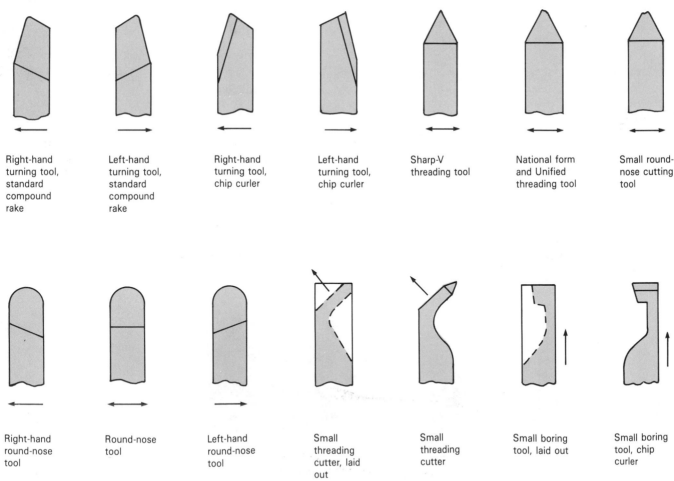

Right-hand turning tool, standard compound rake

Left-hand turning tool, standard compound rake

Right-hand turning tool, chip curler

Left-hand turning tool, chip curler

Sharp-V threading tool

National form and Unified threading tool

Small round-nose cutting tool

Right-hand round-nose tool

Round-nose tool

Left-hand round-nose tool

Small threading cutter, laid out

Small threading cutter

Small boring tool, laid out

Small boring tool, chip curler

Figure 9–11 Top view of suggested tool bit shapes. Arrows indicate direction of tool travel

and its cutting edge will break down. This can happen if your carbon tool steel cutting tool turns blue when you are grinding it, or if you try to cut too quickly. The cutting tool must then be rehardened and retempered. Carbon tool steel tool bits are seldom used in machining today.

Fast finishing steels are similar to carbon tool steels, but in addition to carbon they contain small percentages of tungsten, chromium, or vanadium. The addition of these metals makes a harder steel, giving increased wear resistance to cutting tools made from this steel. Fast finishing steels do not have red hardness, however, which is the ability to keep on cutting even though the cutting edge glows red hot. Fast finishing steels are also seldom used in machining today.

High-speed steels contain either 14% to 22% tungsten or 6% to 91% molybdenum together with

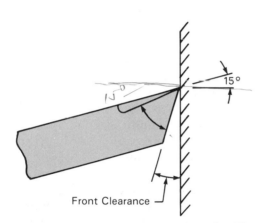

Figure 9–12A Cutter at approximately 15°

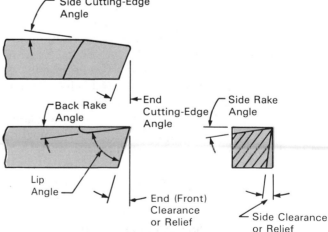

Figure 9–13 The parts of a cutting tool

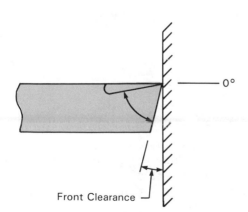

Figure 9–12B Cutter at 0°

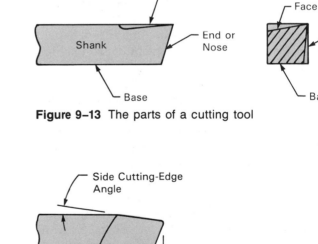

Figure 9–14 The angles of a cutting tool

1.5% to 6% tungsten. Cutting tools made from these steels have red hardness. Some high-speed steels also have a percentage of cobalt which imparts additional red hardness. Cutting tools made from this type of high-speed steel can tolerate heavy cuts where excessive heat will be generated. Depending on the type of high-speed steel from which they are made, cutting tools can withstand temperatures of about 1000 degrees F (540 degrees C). Above this temperature they tend to soften and their cutting edges will break down. During cooling, the tools will reharden themselves. Care must be taken when grinding these steels, as overheating and sudden cooling (quenching) may cause cracks to occur.

Cast material alloys have such trade names as Stellite, Renalloy, or Tantung. Stellite, for example, is an alloy of cobalt, chromium, and tungsten. It contains little or no iron, and is capable of performing at cutting speeds 25% to 80% greater than the maximum speeds for high-speed steel. Stellite tool bits are often used for heavy and intermittent cuts on chilled iron castings.

At one time, cutting tools made of cemented carbides were used exclusively for cast-iron machining. Now, however, they are used to machine all metals. Feeds for cemented carbide cutting tools are customarily lighter than those used for high-speed steel or cast-alloy tools because of the tendency of the cemented carbide tools to chip. Pure tungsten, carburized or impregnated with carbon to form tungsten carbide, is suitable for machining cast iron, aluminum, nonferrous alloys, plastics, and fiber. The addition of tantalum, titanium, and molybdenum carbides results in harder tools which are suitable for all types of steels. During manufacturing, tungsten carbide or a mixture of two or more carbides and a binder, commonly cobalt, are powdered, blended thoroughly, molded under pressure, and sintered.

These cutting tools can be used efficiently at temperatures up to 1600 degrees F (870 degrees C). Carbide tools are harder than high-speed steel tools and have greater wear resistance. They can be used at nearly three times the cutting speeds limiting high-speed steel tools.

Cutting tools made of diamonds compete to an extent with cemented carbides where surface finish and dimensional accuracy requirements are high, but these tools are chipped by hardspots in the material being cut and also by poor handling. Metals, hard rubber, and plastics can be effectively finish turned and bored with diamond tools.

Ceramic (or cemented oxide) tools are made primarily from aluminum oxide. Carboloy cemented oxide grade 0–30, for example, contains about 89% to 90% aluminum oxide and 10% to 11% titanium oxide. Other ceramic cutting tools are made from pure aluminum oxide blended with minute amounts of secondary oxides.

Ceramic tools retain strength and hardness at temperatures in excess of 2000 degrees F (1095 degrees C). Cutting speeds can range from fifty percent to several hundred percent higher than those used with cemented carbide tools. Ceramics are the hardest cutting tool materials currently used in industry with the exception of titanium carbide and diamonds, especially at high temperatures.

Ceramics fracture easily under certain conditions, having rupture strengths about one-half to two-thirds that of carbides. Interrupted cuts, scale removal operations, and milling applications are generally not recommended for these tools. Ceramic cutters usually fail by fracture rather than wear because they lack ductility and have lower tensile strength than other materials.

To summarize, the most widely used cutting tool materials are high-speed steels, cast material alloys, and cemented carbides.

CLEARANCE AND RAKE

The principles of clearance and rake are easiest to explain by using a lathe tool bit as it is used on a lathe. The shape or contour, the amount of clearance, and the type and amount of rake will vary with the machinist. Grinding a tool bit is as individual as brushing your teeth.

Clearance prevents the cutting edge of a cutting

tool from rubbing on the work. If there is no clearance, as in the knife blade in Figure 9–15A, the flank of the cutting tool will rub and will not cut. If there is clearance, Figure 9–15B, the cutting tool will cut. This basic fact applies to any kind of cutting tool.

Clearance is variable depending on the material being cut and the deformation of the material being cut. For example, aluminum is quite soft and tends to deform or bulge slightly as the cutting tool bites into it. For this reason, it is best to grind as much as 30 degrees of clearance under the cutting edge to prevent rubbing. If this much clearance were ground on a steel cutting tool, the cutting edge would vibrate (chatter) and would probably break down very quickly. Table 9–1 (page 340) shows the degrees of clearance to grind for different materials.

The correct amount of clearance will support the cutting edge properly. Too much clearance will cause vibration (chatter) of the cutting edge and possibly the complete breakdown of the cutting edge. Tool bits for lathes must have side clearance, end (front) clearance, and an end cutting edge angle. The primary cutting edge, the cutting edge that does most of the cutting, can be located on the side of the cutting tool, as with a left-hand or right-hand turning tool, or on the end of the cutting tool, as with a cutting-off tool.

The effect of side clearance is illustrated on a lathe tool bit as it moves toward the left. With no side clearance, Figure 9–16A, the tool will rub but will not cut. If there is good side clearance, Figure 9–16B, the cutting edge will cut and be well-supported. If there is too much clearance, Figure 9–16C, the cutting edge will have poor support which will cause it to vibrate (chatter) and possibly to break down completely.

The effect of front or end clearance on a lathe tool bit as it is held at a zero-degree angle is shown in Figure 9–17. If there is no front clearance, Figure 9–17A, the tool will rub and not cut. If there is good front clearance, Figure 9–17B, the tool will cut well and the cutting edge will be well-supported. If there is too much front clearance, Figure 9–17C, the tool will lack support, chatter will result, and the cutting edge might break down.

Figure 9–18 illustrates the effect of front clearance on a lathe tool bit as it is held at a 15-degree angle. The same amount of front clearance is presented to the work as with a cutting tool held at zero degrees, but the point of the tool is thinner. Not as much heat can be carried away from the cutting edge.

As a rule, not too much is said about side cutting-edge angle, or lead angle, Figure 9–19. It usually varies from zero degrees to 20 degrees with an average of about 15 degrees. There are distinct advantages, as follows: The cutting edge tends to take a better shearing cut, and the chip is thinner because of the angular position of the cutting edge. The point of the cutting tool is the weakest part. With lead angle, the cutting edge strikes the work before the point. The shock of cutting is taken up gradually by the whole cutting edge. When the cut extends to the end of the work,

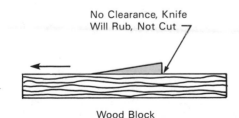

Figure 9–15A Knife blade with no clearance

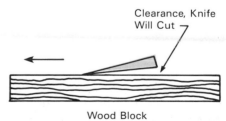

Figure 9–15B Knife blade with clearance

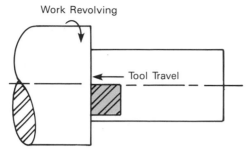

A. No side clearance

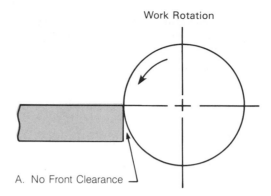

A. No Front Clearance

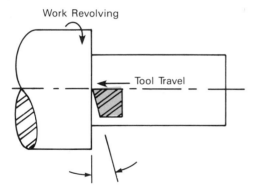

B. Good side clearance

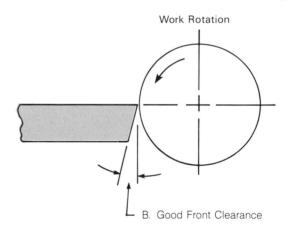

B. Good Front Clearance

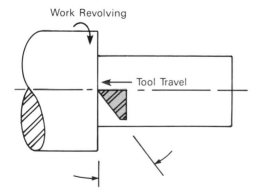

C. Too much side clearance

Figure 9–16 Side clearance

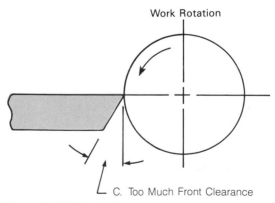

C. Too Much Front Clearance

Figure 9–17 Front clearance

the final section of metal is removed gradually instead of being pushed off in the form of a ring, hoop, or tangled ball of metal. Pressure tends to push the cutting tool away from the work rather than pull it into the work.

An example of a thinner chip when using a tool bit with a 30-degree side cutting-edge angle is illustrated in Figure 9–19A. A mathematical proof is illustrated in Figure 9–19B using an enlargement of the same right angle triangle as in Figure 9–19A, and a feed of 0.010 in. The adjacent side (b) of the right angle triangle is equal to the chip thickness. The hypotenuse (c) is equal to the feed of 0.010 in.

The following formula uses trigonometry for the solution:

$$\text{Cosine of the angle } A = \frac{\text{adjacent side } (b)}{\text{hypotenuse } (c)}$$

$$\text{or Cos 30 degrees} = \frac{b}{c}$$

$$0.866 = \frac{b}{0.010}$$

$$b = 0.866 \times 0.010$$

$$b = 0.00866 \text{ (chip thickness)}$$

The purpose of end cutting-edge angle is to clear the end of the cutting tool from the machined surface of the work as the tool moves. This angle usually varies from 8 degrees to 15 degrees, but in some cases it can be as high as 20 degrees to 30 degrees. If there is no clearance, Figure 9–20A, the cutting tool will bind, shriek, and possibly snap off. With proper clearance, Figure 9–20B, the cutting tool will cut well.

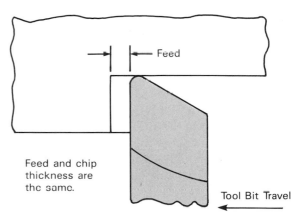

Feed

Feed and chip thickness are the same.

Tool Bit Travel

Tool bit with zero side cutting-edge angle

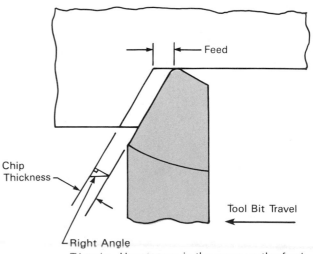

Feed

Chip Thickness

Tool Bit Travel

Right Angle Triangle—Hypotenuse is the same as the feed. Adjacent side is the same as the chip thickness, making the chip thinner.

Tool bit with side cutting-edge angle

Figure 9–19A Proof of a thinner chip

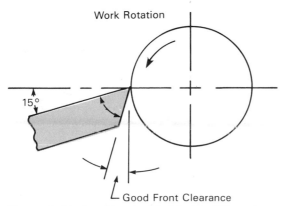

Work Rotation

15°

Good Front Clearance

Figure 9–18 A cutting tool held at a 15° angle

A factory made cut-off or parting tool blade has three clearances, one on the end or nose, and one on each flank, Figure 9–21. If ground from a tool bit, it could have as many as five clearances, Figure 9–22. This type of grooving tool is sometimes referred to as a necking tool for cutting shallow necks as at the end of a screw thread.

Rake refers to the slope of the top of the cutting tool or, in the case of a tool bit, the face. Rake varies depending on the material being cut. It improves the cutting action, forms the chip, and directs the flow of the chip away from the cutting edge. Rake is named after the direction of flow of the chip. For example, if the chip flows to the side of the cutting tool, it is called side rake. If the chip flows toward the back, that is, toward the tool holder, it is called back rake. Some machinists mistakenly refer to back rake as front rake or top rake.

Single-point cutting tools like tool bits may have side rake only with the cutting edge on the side, or back rake only with the cutting edge on the end, or they may have a combination of back and side rake with the primary cutting edge on the side and a secondary cutting edge on the end or nose of the tool. In the latter case, the bulk of the cutting is done on the side and the chips move at an angle toward both the back and the side of the cutting tool.

Two different side rakes are illustrated in Figures 9–23A and 9–23B. The amount and type of side rake ground depend on the material to be cut.

Two different back rakes with the cutting tools held at a zero-degree angle are illustrated in Figures 9–24A and 9–24B. The tools are held at a 15-degree angle in Figures 9–25A and 9–25B. Figure 9–26 shows the tool held at a 15-degree angle, but in this case a combination of both back and side rake are presented to the work. The double or compound rake causes the

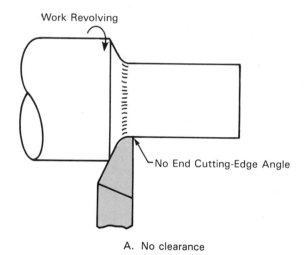

A. No clearance

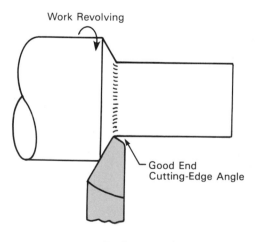

B. Good clearance

Figure 9–20 End cutting-edge angle

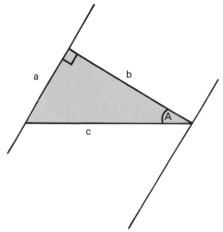

Figure 9–19B The right angle triangle from Figure 9–19A

chip to flow away from the cutting edge at an angle or toward both the back and side of the cutting tool.

To compare back rakes, a side view of tool bits with positive, zero, and negative back rakes is shown in Figure 9–27. These rakes are ground on tool bits depending on what is required for the work at hand. The type of back rake ground depends on the type of material being cut and also on the way the cutting tool is held by the tool holder.

Types of Side Rake

Figure 9–28 illustrates cross sections of tool bits and four different types of side rake.

Zero side rake, Figure 9–28A, is especially necessary for materials like some brasses, some bronzes, and certain brittle plastics.

Standard side rake, Figure 9–28B, is probably one of the most common side rakes. During deep cuts on ductile materials, the chips tend to pile up around the cutter presenting a hazard to the operator. Chip disposal also becomes a problem. This cutter is excellent for gray cast iron.

The chip curler, Figure 9–28C, has one of the best types of rake, especially for extremely deep cuts on ductile materials. The chips form tight little curls that break off against the side of the cutter or sometimes against the tool holder. Keep the width of the chip curler groove narrow to ensure that the chips will curl tightly. Chip disposal is very easy. A chip curler is not a chip breaker.

The chip breaker, Figure 9–28D, causes the chips to fracture in the corner and fall into the chip pan as small chips. As much as 25% more power is required for a chip breaker. This rake is good for tough steels.

Clearance (Relief) Angles

When cutting any material, clearances should always be kept to a minimum, but clearance angles should never be smaller than necessary. Clearance

angles that are too small will cause the cutter to rub. The following points are meant to be a guide only:

1. When cutting hard, tough materials, clearances should be 6 to 8 degrees for high-speed steel cutting tools and 5 to 7 degrees for carbide tools.
2. When cutting medium steels, mild steels, and some cast irons, clearances should be 8 to 12 degrees for high-speed steel cutting tools and 5 to 10 degrees for carbide tools.
3. When cutting ductile materials like copper, brass, bronze, aluminum, ferritic malleable

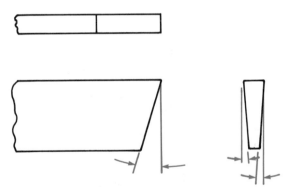

Figure 9–21 Clearances on cut-off or parting tools

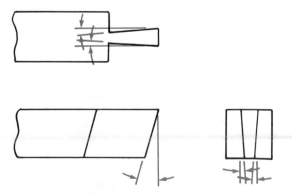

Figure 9–22 Clearances on cut-off or parting tools

iron, etc., clearances should be 12 to 16 degrees for high-speed steel cutting tools and 8 to 14 degrees for carbide tools, partly because of plastic deformation of these metals. This means that softer metals tend to deform or bulge slightly as the cutter strikes them, and they will rub on the side of the cutter. Extra clearance must be ground on the cutter when this occurs.

4. Larger relief angles generally tend to produce better surface finishes because less surface of the flank of the cutting tool rubs against the work. For this reason, clearance angles in

single-point threading tools should be as large as circumstances permit.

5. Problems can occur when machining such materials as stainless steels which tend to work-harden as they are being cut. This can be overcome by increasing the size of the clearance angles. Above all, keep the cutting tool cutting. Don't let cutting edges rub on stainless steels, which tend to work harden if rubbing occurs.

Rake Angles

The following points should be observed in the grinding of rake angles.

1. When rake angles are made larger in a positive direction, cutting forces and cutting temperatures will decrease. For example, if side rake is changed from 5 degrees positive to 5 degrees negative, the cutting force will be about 10% larger.

2. Negative rake angles strengthen cutting edges enabling them to sustain heavier cutting loads and shock loads. Negative rake angles can be recommended for turning very hard materials and for heavy interrupted cuts, especially when using carbide tools.

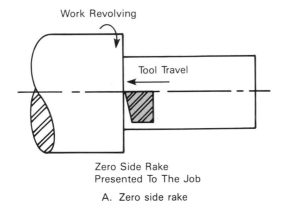

Zero Side Rake
Presented To The Job

A. Zero side rake

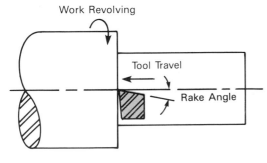

Positive Side Rake
Presented To The Job

B. Positive side rake

Figure 9–23 Side rake in cutting tools

Nose Radius

Nose radius is a very critical part of the cutting edge. It is a slight rounding of the point of the cutting tool. If the tool has a sharp point, surface finish is usually unacceptable and the life of the tool will be short. A nose radius is required to obtain both an acceptable surface finish and a longer tool life.

The quality of surface finish is determined by both the rate of feed and the size of the nose radius. The larger the nose radius, the better the surface finish will be and the faster the feed rate can be. Keep in mind that too large a nose radius will expose more cutting edge than is necessary to contact the work, which will cause more heat to be created.

Nose radii of 1/64 in and 1/32 in can be used successfully on smaller cutters of 1/4-in square up to 1/2-in square. These radii will give an acceptable surface finish, but larger nose radii can also be used. There are no hard and fast rules.

Grinding Tool Bits

Figures 9–29A to D demonstrate both a good method of holding tool bits and a good method of applying tool bits to the grinding wheel. Note that the tools are not being held in a tool holder, for reasons that have already been stated.

Figure 9–29A shows the grinding of front or end clearance and also end cutting-edge angle. Figure 9–29B shows the grinding of side clearance and side

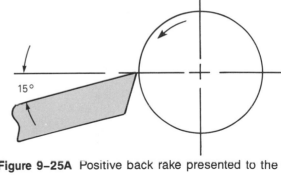

Figure 9–25A Positive back rake presented to the work

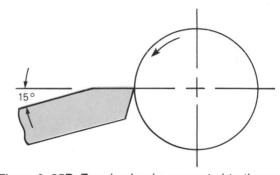

Figure 9–25B Zero back rake presented to the work

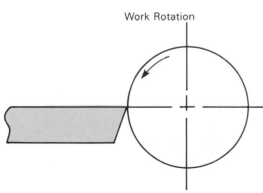

A. Zero back rake

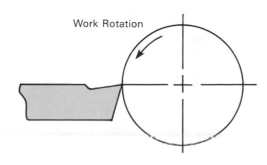

B. Positive back rake

Figure 9–24 Back rake

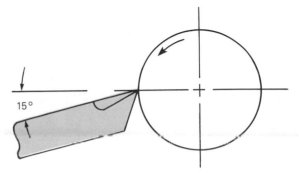

Figure 9–26 A compound back and side rake presented to the work

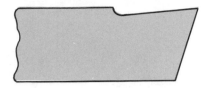

A. Positive back rake

B. Zero back rake

C. Negative back rake

Figure 9–27 Back rake

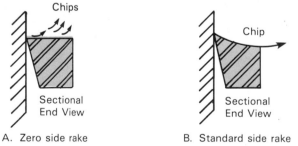

A. Zero side rake

B. Standard side rake

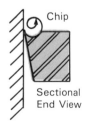

C. Chip curler

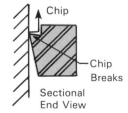

D. Chip breaker

Figure 9–28 Types of side rake

cutting-edge angle, more commonly called lead angle. The grinding of side and/or back rake is shown in Figure 9–29C. The grinding of a chip curling groove on the corner of a grinding wheel is shown in Figure 9–29D.

Finally, when you think your tool bit is correctly ground, examine it closely and critically. If the face or flank or nose show more than one facet on each or any of their surfaces, go back to the grinder and make these surfaces smooth. In other words, make the tool "look pretty," as though it had been ground by a professional. Do not be satisfied with, "it's good enough." Last of all, hone the cutting edge to make it keen and smooth, especially in the case of a finishing tool.

Table 9–1 (page 340) presents suggested cutting angles on high-speed steel tool bits for turning, facing, boring, and taper turning.

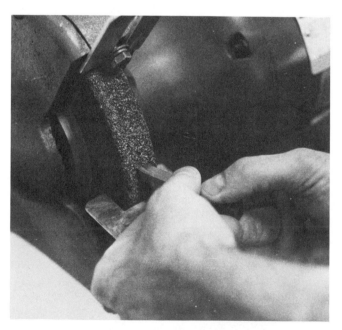

Figure 9–29A Grinding front clearance and end cutting-edge angle

Figure 9–29B Grinding side clearance and side cutting-edge angle (lead angle)

Figure 9–29D Grinding a chip curler groove

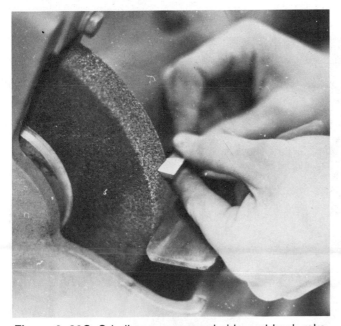

Figure 9–29C Grinding a compound side and back rake

LASER BEAMS – CLADDING AND SURFACE ALLOYING

Among the many exciting aspects of machining technology and manufacturing today is the increasing use of lasers. A laser is a device which amplifies light waves by the stimulated emission of radiation. The intense narrow beam of light produced by a laser can be focused to pinpoint accuracy. One example of the use of a laser is in delicate eye surgery where a detached retina can be "welded" in place without harming other eye tissues. In manufacturing, lasers are being used for cutting and welding metals, metal cladding (hard facing or hard surfacing), surface hardening, drilling, and to some extent for some types of milling cuts. At present, the latter use has some limitations, such as problems with surface finish. Certain basic milling cuts also present difficulties. Laser beams are also being used experimentally in machine vision systems, in precision measuring, and for various lathe operations.

Metal cladding and surface alloying both require intense heat and precision energy control.

Cladding, Figure 1, is also known as hard facing or hard surfacing. It is a welding process in which hard materials, such as nickel, chromium, and manganese, are alloyed and welded to the surface of a softer metal, or substrate, to give that metal increased corrosion and wear resistance. Cladding improves the ability of the part to withstand a harsh environment. An example of the use of the process is the hard surfacing of a bulldozer blade so that it will wear better against harsh abrasion.

The cladding process is an economical technique for altering the surface of a softer part to improve its wear resistance. The relatively inexpensive base material (the blade) is coated with a more expensive hard facing blend or alloy of metals. This is the most inexpensive method of obtaining the unique properties of the hard surfacing alloy. Hard surfacing conserves costly, strategic materials since the bulk of the part is made from relatively inexpensive materials. Most

Figure 1 Laser cladding application *(Courtesy of Combustion Engineering, Inc.)*

commercial hard facing alloys are iron base, nickel base, cobalt base, or carbide. Carbides, primarily tungsten carbide plus other alloys, are the most expensive hard surfacing materials, but they provide the longest life. The iron base, nickel base, and cobalt base materials are less expensive and are entirely satisfactory for most applications.

Conventional cladding processes include tungsten inert gas (TIG) welding, plasma transferred arc, plasma spray, and flame spray. A common problem with these processes is porosity in the hard surfacing material, resulting in poor surface adhesion and cracking. Uneven coating thicknesses require extensive grinding for proper finishing. This is especially true for the hard surfacing obtained by welding procedures. Expensive coating material can be wasted and more rebuilding is often necessary. High heat input can result in lower surface hardness, reduced wear and reduced corrosion resistance. Parts can often become distorted or warped because of the high heat input, especially with the various welding processes.

The process of cladding uses a high-power carbon dioxide laser. The cladding material is introduced in the form of preplaced chips or powder, or it is fed into the interaction zone by means of a powder feeder, Figure 2. As a result, very accurate control of clad thickness and degree of dilution (blending) is obtained. The clad surface is smooth enough to require only 0.005 inch (0.127 mm) to 0.010 inch (0.254 mm) for finish machining.

Laser surface alloying changes the chemical composition of the surface of the substrate or part. The surface is melted to a certain predetermined depth and the desired alloying elements are introduced to the meltpool. The flow of energy to the surface must be intense enough to allow local melting. However, the amount of energy must be accurately controlled to ensure the correct amount of melting without excessive heating of the substrate. If too much heat is applied to the substrate or part, warping or distortion will occur. These requirements are met by the surface alloying performance of the laser as the energy source.

Laser surface alloying, Figure 3, is carried out in the same way as laser cladding, but the processing parameters are adjusted to give greater dilution (or blending). Laser surface alloying combinations of alloy and substrate are nearly endless. Stainless steel (chromium/nickel) surfaces on low carbon steel, and manganese, chromium, and carbon surfaces added to carbon steels are just two of the possible combinations of surface alloys that can be obtained by laser processing.

By coupling the abilities of the laser beam with the abilities of computer control, lasers are taking their place as one of the newest tools in modern manufacturing.

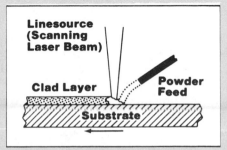

Figure 2 Laser cladding *(Courtesy of Combustion Engineering, Inc.)*

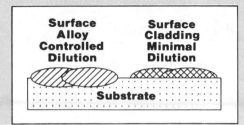

Figure 3 Laser cladding and surface alloying with overlapping *(Courtesy of Combustion Engineering, Inc.)*

Table 9–1 Suggested Cutting Angles for High-Speed Steel Tool Bits

MATERIAL	FRONT RELIEF ANGLE (°)	SIDE RELIEF ANGLE (°)	BACK RAKE ANGLE (°)	SIDE RAKE ANGLE (°)
High-Speed Steels (annealed)	10	6–8	8	12
Tool Steels (annealed) — alloy	10	6–8	8	12
Tool Steels (annealed) — high carbon	10	6–8	8	12
Stainless Steels (chip breaker a must)	10	6–8	8	12
SAE Steels: 1020	10–12	6–8	15	14
1045	10–12	6–8	12	14
1095	10	6–8	8	12
1320 (Mn)	10	6–8	15	15
2317 (Ni)	10	6–8	12	14
3120 (Ni – Cr)	10	6–8	12	14
4340 (Mo)	10	6–8	8	12
Aluminum	15–20	12–15	15–30	15–35
Bakelite — pressure-molded plastics	12	8	0	0
Brass — free cutting	10–12	8–10	0	0
— red and yellow	10–12	8–10	0	0
Bronze — cast, commercial	10–12	8	0–5	0–5
— free cutting	10–12	8	0	0
— hard	10	8	0	0
— phosphor	10	8–10	0	0
Cast Iron — gray	5–10	6–10	5–8	10–12
Celluloid — cast plastics	10–15	8–10	0–5	0
Copper	10–15	10–12	16	20
Fiber	10–15	10–12	0	0
Formica Gear Material	10–15	10	16	10
Monel Metal	10	6–8	8	14
Nickel	10	6–8	8	14
Nickel Silvers	10	6–8	0	0
Rubber — hard	10–15	10–15	0–3	0–7

REVIEW QUESTIONS

1. Explain the difference between roughing and finishing tools.
2. What is the effect of a chip curler compared to a cutting tool with a standard rake with regard to the cutting action?
3. What is meant by the shape of a cutting tool?
4. Some tool holders hold a tool bit at a 15–20 degree angle; other tool holders hold a tool bit at zero degrees. What difference will this make when grinding clearance?
5. Sketch a tool bit and show all its parts.
6. Sketch a tool bit and show all its angles.
7. Should a tool bit be ground in a tool holder?
8. Is it good practice to hone a tool bit? Give reasons for your answer.
9. List the three most widely used materials for cutting tools. Give one advantage and one application for each of these materials.
10. Explain what is meant by the term red hardness.

11. Compare the speeds and feeds for cemented carbide cutting tools as compared to high-speed steel cutting tools.
12. What is the reason for grinding clearance in a cutting tool?
13. What happens if you grind too much clearance?
14. Where do you find the cutting edge on a tool bit?
15. How many clearances are there on a right-hand turning tool?
16. What is the purpose of a front or end clearance on a turning tool?
17. What is the purpose of side cutting-edge angles?
18. What is a more common term for side cutting-edge angle?
19. What angles are usual for a side cutting-edge angle?
20. What is the purpose of an end cutting-edge angle?
21. What angles are usual for an end cutting-edge angle?
22. How many clearances are there on factory-made parting tools?
23. What is rake on a tool bit?
24. What is the purpose of rake on a tool bit?
25. What rakes are there on a lathe tool bit?
26. List three types of back rake and the factors that determine which type is required for a specific job.
27. Compare four types of side rake and give the advantages or disadvantages of each type.
28. What recommendation is made for clearance angles in single-point cutting tools?
29. What is the importance of nose radius?
30. List three considerations in obtaining a properly ground cutting tool.

UNIT 9-3

CARBIDE TOOLING

OBJECTIVES

After completing this unit, the student will be able to:
- explain the difference between indexable carbide insert tools and standard brazed carbide tools and give an advantage of each.
- select the proper carbide cutting tool for the specific job.
- identify tool holders and inserts using identification systems developed by the American Standards Association.
- use carbide cutting tools properly based on the machining recommendations provided.
- identify and correct common problems that occur in the use of carbide cutting tools.

KEY TERMS

Carbide cutting tool
Carburize
Binder
Sinter
Cemented carbide
Alloy steel body
Tool nomenclature
Indexable insert
  cutting tool
Standard brazed
  cutting tool
Tool style
Insert
Negative rake tool
Positive rake tool

Chip breaker plate
Utility insert
Precision insert
Tool holder
  identification system
Boring bar
Indexable inserts
  identification system
Continuous chip
Discontinuous
  (crumbling) chip
Edge wear
Top wear (cratering)
Heat checks or cracks

INTRODUCTION

There are many manufacturers of carbide tooling in the modern machining industry. A few of the more common trade names used are Kennametal, Carboloy, Sandvik, and Wesson, each having differences in composition and design of cutting tools and tool holders.

The essential ingredients of carbide cutters are tungsten (W), tantalum (Ta), columbium (Cb), and titanium (Ti), with a metal-like cobalt (Co) acting as a binder. The invention of WTiC (tungsten-titanium-carbide) was the first commercially successful carbide for cutting steel. Blends of different carbides will give different grades of cutting tools, each one superior for a particular job.

Essentially, pure tungsten is carburized, that is, impregnated with carbon, to form tungsten carbide. It is one of the hardest known substances, next to diamond, but it is brittle and porous. The carbide or a mixture of carbides and a binder (cobalt is often used) is powdered and thoroughly blended. The blend of powders is poured into a small mold and is hydraulically pressed into a small cake under tons of pressure. The cake is ejected from the mold and then sintered in an oxygen-free furnace. The binder holds or cements the carbide particles together, hence the name cemented carbide.

The resulting cemented carbide is extraordinarily hard, but it is also brittle. It is not used directly in a tool holder like a high-speed steel tool bit. It is either brazed to a supporting tough alloy steel body or it is clamped as an insert into a tough steel tool holder specially made for the job of holding the insert. Cutting speeds attainable are three times faster than for high-speed steel cutting tools. With the advent of ceramic cutting tools (aluminum oxide) in recent years, cutting speeds are higher yet, and formerly non-machinable materials such as nickel and white cast iron are now machinable. Carbide cutting tools need power and speed to perform adequately and properly.

TOOL NOMENCLATURE

There are many similarities in nomenclature between high-speed steel tool bits and carbide tools. The body of the tool, which is held in the machine and which supports the cutting edge, is known as the shank. The cutting material, which may be clamped or brazed to the shank, is called the insert or tip. Figure 9–30 illustrates carbide-tool nomenclature.

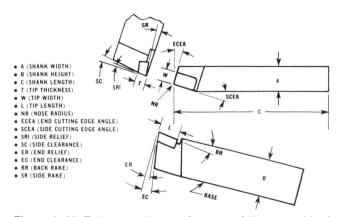

- A (SHANK WIDTH)
- B (SHANK HEIGHT)
- C (SHANK LENGTH)
- T (TIP THICKNESS)
- W (TIP WIDTH)
- L (TIP LENGTH)
- NR (NOSE RADIUS)
- ECEA (END CUTTING EDGE ANGLE)
- SCEA (SIDE CUTTING EDGE ANGLE)
- SRI (SIDE RELIEF)
- SC (SIDE CLEARANCE)
- ER (END RELIEF)
- EC (END CLEARANCE)
- BR (BACK RAKE)
- SR (SIDE RAKE)

Figure 9–30 Tool nomenclature *(Courtesy of Kennametal Inc.)*

BASIC TOOL TYPES

There are two basic tool types used in carbide cutting. They are indexable insert tools and standard brazed tools.

Indexable insert tools are the most economical for practically all types of metal-cutting operations. The inserts provide a number of low cost, indexable cutting edges. After all edges are used, it is more economical for industry to replace the insert than it is to pay a machinist to regrind a brazed tool. Various methods are used to lock the insert in position. A cam-type lock pin, or a clamp, or a combination of lock pin and clamp are some of the methods used.

Standard brazed tools, low in initial cost, are suitable for most general-purpose machining operations. They can be modified easily for special-purpose, short-run applications. These tips are usually silver soldered or brazed to an alloy steel shank. They can be resharpened many times.

These basic tool types are illustrated in Figure 9–31.

Standard Tool Styles

Most carbide tool companies conform to industry standards in styles and sizes of cutting tools for general types of machining operations. Figure 9–32 illustrates the styles of cutting tools available and the operations they will perform.

Indexable insert tools use flat, multiple-edge inserts in triangular, square, round, and diamond shapes. Inserts are clamped in heat-treated holders at negative, positive, or neutral rake angles.

Negative rake tools permit inserts to be turned over, thereby doubling the number of cutting points available. Inserts used with positive rake tools are indexable, but cutting edges are provided on one face only. Up to eight new cutting edges are available by loosening the clamp and rotating the insert or by turning it over. Insert seats or shims provide hard, flat, back-up surfaces for the insert to sit on. Shims also provide positive protection to the holder.

Effective chip control is provided by solid carbide chip-breaker plates which have exceptional strength and resistance to "pick-up." Some inserts also have preformed chip grooves that provide constant chip control over a wide range of feeds on general machining operations.

Selecting Tools

Following are suggestions for when to use negative rake tools and positive rake tools. Use negative rake tools for general-purpose machining of most materials, especially for rough or interrupted cuts. They are also used for hard materials on rigid setups.

Use positive rake tools for machining softer steels and nonferrous metals; for gummy, work-hardening alloys; for slender parts or thin wall tubing that will not stand high cutting forces; and on low-powered machines or setups that lack rigidity.

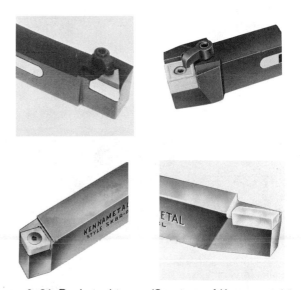

Figure 9–31 Basic tool types *(Courtesy of Kennametal Inc.)*

Kennametal Standard Tool Styles

Kennametal tools conform to industry standards and are made in styles and sizes for all general types of machining operations:

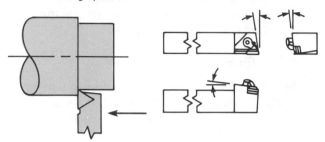

Style A for turning, facing, or boring to a square shoulder.

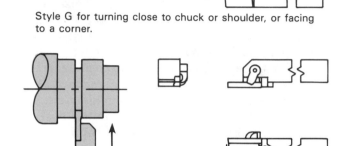

Style G for turning close to chuck or shoulder, or facing to a corner.

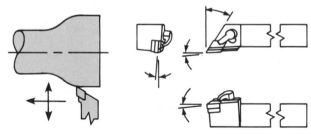

Style GCH for deep grooving.

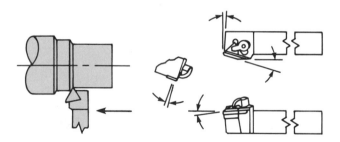

Style B for rough turning, facing, or boring where a square shoulder is not required.

Style J for profiling and finish turning.

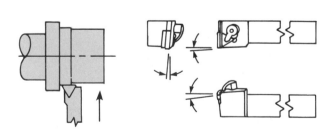

Style F for facing, straddle facing, or turning with shank parallel to work axis.

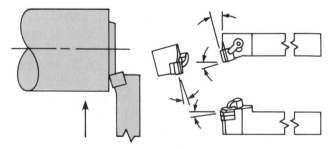

Style K for lead angle facing or turning with shank parallel to work axis.

Figure 9–32 Standard tool styles and machining operations *(Courtesy of Kennametal Inc.)*

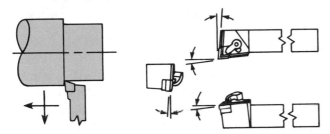

Style L for both turning and facing with same tool. 80° diamond insert.

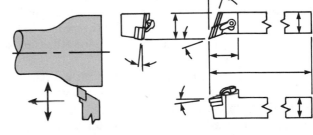

Style Q for profile machining 55° insert.

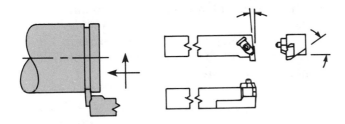

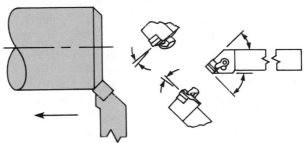

Style NE and Style NS for threading and grooving operations. Inserts available for 60° V, Acme, Buttress and API threads.

Style S for chamfering and facing. 45° lead angle.

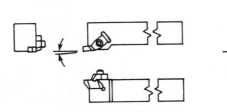

Style U for profile machining 35° insert.

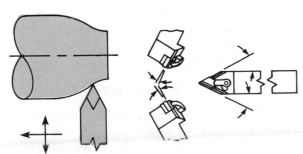

Style P for profile machining. Insert centrally located.

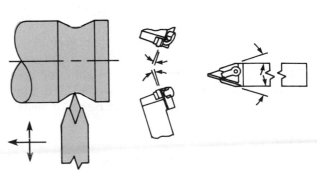

Style V for profile machining. 35° insert.

Figure 9–32 Continued

In selecting tool holders and inserts, the following suggestions are common for carbide cutting. Determine the proper tool style for the work. Select an insert with adequate cutting-edge length. Choose the largest shank possible.

When selecting utility or precision inserts, keep the following in mind. Utility inserts have both the top and the bottom ground for rough machining. Precision inserts have all surfaces ground for general rough and finish machining. Inserts with preformed, chip-control grooves can be used for general rough and finish turning, particularly for ductile materials.

The following cutting tool shapes are available. Square inserts have a strong structural shape (90-degree point angle). They are used in lead angle and chamfering tools. Triangular inserts (60-degree point angle) are used for cutting to a square shoulder, for profiling, chamfering, or plunge turning. Round inserts provide shallow feed marks at high feed rates on finishing passes. They are particularly suitable for cutting cast iron because they decrease the incidence of edge breakout. Diamond inserts (80-degree nose angle) are used for combination turning and facing tools. Inserts with a 55-degree nose angle are for profiling tools. Rectangular inserts are used for heavy-duty Kendex tools on cuts greater than 1/2 in (12.7 mm) in depth. Threading and grooving inserts are also available.

When using a nose radius, use a small nose radius for steel and materials that cut with a continuous chip. Use a larger nose radius for cast iron (discontinuous chip).

Chip-breaker plates are available in various widths to match the size and shape of an insert. Inserts with preformed chip grooves are available for general machining operations. The preformed chip grooves provide chip control over a wide range of feeds and speeds.

Tool Holder Identification System

The identification systems shown in Tables 9–2 and 9–3 have been developed by the American Standards Association and are generally adopted by the carbide cutting tool industry. These charts make identification of tool holders and inserts fast and accurate.

Boring Bars

Boring bars, Figures 9–33 and 9–34, are made with steel shanks or with tungsten-carbide shanks. The latter type takes advantage of the high-rigidity of tungsten carbide, which is three times that of hardened steel.

The following suggestions should be considered when selecting boring bars: use the largest diameter bar for the bore; hold overhang to a minimum; use carbide bars where deflection is excessive; and use carbide bars when overhang-to-bar diameter exceeds 5 to 1.

For best results, the boring bar must be held in the machine as rigidly as possible.

Kennametal Grades

Different grades of carbide cutting tools are available, depending on the machining conditions of the work. In the Kennametal system, for example, three general-purpose grades (K42, K45, and K68) can machine about 90% of the work in most shops.

The KC75 cutter is a coated insert with a micro-thin titanium-carbide coating (0.0002 in or 0.005 mm thick) on a specially alloyed, tough core grade of carbide. This coated insert is another general-purpose grade of cutter that permits higher operating speeds and feeds, resulting in higher metal removal rates than previously attainable.

Table 9–4 (page 351) lists different grades of cutting tools and the typical machining applications for which they are suited.

MACHINING RECOMMENDATIONS

The following tips will be valuable when using carbide cutting tools for carbon and alloy steels.

These steels have much the same cutting characteristics in that both cut with a continuous chip and form a built-up edge on the tool if run at low speeds. As speed increases, a critical point is reached above

Table 9-2 Tool Holder Identification System *(Courtesy of Kennametal Inc.)*

toolholder identification system

This identification system was developed for qualified holders, and has been used in listing the catalog numbers for qualified holders shown in this catalog.

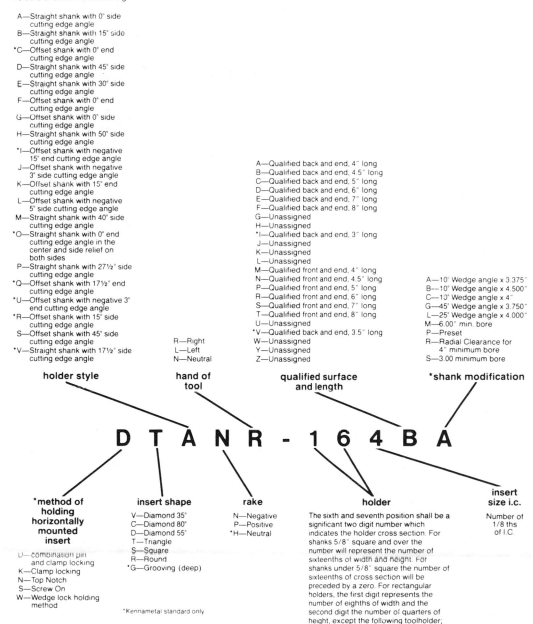

A—Straight shank with 0° side cutting edge angle
B—Straight shank with 15° side cutting edge angle
*C—Offset shank with 0° end cutting edge angle
D—Straight shank with 45° side cutting edge angle
E—Straight shank with 30° side cutting edge angle
F—Offset shank with 0° end cutting edge angle
G—Offset shank with 0° side cutting edge angle
H—Straight shank with 50° side cutting edge angle
*I—Offset shank with negative 15° end cutting edge angle
J—Offset shank with negative 3° side cutting edge angle
K—Offset shank with 15° end cutting edge angle
L—Offset shank with negative 5° side cutting edge angle
M—Straight shank with 40° side cutting edge angle
*O—Straight shank with 0° end cutting edge angle in the center and side relief on both sides
P—Straight shank with 27½° side cutting edge angle
*Q—Offset shank with 17½° end cutting edge angle
*U—Offset shank with negative 3° end cutting edge angle
*R—Offset shank with 15° side cutting edge angle
S—Offset shank with 45° side cutting edge angle
*V—Straight shank with 17½° side cutting edge angle

R—Right
L—Left
N—Neutral

A—Qualified back and end, 4" long
B—Qualified back and end, 4.5" long
C—Qualified back and end, 5" long
D—Qualified back and end, 6" long
E—Qualified back and end, 7" long
F—Qualified back and end, 8" long
G—Unassigned
H—Unassigned
*I—Qualified back and end, 3" long
J—Unassigned
K—Unassigned
L—Unassigned
M—Qualified front and end, 4" long
N—Qualified front and end, 4.5" long
P—Qualified front and end, 5" long
R—Qualified front and end, 6" long
S—Qualified front and end, 7" long
T—Qualified front and end, 8" long
U—Unassigned
*V—Qualified back and end, 3.5" long
W—Unassigned
Y—Unassigned
Z—Unassigned

A—10° Wedge angle x 3.375"
B—10° Wedge angle x 4.500"
C—10° Wedge angle x 4"
G—45° Wedge angle x 3.750"
L—25° Wedge angle x 4.000"
M—6.00" min. bore
P—Preset
R—Radial Clearance for 4" minimum bore
S—3.00 minimum bore

holder style **hand of tool** **qualified surface and length** ***shank modification**

D T A N R - 1 6 4 B A

***method of holding horizontally mounted insert**

D—combination pin and clamp locking
K—Clamp locking
N—Top Notch
S—Screw On
W—Wedge lock holding method

insert shape

V—Diamond 35°
C—Diamond 80°
D—Diamond 55°
T—Triangle
S—Square
R—Round
*G—Grooving (deep)

*Kennametal standard only

rake

N—Negative
P—Positive
*H—Neutral

holder

The sixth and seventh position shall be a significant two digit number which indicates the holder cross section. For shanks 5/8" square and over the number will represent the number of sixteenths of width and height. For shanks under 5/8" square the number of sixteenths of cross section will be preceded by a zero. For rectangular holders, the first digit represents the number of eighths of width and the second digit the number of quarters of height, except the following toolholder; 1¼" x 1½" which is given the number 91

insert size i.c.

Number of 1/8 ths of I.C.

Table 9–3 Indexable Inserts Identification System *(Courtesy of Kennametal Inc.)*

indexable inserts identification system

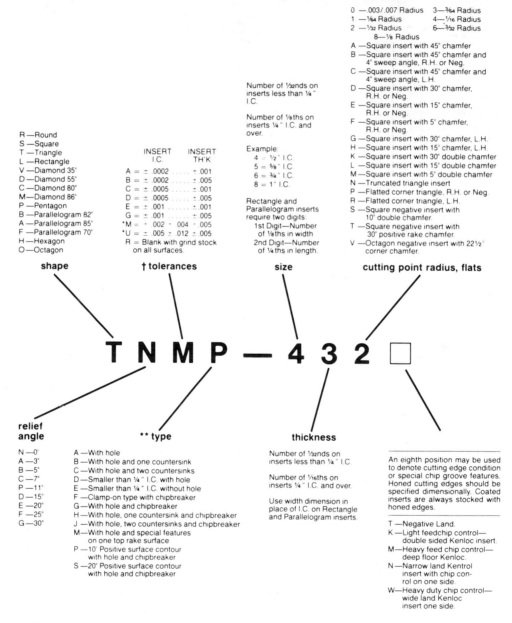

0 —.003/.007 Radius   3—³⁄₆₄ Radius
1 —¹⁄₆₄ Radius   4—¹⁄₁₆ Radius
2 —¹⁄₃₂ Radius   6—³⁄₃₂ Radius
            8—¹⁄₈ Radius
A —Square insert with 45° chamfer
B —Square insert with 45° chamfer and
      4° sweep angle, R.H. or Neg.
C —Square insert with 45° chamfer and
      4° sweep angle, L.H.
D —Square insert with 30° chamfer,
      R.H. or Neg.
E —Square insert with 15° chamfer,
      R.H. or Neg.
F —Square insert with 5° chamfer,
      R.H. or Neg.
G —Square insert with 30° chamfer, L.H.
H —Square insert with 15° chamfer, L.H.
K —Square insert with 30° double chamfer
L —Square insert with 15° double chamfer
M —Square insert with 5° double chamfer
N —Truncated triangle insert
P —Flatted corner triangle, R.H. or Neg.
R —Flatted corner triangle, L.H.
S —Square negative insert with
      10° double chamfer.
T —Square negative insert with
      30° positive rake chamfer.
V —Octagon negative insert with 22½°
      corner chamfer.

Number of ¹⁄₃₂nds on inserts less than ¼″ I.C.

Number of ¹⁄₈ths on inserts ¼″ I.C. and over.

Example:
4 = ½″ I.C.
5 = ⅝″ I.C.
6 = ¾″ I.C.
8 = 1″ I.C.

Rectangle and Parallelogram inserts require two digits:
1st Digit—Number of ⅛ths in width
2nd Digit—Number of ¼ths in length.

R —Round
S —Square
T —Triangle
L —Rectangle
V —Diamond 35°
D —Diamond 55°
C —Diamond 80°
M —Diamond 86°
P —Pentagon
B —Parallelogram 82°
A —Parallelogram 85°
F —Parallelogram 70°
H —Hexagon
O —Octagon

	INSERT I.C.	INSERT TH'K
A =	± .0002	± .001
B =	± .0002	± .005
C =	± .0005	± .001
D =	± .0005	± .005
E =	± .001	± .001
G =	± .001	± .005
*M =	± .002 ± .004	± .005
*U =	± .005 ± .012	± .005

R = Blank with grind stock on all surfaces.

shape   **† tolerances**   **size**   **cutting point radius, flats**

T N M P — 4 3 2 □

relief angle

N —0°
A —3°
B —5°
C —7°
P —11°
D —15°
E —20°
F —25°
G —30°

** type

A —With hole
B —With hole and one countersink
C —With hole and two countersinks
D —Smaller than ¼″ I.C. with hole
E —Smaller than ¼″ I.C. without hole
F —Clamp-on type with chipbreaker
G —With hole and chipbreaker
H —With hole, one countersink and chipbreaker
J —With hole, two countersinks and chipbreaker
M —With hole and special features
      on one top rake surface
P —10° Positive surface contour
      with hole and chipbreaker
S —20° Positive surface contour
      with hole and chipbreaker

thickness

Number of ¹⁄₃₂nds on inserts less than ¼″ I.C.

Number of ¹⁄₁₆ths on inserts ¼″ I.C. and over.

Use width dimension in place of I.C. on Rectangle and Parallelogram inserts.

An eighth position may be used to denote cutting edge condition or special chip groove features. Honed cutting edges should be specified dimensionally. Coated inserts are always stocked with honed edges.

T —Negative Land.
K —Light feedchip control—
      double sided Kenloc insert.
M —Heavy feed chip control—
      deep floor Kenloc.
N —Narrow land Kentrol
      insert with chip con-
      rol on one side.
W —Heavy duty chip control—
      wide land Kenloc
      insert one side.

*Exact tolerance is determined by size of insert.
**Shall be used only when required.
†A & B I.C. Tolerances only in uncoated grades.

which the built-up edge is swept away, cutting is more efficient, tool life is greatly extended, and finish is improved. This critical speed is affected by the hardness of the steel, chip thickness, and, to a lesser degree, the depth of cut. Good cutting practice is usually 50% to 100% above critical speed.

Because of the strong continuous chip, steel tends to **crater** (erode the top face of the tool). A flat chip will curl away from the face of the tool with comparatively light force, but the same amount of steel in a channel-shaped chip has greater structural rigidity and requires more force to deflect. A large nose radius produces a flat ribbon-like chip, tool life is better, and less power is required.

There are two general groups of stainless steels, the hardenable (magnetic) and the austenitic (non-magnetic). Hardenable stainless steels machine much the same as alloy steels of equal hardness. Austenitic stainless steels are work-hardening, yet are soft and gummy in their tendency to tear and build up on the cutting edge. The build-up tendency calls for high speeds, but the work-hardened chip calls for speeds in the lower ranges. A compromise, then, is a feed rate heavy enough to get under the work-hardened surface of the previous stroke or revolution and a speed high enough to avoid excessive build-up.

Gray cast iron machines with a discontinuous or a crumbling chip that has very little tendency to build up along the cutting edge. As a result, machining techniques for gray cast iron are different than those for steel. Depending on feed rate and depth of cut, cutting speeds up to 400 ft/min (122 m/min) are common. An example of normal cutting would be a feed of 0.025 in (0.64 mm), a depth of 0.200 in (5 mm), and a cutting speed of 275 ft/min (83 m/min).

The low-strength chip breaks into a crumbly powder, so a large nose radius permits better finish, faster feed, and longer tool life.

The addition of nickel or steel scrap will produce a higher strength cast iron called high-tensile cast iron. There is little difference in the machining characteristics except that the chip is mechanically stronger and has more tendency to crater the top surface of the tool.

Nonferrous materials such as alloys of copper

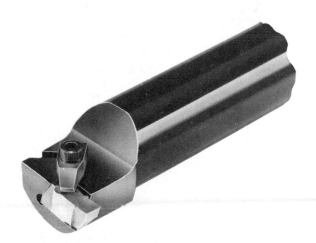

Figure 9–33 Steel shank boring bars *(Courtesy of Kennametal Inc.)*

(brasses and bronzes) machine with a low-strength chip and can be run 500 to 1000 ft/min (150 to 300 m/min) depending on hardness. Aluminum-bronze alloys have some tendency to build up on the cutting edge, but a cutting speed of 250 to 500 ft/min (75 to 150 m/min) will prevent this build-up.

Aluminum and magnesium alloys machine readily with low-tensile chips that exert little pressure on the tool. Rakes and clearances can be as much as 30 degrees, with cutting speeds in excess of 500 ft/min (150 m/min).

None of the common types of plastics present any real machining problems, but when combined with fillers or fibers such as clay, asbestos, cotton, paper, glass, etc., they tend to become abrasive. The more abrasive the filler, the lower the practical cutting speed should be, ranging from 200 to 12000 ft/min (60 to 3660 m/min.)

High-temperature alloys are difficult to machine due to their high-alloy content and must be cut at reduced cutting speeds. There are five basic reasons for the difficulty in machining these super alloys: they have a tendency to work-harden; they have high

strength levels; they are abrasive; high heat is generated; and they have low thermal conductivity.

These alloys can be grouped roughly into three major classifications: iron base, nickel base, and cobalt base. Machining recommendations for these alloys are given in Table 9–5.

Tips for Machining Super Alloys

1. Tool must be kept sharp at all times.
2. Tool should be lightly honed.
3. Carbide tools are almost mandatory, but there are exceptions.
4. Tool overhang must be keep to a minimum.
5. Avoid excessive tool wear to minimize cutting forces.
6. Use positive rake tools wherever possible.
7. Use negative rake tools only where necessary and when surface speeds can be kept in the higher ranges.
8. Machine tools must be kept rigid.
9. Machine tools should be over powered, rather than under powered.
10. The workpiece must be well-clamped.
11. The workpiece must be well-supported to avoid flexing.
12. Depth of cut should be deep enough to avoid glazing.
13. Feed should be positive to avoid dwelling and work hardening.
14. Minimum chip discoloration is desirable.
15. Do not allow the cutting tool to rub on the work. Keep it cutting.

The same good shop practice applied in machining carbon and alloy steels must be used in machining the high-temperature alloys.

Carbides can be used most successfully to machine these metals, but since most machining operations are special cases, no fixed rules will apply. In general, keep in mind that feeds are moderate, speeds and depths of cut are low, and tool life is relatively short. Table 9–6 contains data for suggested starting points. Use positive rake cutters.

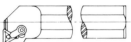
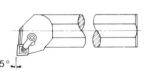

STYLE B-31-00 & B-32-00
0° lead angle for boring to a square shoulder—uses triangular inserts

STYLE B-41-00 & B-42-00
15° lead angle for through boring—uses square inserts

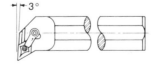

STYLE B-37-00 & B-38-00
5° and 5° reverse lead for boring and facing—uses 80° diamond inserts

STYLE B-36-00 3° reverse lead for profiling—uses 55° diamond inserts.

Figure 9–34 Boring bar styles *(Courtesy of Kennametal Inc.)*

Table 9-4 Cutting Tool Grades and Machining Applications *(Courtesy of Kennametal Inc.)*

Kennametal Grade System

Coated Carbide Grades	Typical Machining Application	Type Cooling
KC950	Productive over a wide range, from medium roughing to finishing cast irons, high-temperature alloys, and stainless, low carbon and alloy steels. Can handle up to 50 to 60 percent of a typical plant's metalcutting operations. Designed with strength to withstand interrupted cuts at ceramic-coated insert speeds.	Multi-layer titanium nitride, aluminum oxide, titanium carbide
KC910	High speed finishing to light roughing applications. May be used in machining carbon or alloy steels, cast irons, some high temperature alloys, and stainless steels. Maximum edge wear and heat resistance for very abrasive or hot machining cuts.	Aluminum-oxide coated
KC850	Finishing to heavy roughing applications ideally suited to conditions that demand *maximum edge strength and wear resistance*. May be used in machining carbon or alloy steels, most cast irons high temperature and stainless steel alloys. Superior thermal and mechanical shock resistance make this an excellent heavy roughing grade for milling and turning all types of steel.	Multi-phase TiN coated
KC810	General purpose grade for finishing to roughing applications in steels, cast irons and stainless steel alloys at conventional coated carbide speeds.	Multi-phase TiN coated
KC250	Semi-finishing to heavy roughing applications in cast irons and stainless steel alloys. Ideally suited to conditions that demand high edge strength and wear resistance such as roughing cuts on stainless steel castings. Also suitable for machining some high temperature alloys and steels at moderate speeds. Superior mechanical shock resistance.	Multi-phase TiN coated
KC210	Finishing to light roughing applications on cast irons, stainless steel alloys, high temperature alloys and heat treated steels. Excellent wear in very abrasive cuts.	Multi-phase TiN coated

Uncoated	Hardness HRA	Typical Machining Applications
K7H	93.5	Precision finishing of steels and alloyed cast irons at high speeds and low to moderate chip loads. Frequently applied in single point threading of steels heat treated to a high hardness.
K45	92.5	May be used in finishing and light roughing of all steels. Excellent crater, edge wear and thermal shock resistance. Frequently applied in grooving where maximum edge wear resistance is required.
K4H	92.0	Excellent threading and form tool grade for steels and cast irons. May also be used for semi-finishing to light roughing of steels and cast irons at moderate speeds and chip loads.
K2884	92.0	General purpose steel *milling* grade that may be used in moderate to heavy chip loads. Excellent edge wear and mechanical shock resistance.
K420	91.3	Heavy to moderate steel machining grade. Superior edge strength and thermal shock resistance. Used in steel milling or turning through severe interruptions at high chip loads.
K21	91.0	For light to heavy roughing of steels at moderate speeds and chip loads. Excellent mechanical and thermal shock resistance.
K11	93.0	Primarily used for precision finishing of cast irons, nonferrous alloys, nonmetals at high speeds and light chip loads. K11 may also be used in finishing hard steels at moderate speeds and light chip loads. Excellent edge wear resistance.
K68	92.5	General purpose turning, milling and threading grade for light roughing to finishing of cast irons, nonferrous alloys and nonmetals at moderate to high speeds and light chip loads. Excellent edge wear resistance for machining all nonferrous alloys and most high temperature alloys.
K6	92.0	Moderate roughing grade for cast irons, nonferrous alloys, nonmetals and most high temperature alloys. High edge strength and good edge wear resistance.
K8735	92.0	Excellent *milling* and broaching grade for gray, malleable and nodular cast irons at high speeds and light chip loads. Also superior resistance to "built-up-edge" in machining all stainless steels and aluminum alloys.
K1	90.0	Excellent mechanical shock resistance for roughing through heavy interruptions when machining stainless steels, most high temperature alloys, cast irons and steels, and rough cast nonferrous alloys. Machining speeds will be low to moderate in most applications.

Table 9–4 Continued

Kyon and Ceramic Grades	Hardness HRA	Typical Machining Applications
Kyon 2000		A new cutting tool material designed for high velocity machining of cast irons and nickel-base alloys with heavier chip loads than solid ceramics or ceramic-coated carbides. Offers excellent thermal shock and impact resistance for use with coolants and milling applications.
K090	94 — 94.5	A black hot pressed ceramic for machining cast irons over 235 Bhn and to 66 HRC, also for steels over 34 HRC to 66 HRC. Excellent edge wear resistance and hot hardness for machining most high temperature alloys.
K060	93 — 94	A white cold pressed ceramic for machining cast irons below 235 Bhn and steels under 34 HRC. Superior edge wear resistance in machining most cast irons below 235 Bhn.

Table 9–5 Machining Recommendations for High-Temperature Alloys *(Courtesy of Kennametal Inc.)*

IRON-BASED ALLOYS	ROUGHING	GENERAL	FINISHING
Speed Feed Depth of cut Grade	30–100 ft/min (9–30 m/min) 0.010–0.040 in (0.25–1.00 mm) 0.125–0.375 in (3.0–9.5 mm) K21–K42	50–125 ft/min (15–38 m/min) 0.008–0.020 in (0.20–0.50 mm) 0.0625–0.250 in (1.5–6.5 mm) K42–K6	75–200 ft/min (23–60 m/min) 0.003–0.010 in (0.08–0.25 mm) 0.031–0.093 in (0.8–2.4 mm) K68–K8

NICKEL-BASED ALLOYS	ROUGHING	GENERAL	FINISHING
Speed Feed Depth of cut Grade	30–90 ft/min (9–27 m/min) 0.010–0.035 in (0.25–0.90 mm) 0.125–0.250 in (3.0–6.5 mm) K21–K42	50–100 ft/min (15–30 m/min) 0.008–0.020 in (0.20–0.50 mm) 0.0625–0.1875 in (1.5–4.8 mm) K42–K6	70–175 ft/min (21–53 m/min) 0.003–0.010 in (0.08–0.25 mm) 0.031–0.093 in (0.8–2.4 mm) K68–K8

COBALT-BASED ALLOYS	ROUGHING	GENERAL	FINISHING
Speed Feed Depth of cut Grade	25–75 ft/min (8–23 m/min) 0.010–0.030 in (0.25–0.75 mm) 0.0625–0.1875 in (1.5–4.8 mm) K21–K42	40–90 ft/min (12–27 m/min) 0.008–0.020 in (0.20–0.50 mm) 0.031–0.125 in (0.8–3.0 mm) K42–K45	60–125 ft/min (18–38 m/min) 0.003–0.010 in (0.08–0.25 mm) 0.031–0.093 in (0.8–2.4 mm) K45–K68

Table 9–6 Recommendations for Machining Super Alloys *(Courtesy of Kennametal Inc.)*

	LEAD ANGLE	NOSE RADIUS in (mm)	GRADE	DEPTH OF CUT in (mm)	FEED in (mm)	SPEED fpm (m/min)
Molybdenum (Mo)						
Roughing	15°–30°	1/64–1/32 (0.38–0.79)	K6–K68	1/16–1/8 (1.59–3.18)	0.010 (0.25) 0.020 (0.50)	100–200 (30–60)
Finishing	15°–30°	1/64–1/32 (0.38–0.79)	K8–K11	1/32–1/16 (0.79–1.59)	0.005 (0.13) 0.015 (0.38)	150–300 (45–90)

Table 9–6 Continued

	LEAD ANGLE	NOSE RADIUS in (mm)	GRADE	DEPTH OF CUT in (mm)	FEED in (mm)	SPEED fpm (m/min)
Tungsten (W)						
Roughing	30°–45°	1/32–3/64 (0.79–1.14)	K6–K68	1/16–1/8 (1.59–3.18)	0.020 (0.50) 0.040 (1.00)	20–60 (8–18)
Finishing	15°–30°	1/64–1/32 (0.38–0.79)	K8–K11	1/32–1/16 (0.79–1.59)	0.015 (0.38) 0.030 (0.76)	50–80 (15–24)
Tantalum (Ta)						
Roughing	15°–30°	1/64–1/32 (0.38–0.79)	K21–K2S	1/16–3/16 (1.59–4.77)	0.015 (0.38) 0.030 (0.76)	60–100 (18–30)
Finishing	15°–30°	1/64–1/32 (0.38–0.79)	K4H–K45	1/64–1/16 (0.38–1.59)	0.005 (0.13) 0.020 (0.50)	80–200 (24–60)
Columbium (Cb)						
Roughing	30°–45°	1/32–3/64 (0.79–1.14)	K6–K68	1/32–1/16 (0.79–1.59)	0.010 (0.25) 0.030 (0.76)	25–60 (7.5–18)
Finishing	30°–45°	1/64–1/32 (0.38–0.79)	K8–K11	1/64–1/32 (0.38–0.79)	0.008 (0.20) 0.015 (0.38)	50–70 (15–21)

TROUBLESHOOTING CARBIDE CUTTING TOOLS

Some of the more common problems that occur in the use of carbide cutting tools and their probable causes follow.

1. Chipping.
   (a.) "Saw toothed" or too keen a cutting edge.
   (b.) Chip breaker too narrow.
   (c.) Chatter.
   (d.) Scale or inclusions (sand, slag, etc.).
   (e.) Incorrect grade.
   (f.) Too much relief.
   (g.) Lack of rigidity.
   (h.) Improper grinding.
2. Cracking or breaking.
   (a.) Feed too heavy.
   (b.) Worn or chipped cutting edges.
   (c.) Improperly applied coolant.
   (d.) Too much rake or relief.
   (e.) Too much overhang.
   (f.) Lack of rigidity in the setup.
   (g.) Speed too slow.
   (h.) Too much variation in depth of cut for size of tip.
   (i.) Incorrect grade.
   (j.) Chatter.
   (k.) Braze or grinding strains.
   (l.) Built-up edges.

3. Chatter.
   (a.) Tool not on center.
   (b.) Insufficient relief or clearance.
   (c.) Too much rake.
   (d.) Too much overhang.
   (e.) Nose radius too large.
   (f.) Feed too high.
   (g.) Lack of rigidity in the setup.
   (h.) Insufficient horsepower.

4. Torn finish.
   (a.) Speed too low.

(b.) Dull tool.
(c.) Chip breaker too narrow.
(d.) Improper grinding.

5. Wear.
   (a.) Speed too high.
   (b.) Feed too light.
   (c.) Incorrect grade.
   (d.) Nose radius too large.
   (e.) Improper grinding.

6. Glaze.
   (a.) Dull tool.
   (b.) Feed too light.
   (c.) Nose radius too large.
   (d.) Insufficient relief.

7. Buildup.
   (a.) Speed too low.
   (b.) Finer finish grind needed.
   (c.) Too little rake.

8. Crater.
   (a.) Speed too high.
   (b.) Feed too high.
   (c.) Incorrect grade of carbide.

N.B. Buildup on the top tool surface, which results in crater wear, is directly related to cutting speed.

Tool Wear

Figure 9–35 illustrates the types of wear that can occur in carbide cutters.

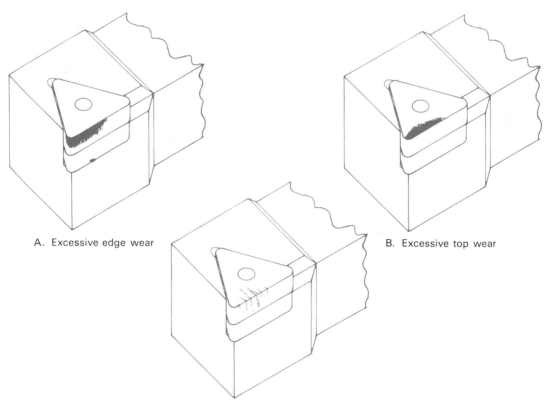

A. Excessive edge wear

B. Excessive top wear

C. Heat checks or cracks

Figure 9–35 Wear on carbide cutting tools *(Courtesy of Kennametal Inc.)*

REVIEW QUESTIONS

1. Name four essential ingredients used in cemented carbide compositions.
2. Briefly describe the procedure for making cemented carbides.
3. Name one of the most common binders used in cemented carbides.
4. What is the most important disadvantage of a cemented carbide and how is it overcome?
5. How does cutting speed compare between carbide tools and high-speed steel cutting tools?
6. Name two basic tool types for carbide tools and explain the differences between them.
7. Name eight different styles of carbide tools and explain their particular uses.
8. What is an indexable insert?
9. What is the advantage of negative rake indexable insert tools as compared to positive rake tools?
10. Name two methods of chip control provided with indexable insert tools.
11. List the shapes available for cutting tool inserts and state the machining operations they are designed for.
12. List the nine identifying factors given in the tool-holder identification system.
13. State five considerations in the selection and use of boring bars.
14. What problems could occur when machining carbon and alloy steels with carbides?
15. What problems could occur in the machining of stainless steels?
16. What is the difference between a continuous and a discontinuous chip?
17. List five reasons why high-temperature alloys are difficult to machine.
18. Give eight recommended tips for machining super alloys successfully.
19. List five causes for a carbide tool to chip.
20. What are five reasons for a carbide tool to crack or break?
21. What causes a torn finish when machining with carbide tools?
22. What causes a carbide tool to chatter? Give six factors.
23. What causes a carbide tool to wear?
24. What causes a glazed finish when machining with carbide tools?
25. What causes a built-up edge on carbide tools?
26. What causes a carbide tool to crater?

UNIT 9-4

SPEED, FEED, AND DEPTH OF CUT

OBJECTIVES

After completing this unit, the student will be able to:

- calculate speeds for different cutting tools.
- calculate approximate feeds for different cutting tools.
- select an approximate depth of cut for different cutting tools.

KEY TERMS

Roughing cut
Cutting speed (CS)
Depth of cut
Finishing cut

Machining efficiency
Feed
Fine feed
Coarse feed

INTRODUCTION

In any machining operation, a machine tool should be made to work hardest during the roughing cut. The cutting speed (CS) should be selected according to the material being cut and the cutting tool material. Depth of cut and the amount of feed should be as much as the machine and cutting tool can stand. Time can be made only during roughing cuts. Finishing is a slower process that cannot be hurried. To remove the most metal quickly and efficiently, a deep cut with a finer feed is better than a lesser depth and a coarse feed.

A thick chip requires more power to remove, and surface finish is rougher.

For greatest machining efficiency:

1. Set the maximum depth of cut.
2. Set the heaviest feed.
3. Set the highest speed.
4. Adjust the feed and speed to give maximum tool life.

There are three basic reasons for machining:

1. Machining removes surplus material from the work.
2. Machining brings the work to a required size and shape.
3. Machining produces a specific surface finish.

CUTTING SPEED

Cutting speed (CS) is the machine shop method of measuring speed, either of a point on the cutting tool moving past the work, or of a point on the work moving past the cutting tool. Cutting speed, measured in surface feet per minute, is the distance in feet that a point on a cutter will move in one minute, or it is the distance in feet that a point on the work will move past the cutter in one minute. In metric measure, cutting speed is measured in meters per minute. Cutting speed depends on: the type of material being cut; the amount of material being removed in a specific time; and the cutting tool material.

Table 9–7 contains suggested cutting speeds for a few of the more common materials machined in the machine shop. Keep in mind that there are no set rules. These figures should be used only as a guide.

The standard formulas for calculating rpm are:

$$rpm = \frac{4 \times CS \text{ (ft/min)}}{D \text{ (inches)}}$$

$$rpm = \frac{CS \text{ (m/min)} \times 1000}{\pi D \text{ (mm)}}$$

FEEDS FOR HIGH-SPEED STEEL AND CARBIDE CUTTING TOOLS

There are no set rules for the feed of cutting tools. There are many variables that will determine what feed you should use; e.g., the condition of the machine, the rigidity of the machine, the rigidity of the setup of the work, the rigidity and strength of the cutting tool, the style and shape of the cutting tool, the material to be cut, and whether the cut is to be a roughing cut or a finish cut. All of these factors will govern what feed you should use. The machine should be made to work hardest during the roughing cut where surface finish is not a particular factor. Use a feed and depth of cut that is as much as the machine and work can stand. If metal has to be removed, remove it as quickly as possible. Don't play with it. Table 9–8 gives suggested feeds. It should be used only as a guide.

Table 9–7 Suggested Cutting Speeds for High-Speed Steel and Carbide Cutting Tools

MATERIAL	HIGH-SPEED STEEL		CARBIDE	
	FT/MIN	m/MIN	FT/MIN	m/MIN
Steel: SAE 1020	80–100	24–30	180–280	60–80
SAE 1050	60–80	18–24	140–230	40–70
SAE 1090	40–80	12–24	90–180	27–55
SAE 3120 (Ni-Cr)	50–80	15–24	110–180	35–55
SAE 4340 (Mo)	50–80	15–24	110–180	35–55
Cast Iron — Gray	70–100	21–30	180–280	55–80
Aluminum, Aluminum Alloys	100–300	30–90	250–800	105–240
Brass, Bronze — free cutting	100–200	30–60	230–500	70–150

Cutting speed (CS) is measured in feet per minute or meters per minute.

Standard formulas for calculating rpm:

$$rpm \text{ (English)} = \frac{4 \times CS \text{ (ft/min)}}{D \text{ (diameter in inches)}}$$

$$rpm \text{ (Metric)} = \frac{CS \text{ (m/min)} \times 1000}{\pi D \text{ (diameter in mm)}}$$

SAMPLE PROBLEMS

EXAMPLE 1. A piece of cold rolled steel (SAE 1020), 2-in diameter, is to be turned on a lathe using a HSS tool bit. What rpm is required?

$$rpm = \frac{4 \times CS}{D}$$
$$= \frac{4 \times 80}{2} \quad \text{(using 80 fpm as the cutting speed for SAE 1020)}$$
$$= 160 \text{ rpm (answer)}$$
$$or = \frac{4 \times 100}{2} \quad \text{(using 100 fpm as the cutting speed for SAE 1020)}$$
$$= 200 \text{ rpm (answer)}$$

Here are two solutions for the same problem. Using 100 fpm is an easier, quicker calculation and, in fact, this cutting speed might be alright to use. In actual practice, however, the cutting speed of 80 fpm will probably be more practical and will give you a longer tool life.

If your selections of speeds available on the machine are 75, 150, 200, 300, 500, 800, 1250, and 2000 rpm, the highest speed to select would be 200 rpm. Keep in mind that you may have to drop your speed to 150 rpm if the tool bit tends to burn at 200 rpm.

EXAMPLE 2. A piece of cold rolled steel (SAE 1020), 2-in diameter, is to be turned on a lathe using a carbide cutting tool. What rpm is required?

$$rpm = \frac{4 \times CS}{D}$$
$$= \frac{4 \times 250}{2} \quad \text{(using 250 fpm as the cutting speed for SAE 1020)}$$
$$= 500 \text{ rpm (answer)}$$
$$or = \frac{4 \times 300}{2} \quad \text{(using 300 fpm as the cutting speed for SAE 1020)}$$
$$= 600 \text{ rpm (answer)}$$

Again, using the same selection of speeds as used in Example 1, the highest speed to select would be 500 rpm as the solution for either of the preceding calculations. The next speed up, 800 rpm, would be too high. The risk of burning the cutting tool would be too great.

EXAMPLE 3. A piece of cold rolled steel (SAE 1020), 50.8 mm in diameter, is to be turned on a lathe using a HSS tool bit. What rpm is required?

$$rpm = \frac{CS \text{ (m/min)} \times 1000}{\pi D \text{ (mm)}}$$
$$= \frac{24 \times 1000}{\pi \times 50.8}$$
$$= \frac{24 \times 1000}{3.14 \times 50.8} \quad \text{(using 24 m/min as the cutting speed for SAE 1020)}$$
$$= 150 \text{ rpm} \quad \text{(answer)}$$
$$or = \frac{30 \times 1000}{\pi \times 50.8}$$
$$= \frac{30 \times 1000}{3.14 \times 50.8} \quad \text{(using 30 m/min as the cutting speed for SAE 1020)}$$
$$= 188 \text{ rpm} \quad \text{(answer)}$$

Using the same speed selection as in Example 1, the best practical speed in both of these calculations would be 150 rpm. The next highest speed of 200 rpm would probably be a little high.

EXAMPLE 4. A piece of cold rolled steel (SAE 1020), 50.8 mm in diameter, is to be turned on a lathe using a carbide cutting tool. What rpm is required?

$$rpm = \frac{CS \text{ (m/min)} \times 1000}{\pi D \text{ (mm)}}$$
$$= \frac{75 \times 1000}{\pi \times 50.8}$$
$$= \frac{75 \times 1000}{3.14 \times 50.8} \quad \text{(using 75 m/min as the cutting speed for SAE 1020)}$$
$$= 470 \text{ rpm} \quad \text{(answer)}$$
$$or = \frac{90 \times 1000}{\pi \times 50.8}$$
$$= \frac{90 \times 1000}{3.14 \times 50.8} \quad \text{(using 90 m/min as the cutting speed for SAE 1020)}$$
$$= 564 \text{ rpm} \quad \text{(answer)}$$

Again, using the same speed selection as in Example 1, 500 rpm would probably be the best practical selection in both of these calculations.

EXAMPLE 5. A 3/4-in, high-speed steel drill is used to drill a hole in cold rolled steel (SAE 1020). What rpm is required?

$$rpm = \frac{4 \times CS}{D}$$

$$= \frac{4 \times 80}{3/4} \quad \text{(using 80 fpm as the cutting speed for SAE 1020)}$$

$$= \frac{4 \times 80 \times 4}{3}$$

$$= \frac{1280}{3}$$

$$= 426 \text{ rpm} \quad \text{(approximate answer)}$$

If your selection of speeds is 200, 400, 450, and 800 rpm, the best speed to select would be 400 rpm.

EXAMPLE 6. A 4-in, high-speed steel milling cutter is cutting cold rolled steel (SAE 1020). What rpm is required?

$$rpm = \frac{4 \times CS}{D}$$

$$= \frac{4 \times 80}{4}$$

$$= 80 \text{ rpm} \quad \text{(answer)}$$

If your selection of speeds is 75, 90, 150, and 200 rpm, the best speed to select would be 75 rpm.

A practical example of setting a tool bit for speed, feed, and depth of cut is shown in the views of the turning operation in Figure 9–36.

Table 9–8 Suggested Feeds per Tooth for Milling Cutting Tools

SUGGESTED FEEDS PER TOOTH FOR HIGH-SPEED STEEL MILLING CUTTERS				
MATERIAL	FEED PER TOOTH — INCHES (mm)			
	Face Mills	End Mills	Slot Drills	Saws
Steel: SAE 1020	0.004 (0.10)	0.001 (0.05)	0.004 (0.10)	0.0003 (0.008)
1050	0.001 (0.02)	0.002 (0.05)	0.003 (0.08)	0.0002 (0.005)
1090	0.001 (0.02)	0.002 (0.05)	0.003 (0.08)	0.0002 (0.005)
Cast Iron — Gray	0.006 (0.15)	0.004 (0.10)	0.005 (0.12)	0.0004 (0.010)
Stainless Steel	0.004 (0.10)	0.001 (0.05)	0.003 (0.08)	0.0002 (0.005)
Aluminum	0.004 (0.10)	0.009 (0.45)	0.006 (0.15)	0.0004 (0.010)
Brass, Bronze — free cutting	0.007 (0.35)	0.004 (0.10)	0.005 (0.12)	0.0003 (0.008)
SUGGESTED FEEDS PER TOOTH FOR CARBIDE MILLING CUTTERS				
MATERIAL	FEED PER TOOTH — INCHES (mm)			
	Face Mills	End Mills	Slot Drills	Saws
Steel: SAE 1020	0.012 (0.30)	0.003 (0.08)	0.004 (0.10)	0.0003 (0.008)
1050	0.010 (0.25)	0.002 (0.05)	0.003 (0.08)	0.0002 (0.005)
1090	0.008 (0.20)	0.002 (0.05)	0.003 (0.08)	0.0002 (0.005)
Cast Iron — Gray	0.015 (0.40)	0.004 (0.10)	0.005 (0.12)	0.0004 (0.010)
Stainless Steel	0.010 (0.25)	0.002 (0.05)	0.003 (0.08)	0.0002 (0.005)
Aluminum	0.018 (0.45)	0.005 (0.12)	0.006 (0.15)	0.0004 (0.010)
Brass, Bronze — free cutting	0.015 (0.40)	0.004 (0.10)	0.005 (0.12)	0.0003 (0.008)

Figure 9–36A Turning 2″ diameter, cold rolled steel at 200 rpm with a 5/16″ high-speed steel tool bit and a chip curler groove. Depth of cut is 1/4″ (6.4 mm), feed is 0.006″/rev (0.15 mm/rev). This combination produces a long, curling chip.

Figure 9–36B Doubling the feed to 0.012″/rev (0.30 mm/rev) forces the chip to curl against the side of the cutter and break into short chips shaped like a number nine.

REVIEW QUESTIONS

1. When should a machine tool be made to work hardest?
2. Considering the time factor in machining, what is the difference between roughing and finishing?
3. How is the greatest amount of metal removed quickly and efficiently?
4. What is cutting speed?
5. What is the formula for calculating rpm in English (U. S. Customary) measure? In metric measure?
6. List four factors that govern the selection of the feed of a cutting tool for a specific job.
7. A piece of cold rolled steel 3 inches in diameter is turned on a lathe with a HSS tool bit. What rpm is required?
8. A piece of cold rolled steel 75 mm in diameter is turned on a lathe with a HSS tool bit. What rpm is required?
9. A 7/8-in HSS drill is used to drill a hole in cold rolled steel. What speed should be selected for the drill?
10. A 20-mm HSS drill is used to drill a hole in cold rolled steel. What speed should be selected for the drill?
11. A 2½-in HSS milling cutter is cutting cold rolled steel. What is the rpm of the cutter?
12. If the speed of the cutting tool is too high, what may happen to the cutting tool?

UNIT 9-5

CUTTING FLUIDS

OBJECTIVES

After completing this unit, the student will be able to:

- list the functions of a cutting fluid.
- describe the basic types of cutting fluids.
- use the proper cutting fluid for the job at hand.
- use cutting fluids safely.

KEY TERMS

Cutting fluid	Mineral-lard oil
Built-up edge	Sulfurized oil
Soluble oil	Chlorinated oil
Aqueous cutting fluid	Chemical (synthetic)
Mineral oil	cutting fluid
Lard oil	Wetting agent

INTRODUCTION

Cutting fluids make it possible to cut metals at a higher rate of speed. Some cutting fluids form a metallic film on the metal surface. This prevents the chip from sticking to the cutting edge, a phenomenon commonly called a built-up edge. The surface finish of most metals can be improved considerably by the use of the proper types of cutting fluids.

Application of cutting fluids should be directed at the tool face, if possible, so that the cutting fluid actually works between the chip and the cutting tool face.

FUNCTIONS OF A CUTTING FLUID

A cutting fluid is used to reduce the temperature of a cutting action; to reduce the friction of chips sliding along the tool face; to prevent rust; and to flush away chips. Cutting fluids have antiseizure and antiwelding properties that inhibit the formation of a built-up edge and other adhesions. The hygienic properties of cutting fluids act to prevent dermatitis, which is a skin rash. This property also prevents bacterial growth that can cause unpleasant odors. Cutting fluids also have the ability to resist foaming, smoking, and misting.

BASIC TYPES OF CUTTING FLUIDS

Water is an ideal coolant, but to avoid rusting, it is mixed in various proportions with soluble or emulsifying oils, sometimes called aqueous cutting fluids. "All-purpose" oils now on the market are mixed with water and are even used as coolants and machine lubricants on screw cutting machines. Mixes of 1:5 (one part oil to five parts water) to 1:100 (one part oil to 100 parts water) are possible. A 1:5 mix is very effective for drilling, turning, milling, etc., while a 1:20 mix is very effective for grinding operations. Soluble oils act mainly as coolants and have some lubricant qualities as well. In order to blend properly, it is important to *always add the oil to the water*, not the water to the oil. It is also preferable to use warm water above 40 degrees F (4.4 degrees C) but not boiling. This will ensure that proper blending will take place. Proportions of the mixture should be determined by test. To prevent corrosion when a soluble oil is used, there should be sufficient air circulation to evaporate the water from the machined surface, thus leaving the oil to form a protective coating. If rapid drying is prevented either by lack of air circulation or high humidity, rusting is likely to occur.

Straight mineral oils are used mainly for light machining operations on certain free machining steels and some nonferrous metals. These oils combine cooling and lubricating qualities. They are used primarily for blending with base cutting oils to obtain specific properties.

Lard oil is for general use. At one time, it was used extensively for heavier machining operations, but it is expensive, has a tendency toward turning rancid, has an objectionable odor, and contributes to dermatitis and the formation of bacteria. It has been replaced largely by mineral-lard oils and to a considerable degree by sulfurized oils, except where sulfur may tarnish the metal.

Mineral-lard oils are for general use, but they are used especially for automatic machines to obtain a better finish than that obtained with straight mineral oils. These oils are noncorrosive to copper and its alloys (brasses and bronzes). These oils have greater lubrication value than straight mineral oils and are much less expensive than straight lard oil. Mixtures may contain 10% to 40% of lard oil. As a general rule, the percentage of lard oil is increased with the hardness of stock.

Sulfur additions to mineral and mineral-lard oils, called sulfurized oils, increase cooling and lubricating qualities, and the strength of the oil film that is formed is very high. These oils are used for machining carbon and alloy steels such as stainless steels and also other high-nickel alloys. Brasses and other nonferrous alloys

are blackened by high-sulfur oils. Sulfur is blended with mineral oils to produce metallic oxides or metallic film lubrication instead of fluid film lubrication. Chlorine is added to mineral oils, called chlorinated oils, to obtain cutting fluids suitable for high chip-bearing pressures.

Chemical or synthetic cutting and grinding fluids are pure solutions of organic and inorganic material in water. They contain little or no petroleum products. There are two types of these fluids:

1. Some have added lubricants and wetting agents. These are recommended for tough machining operations such as tapping, boring, drilling, and milling. These fluids have excellent rust inhibitors, high heat conductivity, and excellent lubricity. They produce good finishes that do not require degreasing. A tendency to foam makes them unsuitable for disc-wheel-type surface grinders.
2. The other type of fluids has no lubricants and wetting agents. These fluids do not foam readily, are excellent coolants and rust inhibitors. They are clear solutions and are good for surface grinding operations. These fluids are not recommended for heavier machine operations. A lack of lubricity is a disadvantage. Both types of fluids are expensive.

Aqueous solutions are usually coolants only, but they also serve to wash away chips. They have been replaced largely by modern cutting fluids. Soda water mixes are the cheapest of all coolants. They are very effective for cooling but have practically no lubricating qualities. They will cause steel to rust. A popular mixture still used in some shops is one pound of Sal Soda (carbonate of soda), one quart of lard oil, one quart of soft soap, and ten gallons of water, boiled together for half an hour.

The most common gaseous cutting fluid is compressed air. Other types are carbon dioxide, liquid argon, and liquid nitrogen.

Casite is an old cutting fluid that once was used by many machinists. It was a 50/50 mixture of carbon tetrachloride and kerosene. Carbon tetrachloride was also used as a cleaning fluid on machinery and machine parts. **Carbon tetrachloride is dangerous and its use can result in death. It should not be used either as a cutting fluid or as a cleaning fluid in a machine shop.**

USES OF CUTTING FLUIDS

There are no fixed rules or formulas for selecting cutting fluids except the rule of an actual test with a given cutting tool or operation or material. Following are a few suggestions for the successful use of cutting fluids:

For general use with high-speed steel cutting tools used in turning, boring, drilling, and milling of medium- and low-strength steels, a blend of one part soluble oil and ten parts water is good.

In the cutting of tool steels and tough alloy steels, a heavier soluble oil mix of one part oil to five parts water is recommended for turning, milling, drilling, and reaming. A mineral-lard oil would also give excellent results in this case.

For tougher operations like tapping, threading, and broaching of tool steels and high-strength steels, sulfo-chlorinated oils are recommended. Sulfurized mineral oils and mineral-lard oils are also recommended for tough, stringy, low-carbon steels to reduce the tearing effect that some of these steels are subject to.

With copper and its alloys, a soluble oil mix of 1:10 is recommended for most operations. For more severe operations, use a heavier soluble oil mix of 1:5 or a mineral-lard oil blend of 5% to 10% lard oil.

Sulfo-chlorinated oils are not recommended because they will stain most brasses, bronzes, and copper. Brasses and bronzes are usually machined dry, but particularly in drilling and reaming, a soluble oil blend of 1:10 is good practice, and for tapping and threading, a slightly heavier blend of 1:5 will do the job.

Copper is a soft, sticky metal to cut, but it can be worked well with a 50/50 blend of lard oil and turpentine or better yet, a 50/50 blend of mineral-lard oil and turpentine.

Babbitt is a stock bearing metal composed basically of lead, tin, and antimony. It is usually cut dry, but bushings are sometimes difficult to bore dry. The material has a tendency to roll around the tool bit in a hard ball. A 50/50 mixture of kerosene and lard oil or kerosene and mineral-lard oil works very well.

Aluminum alloys are frequently cut dry, but a 50/50 blend of kerosene and lard oil or kerosene and mineral-lard oil works well. In addition, a thin soluble oil mix of 1:20 to 1:30 will also work well.

Cast iron is usually worked dry. Sometimes a few drops of a thin soluble oil mix will prevent a drill from screeching in a drill jig. A flood of soluble oil mix or strong soda water will wash away chips and hold the dust down, but an objectionable mess will be created and the combination of oil, chips, and sand will form an abrasive mixture that will quickly dull most cutting tools. Cast iron sometimes hand reams easily with a tallow and graphite mix.

A common mixture for grinding on surface grinders or valve grinders, etc., is a soluble oil mix of 1:20. Light mineral oils, about an SAE 10, will give a fine surface finish with vitrified bonded wheels but not with rubber-bonded or shellac-bonded wheels.

A dangerous situation should be noted here: Under certain conditions, an oil vapor mist composed of compressed air and mineral oil can be ignited.

Soda water is still used by many shops because it is effective and inexpensive.

After the grinding operation is completed, it is best to allow the wheel to run until it is reasonably dry or until most of the fluid is thrown off.

As a general rule, cutting fluids are not used with cemented carbide cutting tools unless a great quantity of cutting fluid can be applied to ensure uniform temperatures that will prevent carbide inserts from cracking. They are sensitive to sudden changes in temperature. Thermal shocks cause thermal cracks near the cutting edge.

The effectiveness of cutting fluids lessens as cutting speed is increased.

The flow of the chip interrupts the flow of coolant to the cutting edge. One method of applying coolant is to direct the coolant flow to the work from behind rather than at the cutter or to apply the coolant from below the cutter so that the coolant actually works its way effectively between the chip and the face of the cutter.

The danger of thermal cracking is great when milling. Carbide milling operations should be performed dry.

REVIEW QUESTIONS

1. What are the functions of a cutting fluid?
2. What is a built-up edge?
3. Name the types of cutting fluids.
4. What is the difference between a soluble cutting fluid and an insoluble cutting fluid?
5. How are emulsifying oils mixed with water?
6. Name two problems encountered when using lard oil.
7. List three advantages of mineral-lard oil as a cutting fluid compared to either mineral oil or lard oil alone.
8. What is the purpose of sulfurized oils?
9. What should sulfurized oil not be used for?
10. What is synthetic cutting fluid used for?
11. Why do we use the term *cutting fluid* rather than *cutting oil*?
12. What is an aqueous solution?
13. What is the most common gaseous cutting fluid?
14. Why is cast iron usually worked dry?
15. What dangerous situation can be encountered when using oil vapor mist?
16. How are cutting fluids used with carbide tooling?
17. What is the danger of using cutting fluids with carbide tooling?
18. Describe one method of applying coolant to the cutting edge of a tool.

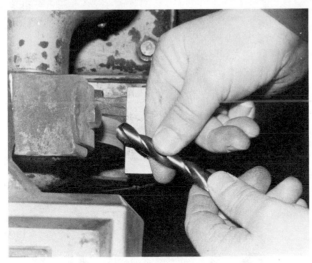

Top view showing a manual method of sharpening a twist drill. The forefinger supporting the drill rests on the tool rest and raises the drill to ensure clearance being ground behind the cutting lips. This view shows the point angle being formed. The potential danger is that the forefinger will be hit by the wheel. *Courtesy of Cominco Ltd..*

Quick change tooling systems allow tools to be changed even while the spindle is rotating. *Courtesy of SKF & Dormer Tools Ltd.*

Spiral flute taps force the swarf out of the hole and prevent the chips from clogging at the bottom or in the flutes which can result in tap breakage or damaged threads. *Courtesy of SKF & Dormer Tools Ltd.*

Selecting the proper chipbreaking geometry with carbide inserts results in improved tool life and surface finish. *Courtesy of SKF & Dormer Tools Ltd.*

A heavy-duty turning operation using a triangular indexable insert tool. *Courtesy of Kennametal Inc.*

Adequate supply of the correct lubricant will improve surface finish and greatly aid in heat dissipation. *Courtesy of SKF & Dormer Tools Ltd.*

CHAPTER TEN

MILLING AND MILLING OPERATIONS

A milling operation is one in which the work is fed into a revolving cutting tool. The cutting tool, which can be a single-point tool or a multitooth tool, can be held in either a horizontal or vertical position.

Manufacturers provide milling machines ranging from manual horizontal and vertical milling machines to the complex CNC machining centers. These machining centers make up a major portion of the Flexible Manufacturing Systems in use today. Milling machines handle a great variety of machining operations, including horizontal milling, vertical milling, angular milling, and form milling. The milling machine, in fact, has replaced the metal shaper which is hard, if not impossible, to find in modern machine shops. For example, vertical milling machines perform typical metal shaper work much faster and easier and with much more accuracy.

MILLING MACHINE SAFETY

The revolving milling cutter is the primary danger in milling. However, the work breaking loose and the chips are also potential dangers.

1. Permission should be obtained to operate the machine, at the discretion of the instructor.
2. Always roll up loose sleeves and remove ties, rings, watches, and so on, before operating the machine.

3. Always wear specified eye protection.
4. Make all adjustments with the power off and the brake full on.
5. Be certain that the work is held securely in the vise, or clamped securely to the table.
6. Be sure that the cutter is held securely in the spindle or on the arbor.
7. Be sure of the correct spindle rotation. Never run a cutter backward.
8. Try to avoid "climb" milling unless the machine is equipped with a backlash eliminator. One of the occasions for "climb" milling would be the cutting of thin sheet metal.
9. Check spindle rotation, spindle speed, depth of cut, and feed before cutting the work.
10. When changing milling cutters or adjusting the work, the machine must be shut off and the brake must be full on.
11. Never walk away from the milling machine when it is operating.
12. When the work is complete, shut off the power and clean the machine.

UNIT 10-1

BASIC MILLING OPERATIONS

OBJECTIVES

After completing this unit, the student will be able to:

- define the basic milling operations.
- identify the basic cutters in common use for milling operations.
- name the parts of various common milling cutters.
- select the proper milling cutter for the specific job.
- set up milling cutters in tool holding devices.
- set up work on the milling machine table using vises and clamps.

KEY TERMS

Milling

Gang milling

Plain milling

Plain metal slitting saw

Side milling cutter

Face milling

Angular milling

Form milling

Rotating cutter

Straight or spiral tooth

Single-angle milling cutter

Double-angle milling cutter

Convex cutter

Column-and-knee
  universal milling
  machine
Indexable carbide
  inserts
Form relieved tooth
Straight teeth or
  helical teeth
Left-hand helical tooth
Right-hand helical
  tooth
Primary clearance
  land (angle)
Secondary clearance
  land (angle)

Concave cutter
Face milling cutter
Shell mill
End mill
Shell end mill
Woodruff keyseat
  cutter
Milling arbor
End milling cutter
  holder (adapter)
Shell end mill
  adapter
Dormer fast lock
  holder
Style "E" adapter

INTRODUCTION

Milling is a machining operation in which the work is clamped to the work table and is moved or fed into a revolving cutter that is held on a horizontal arbor, in a horizontal spindle, or in a vertical spindle.

Figure 10–1 shows a gang milling operation. Seven cutters are mounted on one horizontal arbor, and each cutter makes its own particular cut on the work.

In milling, there are four basic cutting operations: plain milling, face milling, angular milling, and form milling. Other types of milling are combinations of one or more of these basic cuts. All operations are performed by rotating cutters that can have one cutting tooth, as in a "fly" cutter, or many teeth.

Plain milling is any cutting done on a flat surface parallel to the axis of the cutter. **Face milling** is the cutting of a flat surface at right angles to the axis of the cutter (the cutter axis can be horizontal or vertical). **Angular milling** is the cutting of a flat surface inclined to the axis of the cutter. Do not confuse this operation with angular indexing which is covered later in this chapter. **Form milling** is the cutting of a surface having a regular or an irregular outline; that is, a surface other than flat.

THE PARTS OF THE MILLING MACHINE

The essential parts of the column-and-knee-type milling machine, Figure 10–2, are the base, column, overarm, arbor, arbor supports, worktable, saddle, and knee.

MILLING CUTTERS

There is a great variety of cutters for milling machines. All of them are rotary-type cutters which can usually be adapted to either horizontal or vertical milling machines. Most cutters are made of high-speed steel or carbide, or have special alloy steel bodies to hold indexable carbide inserts. The carbide inserts permit the use of cutting speeds two to three times higher than those used for high-speed steel cutters.

There are generally two types of cutting teeth: the straight or spiral tooth as found on a slitting saw or a plain milling cutter; and the form relieved tooth as found on convex or concave cutters and involute gear cutters. The straight or spiral tooth cutter requires

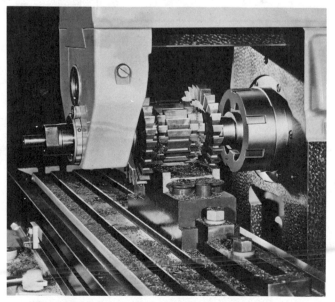

Figure 10–1 Gang milling operation *(Courtesy of Cincinnati Milacron)*

sharpening on the outside diameter or periphery of the tooth to provide clearance behind the cutting edge. Metal slitting saws, side milling cutters, plain milling cutters with straight and spiral teeth, end mills, shell end mills, face mills, and angle cutters all fall into this category. Form relieved teeth are ground on the front faces of the teeth. This group of cutters includes involute gear cutters, concave or convex cutters, and hobbing-type gear cutters, all of which have a built-in clearance from the factory.

Straight teeth or helical teeth are sharpened by grinding the outer periphery of each tooth. Details of a typical straight milling cutter tooth are shown in Figure 10–3, including the names of the parts of the tooth. The primary clearance angle is 10 to 12 degrees,

which is normal behind a cutting edge. After repeated sharpenings, the land becomes too wide, causing the cutter to rub on the work as shown in Figure 10–4. A secondary clearance angle must be ground, creating a second land. This is normal grinding practice. Figure 10–5 shows the effect of wheel shape on the milling cutter tooth. To attain a 10- to 12-degree clearance, less material would be removed with a cup wheel, although a straight wheel is also used. Form relieved teeth are sharpened on the face of the cutter using a dish-shaped grinding wheel. Rake angle is the same as on the original tooth. Form relieving means that the portion of the tooth behind the cutting face is always slightly smaller than the cutting face. This is a factory built-in clearance to prevent the cutter from

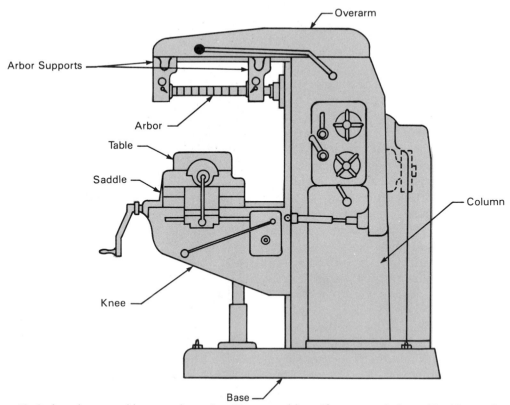

Figure 10–2 A column-and-knee universal milling machine *(Courtesy of Gate Machinery Co. Ltd.)*

rubbing. The milling machine operator must be aware of this information so that the best performance is obtained from milling cutters.

A common milling cutter is the plain metal slitting saw, Figure 10–6. It can be used to cut slots as thin as 0.006 in (0.15 mm) to 3/16 in (4.76 mm) wide. Plain milling cutters are slightly hollow ground. They are thinner at the center than at the periphery of the teeth. This clearance is built in at the factory to prevent the cutter from binding in the cut. The parts of the slitting saw are given in Figure 10–7. Side milling cutters, Figure 10–8, are used for slotting, side milling, and straddle milling. Sizes start at 1/4 in (6.35 mm) thick and range up to 1 in (25.4 mm). Cutting is done on both the outer periphery and the sides, making this a very versatile and widely used cutter. The parts of the side milling cutter are given in Figure 10–9. Plain milling cutters, Figure 10–10, usually range from 3/16 in (4.76 mm) wide up to 4 in (100 mm) wide for light-duty cutters, and up to 6 in (150 mm) wide for heavy-duty cutters. Some cutters are designed to take light cuts, and others moderate or heavy cuts on flat surfaces. Cutters over 3/4 in wide have 25-degree left-

or right-hand helical teeth, with some cutters having as high as a 52-degree helix. All of these cutters have a 10 to 12 degree built-in, positive, radial rake angle, which is shown in Figure 10–11. To determine which is a left-hand or right-hand helix, look at the cutter from the end. A left-hand helical tooth slopes up and curves to the left. A right-hand helical tooth slopes up and curves to the right. A helical tooth gives an easier shearing cut than a straight tooth. A wide milling cutter is sometimes referred to as a slabbing cutter.

Angular cutters are either single angle, Figure 10–12 (page 374), or double angle, Figure 10–13 (page 374). Single-angle milling cutters are made in a left- or right-hand cut with included angles of 45 or 60 degrees. Cutting teeth are on both the periphery and the face. These cutters are used for cutting dovetails and angular surfaces. Double-angle milling cutters are used for cutting angular grooves such as notches, serrations, and bevels. They have included angles of 45, 60, or 90 degrees. The details of both angular cutters are shown in Figure 10–14 (page 375).

Formed cutters like the convex cutter, Figure 10–15 (page 375), will produce a true convex radius. A con-

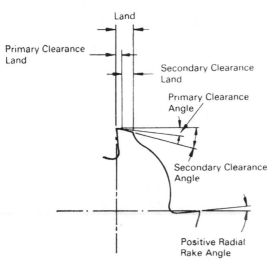

Figure 10–3 Details of a typical straight milling cutter tooth *(Courtesy of SKF & Dormer Tools Ltd.)*

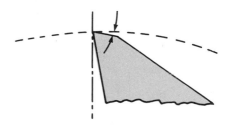

A. 10°–12° clearance on a new tooth

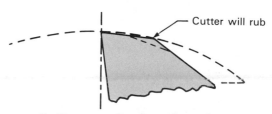

B. Clearance after many sharpenings

Figure 10–4 Clearance ground on a straight milling cutter tooth *(Courtesy of Norton Company)*

cave cutter will produce a true concave radius. Details of both cutters are shown in Figure 10–16 (page 376). Corner rounding cutters, Figure 10–17 (page 376), are made in both right- and left-hand cuts. The left-hand cut is illustrated. The involute gear cutter is shaped exactly like a gear tooth, and like other formed cutters, the teeth are form relieved. Involute gear cutters are discussed in greater detail in the gearing section of this chapter.

Face milling cutters or shell mills, Figure 10–18 (page 376), are very common types of carbide cutters.

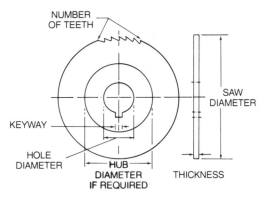

Figure 10–7 Parts of a plain metal slitting saw *(Courtesy of The Cleveland Twist Drill Company)*

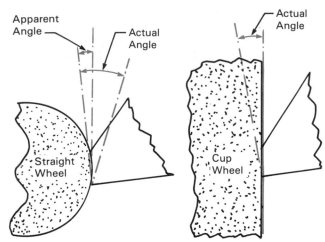

Figure 10–5 Effect of wheel shape on the relief angle *(Courtesy of Norton Company)*

Figure 10–8 A side milling cutter *(Courtesy of The Cleveland Twist Drill Company)*

Figure 10–6 A plain metal slitting saw *(Courtesy of The Cleveland Twist Drill Company)*

Side and face milling cutter nomenclature

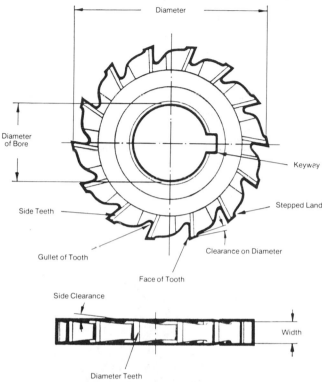

Figure **10–9** Parts of a side milling cutter *(Courtesy of SKF & Dormer Tools Ltd.)*

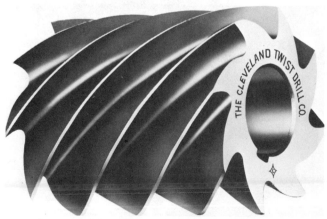

Figure **10–10** A plain milling cutter *(Courtesy of The Cleveland Twist Drill Company)*

The special alloy steel body is designed to hold a number of indexable inserts. They are designed especially for CNC milling, but they can be adapted for use on all milling machines that have enough power to use them. Carbide cutters need power and speed to perform adequately. They can be used in either the horizontal or vertical position for planing large flat surfaces, Figure 10–19 (page 377).

End mills are designed for the accurate cutting of slots, pockets, keyways, and intricate contours, which are formed to close tolerances. They are made in many flute styles, lengths of cuts, and shapes with left- or right-hand helical flutes. There are also different styles of shank — plain, threaded, or with

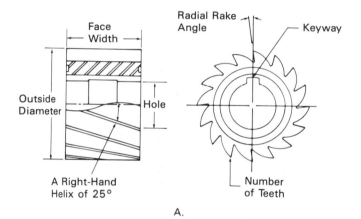

A.

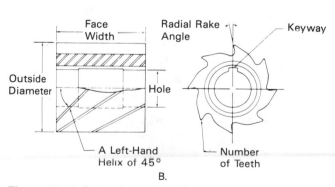

B.

Figure **10–11** Parts of a plain milling cutter with a right-hand helix (A.) and a left-hand helix (B.) *(Courtesy of The Cleveland Twist Drill Company)*

special locking arrangements, depending on the particular application. Tool holders or adapters are made for each shank style.

The single-end, high-speed steel, two-flute end mill, Figure 10–20, is suitable for cutting keyways accurately. This type of mill has center cutting capability, that is, the end cutting edges go right to the center of the tool. Plunge cutting, or cutting straight into the work, can be done as well as slotting cuts and general-purpose milling.

The single-end, high-speed steel, four-flute end mill, Figure 10–21, does not have plunge cutting capability since the cutting edges on the end do not go all the way to the center of the tool. However, these end mills are also made with center cutting and plunge cutting capabilities. The names of the parts of an end mill and the details of the teeth are shown in Figure 10–22 (page 378).

To illustrate the difference in center cutting and noncenter cutting capability, Figure 10–23A (page 379) shows the end view of a two-flute end mill that is center cutting. The cutting edges go right to the center. The four-flute end mill shown in Figure 10–23B is not center cutting since the cutting edges do not go right to the center.

Shell end mills, Figure 10–24 (page 379), are used for face milling or end milling. Shell end mills are less expensive than other end mills of comparable size because there is no shank. A number of different size shell end mills can be used on one arbor or adapter.

Figure 10–12 A single-angle, right-hand cut, milling cutter *(Courtesy of The Cleveland Twist Drill Company)*

Figure 10–13 A double-angle milling cutter *(Courtesy of The Cleveland Twist Drill Company)*

For example, 1 1/4-in, 1 3/8-in, and 1 1/2-in diameter cutters can all fit the same adapter.

Woodruff keyseat cutters, Figure 10–25 (page 379), are used for cutting keyseats to fit Woodruff keys. They are made in a shank style, as illustrated, and also in an arbor style.

A typical machining operation for end mills, Figure 10–26 (page 380), is a slotting operation using a high-speed steel, four-flute end mill in the vertical position. A surfacing operation, Figure 10–27 (page 380), is performed by a premium steel, SPIRA-LOC, helical-blade, multiple-flute end mill in the horizontal position.

MILLING CUTTER HOLDING DEVICES

There is a great variety of holders for adapting different types of cutters for different cuts. Figures 10–28 to 10–31 show a few of the more commonly used holders. The milling arbors, Figure 10–28 (page 380), hold and rotate cutters on a horizontal milling machine. They are complete with support bearings and spacing collars.

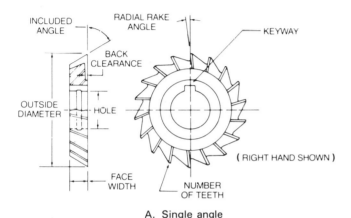

A. Single angle

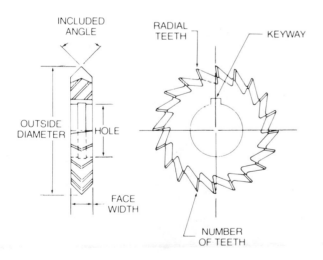

B. Double angle

Figure 10–14 Parts of a single-angle cutter (A.) and a double-angle cutter (B.) *(Courtesy of The Cleveland Twist Drill Company)*

Figure 10–15 A convex cutter *(Courtesy of The Cleveland Twist Drill Company)*

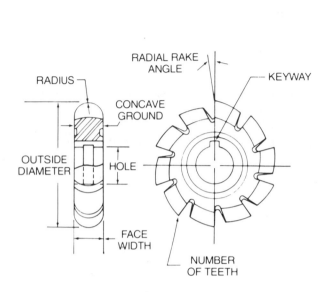

A. Convex cutter

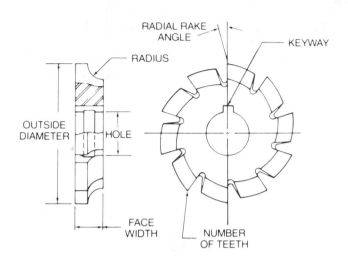

Left hand

Figure 10–17 Parts of a corner rounding cutter *(Courtesy of The Cleveland Twist Drill Company)*

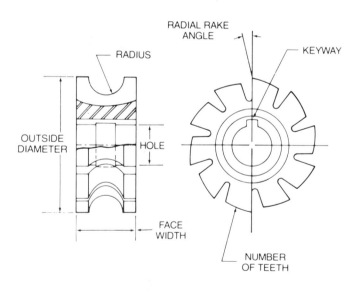

B. Concave cutter

Figure 10–16 Parts of a convex cutter (A.) and a concave cutter (B.) *(Courtesy of The Cleveland Twist Drill Company)*

Figure 10–18 A face milling cutter (shell mill) *(Courtesy of Kennametal Inc.)*

Figure 10–19 Planing a large, flat surface using a carbide face milling cutter *(Courtesy of Kennametal Inc.)*

End milling cutter holders or adapters are used for holding and driving different types of end milling cutters. The shell end mill adapter (style "C") will hold a number of different sizes of shell end milling cutters. A sectional sketch of the style "C" adapter is shown in Figure 10–29 (page 380). A Dormer fast lock holder, Figure 10–30 (page 380), is used for threaded shank end milling cutters. The style "E" holder is used for holding taper shank end milling cutters. A sectional sketch of the style "E" holder is shown in Figure 10–31 (page 381). All of these holders are adaptable to both vertical and horizontal milling machines.

WORK HOLDING METHODS

The work must be held securely on the milling machine table when cutting is taking place. Two of the most common methods of holding work are vises and clamps.

Figure 10–20 A single-end two-flute end mill *(Courtesy of The Cleveland Twist Drill Company)*

Figure 10–21 A single-end four-flute end mill *(Courtesy of The Cleveland Twist Drill Company)*

When using a vise, it is best, but not always possible, to work against the jaws of the vise rather than having the jaws in line with the cut. It is also best to grip the work in the center of the vise jaws. If the work is off center, unequal pressure is exerted on the work which could move under pressure of the cut. The unequal pressure is also hard on the vise itself. Make sure that all work is properly supported when held in a vise. The work must be supported from below,

Figure 10–32A, or the work will be forced downward, Figure 10–32B.

The work may be clamped directly to the table. It is always best to use a stop as well as a clamp. If the work surface is finished, use a protective shim between the clamp and the work. Always use washers between nuts and clamps. In proper clamping procedure, both ends of the work should be clamped, Figure 10–33.

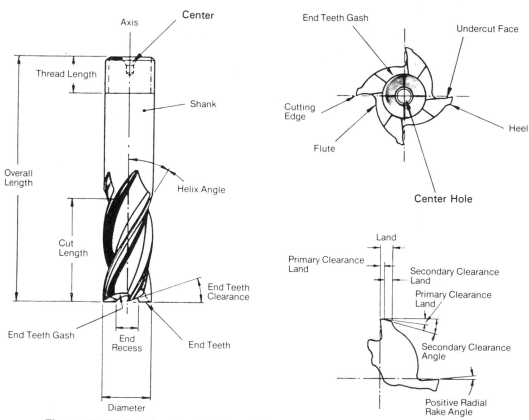

Figure 10–22 Details of end mill teeth *(Courtesy of SKF & Dormer Tools Ltd.)*

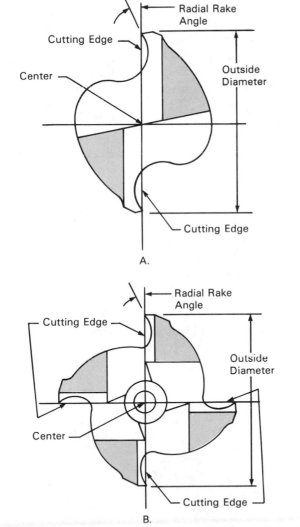

A.

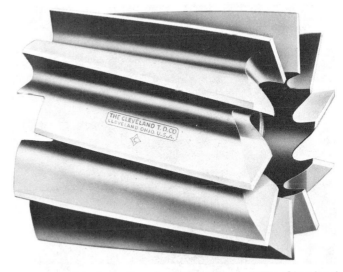

Figure 10–24 A shell end mill *(Courtesy of The Cleveland Twist Drill Company)*

B.

Figure 10–23 End view of a center cutting, two-flute end mill (A.) and a four-flute end mill (B.) *(Courtesy of SKF & Dormer Tools Ltd.)*

Figure 10–25 Woodruff keyseat cutter *(Courtesy of The Cleveland Twist Drill Company)*

Figure 10–26 Slotting operation using a high-speed steel end milling cutter *(Courtesy of The Cleveland Twist Drill Company)*

Figure 10–27 Surfacing operation using a SPIRA-LOC, helical-blade end milling cutter *(Courtesy of The Cleveland Twist Drill Company)*

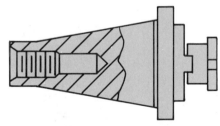

Figure 10–29 Shell end mill arbor, style "C"

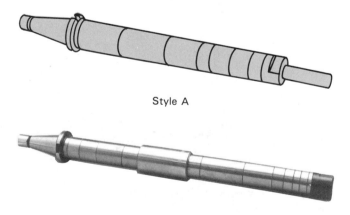

Style A

Figure 10–28 Milling arbors *(Photo courtesy of DoALL Company)*

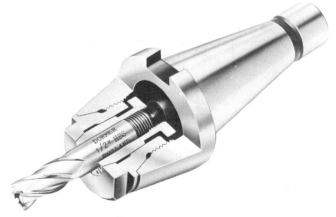

Figure 10–30 Dormer fast lock holder *(Courtesy of SKF & Dormer Tools Ltd.)*

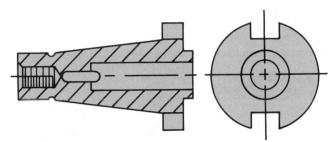

Figure 10-31 Style "E" adapter

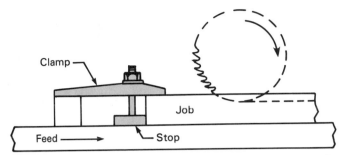

Figure 10-33 Work must be clamped securely on both ends

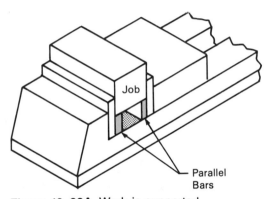

Figure 10-32A Work is supported

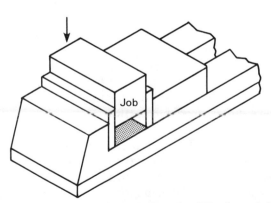

Figure 10-32B Work is not supported, will be forced down

REVIEW QUESTIONS

1. List four precautions to be taken when performing milling operations.
2. What is milling?
3. Briefly describe each of the four basic milling operations.
4. Name the essential parts of a horizontal milling machine.
5. What is the difference in the sharpening of a straight or spiral tooth cutter as compared to sharpening a form relieved tooth cutter?
6. Explain the term form relieving. Who is responsible for form relieving a cutter originally?
7. What is the difference between a plain milling cutter and a side milling cutter?
8. What two requirements must be met if indexable carbide insert milling cutters are to perform adequately?
9. For what purpose are end mills used?
10. List three commonly used holders for milling cutters.
11. Give three considerations for holding work in a vise on a milling machine table.
12. Give three considerations for clamping work to the milling machine table.

UNIT 10-2

MILLING MACHINES

OBJECTIVES

After completing this unit, the student will be able to:

- identify the various types of milling machines.
- set up tools on a jig boring machine.
- set up work on a jig boring machine.
- apply the coordinate system to locate holes on the work for jig boring.
- set up and perform jig boring work on a vertical milling machine.

KEY TERMS

Column-and-knee-type milling machine
Bed-type milling machine
Plain milling machine
Universal milling machine
Vertical milling machine
Jig boring machine
Die
Jig
Fixture
Gage
Single-point boring bar
Dial indicator
Locating microscope
Coordinate system
Order of operations
Reference bar
Coordinate boring and milling machine
Circular dividing table with digital readout

GENERAL MILLING MACHINES

There are two general categories of milling machines: the column and knee type and the bed type.

There are three styles of column-and-knee-type milling machines. The plain milling machine, Figure 10–34, has a horizontal spindle driving a horizontal arbor or shaft, which in turn drives a rotary cutter. The work table has horizontal and traverse movements, but there is no table swiveling arrangement.

The universal milling machine, Figure 10–35, is identical to the plain milling machine except that there is a two-part saddle containing a swiveling arrangement so that the table can be swiveled or rotated. The swiveling table gives this machine the capability of being able to produce a helical gear which is described in the section on gearing in this chapter. This is the only difference between a universal and a plain milling machine.

The third style is the very versatile vertical milling machine which has a vertical spindle. It normally drives a cutter holding device, which drives a vertical milling cutter, a slot drill, or a face mill. Some rotary cutters can also be held on this machine; for example, a gear cutter used for cutting a gear. Figure 10–36 illustrates a face milling operation using a shell end milling cutter. The cut being made by the face of the cutter is at right angles to the axis of the cutter. In angular milling, the cutting of both a right-hand and a left-hand bevel can be done without moving the work. The head can be tilted in either direction. Scroll plates can be cut, as well as spur gears and helical gears.

All column-and-knee-type machines have a vertical adjustment of the worktable. The vertical milling machine has, in addition, some vertical adjustment of the spindle. Column-and-knee machines are probably the most common machines in school shops, in job shops, and in tool and die shops because they are, by far, the most versatile type of milling machine.

A bed-type machine, Figure 10–37 (page 386), is much more rugged than the column-and-knee-type machine. It has a vertical adjustment of the spindle on a solid worktable base. It is not as easily and

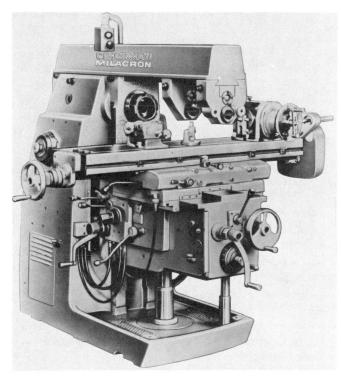

Figure 10–35 A universal (horizontal) milling machine *(Courtesy of Cincinnati Milacron)*

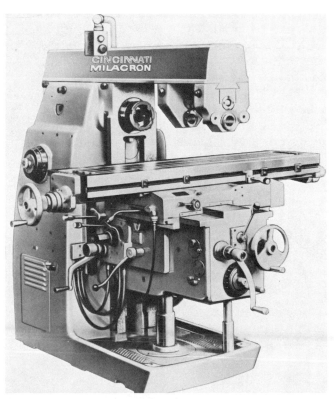

Figure 10–34 A plain (horizontal) milling machine *(Courtesy of Cincinnati Milacron)*

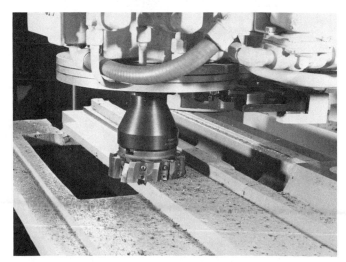

Figure 10–36 Face milling using a carbide face mill *(Courtesy of Cincinnati Milacron)*

LASER SURFACE HARDENING

CO_2 lasers can be used to surface harden a wide variety of metal parts. They can effectively deliver power in the range of 0.5 to 8 kilowatts. The stability and accuracy of the beam makes the laser ideal for the hardening process.

The laser output beams, which normally are "doughnut" shaped, can easily be changed by optical means into square spots or into more complex shapes for special purposes. With this kind of flexibility, it is easy to heat treat flat surfaces and machine parts such as bearing races, gears, shafts, cylinders, and others. Large laser spots can be used to enhance productivity and provide high coverage rates as well as fewer overlap zones (areas which can be troublesome in many cases). The large power range of the laser also makes it possible to vary the depth of case hardening to an unusual degree. The precise power control and extreme stability result in very uniform hardening depths.

Heat treating is the transformation or changing of ferrous surfaces. In surface hardening, heat treating improves mechanical properties of highly stressed, ferrous machine parts. Surface hardening increases wear resistance and fatigue life through structural changes induced in the metal surface during the process.

The surface hardening process is basically the same as the conventional through-hardening process of ferrous materials. In both processes, increased hardness and strength are obtained by heating the material sufficiently to transform it to austenite, followed by rapid cooling or quenching to room temperature. Surface hardening, however, differs from the through-hardening process in that only a thin surface layer is heated to the transformation temperature prior to quenching. As a result, the interior of the workpiece remains essentially unaffected. If the surface heating is done very rapidly, the use of a quenching medium (such as oil, water, or cold air) may not be necessary. That is, when the heat source is removed, the surface cools very rapidly by heat conduction to the still cool interior (or, the cool interior quickly absorbs the heat from the surface). This process is known as self-quenching and is generally used in laser surface hardening, Figure 1. A laser can generate very intense heat at the workpiece surface, making it ideal for surface hardening.

Laser surface hardening is high quality work and takes place quickly. A broad area laser beam can surface harden a ferrous metal surface at rates up to 35 square inches per minute (225 square centimeters/min). The depth of hardening can be varied as much as 0.120 inch to 0.160 inch (3 to 4 mm) in some alloy steels. No quenching media are required. The precise control of the size and shape of the laser spot is achieved through the use of optical systems designed for this purpose.

Figure 2 shows how the "doughnut" shaped beam is transformed into a square laser spot by means of an optical integrator. By moving this spot over a surface, hardening is accomplished.

Figure 1 Laser surface hardening of a cast iron clutch plate *(Courtesy of Combustion Engineering, Inc.)*

The accuracy of the process is maintained by controlling the laser power and spot translation/speed (speed of moving from one place to another). The depth of hardening and the maximum hardness obtained depend upon the power, processing speed, and type of material. Figure 3 shows the depth of case hardening in thousandths of an inch obtainable in an SAE 4140 steel which has very good hardenability. Table 1 lists steels which are suitable for laser surface hardening, as well as other steel alloys which are not well-suited for the process.

The advantages of laser surface hardening over more conventional methods are numerous. There is less workpiece distortion or warping because of the low energy input required. The process is easy to control because of the stable laser power, processing speed, and laser beam accuracy. Surface hardening can be done easily in localized patterns by computer process control. No specialized controlled atmosphere (oxygen-free atmosphere) or vacuum is needed. Large workpieces are easily processed. There is no need for quenching media and tanks because of the concept of self-quenching. Inaccessible areas are hardened by the use of special optical systems which can guide laser beams easily. Processing rates and productivity are high.

Surface hardening by means of a laser beam, with its ease and accuracy of control, its convenience and flexibility is also finding its proper place in modern manufacturing.

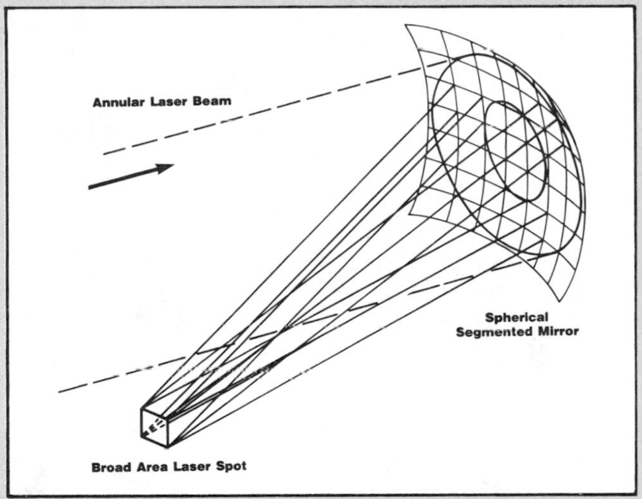

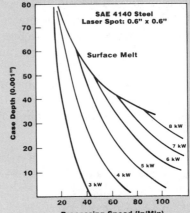

Figure 3 Laser surface hardening of SAE 4140 steel *(Courtesy of Combustion Engineering, Inc.)*

Figure 2 Forming a square laser spot with uniform power density using an optical integrator *(Courtesy of Combustion Engineering, Inc.)*

Table 1 Hardenability of ferrous materials by laser beam *(Courtesy of Combustion Engineering, Inc.)*

GOOD	MARGINAL	NOT HARDENABLE
Medium Carbon Steel (1045)	Annealed Carbon Steel	Low Carbon Steel (1010)
High Carbon Steel (1080)	Spheroidized Carbon Steel	Austenitic Stainless Steel
Tool Steel (52100)	SAE 1020	Non-ferrous Alloys and Metals
Low Alloy Steel (4140)	Ferritic Nodular Cast Iron	Wrought Iron
Cast Iron		

quickly adaptable for a variety of work. Bed-type machines are used primarily for manufacturing large numbers of duplicate pieces and are usually found in production shops.

JIG BORING MACHINES

A **jig boring machine**, Figure 10–38, in one sense, is a glorified drilling machine. It has all of the essential elements of a vertical spindle milling machine, but it is built much lower to the floor. It is much more rigid in construction, making it more free from vibration. It is extremely accurate, to within 0.0001 in (0.002 mm), or even closer in some machines. A jig boring machine is rugged for heavy cuts, yet sensitive for the lighter, more delicate cuts. The machine usually has a wide range of spindle speeds, 200 rpm to 3000 rpm being one example. The main purpose of this machine is to accurately locate, drill, ream, and bore the numerous holes necessary for dies, jigs, fixtures, gages, and other precision parts.

A **die** is one of a pair of cutting or shaping tools operated by being pressed or driven toward each other. The smaller tool that enters the other is called a punch,

and the larger tool, a die. A die can also be a hollow, internally threaded tool used for cutting screw threads on bolts or studs.

A **jig** is a work-holding device usually used on a drill press. The jig may be moved around on the table to bring each drill bushing under the drill. It is not fastened to the machine on which it is used. A clamp is usually used on the work.

A **fixture** is also a work-holding device, but it is clamped to the table. The drill is moved to the drill bushing as with a radial drill, or to the cutting tool by moving the table, as with a milling machine.

A **gage** is an instrument used for measuring or testing mechanical parts to determine whether or not dimensions are correct or are within specified limits.

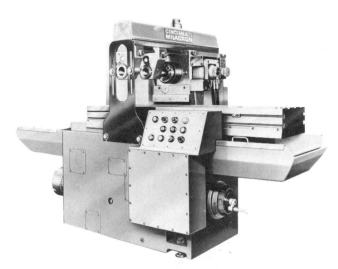

Figure 10–37 A bed-type milling machine *(Courtesy of Cincinnati Milacron)*

Figure 10–38 A jig boring machine *(Courtesy of Cominco Ltd. and American SIP Corp.)*

Tools Used in Jig Boring

The cutting tools used in jig boring are center drills, drills, reamers, and counterbores, but the most important cutting tool used is the single-point boring bar. One style of **single-point boring bar**, Figure 10–39, is used to bore holes accurately to exact diameter. Note the micrometer dial used to move the cutter outward from center.

A **dial indicator**, Figure 10–40, is used to set the edge of the work parallel to the longitudinal line of travel or parallel to the cross slide line of travel.

A **locating microscope**, Figure 10–41, is used to locate either the edge of the work or a reference line marked on the work.

Setting Up the Work

The T-slots in the worktable are parallel to the longitudinal line of travel of the table. One method of setting the work parallel to the line of travel is to set steel blocks in the T-slots to act as stops for the work to rest against, Figure 10–42. The work is then set on parallel bars and clamped securely, Figure 10–43. This method of setup ensures two things: (1.) the work will be exactly parallel to the line of travel of the worktable, and (2.) there is room under the work to allow the cutter to drill through without damaging the table.

Another, perhaps more common, method of setting the work parallel to the line of table travel is to run a dial indicator along the edge of the work. If the dial reads the same at both ends of the edge, the work

Figure 10–39 A single-point boring bar *(Courtesy of Cominco Ltd. and American SIP Corp.)*

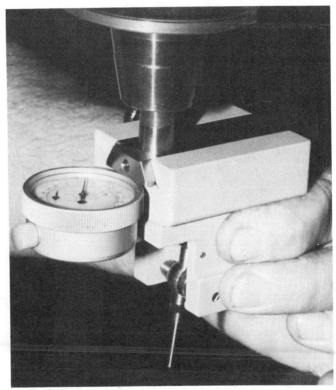

Figure 10–40 A dial indicator *(Courtesy of Cominco Ltd. and American SIP Corp.)*

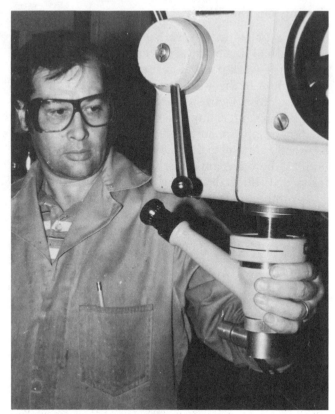

Figure 10–41 A locating microscope *(Courtesy of Cominco Ltd. and American SIP Corp.)*

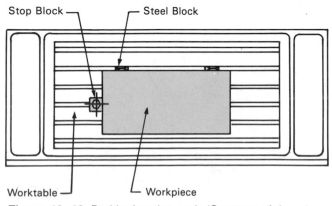

Figure 10–42 Positioning the work *(Courtesy of American SIP Corp.)*

is parallel. More modern jig borers use a locating microscope, Figure 10–44A, to either locate the edge of the work or to locate a reference line that has been marked on the work. The microscope is clamped into the spindle. When you look through the microscope and focus the eyepiece to get a clear image, you are looking down the exact center of the spindle of the machine. The image, Figure 10–44B, is similar to that seen when looking through a rifle scope. By setting the cross hairs on the edge of the work, you know exactly where the center of the spindle is with respect to the edge of the work. It is also essential that the edge of the work be sharp, not bevelled, when using the microscope. The microscope can also be used to locate a reference line.

The Coordinate System

The **coordinate system** is the most efficient method of locating holes for jig boring. Points are located from two reference lines, or **axes**, or from two edges which are also called axes. The table is moved by two precision lead screws working at right angles to each other. The line or edge parallel to the longitudinal line of travel is called the X-axis. The line or edge parallel to the cross slide line of travel is called the Y-axis.

Any point on the work may be used as a reference origin or starting point. It can be a hole, a dowel or pin, or an intersection of two edges or reference lines. Establish this point with relation to the centerline of the spindle, then set and lock the lead screw dials at zero. Set the longitudinal and cross slide scales to the nearest inch graduation, or to the nearest millimeter graduation in the case of a metric machine. As an example, suppose that the cross slide scale was set on 3 in (76.2 mm) and the longitudinal scale was set on 5 in (127 mm). Figures 10–45A and 10–45B show what would happen. Working from the X and Y axes, the revised settings would be the actual settings for the longitudinal and the cross slide scales and dials.

SUGGESTED ORDER OF OPERATIONS

Once the settings are known, the suggested order of operations is as follows:

1. Spot drill all holes using a center drill. A center drill is much more rigid than a longer twist drill.

2. Drill and bore all holes to rough dimensions; i.e., undersized diameters.

3. Check the setting of the work to make sure that the work has not moved.

4. Finish bore all holes to exact diameter using the single-point boring tool.

USING A VERTICAL MILLING MACHINE FOR JIG BORING

The same work described for the jig boring machine can also be done on a vertical milling machine, Figure 10–46. Somewhat less accuracy is obtained, but the principles are exactly the same. The work can be set on the table in the same way as on the jig boring machine; i.e., using parallel bars and clamps. The work can be set square and true using a dial indicator clamped to the spindle. The two edges of the work can be dialed and the work can be set square and true with the longitudinal axis and cross slide axis. The coordinate positions can be established by using a reference bar. A piece of 1/2-in (12-mm) cold rolled steel gripped in a drill chuck will do. Set this bar so that it just touches the edge of the work; raise the bar clear of the work; set the lead screw dial on zero; advance the table an additional 1/4 in (6 mm), or half the diameter of the reference bar. The center of the spindle will be over the edge of the work. Reset the lead screw dial to zero. Make your adjustments for drilling and boring by using the sizes stated on the original sketch. Repeat this same procedure for both the X and Y axes. The suggested order of drilling and boring operations can be exactly as done on a jig boring machine. It must be emphasized that the principles of jig boring can be taught on a vertical milling machine, but in no way

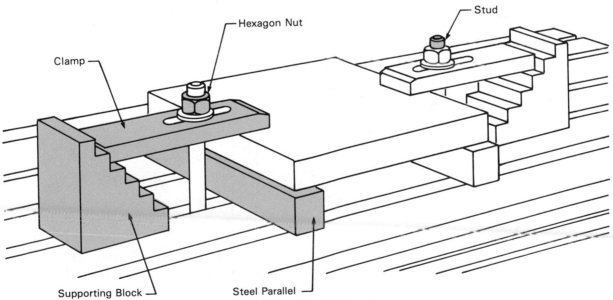

Figure 10–43 Clamping the work *(Courtesy of American SIP Corp.)*

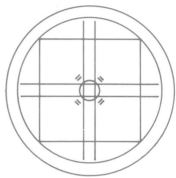

Figure 10–44B The view seen through the microscope *(Courtesy of American SIP Corp.)*

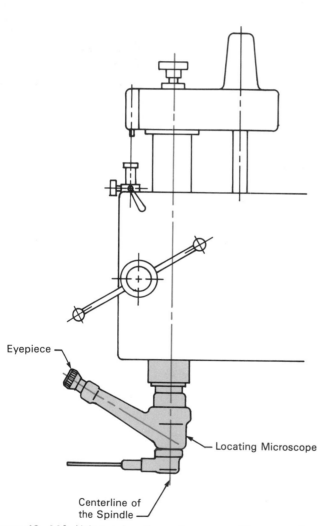

Eyepiece

Locating Microscope

Centerline of
the Spindle

Figure 10–44A Using a locating microscope *(Courtesy of American SIP Corp.)*

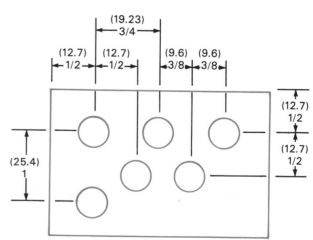

Figure 10–45A Original sketch [measurements in English (U.S. Customary) and metric units]

is this machine as accurate as a jig boring machine.

One of the newest machines on the market today is called a coordinate boring and milling machine, Figure 10–47. It combines all of the sensitivity, delicacy, and accuracy of a jig boring machine with the ruggedness of a milling machine. Its rugged design and versatility enable the machine to handle roughing and semifinishing operations as well as the very sensitive, smallest jig boring operations, to within an accuracy of 0.00006 in (0.0015 mm). The operations of drilling, boring, surface milling, side milling, and slot milling can all be performed with equal ease. Figure 10–48 illustrates two common boring operations performed by jig boring machines.

A very useful attachment for the boring machine is the circular dividing table with a digital readout, Figure 10–49. In principle, it is the same as any other circular dividing table, but with the addition of the digital readout it becomes accurate to within one second of arc (one degree is 60 seconds of arc.).

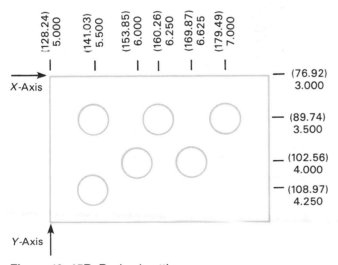

Figure 10–45B Revised settings

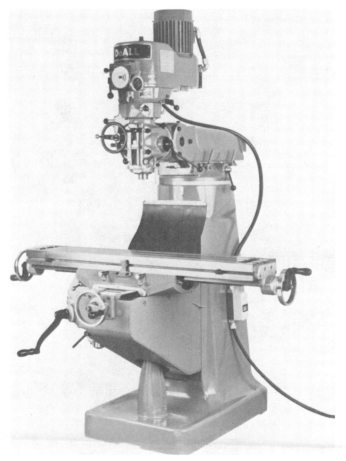

Figure 10–46 A vertical milling machine *(Courtesy of DoALL Company)*

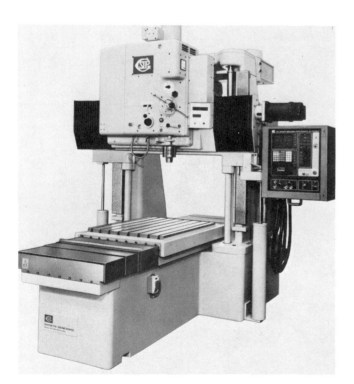

Figure 10–47 A coordinate boring and milling machine *(Courtesy of American SIP Corp.)*

Figure 10–48A A typical jig boring operation *(Courtesy of American SIP Corp.)*

Figure 10–49 A circular dividing table with digital readout; accuracy is 1 second of arc *(Courtesy of American SIP Corp.)*

Figure 10–48B A boring operation on a jig boring machine *(Courtesy of American SIP Corp.)*

REVIEW QUESTIONS

1. Name the three styles of column-and-knee-type milling machines.
2. How does a plain milling machine differ from a universal milling machine?
3. How does a vertical milling machine differ from a horizontal milling machine?
4. What is the difference between a column-and-knee-type and a bed-type milling machine?
5. In what types of machine shops are column-and-knee-type milling machines most commonly used? Bed-type machines?
6. List three ways in which a jig boring machine differs from a vertical spindle milling machine.
7. What is the primary purpose of a jig boring machine?
8. A jig and a fixture are both work holding devices. Describe how they differ in the way in which they hold the work.
9. What is the most important cutting tool used in jig boring?
10. What is the purpose of a dial indicator in jig boring?
11. Briefly describe a method of setting the work parallel to the line of table travel using a locating microscope.

12. How is the coordinate system method used to locate holes in the work for jig boring?
13. List the four steps in the suggested order of operations for jig boring.

14. How can the coordinate positions be established on a vertical milling machine being used for jig boring?

UNIT 10-3

MILLING OPERATIONS

OBJECTIVES

After completing this unit, the student will be able to:

- describe the basic milling operations.
- compare climb milling and conventional milling and state the advantages of each form of milling.
- select the proper speeds and feeds for milling operations.
- set up properly for milling operations.

KEY TERMS

Plain milling
Slab milling
Face milling
Angular milling
Gang milling

Climb (down) milling
Conventional (up) milling
Cutting speed
Feed

BASIC MILLING OPERATIONS

Plain milling, Figure 10–50, is the cutting of a flat surface parallel to the axis of the cutter. Sometimes plain milling is referred to as **slab milling**.

In judging the slope of the teeth, always look at a cutter from one end. The cutter in Figure 10–51 has a right-hand helix; i.e., the helix slopes up and curves to the right. A right-hand helix tends to pull the cutter toward the spindle, exerting pressure toward the spindle rather than away from it. This is good milling practice.

Face milling is the cutting of a flat surface at right angles to the axis of the cutter. The face milling cutter, in some cases a shell end milling cutter, can be held directly in the spindle nose of a horizontal milling machine, or directly in the spindle nose of a vertical milling machine, Figure 10–52. The face milling cut can be either vertical or horizontal.

Angular milling is the cutting of a flat surface at an inclination to the axis of the cutter. Figure 10–53 shows a dovetailing cutter at work.

Form milling, Figures 10–54 and 10–55, is the cutting of a surface having a regular or an irregular outline; i.e., a surface other than flat. A gear cutter cutting a gear tooth is an example of form milling.

Two other milling operations are actually combinations of the basic milling cuts. These operations are straddle milling and gang milling.

The photo of **straddle milling**, Figure 10–56, shows two inserted tooth cutters on each side of the work, each cutter making its own particular cuts. A combination of horizontal and vertical cuts is being made.

Gang milling is when a number of different types of cutters are all mounted on the same horizontal arbor, each cutter performing its own particular operation. It should be emphasized that because rotary milling cutters are slightly hollow ground, proper shimming should be used when cutters are set side by side on the same arbor.

Rotation of the Cutter

The direction of rotation of the cutter is most important, Figure 10–57. Never run a cutter backward, Figure 10–57B. If you do so, you will ruin the cutter.

CONVENTIONAL MILLING AND CLIMB MILLING

It is important to know what kind of milling to use, conventional or climb milling, Figure 10–58. Climb milling should be used only if the milling machine is equipped with a **backlash eliminator**, which eliminates any play or looseness in the milling machine lead screw. If you ignore this fact, your work can suddenly be pulled into the cutter, resulting in ruined work and/or a broken cutter.

In climb or down milling, chips are cut at maximum thickness at the start of a cut and decreased to zero thickness at the end of the cut. This can be used

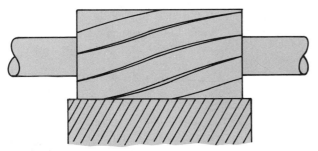

Figure 10–50 Plain milling *(Courtesy of Kearney and Trecker)*

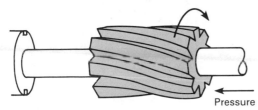

Figure 10–51 Good milling practice *(Courtesy of Kearney and Trecker)*

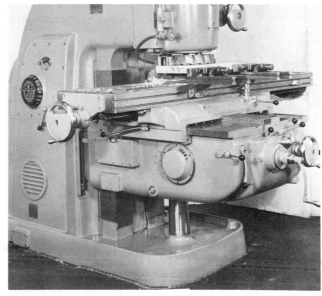

Figure 10–52 Face milling on a vertical milling machine *(Courtesy of Cincinnati Milacron)*

for most milling operations except that machines must be rigid and must have backlash eliminators. These are essential for climb milling.

There are several advantages of climb milling, Figure 10–59, over conventional milling. Fixtures and holding devices are simpler and less costly because of downward pressure exerted by the cutter. In slitting thin sheet, climb milling is preferred. Cutters with higher rake angles can be used. Less power is needed. Chips are less likely to be carried by the cutter teeth, reducing the possibility of marring the surface. Chip disposal is easier. Chips pile up behind the cutter instead of in front. The cutter wears less because chips are thicker at the start of the cut.

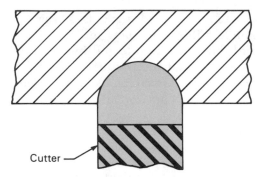

Figure 10–54 A convex cutter

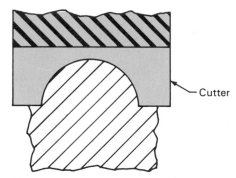

Figure 10–55 A concave cutter

Figure 10–53 Angular milling with a dovetailing cutter (*Courtesy of Cincinnati Milacron*)

Figure 10–56 Straddle milling (*Courtesy of Cincinnati Milacron*)

In conventional or up milling, the chips are cut with no thickness at the start of the cut and increased to maximum thickness at the end of the cut.

There are some advantages to conventional milling over climb milling. There is a lower impact on the cutter at the start of the cut. The direction of the milling force compensates for backlash of the feed mechanism. This is preferred over milling on the surfaces where depth of cut varies (say by 20%) and on milling castings or forgings with rough surfaces due to sand or scale.

It should be noted that the arbor, collars, and cutter must always be wiped clean before assembly. If this is not done, the cutter will run out of true. Due to the slimness and length of the arbor, and also because there are so many collars, a small run-out of the cutter is practically inevitable.

SPEEDS AND FEEDS

There are a number of factors to be considered before choosing a speed or a feed in milling. Two major

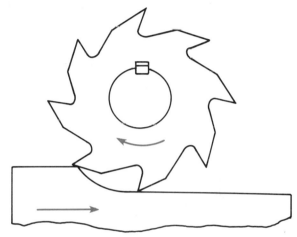

Figure 10–58A Conventional milling *(Courtesy of Cincinnati Milacron)*

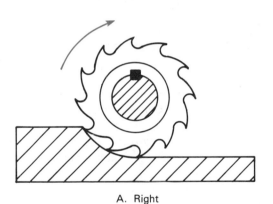

A. Right

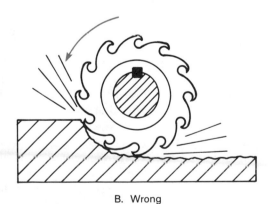

B. Wrong

Figure 10–57 Rotation of the cutter *(Courtesy of Kearney and Trecker)*

Figure 10–58B Climb milling *(Courtesy of Cincinnati Milacron)*

factors are the hardness and the toughness of the material being cut and also the hardness and the toughness of the cutter being used. Other factors are the rigidity and general condition of the machine being used and also the rigidity of the setup of the work. Because of these factors, it is difficult to give hard and fast rules for speeds and feeds, but there are some guidelines that can be used. Keep in mind that adjustments might have to be made to both speeds and feeds by lowering both to minimum recommendations

or even lower. The recommended guidelines are based on ideal conditions, especially in rigidity and the condition of the machine and setup of the work. It is assumed that cutters are sharpened properly.

Suggestions for cutting speeds for both high-speed steel and carbide cutters are listed in Table 9–7. Standard formulas for calculating rpm in both English (U.S. Customary) and metric measure are also included. Suggestions for rpm for different diameters in both English (U.S. Customary) and metric measure are contained in Table 10–1, as calculated according to cutting speeds.

It is much easier to suggest speeds than it is to suggest feeds. The best practice is to start with a slow feed and then make adjustments according to the way the cutter performs on a certain piece of work. Recommended feeds are given in Table 9–8 for both high-speed steel cutters and carbide cutters. Using a cautious approach, it might be best to treat your cutter as though it has one tooth instead of six or twelve teeth when making your initial calculation for feed. When cutting is started, it is then easy to double the amount of feed and double it again if you are not removing metal quickly enough. Remember that time can be made only during roughing cuts. Finishing is a much slower process. Surface finish is important only after a finishing cut, not after a roughing cut. Feeds are normally slower during finish cuts than they are during roughing cuts.

In addition to the speed and feed charts, Table 10–2 contains formulas for making speed and feed calculations.

MILLING HINTS

When you are setting up for milling, choose the correct cutter for the work. Make sure that your cutter is sharp. It is poor practice and dangerous to use a dull cutter. Make sure that your cutter is rotating in the proper direction and use climb milling only if your machine has a backlash eliminator to take up all of the play or looseness in the lead screw. Set the correct feeds and speeds and make sure that the work is securely clamped to the worktable. Inspect the taper on the end of the arbor. It should be thoroughly clean and free of nicks or bruises. The spacing collars and bearings

F = feed rate ϕ = angle of tooth spacing
d = depth of cut θ = clearance angle
f_T = feed per tooth α = Rake angle

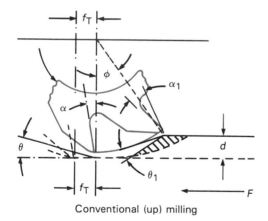

Conventional (up) milling

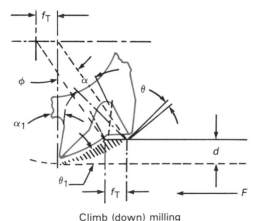

Climb (down) milling

Figure 10–59 Advantages of climb milling *(Courtesy of Cincinnati Milacron)*

Table 10-1 Rpm for Different Diameters Based on Cutting Speed *(Courtesy of SKF & Dormer Tools, Ltd.)*

PERIPHERAL SPEED						
Ft./min.	20	30	50	70	80	100
m/min.	6	9	15	21	24	30
Diam. in.	REVOLUTIONS PER MINUTE					
1/64	4,897	7,346	12,243	17,140	19,588	24,485
1/32	2,449	3,673	6,121	8,570	9,794	12,243
3/64	1,629	2,443	4,072	5,701	6,515	8,144
1/16	1,222	1,833	3,056	4,278	4,889	6,112
5/64	978	1,467	2,445	3,424	3,913	4,891
3/32	814	1,222	2,036	2,851	3,258	4,072
7/64	698	1,047	1,746	2,444	2,793	3,492
1/8	611	917	1,528	2,139	2,445	3,056
5/32	489	734	1,223	1,712	1,956	2,445
3/16	407	611	1,019	1,426	1,630	2,037
7/32	349	524	873	1,222	1,397	1,746
1/4	306	458	764	1,070	1,222	1,528
5/16	244	367	611	856	978	1,222
3/8	204	306	509	713	815	1,019
7/16	175	262	437	611	698	873
1/2	153	229	382	535	611	764
9/16	136	204	340	475	543	679
5/8	122	183	306	428	489	611
11/16	111	167	278	389	444	556
3/4	102	153	255	356	407	509
13/16	94	141	235	329	376	470
7/8	87	131	218	306	349	437
15/16	81	122	204	285	326	407
1	76	115	191	267	306	382
1, 1/8	68	102	170	238	272	340
1, 1/4	61	92	153	214	244	306
1, 3/8	56	83	139	194	222	278
1, 1/2	51	76	127	178	204	255
1, 5/8	47	71	118	165	188	235
1, 3/4	44	65	109	153	175	218
1, 7/8	41	61	102	143	163	204
2	38	57	95	134	153	191
2, 1/4	34	51	85	119	136	170
2, 1/2	31	46	76	107	122	153
2, 3/4	28	42	69	97	111	139
3	25	38	64	89	102	127
3, 1/2	22	33	55	76	87	109
4	19	29	48	67	76	95
4, 1/2	17	25	42	59	68	85
5	15	23	38	53	61	76
5, 1/2	14	21	35	49	56	69
6	13	19	32	45	51	64
6, 1/2	12	18	29	41	47	59
7	11	16	27	38	44	55
7, 1/2	10	15	25	36	41	51
8	9	14	24	33	38	48

PERIPHERAL SPEED						
Ft./min.	20	30	50	70	80	100
m/min.	6	9	15	21	24	30
Diam. mm.	REVOLUTIONS PER MINUTE					
0.5	3,878	5,817	9.695	13,573	15,512	19,389
1.0	1,939	2,908	4,847	6,786	7,756	9,695
1.5	1,293	1,939	3,232	4,524	5,171	6,463
2.0	971	1,456	2,427	3,397	3,883	4,854
2.5	776	1,165	1,941	2,717	3,105	3,882
3.0	647	970	1,617	2,264	2,587	3,234
3.5	554	832	1,386	1,940	2,218	2,772
4.0	485	728	1,213	1,698	1,940	2,425
4.5	431	647	1,078	1,509	1,724	2,156
5.0	388	582	970	1,359	1,552	1,940
6.0	323	485	809	1,132	1,294	1,617
7.0	277	416	693	970	1,109	1,386
8.0	243	364	606	849	970	1,213
9.0	216	323	539	755	862	1,078
10.0	194	291	485	679	776	970
11.0	176	265	441	617	706	882
12.0	162	243	404	566	647	809
13.0	149	224	373	522	597	746
14.0	139	208	346	485	554	693
15.0	129	194	323	453	517	647
16.0	121	182	303	424	485	606
17.0	114	171	285	399	457	571
18.0	108	162	269	377	431	539
19.0	102	153	255	357	409	511
20.0	97	146	243	340	388	485
22.0	88	132	221	309	353	441
24.0	81	121	202	283	323	404
26.0	75	112	187	261	299	373
28.0	69	104	173	243	277	346
30.0	65	97	162	226	259	323
35.0	55	83	139	194	222	277
40.0	49	73	121	170	194	243
45.0	43	65	108	151	172	216
50.0	39	58	97	136	155	194
60.0	32	49	81	113	129	162
70.0	28	42	69	97	111	139
80.0	24	36	61	85	97	121
90.0	22	32	54	75	86	108
100.0	19	29	49	68	78	97
115.0	17	25	42	58	66	83
130.0	15	22	38	51	59	73
145.0	13	20	33	46	53	66
160.0	12	18	30	42	48	60
180.0	11	16	27	37	42	53
200.0	10	14	24	33	38	48

R.P.M. for Peripheral Speeds not given, can be obtained by simple addition or subtraction, e.g.: 150 ft./min. 100 + 50 = 1,146 R.P.M. (for 1/2" dia.) 60 ft./min. = 80 — 20 = 4,886 R.P.M. (for 3/64" dia.)

R.P.M. for Peripheral Speeds not given, can be obtained by simple addition or subtraction, e.g.: 45 m/min. = 15 + 30 = 1,455 R.P.M. (for 10 mm dia.) 18 m/min. = 24 — 6 = 3.878 R.P.M. (for 1.5 mm dia.)

on the arbor must also be thoroughly clean and free of nicks and bruises. Set the work and cutter as close to the column of the machine as possible.

Most cutter failures are caused by vibration. This can be caused by loose bearings, a loose table, or a loose knee. The cutter may not be ground properly or it may be dull. The land may be too wide, causing the cutter to rub on the work instead of cutting it. Speed of the cutter may be too fast or the feed may be too heavy. There may not be enough coolant.

Table 10-2 Formulas for Speed and Feed Calculations *(Courtesy of SKF & Dormer Tools, Ltd.)*

Where D = Diameter of Cutter.
T = Number of Teeth in Cutter.
π = 3.1416 (approx. $^{22}/_7$).
R.P.M. = Revolutions per minute of Cutter.
ft./min. =
or m/min. = Peripheral Speed of Cutter.
F = Feed of Cutter per Minute, or Table Travel.
F.R. = Feed per Revolution of Cutter.
F.T. = Feed per Tooth per Revolution, i.e. Chip and Tooth Load.

TO FIND	KNOWN	FORMULA ENGLISH	FORMULA METRIC
Speed of Cutter	(a) Diameter of Cutter (b) Revolutions per minute of cutter	$\dfrac{D \times \pi \times R.P.M.}{12}$ = ft./min. .2618 D x R.P.M. = ft./min.	$\dfrac{D \times \pi \times R.P.M.}{1000}$ = m/min.
Revolutions per Minute of Cutter	(a) Diameter of Cutter (b) Peripheral Speed in feet per minute. (m/min.)	$\dfrac{ft./min.}{\left(\dfrac{D \times \pi}{12}\right)}$ $\dfrac{ft./min.}{.2618D}$ R.P.M. $3.82 \times \dfrac{ft./min.}{D}$ = R.P.M. or $4 \times \dfrac{ft.min.}{D}$ = R.P.M. app.	$\dfrac{\text{Metres/min.} \times 1000}{\text{'D'} \times \pi}$ = R.P.M.
Feed per Revolution of Cutter	(a) Feed of Cutter per minute (b) Revolution per minute of cutter	$\dfrac{F}{R.P.M}$ = F.R.	
Feed per Tooth per Revolution	(a) Feed of Cutter per minute (b) Revolutions per minute of cutter (c) Number of teeth in cutter	$\dfrac{F}{T \times R.P.M.}$ = F.T. also $\dfrac{F.R.}{T}$ = F.T.	
Feed of Cutter per Minute	(a) Feed per tooth per revolution (b) Number of Teeth in cutter (c) Revolutions per minute of cutter	F.T. x T. = R.P.M. = F.	
Feed of Cutter per Minute	(a) Feed per revolution of cutter (b) Revolutions per minute of cutter	F.R. x R.P.M. = F.	
Number of Teeth	(a) Feed of Cutter per minute (b) Feed per tooth per revolution (c) Revolutions per minute of cutter	$T = \dfrac{F}{F.T. \times R.P.M.}$	

REVIEW QUESTIONS

1. What is the difference between plain milling and face milling?
2. Briefly define angular milling and form milling.
3. What is the difference between straddle milling and gang milling?
4. In the process of climb (down) milling, at what point are the chips at maximum thickness?
5. What two machine requirements must be met before climb milling can be used?
6. List four advantages of climb milling over conventional milling.
7. What is the major difference between conventional (up) milling and climb milling?
8. List two advantages of conventional milling over climb milling.
9. What are four factors to be considered in selecting the cutting speed for a milling operation?
10. What is the recommended practice for determining the proper feed for milling?
11. List six checks to make during the set up for milling.
12. What are four possible causes of cutter failure?

UNIT 10-4

INDEXING

OBJECTIVES

After completing this unit, the student will be able to:

- define the following terms: plain or direct indexing, simple indexing, angular indexing, and differential indexing.
- set up and perform direct indexing.
- use the universal dividing head for simple indexing and angular indexing.
- set up and use the universal dividing head for differential indexing.

KEY TERMS

Indexing
Plain or direct indexing
Simple indexing
Dividing head
Index plate
Universal dividing head

Angular indexing
Differential indexing
Index crank
Plain or direct index head
Hole circle
Gear ratio
Spindle gear
Worm gear
Idler gear

INTRODUCTION

Indexing is a method used to accurately divide the circumference of a circle into a number of equal parts, as in the cutting of gear teeth on a gear, the cutting of lines on a graduated collar (as on a lathe), or the cutting of lines on a protractor. The types of indexing are plain or direct indexing, simple indexing, angular indexing, and differential indexing.

Plain or direct indexing is when the index crank is connected directly to the spindle. If two equally spaced cuts were required, the spindle would be moved a half turn for each cut. **Simple indexing** increases the number of equal divisions attainable. The index crank turns 40 times to the spindle's one turn. The bulk of indexing is simple indexing. **Angular indexing** is the division of a circle into degrees and minutes, working with the ratio of 40:1. Since it is impossible to index all numbers, we must differential index. **Differential indexing** is a method used to index those numbers that cannot be simple indexed. Differential indexing is a combination of two movements, the simple movement of the index crank and the movement (or rotation) of the index plate itself either in a clockwise or a counterclockwise direction to achieve the needed number of divisions.

DIRECT INDEXING

The plain or direct index head, Figure 10–60, is the simplest of the dividing mechanisms. The index crank is connected directly to the spindle. The crank and spindle are rotated directly to the required number of equal parts of the circle desired, as marked on the index plate. The number of equal divisions that can be obtained on this type of index head is limited by the number of holes on the index plate. Common direct indexing plates contain 24-hole (Figure 10–61), 30-hole, 36-hole, 48-hole, and/or 60-hole circles. The most common index plate has a 24-hole circle. The direct index head can be used for average work on a milling machine.

EXAMPLE OF DIRECT INDEXING. To cut a square, the work would have to be revolved through four equal cuts.

Using a 24-hole circle:

$$\frac{24}{N} = \frac{24}{4} = 6 \text{ holes for each of four cuts}$$

This index crank would be moved 6 holes for each cut. A 24-hole circle will divide into 2, 3, 4, 6, 8, 12, and 24 equal cuts. Each of these numbers is a multiple of 24.

A 30-hole circle will divide into 2, 3, 5, 6, 10, 15, and 30 equal cuts, each a multiple of 30.

A 36-hole circle will divide into 2, 3, 4, 6, 9, 12, 18, and 36 equal cuts, each a multiple of 36.

Figure 10–60 The plain or direct index head *(Courtesy of Cincinnati Milacron)*

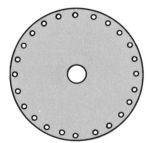

Figure 10–61 A 24-hole index plate

THE UNIVERSAL DIVIDING HEAD

The universal dividing head is by far the most common dividing head in most machine shops. Basically, the index crank revolves a worm which turns the spindle in the ratio of 40:1; i.e., the index crank will turn 40 times and the spindle will turn once. The spindle has a tapered hole, a Brown and Sharpe taper, and sometimes an external thread to take a threaded chuck. (The Brown and Sharpe Company did the bulk of the early development on the milling machine.) The spindle is set in a swiveling block that can swivel from −10 to +100 degrees. The footstock and an adjustable center rest will support work between centers. The universal dividing head is used for direct indexing, simple indexing, differential indexing, and helical milling.

One example of an index plate containing 20-, 19-, 18-, 17-, 16-, and 15-hole circles is illustrated in Figure 10–62. The numbered hole is the starting point or zero hole.

Index Plates

Brown and Sharpe normally supplies three plates:

Plate #1 — 15-, 16-, 17-, 18-, 19-, and 20-hole circles.
Plate #2 — 21-, 23-, 27-, 29-, 31-, and 33-hole circles.
Plate #3 — 37-, 39-, 41-, 43-, 47-, and 49-hole circles.

Cincinnati Milacron normally supplies one plate drilled on both sides:

Side #1 — 24-, 25-, 28-, 30-, 34-, 37-, 38-, 39-, 41-, 42-, and 43-hole circles.
Side #2 — 46-, 47-, 48-, 51-, 53-, 54-, 57-, 58-, 59-, 62, and 66-hole circles.

Elliott Machine Tool Company (England) usually supplies two plates:

Plate #1 — 15-, 18-, 20-, 23-, 27-, 31-, 37-, 41-, and 47-hole circles.
Plate #2 — 16-, 17-, 19-, 21-, 29-, 33-, 39-, 43-, and 49-hole circles.

Simple Indexing

The ratio of the universal dividing head is 40:1, or 40 turns of the index crank to one turn of the spindle.

EXAMPLES OF SIMPLE INDEXING

Cutting a Square

Cutting a square would require four equally spaced cuts.

$$\frac{40}{N} = \frac{40}{4} = 10 \text{ turns of the crank for each cut}$$

Cutting Ten Equal Cuts

$$\frac{40}{N} = \frac{40}{10} = 4 \text{ turns of the crank for each cut}$$

CUTTING SIX EQUAL CUTS

$$\frac{40}{N} = \frac{40}{6} = 6\ 4/6 \text{ turns of the crank or } 6\ 2/3 \text{ turns for each cut}$$

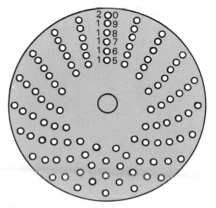

Figure 10–62 A typical index plate

To cut six equal cuts (a hexagon), rotate the index crank 6 2/3 turns for each cut. Convert the two-thirds of a turn to an appropriate hole circle. Use any hole circle divisible by three; e.g., a 15-, 18-, 21-, 27-, 33-, or 39-hole circle.

If the 39-hole circle is selected:

2/3 = 26/39 = 26 holes on the 39-hole circle
(actually 26 spaces, the spaces between the holes)

or:

6 2/3 turns = 6 26/39 turns = 6 turns plus 26 holes on the 39-hole circle for each of 6 cuts

The hole numbered 39 is zero. Starting from zero, count off 26 holes (or spaces). Set the sector arms at zero and also at the 26th hole and lock them so that they will maintain this angle or 2/3 of a turn, Figure 10–63. Make the first cut and then return the cutter to the starting point. To make the second cut, start at zero, crank six turns plus the additional 26 holes, lock the crank in the 26th hole, and move the sector arms until the starting arm rests against the 26th hole. This now becomes your starting point for the second cut. Make the second cut. Return the cutter to the starting point. Again crank six turns plus 26 holes and move the sector arms to the 26th hole. Make the third cut. Repeat the procedure until all six cuts are done.

CUTTING A 60-TOOTH GEAR

To cut a 60-tooth gear, 60 equally spaced cuts are required.

$\dfrac{40}{N} = \dfrac{40}{60} = \dfrac{4}{6} = \dfrac{2}{3}$ of a turn for each of the 60 cuts

Convert 2/3 to an appropriate hole circle; e.g., 15-, 18-, 21-, 27-, 33-, or 39-hole circle. If you use the 39-hole circle:

$\dfrac{2}{3} = \dfrac{26}{39}$ or 26 holes on the 39-hole circle

Set the index crank in the hole numbered 39. This is zero. Count 26 holes and set the sector arms. Crank through 26 holes on the 39-hole circle for each cut. Don't forget to move the sector arms at the start of each cut. Table 10–3 contains a simple indexing chart.

Angular Indexing

In angular indexing, 40 turns of the index crank will move the spindle one turn or 360 degrees. If the crank is turned once, the spindle moves 1/40 turn or $1/40 \times 360$ degrees = 9 degrees. If you turn the crank 1/9 turn, the spindle will move 1 degree.

EXAMPLES. Index for 36 degrees:

$\dfrac{36}{9}$ = 4 turns of the crank to move the spindle 36 degrees

Index for 5 1/2 degrees:

$\dfrac{5\ 1/2}{9}$ (convert to a whole number)
    (convert to an existing hole circle)

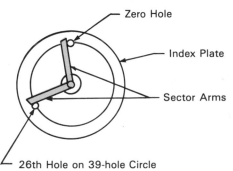

Figure 10–63 One example of two-thirds of a turn of the index crank

Table 10–3 A Simple Indexing Chart *(Courtesy of Gate Machinery Company Ltd.)*

STANDARD PLATES—INDEX PLATE 1 , 2   HOLE CIRCLES—15, 18, 20, 23, 27, 31, 37, 41, 47 , 16, 17, 19, 21, 29, 33, 39, 43, 49

No. of divisions	Hole circle	revolutions of the index-arm	the holes in the circle	No. of divisions	Hole circle	revolutions of the index-arm	Holes in the circle	No. of divisions	Hole circle	revolutions of the index-arm	Holes in the circle	No. of divisions	Hole circle	the holes in the circle	No. of divisions	Hole circle	Holes in the circle	No. of divisions	Hole circle	the holes in the circle	No. of divisions	Hole circle	Holes in the circle
2	—	20	—	18	18	2	4/18	38	19	1	1/19	62	31	20/31	105	21	8/21	168	21	5/21	264	33	5/33
3	39	13	13/39	19	19	2	2/19	39	39	1	1/39	64	16	10/16	108	27	10/27	170	17	4/17	270	27	4/27
	27	13	9/27	20	—	2	—	40	—	1	—	65	39	24/39	110	33	12/33	172	43	10/43	280	49	7/49
4	—	10	—	21	21	1	19/21	41	41	—	40/41	66	33	20/33	115	23	8/23	180	27	6/27	290	29	4/29
5	—	8	—	22	33	1	27/33	42	21	—	20/21	68	17	10/17	116	29	10/29		18	4/18	296	37	5/37
6	39	6	26/39	23	23	1	17/23	43	43	—	40/43	70	49	28/49	120	39	13/39	184	23	5/23	300	15	2/15
	27	6	18/27	24	39	1	26/39	44	33	—	30/33	72	27	15/27		27	9/27	185	37	8/37	360	18	2/18
7	49	5	35/49		27	1	18/27	45	27	—	24/27	74	37	20/37	124	31	10/31	188	47	10/47		27	3/27
	21	5	15/21	25	20	1	12/20	46	23	—	20/23	75	15	8/15	128	16	5/16	190	19	4/19	400	20	2/20
8	—	5	—	26	39	1	21/39	47	47	—	40/47	76	19	10/19	130	39	12/39	195	39	8/39			
9	27	4	12/27	27	27	1	13/27	48	18	—	15/18	78	39	20/39	132	33	10/33	196	49	10/49			
	18	4	8/18	28	49	1	21/49	49	49	—	40/49	80	20	10/20	135	27	8/27	200	20	4/20			
10	—	4	—		21	1	9/21	50	20	—	16/20	82	41	20/41	136	17	5/17	205	41	8/41			
11	33	3	21/33	29	29	1	11/29	*51	51	—	40/51	84	21	10/21	140	49	14/49	210	21	4/21			
12	39	3	13/39	30	39	1	13/39	52	39	—	30/39	85	17	8/17	144	18	5/18	215	43	8/43			
	27	3	9/27		27	1	9/27	*53	53	—	40/53	86	43	20/43	145	29	8/29	216	27	5/27			
13	39	3	3/39	31	31	1	9/31	54	27	—	20/27	88	33	15/33	148	37	10/37	220	33	6/33			
14	49	2	42/49	32	20	1	5/20	55	33	—	24/33	90	27	12/27	150	15	4/15	230	23	4/23			
	21	2	18/21	33	33	1	7/33	56	49	—	35/49	92	23	10/23	152	19	5/19	232	29	5/29			
15	39	2	26/39	34	17	1	3/17	*57	57	—	40/57	94	47	20/47	155	31	8/31	235	47	8/47			
	27	2	18/27	35	49	1	7/49	58	29	—	20/29	95	19	8/19	156	39	10/39	240	18	3/18			
16	20	2	10/20		21	1	3/21	*59	59	—	40/59	98	49	20/49	160	20	5/20	245	49	8/49			
17	17	2	6/17	36	27	1	3/27	60	39	—	26/39	100	20	8/20	164	41	10/41	248	31	5/31			
18	27	2	6/27	37	37	1	3/37		27	—	18/27	104	39	15/39	165	33	8/33	260	39	6/39			

Numerator and denominator must be treated equally.

$$\frac{5 \ 1/2 \ \times \ 2}{9 \ \times \ 2} = \frac{11}{18}$$ or 11 holes on the 18-hole circle

Eleven holes on the 18-hole circle would move the spindle 5 1/2 degrees.

Differential Indexing

It is impossible to simple index for all numbers. The fraction 40/N cannot always be reduced to a factor of one of the available hole circles. Differential indexing must be used. The necessary division is obtained by the combination of two movements, the simple indexing movement of the index crank and the movement of the index plate itself.

A plan view of a simplified version of a dividing head, Figure 10–64, shows that the crank turns a worm shaft, which turns a worm, which turns a 40-tooth worm wheel and then the spindle.

The index plate is attached to a sleeve that connects it to equal gears and the worm shaft. In simple indexing, the index plate is locked to prevent it from turning. In differential indexing, the index plate is unlocked leaving it free to turn.

A gear train is set up to join the outer end of the spindle to the worm shaft (the dividing head is actually geared to itself). When the crank is turned to move the spindle, the gear train will cause the index plate to turn at the same time.

How much the index plate will turn depends on the ratio of the outer spindle gear and the gear on the worm shaft. Whether the index plate revolves clockwise (as does the index crank), or counterclockwise, depends on whether one or two idler gears are used between the spindle gear and the worm gear.

If the spindle gear and the worm gear are equal and if two idler gears are used, the index plate will move counterclockwise (as the index crank is turned

clockwise). Forty turns of the index crank would cause the index plate to move back one revolution. If you stopped the index crank at each turn and took a cut on a gear blank, a 41-tooth gear would result.

If one idler is used, the index plate will move clockwise. Again, if you turned the index crank 40 times, the index plate would move forward one revolu-

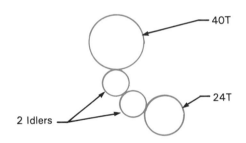

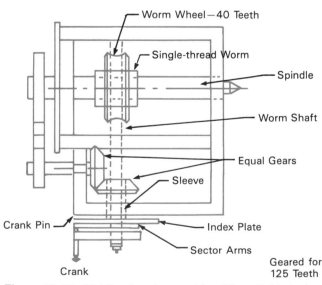

Figure 10–64 Dividing head geared for differential indexing of 125 teeth

tion. If a cut were made at each revolution of the crank, a 39-tooth gear would result.

EXAMPLE 1. Index for 125 divisions:

$$\frac{40}{N} = \frac{40}{125} = \frac{8}{25}$$

The only simple indexing possible would be 40 holes on a 125-hole circle or 8 holes on a 25-hole circle. These circles are not available on most machines. Differential indexing is indicated.

N = number of divisions required = 125 (cannot be simple indexed).

A = 120 = an approximate number close to 125 that can be simple indexed. (A can be any approximate number as long as it can be simple indexed.)

Finding the required gear ratio for differential indexing:

$$R = (A - N) \times \frac{40}{A} \text{ (if A is greater than N)}$$

or

$$R = (N - A) \times \frac{40}{A} \text{ (if A is less than N)}$$

$$R = (N - A) \times \frac{40}{A}$$

$$R = (125 - 120) \times \frac{40}{120} = 5 \times \frac{1}{3} = \frac{5}{3} = \frac{\text{gear on spindle (driver)}}{\text{gear on worm (driven)}}$$

$$R = \frac{5}{3} = \frac{5 \times 8}{3 \times 8} = \frac{40}{24} \frac{\text{(spindle gear)}}{\text{(worm gear)}}$$

Simple indexing for 120:

$$\frac{40}{N} = \frac{40}{120} = \frac{1}{3} \text{ turn of index crank for each cut, plus the movement of the index plate in the ratio of 5:3}$$

$$\frac{1}{3} \text{ turn} = \frac{13}{39} \text{ or 13 holes on the 39-hole circle}$$

If A is greater than N, the index plate and crank must revolve in the same direction or clockwise. One idler gear must be used between the 40-tooth and 24-tooth gears.

If A is less than N, the index plate and crank must revolve in opposite directions — the index plate counterclockwise, the crank clockwise — and the two idler gears must be used between the 40-tooth and 24-tooth gears.

Note that the size of the idler gears does not matter. Their only purpose is to regulate the direction of rotation.

Final solution for 125 divisions: Set the sector arms for 13 holes on the 39-hole circle (other hole circles can also be used). Place a 40-tooth gear on the spindle and a 24-tooth gear on the worm shaft, with two idler gears between them.

EXAMPLE 2. Index for 133 divisions:

$$N = 133$$
$$A = 140$$

Gear ratio required:

$$(A - N) \times \frac{40}{A} = (140 - 133) \times \frac{40}{140} = 7 \times \frac{2}{7} = \frac{2}{1} = \frac{2 \times 24}{1 \times 24} = \frac{48 \text{ (spindle)}}{24 \text{ (worm)}}$$

A is greater than N: therefore, the index plate and crank must revolve in the same direction or clockwise. One idler gear must be used between the 48-tooth gear on the spindle and the 24-tooth gear on the worm shaft.

Simple indexing for 140:

$$\frac{40}{140} = \frac{2}{7} \text{ turn for each cut} = \frac{14}{49} \text{ or 14 holes on the 49-hole circle}$$

Table 10–4 contains a chart of the change gears and movements for differential indexing.

Table 10–4 Change Gears and Movements for Differential Indexing *(Courtesy of Gate Machinery Company Ltd.)*

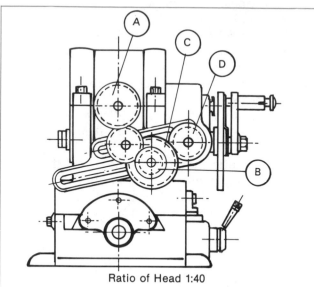

Ratio of Head 1:40

Pitch Circles Provided: 15, 16, 17, 18, 19, 20, 21, 23, 27, 29, 31, 33, 37, 39, 41, 43, 47, 49 holes.

Dividing number	Pitch circle	Revolution of the index arm	Wheel on the spindle of the dividing head — A	Wheel on the bolt — B	Wheel on the bolt — C	Wheel on the shaft of the dividing disc — D	Number of intermediate wheels
51	17	14/17	48	—	—	24	2
53	49	35/49	72	24	40	56	—
57	49	35/49	40	—	—	56	2
59	33	22/23	32	—	—	48	1
61	33	22/33	32	—	—	48	2
63	33	22/33	48	—	—	24	2
67	49	28/49	48	—	—	28	1
69	20	12/20	56	—	—	40	2
71	27	15/27	40	—	—	72	1
73	49	28/49	48	—	—	28	2
77	20	10/20	48	—	—	32	1
79	20	10/20	24	—	—	48	1

Table 10-4 Continued

Dividing number	Pitch circle	Revolution of the index arm	Wheel on the spindle of the dividing head — A	Wheel on the bolt — B	Wheel on the bolt — C	Wheel on the shaft of the dividing disc — D	Number of intermediate wheels
81	20	10/20	24	—	—	48	2
83	20	10/20	48	—	—	32	2
87	15	7/15	24	—	—	40	2
89	27	12/27	32	—	—	72	1
91	39	18/39	48	—	—	24	2
93	27	12/27	32	—	—	24	2
96	49	21/49	32	—	—	28	2
97	20	8/20	48	—	—	40	1
99	20	8/20	32	40	28	56	—
101	20	8/20	48	40	24	72	1
102	20	8/20	32	—	—	40	2
103	20	8/20	48	—	—	40	2
106	43	16/43	48	24	24	86	—
107	20	8/20	64	32	56	40	1
109	16	6/16	28	—	—	32	2
111	39	13/39	72	—	—	24	1
112	33	11/33	64	—	—	24	1
113	39	13/39	56	—	—	24	1
114	39	13/39	48	—	—	24	1
117	33	11/33	24	—	—	24	1
118	39	13/39	32	—	—	48	1
119	39	13/39	24	—	—	72	1
121	39	13/39	24	—	—	72	2
122	39	13/39	32	—	—	48	2
123	39	13/39	24	—	—	24	2
125	39	13/39	40	—	—	24	2
126	39	13/39	48	—	—	24	2
127	39	13/39	56	—	—	24	2
129	39	13/39	72	—	—	24	2
131	20	6/20	28	—	—	40	1
133	49	14/49	48	—	—	24	1
134	49	14/49	48	—	—	28	1

Table 10-4 Continued

Dividing number	Pitch circle	Revolution of the index arm	Wheel on the spindle of the dividing head A	Wheel on the bolt B	Wheel on the bolt C	Wheel on the shaft of the dividing disc D	Number of intermediate wheels
137	49	14/49	24	—	—	28	1
138	49	14/49	32	—	—	56	1
139	49	14/49	24	48	32	56	—
141	18	5/18	40	—	—	48	1
142	49	14/49	32	—	—	56	2
143	49	14/49	24	—	—	28	2
146	49	14/49	48	—	—	28	2
147	49	14/49	48	—	—	24	2
149	49	14/49	72	—	—	28	2
151	20	5/20	72	—	—	32	1
153	20	5/20	56	—	—	32	1
154	20	5/20	48	—	—	32	1
157	20	5/20	24	—	—	32	1
158	20	5/20	24	—	—	48	1
159	20	5/20	28	56	32	64	—
161	20	5/20	28	56	32	64	1
162	20	5/20	24	—	—	48	2
163	20	5/20	24	—	—	32	2
166	20	5/20	48	—	—	32	2
167	20	5/20	56	—	—	32	2
169	20	5/20	72	—	—	32	2
171	21	5/21	40	—	—	56	2
173	27	6/27	64	32	56	72	—
174	27	6/27	32	—	—	24	1
175	27	6/27	64	32	40	72	—
176	27	6/27	64	24	24	72	—
177	27	6/27	48	—	—	72	1
178	27	6/27	32	—	—	72	1
179	27	6/27	32	48	24	72	—
181	27	6/27	32	48	24	72	1
182	27	6/27	32	—	—	72	2
183	27	6/27	32	—	—	48	2

Table 10-4 Continued

Dividing number	Pitch circle	Revolution of the index arm	Wheel on the spindle of the dividing head A	Wheel on the bolt B	Wheel on the bolt C	Wheel on the shaft of the dividing disc D	Number of intermediate wheels
186	27	6/27	64	—	—	48	2
187	27	6/27	56	24	40	72	1
189	27	6/27	64	—	—	32	2
191	20	4/20	72	—	—	40	1
192	20	4/20	64	—	—	40	1
193	20	4/20	56	—	—	40	1
194	20	4/20	48	—	—	40	1
197	20	4/20	24	—	—	40	1
198	20	4/20	32	40	28	56	—
199	20	4/20	32	64	40	100	—
201	20	4/20	24	60	24	72	—
202	20	4/20	48	40	24	72	1
203	20	4/20	24	—	—	40	2
204	20	4/20	32	—	—	40	2
206	20	4/20	48	—	—	40	2
207	20	4/20	56	—	—	40	2
208	20	4/20	64	—	—	40	2
209	20	4/20	72	—	—	40	2
211	16	3/16	28	—	—	64	1
212	43	8/43	48	24	24	86	—
213	27	5/27	40	—	—	72	1
214	20	4/20	64	32	56	40	1
217	21	4/21	64	—	—	48	2
218	16	3/16	56	—	—	64	2
219	21	4/21	48	—	—	28	2
221	17	3/17	24	—	—	24	1
222	18	3/18	72	—	—	24	1
223	43	8/43	64	24	48	86	1
224	18	3/18	64	—	—	24	1
225	27	5/27	40	—	—	24	2
226	18	3/18	56	—	—	24	1
227	49	8/49	72	28	64	56	—

Table 10-4 Continued

Dividing number	Pitch circle	Revolution of the index arm	Wheel on the spindle of the dividing head A	Wheel on the bolt B	Wheel on the bolt C	Wheel on the shaft of the dividing disc D	Number of intermediate wheels
228	18	3/18	48	—	—	24	1
229	18	3/18	44	—	—	24	1
231	18	3/18	48	—	—	32	1
233	18	3/18	56	—	—	40	1
234	18	3/18	24	—	—	24	1
236	18	3/18	32	—	—	48	1
237	18	3/18	24	—	—	48	1
238	18	3/18	24	—	—	72	1
239	18	3/18	32	64	24	72	—
241	18	3/18	32	64	24	72	1
242	18	3/18	24	—	—	72	2
243	18	3/18	32	—	—	64	2
244	18	3/18	32	—	—	48	2
246	18	3/18	24	—	—	24	2
247	18	3/18	56	—	—	48	2
249	18	3/18	48	—	—	32	2
250	18	3/18	40	—	—	24	2
251	18	3/18	64	32	44	48	1
252	18	3/18	48	—	—	24	2
253	33	5/33	40	—	—	24	1
254	18	3/18	56	—	—	24	2
255	18	3/18	72	24	40	48	1
256	18	3/18	64	—	—	24	2
257	49	8/49	64	28	48	56	1
258	43	7/43	64	—	—	32	2
259	49	7/49	72	—	—	24	1
261	29	4/29	72	24	64	48	—
262	20	3/20	28	—	—	40	1
263	49	8/49	72	28	64	56	1
265	49	7/49	72	24	40	56	—
266	49	7/49	64	—	—	32	1
267	27	4/27	32	—	—	72	1
268	49	7/49	48	—	—	28	1

Table 10-4 Continued

Dividing number	Pitch circle	Revolution of the index arm	Wheel on the spindle of the dividing head A	Wheel on the bolt B	Wheel on the bolt C	Wheel on the shaft of the dividing disc D	Number of intermediate wheels
269	20	3/20	28	40	32	64	1
271	49	7/49	72	—	—	56	1
272	49	7/49	64	—	—	56	1
273	49	7/49	24	—	—	24	1
274	49	7/49	48	—	—	56	1
275	49	7/49	40	—	—	56	1
276	49	7/49	32	—	—	56	1
277	49	7/49	24	—	—	56	1
278	49	7/49	24	48	32	56	—
279	27	4/27	32	—	—	24	2
281	49	7/49	24	56	24	72	1
282	43	6/43	56	24	24	86	—
283	49	7/49	24	—	—	56	2
284	49	7/49	32	—	—	56	2
285	49	7/49	40	—	—	56	2
286	49	7/49	48	—	—	56	2
287	49	7/49	24	—	—	24	2
288	49	7/49	32	—	—	28	2
289	49	7/49	72	—	—	56	2
291	15	2/15	48	—	—	40	1
292	49	7/49	48	—	—	28	2
293	15	2/15	56	40	32	48	—
294	49	7/49	48	—	—	24	2
295	15	2/15	32	—	—	48	1
297	33	4/33	56	24	48	28	—
298	49	7/49	72	—	—	28	2
299	23	3/23	24	—	—	24	1
301	43	6/43	48	—	—	24	2
302	16	2/16	72	—	—	32	1
303	15	2/15	48	40	24	72	1
304	16	2/16	48	—	—	24	1
305	15	2/15	32	—	—	48	2
306	15	2/15	32	—	—	40	2

Table 10-4 Continued

Dividing number	Pitch circle	Revolution of the index arm	Wheel on the spindle of the dividing head A	Wheel on the bolt B	Wheel on the bolt C	Wheel on the shaft of the dividing disc D	Number of intermediate wheels
307	15	2/15	56	40	48	72	1
308	16	2/16	48	—	—	32	1
309	15	2/15	48	—	—	40	2
311	16	2/16	72	24	24	64	—
313	16	2/16	28	—	—	32	1
314	16	2/16	24	—	—	32	1
315	16	2/16	40	—	—	64	1
316	16	2/16	32	—	—	64	1
317	16	2/16	24	—	—	64	1
318	16	2/16	24	48	28	56	—
319	29	4/29	72	24	64	48	1
321	16	2/16	24	64	24	72	1
322	23	3/23	64	—	—	32	2
323	16	2/16	24	—	—	64	2
324	16	2/16	32	—	—	64	2
325	16	2/16	40	—	—	64	2
326	16	2/16	24	—	—	32	2
327	16	2/16	28	—	—	32	2
329	16	2/16	72	24	24	64	1
331	16	2/16	48	24	44	64	1
332	16	2/16	48	—	—	32	2
333	27	3/27	72	—	—	24	1
334	16	2/16	56	—	—	32	2
335	33	4/33	40	44	48	72	1
336	16	2/16	64	—	—	32	2
337	43	5/43	56	32	40	86	—
338	16	2/16	72	—	—	32	2
339	27	3/27	56	—	—	24	1
341	43	5/43	40	32	24	86	—
342	27	3/27	64	—	—	32	1
343	15	2/15	86	24	64	40	1
345	27	3/27	40	—	—	24	1
346	27	3/27	64	32	56	72	—

Table 10-4 Continued

Dividing number	Pitch circle	Revolution of the index arm	Wheel on the spindle of the dividing head A	Wheel on the bolt B	Wheel on the bolt C	Wheel on the shaft of the dividing disc D	Number of intermediate wheels
347	43	5/43	40	32	24	86	1
348	27	3/27	32	—	—	24	1
349	27	3/27	48	24	44	72	—
350	27	3/27	64	32	40	72	—
351	27	3/27	24	—	—	24	1
352	27	3/27	64	24	24	72	—
353	27	3/27	56	—	—	72	1
354	27	3/27	48	—	—	72	1
355	27	3/27	40	—	—	72	1
356	27	3/27	32	—	—	72	1
357	27	3/27	24	—	—	72	1
358	27	3/27	24	48	32	72	—
359	43	5/43	100	32	48	86	1
361	19	2/19	64	—	—	32	1
362	27	3/27	32	56	28	72	1
363	27	3/27	24	—	—	72	2
364	27	3/27	32	—	—	72	2
365	20	2/20	56	24	48	32	—
366	27	3/27	32	—	—	48	2
367	27	3/27	56	—	—	72	2
368	27	3/27	64	24	24	72	1
369	41	4/41	64	28	56	32	—
371	21	2/21	64	24	56	32	—
372	27	3/27	64	—	—	48	2
373	20	2/20	72	32	48	40	—
374	27	3/27	56	32	64	72	1
375	27	3/27	40	—	—	24	2
377	29	3/29	24	—	—	24	1
378	27	3/27	64	—	—	32	2
379	20	2/20	72	40	56	48	—
381	27	3/27	56	—	—	24	2
382	20	2/20	72	—	—	40	1

REVIEW QUESTIONS

1. Explain the term indexing.
2. Define each of the following forms of indexing: plain or direct indexing, simple indexing, and angular indexing.
3. Briefly describe how the direct index head is used. What determines the limit of equal divisions that can be obtained using this head?
4. What is meant by the term hole circle?
5. What hole circles are used in direct indexing?
6. If you had to cut a hexagon, what hole circles would be used and how many holes would be used on each of the hole circles?
7. How does a universal dividing head differ from a direct index head?
8. What is the ratio of the universal dividing head and what does the ratio mean?
9. If you were to cut a 60-tooth gear using a universal dividing head, how would you set up the indexing for it?
10. How would you set the sector arms for simple indexing a 60-tooth gear?
11. In angular indexing, how would you index for 27 degrees?
12. Why is differential indexing necessary?
13. Explain how differential indexing is accomplished.
14. Set up the differential indexing for 126 divisions.

UNIT 10-5

GEARING

OBJECTIVES

After completing this unit, the student will be able to:

- define the term gear.
- describe the basic types of gears: spur gear, gear and pinion, rack and pinion, bevel gear, miter gear, herringbone gears, and worm gear.
- explain the relationship between the pitch circle, pitch diameter, root diameter, and outside diameter of a gear.
- identify the parts of a gear tooth.
- use the two gear measurement systems: circular pitch system and diametral pitch system.
- cut a spur gear.

KEY TERMS

Gear	Gear blank
Gear tooth	Pitch circle
Spur gear	Pitch diameter
Pinion	Root diameter
Gear and pinion	Outside diameter
Gear rack	Addendum
Rack and pinion	Dedendum
Bevel gear	Clearance
Truncated cone	Tooth thickness
Miter gear	Working depth
Helical gear	Whole depth
Herringbone gear	Circular pitch system
Worm	Diametral pitch system
Worm gear	Diametral pitch

TYPES OF GEARS

A **gear** is a toothed wheel or cog that is used to transmit nonslip motion between two shafts. Gears are made in a variety of shapes and are used in many combinations.

A **spur gear**, Figure 10–65, is a cylinder. Gear teeth are cut on the outside of the cylinder parallel to the axis of the gear and in line with the shaft. The shafts are parallel to each other. The larger spur gear is usually referred to as a gear and the smaller as a pinion, hence the term *gear and pinion*.

A spur gear in combination with a gear rack is shown in Figure 10–66. If the gear is large, it is often referred to as a bull gear and rack, or just gear and rack. A small gear and rack is usually referred to as a **rack and pinion**. An internal spur gear is illustrated in Figure 10–67.

Bevel gears, Figure 10–68, are shaped like a truncated cone. The teeth are not parallel to the axis of the gear. The gear teeth are smaller at one end than at the other. The two shafts are at an angle to each other,

while the two bevel gears in Figure 10–69 are shown driving shafts at right angles to each other. In cases where the two mating bevel gears are the same size, they are usually referred to as miter gears.

Helical teeth are stronger and run more quietly than spur gear teeth. The teeth of a helical gear are cut parallel to the axis of the shaft but lie at an angle to the shaft. The two shafts illustrated, Figure 10–70,

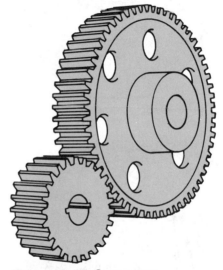

Figure 10–65 Spur gears

Figure 10–66 Rack and pinion

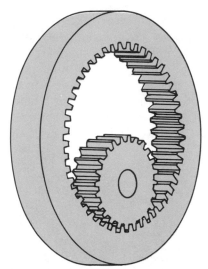

Figure 10–67 Internal spur gear

Figure 10–68 Bevel gears

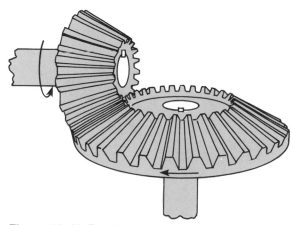

Figure 10–69 Bevel gears

are parallel to each other. Shafts can also be at right angles to each other, Figure 10–71, or at any other desired angle. Since the teeth are at an angle, a directional thrust occurs.

Herringbone gears, Figure 10–72, are like having two helical gears back to back. These gears are stronger, more quiet, and smoother running than spur gears. End thrust, as set up by helical gears, is eliminated. Shafts are parallel to each other.

A worm and worm gear, Figure 10–73, are both illustrated. The worm gear teeth are at an angle to accommodate the lead of the worm. Shafts are at right angles to each other. This combination of gear and worm is a common method of reducing speed in one of the shafts.

GEAR DIAMETERS

A spur gear is the simplest type of gear to cut, but a thorough understanding of gearing is necessary before any cutting is attempted.

The two circles, Figure 10–74, represent two discs touching each other. If one is revolved, friction will cause the other to revolve, unless too much pressure

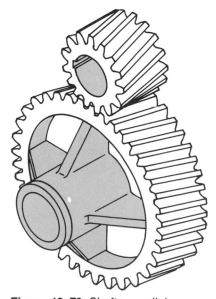

Figure 10–70 Shafts parallel

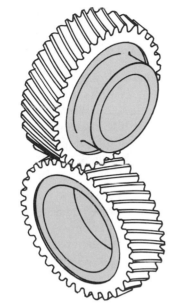

causes slippage. The two discs represent the most important circle in gearing, the **pitch circle**. It is on the diameter of the pitch circle that all calculations are based and on which all measuring of gear teeth is done. The diameter of the pitch circle is called the **pitch diameter (Pd.)**. If the two discs have teeth, Figure 10–75, no slippage will occur.

Figure 10–71 Shafts at right angles

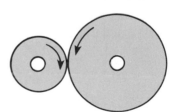

Figure 10–73 Worm and worm gear

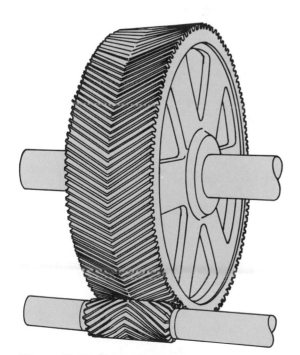

Figure 10–74 Pitch circle

Figure 10–72 Herringbone gears

Figure 10–75 Mating gears

If you were to cut full depth teeth into these discs in order to get the teeth to mesh, the discs would have to be moved closer together, thereby changing the pitch diameter. To develop a gear tooth properly, part of the tooth depth should be cut into the disc and part of the tooth depth should be added to the disc, thereby creating two more diameters, a root diameter (Dr) and an outside diameter (Do), Figure 10–76.

The **root diameter (Dr)**, is the diameter of the circle that forms the bottom of the gear teeth. The **outside diameter (Do)** is the diameter of the circle that forms the top of the gear teeth. It is also the diameter of the gear blank before the teeth are cut.

GEAR TOOTH PARTS

The part of the tooth above the pitch circle is added to the pitch diameter. It is called the addendum (a Latin term meaning "to add to"), *a* in Figure 10–77. The part of the tooth below the pitch circle is deducted from the pitch diameter. It is called the dedendum, (a Latin term meaning "to deduct"), *b* in Figure 10–77. The top of a gear tooth must have clearance at the root of the mating tooth. The **clearance**, *c* in Figure 10–77, is a part of the dedendum, which means that the dedendum is larger than the addendum. Clearance is usually an amount equal to 1/20 of the circular pitch (Pc), which is explained in Figure 10–78. Clearance

causes a slight backlash in the gears that is actually "play" or "looseness" between mating gear teeth. Without backlash, mating teeth would rub and would be quite noisy.

Tooth thickness, Figure 10–79A, is actually the chordal thickness as measured on the pitch circle. **Working depth**, Figure 10–79B, is the actual depth a tooth will penetrate into its mating slot. **Whole depth**, Figure 10–79C, is the total depth of the mating slot. This includes the clearance. Whole depth is also the depth of cut at which the cutter is set.

The **Circular Pitch System** is an actual measure from center to center of two adjacent teeth, measured on the pitch circle. One common use for this measuring system is in making patterns for large cast gears.

The **Diametral Pitch System**, commonly called

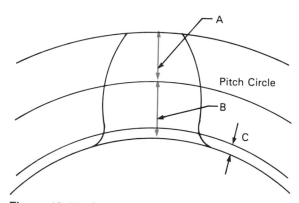

Figure 10–77 Gear tooth

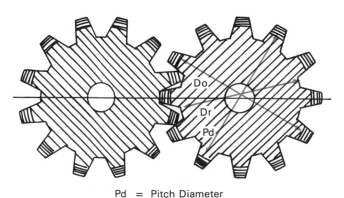

Pd = Pitch Diameter
Dr = Root Diameter
Do = Outside Diameter

A.   B.

Figure 10–76 Developing gear teeth

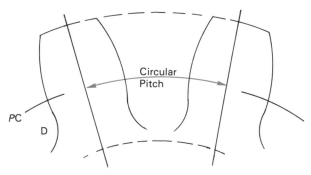

Figure 10–78 Circular pitch

Pitch (P), is the most common measuring system in gearing. The **diametral pitch** of a gear is the number of teeth for each inch of pitch diameter. If, for example, the pitch diameter of a 30-tooth gear is 3 inches, the diametral pitch would be 10. If the pitch diameter of a 36-tooth gear is 3 inches, the diametral pitch would be 12.

CUTTING A SPUR GEAR

A spur gear can be cut on a horizontal milling machine with an involute gear cutter, Figure 10–80. In this method, the gear blank is mounted on a mandrel and is suspended between centers, one center

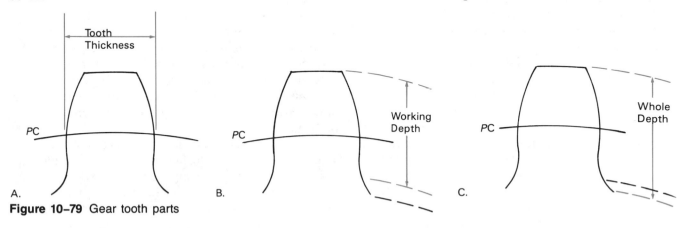

Figure 10–79 Gear tooth parts

Figure 10–80 Cutting a spur gear on a horizontal milling machine *(Courtesy of Cincinnati Milacron)*

LASER BEAM CUTTING AND MACHINING

CO_2 lasers can all be used as efficient metalworking tools for various industrial cutting applications. They can deliver a very stable energy output over a very wide range of power. The beam can be focused to a very intense spot. It is possible to make high-quality cuts without the need for extra devices for smoothing the cuts, such as grinders.

At this time, the laser has been developed primarily for use as a cutting tool similar to the use of a gas cutting torch to cut metal sheets. Smooth cuts, much smoother than those made with a gas cutting torch, are obtainable in materials of 3/4 inch or more in thickness. The high output power of these lasers means that the focusing optics are often of the reflective type like mirrors, rather than the transmissive or lens type.

In the laser cutting process, Figure 1, the material is exposed to the focused laser beam. The rate of energy transfer to the material is so high that the material cannot dissipate it at a comparable rate. The temperature of the interaction zone of the material rises rapidly to the point where it exceeds the vaporization temperature of the material. A cavity forms on the surface of the workpiece. If the power is high enough, the cavity will completely penetrate the workpiece. By directing a high velocity gas stream into this cavity, the molten and vaporized matter will be ejected from the back of the workpiece, Figure 2. The cutting action is generated by the translation (moving) of the laser induced cavity and the assist gas jet nozzle along the path of the desired cut. The assist gas used, Figure 3, can be an inert gas such as argon or helium to minimize the oxide formation on the cut surfaces. However, oxygen is often added to the assist gas stream to increase the rate of cutting. Oxides may be removed more easily from the cut than unoxidized molten material.

There are many advantages to laser cutting. The power density of the focused beam is so high that the cut material shows a narrow, straight-edged kerf with a minimal heat affected zone adjacent to the cut surface. As a result, distortion of the workpiece is low. The laser beam does not exert any measurable mechanical forces. Thus, there is no mechanical distortion of the workpiece and materials can be cut regardless of their hardness. Compensation for tool wear is not required. Since laser cutting does not rely on chemical reactions, there are no combustion products to contaminate the material. The fact that

Figure 1 The laser cutting process *(Courtesy of Combustion Engineering, Inc.)*

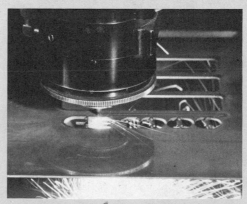

Figure 2 Cutting with a laser *(Courtesy of C-E Industrial Lasers, Inc.)*

the laser beam is essentially a beam of light means that it can be easily controlled and directed. Also, it readily allows for the automation of the cutting process by the use of computers. Cutting in normally inaccessible places is also possible with laser beams because they can be easily directed by optical systems. Another important advantage again relates to the absence of combustion products. That is, one laser beam can be utilized for cutting materials as different as steel plate and textiles. Materials that can be laser cut include polymers, composites, paper, wood, rubber, carbon steel, stainless steel, tool steel, copper alloys, titanium, and aluminum.

Slotting and the drilling of holes are natural jobs for laser beam machining. However, some extra attention must be given to surface finishing if required following these operations. A new machine being developed for manufacturing applications is called a laser processing center,

Figure 4. It is a multiple axis, laser machining center similar to a milling machine type of machining center. However, since it is still in its infancy, its use as a true machining center is limited. There is, however, a program called laser assisted machining (LAM) which is being used. LAM consists of a laser beam moving in front of a conventional milling cutter to soften the material ahead of the cutting tool. This method will extend the life of any conventional cutting tool.

Laser beam cutting on a lathe is in limited use in manufacturing. At the present time, a drawback of the process is that only small amounts of metal can be removed at one time. Metal is removed on a lathe utilizing two laser beams.

Laser cutting, slotting, and drilling will certainly find a place in modern manufacturing following further development.

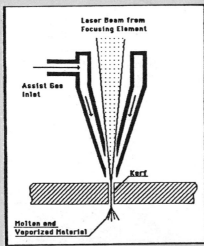

Figure 3 Cutting with a high power CO_2 *(Courtesy of Combustion Engineering, Inc.)*

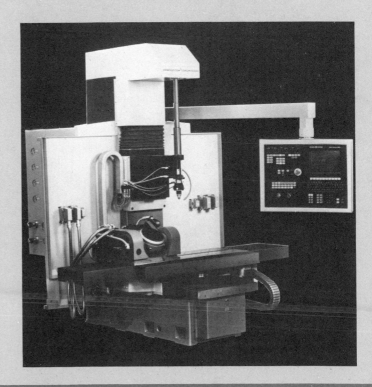

Figure 4 A laser processing center *(Courtesy of C-E Industrial Lasers, Inc.)*

in the universal dividing head spindle, the other supported by the footstock (or tailstock) spindle. In selecting the proper gear cutter, the diametral pitch, the number of the cutter, the range of teeth it will cut, and the whole depth of the cut, are all marked on the side of the cutter. Mount the gear cutter on the milling machine arbor above the gear blank. Set the dividing head to index the required number of teeth as is explained in the indexing section.

A common trade method for centering the cutter and gear blank follows:

1. Center the cutter on the gear blank as closely as possible.

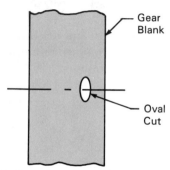

Figure 10–81A Top view of the gear blank showing the oval cut

2. With the cutter revolving, raise the gear blank until it is barely cut by the cutter.

3. Move the gear blank sideways in both directions causing a small oval-shaped cut to be made across the top of the work, Figure 10–81A.

4. Center the cutter on the oval cut and make two short longitudinal cuts forward and backward, Figure 10–81B.

By closely examining these two cuts with respect to the oval cut, it is possible to set the gear blank on center within approximately 0.002 in (0.05 mm), which is close enough for most practical work.

A second method of centering is to set the side of the gear cutter gently against the side of the footstock center, Figure 10–81C. If the center is 0.750 in (19.05 mm) in diameter and the gear cutter is 0.250 in (6.35 mm) thick, the difference between the two is 0.500 in (12.7 mm). This is the amount the cutter must be moved to place it exactly on center. Use the cross feed dial to move the cutter the exact distance.

Once the gear blank has been centered, the full depth of the cut must be set.

1. Lower the gear blank to clear the cutter, and move the gear blank longitudinally beyond the oval cut.

2. Raise the blank until the revolving cutter just touches the top of the gear blank.

3. Move the gear blank longitudinally until the cutter is clear and at the starting point of the cut.

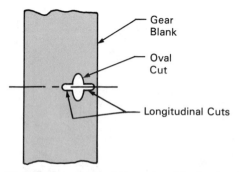

Figure 10–81B Top view of the gear blank showing the oval cut and the two longitudinal cuts

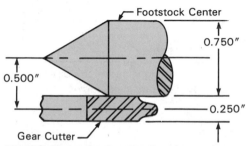

Figure 10–81C Top view showing the gear cutter against the side of the footstock center

4. Set the vertical dial to zero and raise the work to the full depth of the cut, which is the amount of the addendum plus the dedendum.

To check the accuracy of the indexing, it is wise, particularly for a beginner, to index each tooth and move the gear blank into the revolving cutter to lightly cut the corner of each tooth once around the blank. This will ensure that the indexing is correct. It would be frustrating and time consuming to proceed with the cutting of a gear and possibly end up with a half tooth on the final cut. This procedure of slightly cutting the blank does take valuable time, so if the machinist is thoroughly familiar with his equipment and is well experienced, it is not usually done. Once the dividing head is set, the machinist must be extremely careful not to accidentally strike and move the sector arms during each indexing movement.

A method for cutting a large spur gear by using a circular milling table is shown in Figure 10–82. The rules and formulas for spur gear calculations are given in Table 10–5.

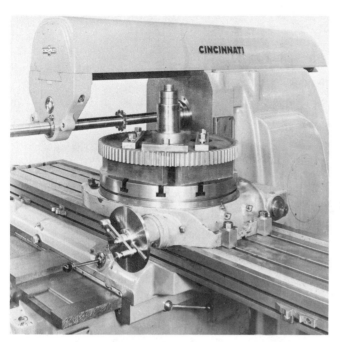

Figure 10–82 Cutting a large spur gear using a circular milling table *(Courtesy of Cincinnati Milacron)*

Table 10–5 Rules and Formulas for Spur Gear Calculations *(Courtesy of Cincinnati Milacron)*

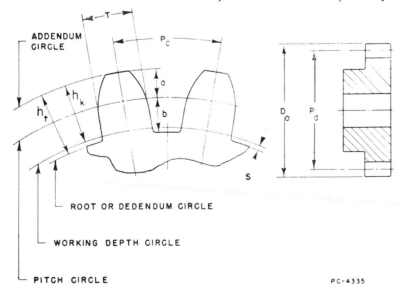

PC-4335

The following symbols are used in conjunction with the formulae for determining the proportions of spur gear teeth.

P = Diametral pitch.
P_c = Circular pitch.
P_d = Pitch diameter.
D_o = Outside diameter.
N = Number of teeth in the gear.
T = Tooth thickness.
a = Addendum.
b = Dedendum.
h_k = Working depth.
h_t = Whole depth.
S = Clearance.
C = Center distance.
L = Length of rack.

Table 10–5　Continued

TO FIND	RULE	FORMULA
Diametral pitch P	Divide 3.1416 by the circular pitch.	$P = \dfrac{3.1416}{P_c}$
Circular pitch P_c	Divide 3.1416 by the diametral pitch.	$P_c = \dfrac{3.1416}{P}$
Pitch diameter P_d	Divide the number of teeth by the diametral pitch.	$P_d = \dfrac{N}{P}$
Outside diameter D_o	Add 2 to the number of teeth and divide the sum by the diametral pitch.	$D_o = \dfrac{N + 2}{P}$
Number of teeth N	Multiply the pitch diameter by the diametral pitch.	$N = P_d\,P$
Tooth circular thickness T	Divide 1.5708 by the diametral pitch.	$T = \dfrac{1.5708}{P}$
Addendum a	Divide 1.0 by the diametral pitch.	$a = \dfrac{1.0}{P}$
Dedendum b	Divide 1.157 by the diametral pitch.	$b = \dfrac{1.157}{P}$
Working depth hk	Divide 2 by the diametral pitch.	$hk = \dfrac{2}{P}$
Whole depth ht	Divide 2.157 by the diametral pitch.	$ht = \dfrac{2.157}{P}$
Clearance S	Divide 0.157 by the diametral pitch.	$S = \dfrac{0.157}{P}$
Center distance C	Add the number of teeth in both gears and divide the sum by two times the diametral pitch.	$C = \dfrac{N_1 + N_2}{2P}$
Length of rack L	Multiply the number of teeth in the rack by the circular pitch.	$L = N\,P_c$

REVIEW QUESTIONS

1. What is a gear?
2. Describe a spur gear.
3. What is a gear and pinion?
4. What is a rack and pinion?
5. Describe a bevel gear.

6. What are miter gears?
7. What is the advantage of helical gear teeth over spur gear teeth?
8. Describe a helical gear.
9. What are herringbone gears and what advantages do they provide?
10. What type of gearing system is used to reduce the speed of one of two shafts?
11. What is the most important circle in gearing? Define the term.
12. What is the pitch diameter?
13. What is the diameter of the gear blank called?

14. What is the root diameter?
15. Describe addendum and dedendum.
16. What is the reason for clearance?
17. What is the tooth thickness?
18. What is the difference between working depth and whole depth?
19. Where is the measurement made in the circular pitch system?
20. What is the recommended method a beginner is to use to check the accuracy of the indexing for cutting a gear?

UNIT 10-6

HELICAL GEARING

OBJECTIVES

After completing this unit, the student will be able to:
- define the term helix.
- describe a helical gear.
- identify a left-hand helix and a right-hand helix.
- determine the helix angle for setting the milling machine table.
- calculate the change gears for the required lead.
- set up to cut a helical gear.

KEY TERMS

Helical milling
Helix
Spiral
Helical gear
Left-hand helix

Right-hand helix
Lead of the helix
Helix angle
Angle of the milling machine table

TERMINOLOGY

Helical milling is the process of cutting helical grooves such as the flutes in a twist drill, the teeth in helical gears, or the worm thread on a shaft.

A more common shop term is *spiral* rather than *helix*, however, both *Machinery's Handbook* and *American Machinist's Handbook* prefer the use of the term *helix*.

A helix, according to the Oxford Dictionary, is anything of a spiral form or coiled form, whether in one plane (like a watch spring) or advancing around an axis (like a corkscrew), but usually the latter. Webster's Dictionary simply states that a helix is a curve formed on any cylinder by a straight line in a plane that is wrapped around the cylinder (like a screw thread). Oxford further states that a helix in geometry is distinct from a spiral which is applied only to plane curves.

A spiral, according to both the Oxford Dictionary and Webster's Dictionary, is a continuous curve traced by a point moving around a fixed point in the same plane (a watch spring) while steadily increasing (or diminishing) its distance from this point. It is also described as a curve traced by a point moving around and simultaneously advancing along a cylinder or cone.

Helical gears differ from spur gears in that the teeth are not parallel to the axis of the gear. The teeth are cut at an angle. In fact, they wind around the pitch cylinder like a screw thread. A complete turn of the helix is seldom seen on a helical gear because the gear is too short.

THE HELIX IN PRACTICE

A helical cut can be either left-hand or right-hand. As with all helical cuts or grooves, always look at the helix from the end. The teeth on the plain milling cutter, Figure 10–83A, slope up and to the left, making this a left-hand helix. Figure 10–83B shows a right-hand helix.

A helical curve is illustrated in Figure 10–84A. The helical cut in Figure 10–84B shows that the cut is actually curved, not straight.

CUTTING A HELIX

1. To cut a helix on a universal milling machine, the table must be set at an angle (to cut the helix angle).

2. Indexing must take place for each gear tooth.

3. As the gear blank is being cut, it must be made to revolve, in order to cut the helix.

4. The revolution of the gear blank is achieved by connecting the table leadscrew to the worm shaft of the dividing head.

5. The lockpin behind the index plate must be disconnected so that the index plate is free to rotate the dividing head spindle.

6. After indexing is done, as the worktable moves, the leadscrew (geared to the worm shaft of the dividing head by a train of gears) causes the index plate to revolve, which causes the dividing head spindle to revolve in a certain ratio. Thus the work revolves as it is being cut.

The lead of the helix is the longitudinal distance the helix advances axially in one complete revolution of the work. The work must rotate one turn while the table travels lengthwise the distance of the lead.

Figure 10–85 shows how the helix angle is determined.

$$\text{Tangent of helix angle} = \frac{\text{work circumference}}{\text{lead of helix}}$$
$$= \frac{3.1416 \times \text{diameter}}{\text{lead of helix}}$$

EXAMPLE. To what angle must the milling machine table be swiveled to cut a helix angle having a lead of 10.882 inches on a piece of work 2 inches in diameter?

$$\text{Tangent of the helix angle} = \frac{(3.1416)(2)}{(10.882)}$$
$$= 0.57739$$
$$\text{Therefore, the helix angle} = 30 \text{ degrees.}$$

DETERMINING THE DIRECTION TO SET THE TABLE

Is the helix left-hand or right-hand? View the work from the end: if the helix slopes up and curves to the

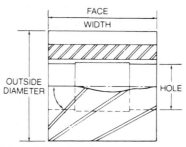

Figure 10–83A Left-hand helix of 45° on a plain milling cutter *(Courtesy of The Cleveland Twist Drill Company)*

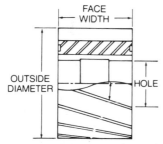

Figure 10–83B Right-hand helix of 25° on a plain milling cutter *(Courtesy of The Cleveland Twist Drill Company)*

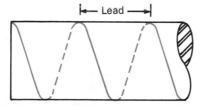

Figure 10–84A A helical curve

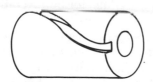

Figure 10–84B A helical cut

right, it is right-hand; if the helix slopes up and curves to the left, it is left-hand.

To cut a left-hand helix, face the machine and swivel the table clockwise. To cut a right-hand helix, face the machine and swivel the table counterclockwise.

Calculating Change Gears for the Required Lead

$$\frac{\text{Lead of helix to be cut}}{\text{Lead of the machine}} = \frac{\text{Product of driven gears}}{\text{Product of driver gears}}$$

This formula may be inverted if preferred.

EXAMPLE 1. Calculate the change gears required to produce a helix having a lead of 25 inches on a piece of work. The available change gears are: 24, 24, 28, 32, 40, 44, 48, 56, 64, 72, 86, and 100 teeth.

In establishing the lead of the machine, assume that the index-head worm shaft is geared to the table leadscrew with equal gears; e.g., both are 24-tooth gears. The index-head ratio is 40:1. A standard mill-

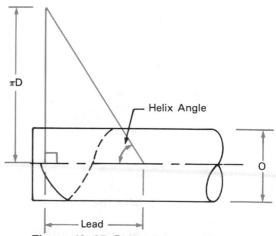

Figure 10–85 Determining a helix angle

ing machine leadscrew has 4 threads per inch. If the leadscrew revolves once, the table moves 1/4 in, and the index head spindle will rotate 1/40 of a revolution. To rotate the index-head spindle one turn, the leadscrew would have to revolve 40 times. The table would have to travel 40 × 1/4 = 10 inches, which means that the machine would have a lead of 10 inches.

$$\text{Gear ratio} = \frac{\text{lead of the helix (driven)}}{\text{lead of the machine (driver)}}$$
$$= \frac{25}{10} = \frac{25}{10} \times \frac{4}{4} = \frac{100 \text{ (driven)}}{40 \text{ (driver)}}$$

EXAMPLE 2. Calculate the change gears required to produce a helix having a lead of 27 inches. The gears are the same as in Example 1.

$$\text{Gear ratio} = \frac{27}{10}$$

There are no gears in the set that are multiples of both 27 and 10. Simple gearing cannot be used. We must go to compound gearing.

$$\text{Gear ratio} = \frac{27}{10} = \frac{3 \times 9 \text{ (driven)}}{2 \times 5 \text{ (drivers)}}$$
$$\frac{3 \times 16}{2 \times 16} = \frac{48}{32} \text{ and } \frac{9 \times 8}{5 \times 8} = \frac{72}{40}$$

$$\text{Gear ratio} = \frac{48 \times 72 \text{ (driven)}}{32 \quad 40 \text{ (drivers)}}$$

Gear on the worm = 72 (driven)
1st gear on stud = 32 (driver)
2nd gear on stud = 48 (driven)
Gear on leadscrew = 40 (driver)

This order of gearing can be changed. The two driven gears can be interchanged and/or the two driver gears can be interchanged, but a driver cannot be interchanged with a driven.

The setup required to cut a right-hand helix needs one idler in order that the gear blank will rotate in the proper direction. The setup required to cut a left-hand helix needs two idlers in order that the gear blank will rotate in the proper direction.

EXAMPLE 3. Following is an example of the main calculations necessary to produce a mating pair of helical gears.

	Gear	Pinion
DP	12	12
PD	3.600	2.400
OD	3.766	2.566
N	42	28
Cutter number	#3	#4
Indexing	20 holes on 21-hole circle	1 turn, 9 holes on 21-hole circle
Full depth	0.1797	0.1797
Helix angle	13 degrees, 33 min (R.H.)	13 degrees, 33 min (L.H.)
Direction to swing table	Counterclockwise	Clockwise
Lead	46.926	31.285
Change gear (Worm)	100	86
(1st stud)	32	40
(2nd stud)	72	64
(Leadscrew)	48	44

Table 10–6 is a chart of leads for helical milling in inch units showing the gear changes and positions necessary. The nomograph given in Table 10–7 is a quick way of finding the angular setting of the machine table. If the diameter of the work is 4 in and the lead is 34.5 in, a line drawn through these two points will give a table setting of 20 degrees. If the work diameter is 100 mm and the lead is 880 mm, a line drawn through these two points will give a table setting of 27.5 degrees. Table 10–8 (page 430) is a chart of leads (in metric units) for helical milling, showing the gear changes and positions necessary.

In Figure 10–86 (page 432), a helical gear rack is being cut using a rack milling attachment. Figure 10–87 (page 433) shows the cutting of a helical gear on one end of a shaft. The rules and formulas for helical gear calculations are given in Table 10–9 (page 434).

Table 10–6 Leads for Helical Milling: Inch Units *(Courtesy of Gate Machinery Company Ltd.)*

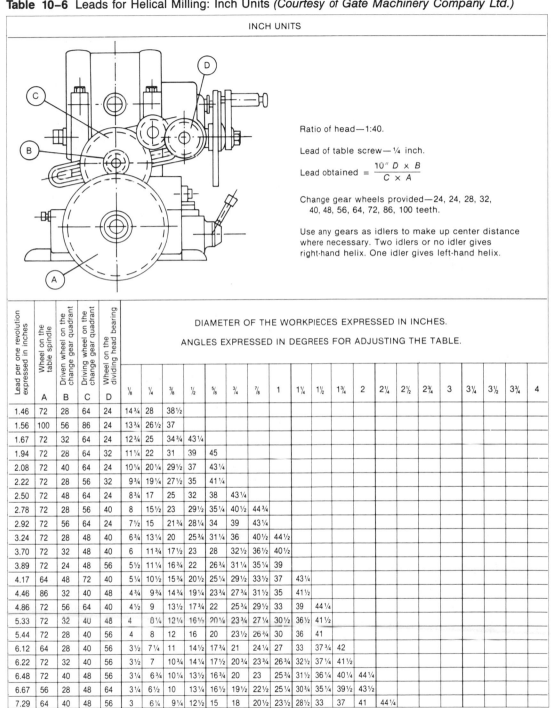

INCH UNITS

Ratio of head—1:40.

Lead of table screw—¼ inch.

$$\text{Lead obtained} = \frac{10'' D \times B}{C \times A}$$

Change gear wheels provided—24, 24, 28, 32, 40, 48, 56, 64, 72, 86, 100 teeth.

Use any gears as idlers to make up center distance where necessary. Two idlers or no idler gives right-hand helix. One idler gives left-hand helix.

DIAMETER OF THE WORKPIECES EXPRESSED IN INCHES.

ANGLES EXPRESSED IN DEGREES FOR ADJUSTING THE TABLE.

Lead per one revolution expressed in inches	Wheel on the table spindle A	Driven wheel on the change gear quadrant B	Driving wheel on the change gear quadrant C	Wheel on the dividing head bearing D	⅛	¼	⅜	½	⅝	¾	⅞	1	1¼	1½	1¾	2	2¼	2½	2¾	3	3¼	3½	3¾	4
1.46	72	28	64	24	14¾	28	38½																	
1.56	100	56	86	24	13¾	26½	37																	
1.67	72	32	64	24	12¾	25	34¾	43¼																
1.94	72	28	64	32	11¼	22	31	39	45															
2.08	72	40	64	24	10¼	20¼	29½	37	43¼															
2.22	72	28	56	32	9¾	19¼	27½	35	41¼															
2.50	72	48	64	24	8¾	17	25	32	38	43¼														
2.78	72	28	56	40	8	15½	23	29½	35¼	40½	44¾													
2.92	72	56	64	24	7½	15	21¾	28¼	34	39	43¼													
3.24	72	28	48	40	6¾	13¼	20	25¾	31¼	36	40½	44½												
3.70	72	32	48	40	6	11¾	17½	23	28	32½	36½	40½												
3.89	72	24	48	56	5½	11¼	16¾	22	26¾	31¼	35¼	39												
4.17	64	48	72	40	5¼	10½	15¾	20½	25¼	29½	33½	37	43¼											
4.46	86	32	40	48	4¾	9¾	14¾	19¼	23¾	27¾	31½	35	41½											
4.86	72	56	64	40	4½	9	13½	17¾	22	25¾	29½	33	39	44¼										
5.33	72	32	40	48	4	8¼	12¼	16½	20¼	23¾	27¼	30½	36½	41½										
5.44	72	28	40	56	4	8	12	16	20	23½	26¾	30	36	41										
6.12	64	28	40	56	3½	7¼	11	14½	17¾	21	24¼	27	33	37¾	42									
6.22	72	32	40	56	3½	7	10¾	14¼	17½	20¾	23¾	26¾	32½	37¼	41½									
6.48	72	40	48	56	3¼	6¾	10¼	13½	16¾	20	23	25¾	31½	36¼	40¼	44¼								
6.67	56	28	48	64	3¼	6½	10	13¼	16½	19½	22½	25¼	30¾	35¼	39½	43½								
7.29	64	40	48	56	3	6¼	9¼	12½	15	18	20½	23½	28½	33	37	41	44¼							

Table 10–6 Continued

INCH UNITS																								
Lead per one revolution expressed in inches	Wheel on the table spindle	Driven wheel on the change gear quadrant	Driving wheel on the change gear quadrant	Wheel on the dividing head bearing	DIAMETER OF THE WORKPIECES EXPRESSED IN INCHES. ANGLES EXPRESSED IN DEGREES FOR ADJUSTING THE TABLE.																			
	A	B	C	D	⅛	¼	⅜	½	⅝	¾	⅞	1	1¼	1½	1¾	2	2¼	2½	2¾	3	3¼	3½	3¾	4
7.41	72	40	48	64	3	6	9	12	14¾	17¾	20¼	22¾	28¼	32½	36½	40¼	43¾							
7.62	56	32	48	64	2¾	5¾	8¾	11½	14½	17¼	19¾	22¼	27½	32	36	39½	43							
8.33	72	40	32	48	2½	5¼	8	10½	13¼	15¾	18¼	20½	25½	29½	33½	37	40½	43½						
8.95	56	28	48	86	2½	5	7½	10	12½	14¾	17	19¼	24	28	31¾	35¼	38½	41¼	44					
9.33	72	48	40	56	2¼	4¾	7¼	9½	11¾	14	16¼	18½	23	27	30½	34	37¼	40¼	43					
9.52	56	40	48	64	2¼	4½	7	9¼	11½	13¾	16	18¼	22½	26½	30	33½	36½	39½	42½	45				
10.29	56	32	40	72	2	4¼	6½	8¾	10¾	12¾	15	17¼	21	24¾	28¼	31½	34½	37½	40	42½	45			
10.37	72	56	48	64	2	4¼	6½	8½	10½	12¾	14¾	17	20¾	24½	28	31¼	34¼	37¼	39¾	42¼	44¾			
10.50	64	56	40	48	2	4¼	6¼	8½	10½	12½	14½	16¾	20½	24¼	27¾	31	34	36¾	39½	42	44¼			
10.67	72	48	40	64	2	4	6¼	8¼	10¼	12¼	14¼	16½	20¼	24	27¼	30½	33½	36½	39	41½	43¾			
10.94	64	40	32	56	2	4	6	8¼	10¼	12	14	16¼	20	23½	26¾	30	33	35¾	38¼	40¾	43			
11.11	72	40	32	64	2	4	6	8	10	11¾	13¾	16	19¾	23	26½	29½	32½	35½	38	40¼	42½	44¾		
11.66	72	48	32	56	1¾	3¾	5¾	7½	9½	11¼	13¼	15¼	18¾	22	25¼	28½	31¼	34	36½	39	41¼	43½		
12.00	48	32	40	72	1¾	3¾	5½	7¼	9¼	11	12¾	15	18¼	21½	24¾	27¾	30½	33¼	35¾	38	40¼	42½	44¾	
13.12	64	48	32	56	1½	3½	5¼	6¾	8½	10¼	11¾	13½	16¾	20	22¾	25¾	28¼	31	33¼	35¾	37¾	40	42	43¾
13.33	72	48	28	56	1½	3¼	5	6½	8¼	10	11½	13¼	16½	19½	22½	25½	28	30½	33	35¼	37½	39½	41½	43¼
13.71	56	48	40	64	1½	3¼	4¾	6½	8	9¾	11½	13	16	19	22	24¾	27¼	30	32¼	34½	36½	38¾	40¾	42½
15.24	72	48	28	64	1½	3	4½	5¾	7¼	8¾	10¼	11¾	14½	17¼	20	22½	25	27¼	29½	31¾	34	35¾	37¾	39½
15.56	72	56	32	64	1¼	2¾	4¼	5¾	7¼	8¾	10	11½	14¼	17	19½	22	24½	27	29	31¼	33¼	35¼	37	39
15.75	40	72	64	56	1¼	2¾	4¼	5½	7	8½	9¾	11¼	14	16¾	19¼	21¾	24¼	26½	28¾	31	33	35	36¾	38½
16.87	64	48	32	72	1¼	2½	4	5¼	6¾	7¾	9¼	10½	13¼	15¾	18½	20½	22¾	25	27	29¼	31¼	33¼	35	36½
17.14	56	48	32	64	1¼	2½	4	5¼	6½	7¾	9	10¼	13	15½	17¾	20¼	22¼	24¾	26¾	29	30¾	32¾	34½	36
18.75	48	40	32	72	1	2¼	3½	4¾	6	7¼	8¼	9½	12	14¼	16¼	18½	20¾	22¾	25	26¾	28½	30¼	32	33¾
19.29	56	48	32	72	1	2¼	3½	4½	5¾	6¾	8	9¼	11½	13¾	16	18¼	20¼	22¼	24	26	28	29¾	31½	33
19.59	56	48	28	64	1	2¼	3¼	4½	5¾	6¾	8	9¼	11½	13½	15¾	18	20	22	23¾	25¾	27½	29¼	31	32¾
19.69	64	56	32	72	1	2¼	3¼	4½	5¾	6¼	8	9	11½	13½	15¾	17¾	20	21¾	23¾	25½	27½	29¼	31	32½
21.43	56	40	24	72	1	2	3¼	4¼	5¼	6¼	7½	8½	10½	12½	14½	16½	18½	20¼	22	23¾	25½	27¼	29	30¼
22.50	64	56	28	72	1	2	3	4	5	6	7	8	10	12	13¾	15¾	17½	19¼	21	22¾	24½	26	27¾	29¼
23.33	48	56	32	64	1	2	3	4	5	5¾	6¾	7¾	9¾	11½	13¼	15¼	17	18¾	20¼	22	23½	25¼	27	28¼
26.25	64	56	24	72	1	1¾	2¾	3½	4¼	5	6	7	8½	10¼	12	13½	15	16¾	18¼	19¾	21¼	22¾	24¼	25½
26.67	48	56	28	64	¾	1¾	2¾	3½	4¼	5	6	6¾	8½	10	11¾	13¼	14¾	16½	18	19½	21	22¼	23¾	25¼
28.00	40	56	32	64	¾	1¾	2½	3¼	4	4¾	5¾	6½	8	9½	11¼	12¾	14¼	15¾	17¼	18¾	20	21½	22¾	24
30.86	40	48	28	72	¾	1½	2¼	3	3¾	4½	5	5¾	7¼	8¾	10	11½	13	14¼	15½	17	18½	19½	21	22
31.50	40	56	32	72	¾	1½	2¼	3	3½	4¼	5	5¾	7¼	8¼	10	11¼	12¾	14	15¼	16½	18	19¼	20½	21¾
36.00	40	64	32	72	½	1¼	2	2½	3¼	3¾	4½	5	6¼	7½	8¾	10	11	12¼	13½	14¾	16	17	18¼	19¼
41.14	40	64	28	72	½	1	1¾	2¼	2¾	3¼	4	4½	5½	6½	7¾	8¾	9¾	10¾	11¾	13	14	15	16	17
45.00	32	56	28	72	½	1	1½	2	2½	3	3½	4	5	6	7	8	9	10	11	11¾	12¾	13¾	14¾	15½
48.00	40	64	24	72	½	1	1½	2	2½	3	3¼	3¾	4¾	5½	6½	7½	8½	9¼	10¼	11¼	12	13	13¾	14½
51.43	32	64	28	72	½	¾	1½	1¾	2¼	2¾	3	3½	4½	5¼	6	7	7¾	8¾	9½	10½	11¼	12	12¾	13¾
60.00	32	64	24	72	½	¾	1¼	1½	2	2¼	2½	3	3¾	4½	5¼	6	6¾	7½	8¼	9	9½	10¼	11	11¾
68.57	28	64	24	72	¼	¾	1	1¼	1¾	2	2¼	2½	3¼	4	4¼	5¼	5¾	6½	7¼	8	8½	9	9¾	10¼

Table 10–7 Nomograph for Finding Angular Settings of Machine Table for Helical Milling *(Courtesy of Gate Machinery Ltd.)*

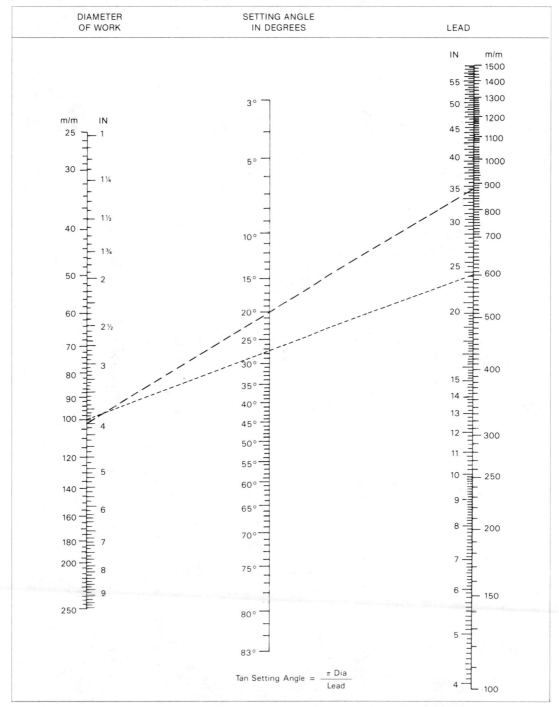

DIAMETER OF WORK | SETTING ANGLE IN DEGREES | LEAD

$$\text{Tan Setting Angle} = \frac{\pi \text{ Dia}}{\text{Lead}}$$

Table 10–8 Leads for Helical Milling (Metric Units) *(Courtesy of Gate Machinery Company Ltd.)*

This table of leads is suitable for the milling of spiral reamers, cutters, etc., but does not give all possible combinations.
Ratio of head — 1 : 40
Change gear wheels provided — 24, 24, 28, 32, 40, 48, 56, 64, 72, 86, and 100 teeth.

Use any gears as idlers to make up center distance where necessary. Two idlers or no idler gives right-hand helix; one idler gives left-hand helix.

$$\text{Lead obtained} = \frac{B \times D \times 200 \text{ mm}}{A \times C}$$

Lead of table screw — 5 mm

Lead per one revolution expressed in mm	Wheel on the table spindle (A)	Driven (B)	Driving (C)	Wheel on the dividing head (D)	Lead per one revolution expressed in mm	Wheel on the table spindle (A)	Driven (B)	Driving (C)	Wheel on the dividing head (D)	Lead per one revolution expressed in mm	Wheel on the table spindle (A)	Driven (B)	Driving (C)	Wheel on the dividing head (D)	Lead per one revolution expressed in mm	Wheel on the table spindle (A)	Driven (B)	Driving (C)	Wheel on the dividing head (D)
13.4	100	24	86	24	52.5	64	28	40	24	86.0	100	28	56	86	127.0	72	32	28	40
15.62	100	28	86	24	53.56	64	40	56	24	86.84	86	28	48	64	128.0	100	24	24	64
16.0	100	24	72	24	54.0	100	72	64	24	87.5	64	28	24	24	129.0	100	48	64	86
17.86	100	32	86	24	54.86	100	64	56	24	88.88	72	28	56	64	130.68	100	28	24	56
18.6	86	24	72	24	55.82	86	48	56	28	89.3	86	24	40	64	133.78	100	56	72	86
20.58	100	24	56	24	56.0	100	28	24	24	89.6	100	32	40	56	134.4	100	48	40	56
21.0	100	28	64	24	56.9	100	64	72	32	90.0	100	40	64	72	135.0	63	24	40	72
22.32	100	40	86	24	57.34	100	24	72	86	90.74	72	28	48	56	136.5	72	32	56	86
24.0	100	24	48	24	58.34	72	56	64	24	91.44	100	64	56	40	137.0	100	32	40	86
25.0	72	24	64	24	60.0	56	28	40	24	93.34	48	32	40	28	139.36	72	28	48	86
26.8	100	48	86	24	62.02	86	48	72	40	95.24	72	24	32	40	140.0	40	24	24	28
28.0	100	28	48	24	62.5	64	40	56	28	95.7	86	24	28	48	143.52	86	24	28	72
29.16	72	28	64	24	63.0	64	72	100	28	96.0	100	24	48	48	144.0	100	24	24	72
30.0	100	40	72	24	63.5	72	40	56	32	97.22	72	28	32	40	146.52	86	28	32	72
32.0	100	32	48	24	63.8	56	64	86	24	98.0	100	28	32	56	147.44	200	24	28	86
33.34	72	28	56	24	64.0	56	64	100	28	99.56	100	64	72	56	148.84	86	24	24	64
33.6	100	28	40	24	64.5	64	86	100	24	100.0	56	28	24	24	150.0	64	24	24	48
34.28	100	40	56	24	65.12	86	28	24	24	100.8	100	28	40	72	152.4	72	24	28	64
35.0	100	40	64	28	66.16	86	64	72	32	102.1	64	56	48	28	154.28	56	24	40	72
36.0	100	48	64	24	66.98	100	72	86	40	102.86	40	24	28	24	155.56	48	28	24	32
36.46	100	56	86	28	67.2	100	24	40	56	103.7	72	32	24	28	156.3	86	48	40	56
37.2	86	32	56	28	68.06	72	56	64	28	104.52	72	28	64	86	157.5	64	28	40	72
37.5	64	24	48	24	68.58	100	24	28	40	105.0	40	28	32	24	160.0	100	40	32	64
38.4	100	32	40	24	70.0	100	56	64	40	106.66	100	32	40	40	162.8	86	40	32	56
39.08	86	28	40	24	71.44	100	64	86	48	107.5	100	40	64	86	164.58	100	32	28	72
41.14	100	24	28	24	72.0	100	24	48	72	108.0	100	24	32	72	166.66	48	24	24	40
41.66	72	40	64	24	72.92	64	28	48	40	108.88	72	28	40	56	168.0	100	28	24	72
42.0	100	56	64	24	73.72	100	24	56	86	109.72	100	24	28	64	170.64	72	40	56	86
42.86	64	32	56	24	74.66	100	56	72	48	111.62	86	24	32	64	172.0	100	24	24	86
43.42	86	56	72	24	75.0	48	24	32	24	112.0	100	24	24	56	173.64	86	28	24	64
43.76	64	28	46	24	76.18	72	32	28	24	114.28	72	24	28	48	175.0	32	24	24	28
44.8	100	32	40	28	76.44	100	32	72	86	115.2	100	32	40	72	178.58	56	24	48	100
45.0	64	24	40	24	76.8	100	24	40	64	116.28	86	32	64	100	180.0	40	24	32	48
45.72	100	40	56	32	77.78	72	56	64	32	119.04	72	24	56	100	182.3	64	28	48	100
46.66	100	40	48	28	78.4	100	56	40	28	119.6	86	40	56	72	184.28	56	24	40	86
47.62	100	64	86	32	80.0	48	32	40	24	120.0	56	28	40	48	185.2	72	32	48	100
48.0	100	48	56	28	81.4	86	40	32	28	121.54	72	28	64	100	186.66	48	28	40	64
49.0	100	56	64	28	83.34	56	40	48	28	122.5	64	28	40	56	188.12	64	28	40	86
50.0	56	28	48	24	84.0	100	56	64	48	123.44	100	24	28	72	190.48	48	32	28	40
51.86	72	32	48	28	85.06	86	32	56	64	125.0	64	40	24	24	191.38	86	32	28	72

Table 10-8 Continued

This table of leads is suitable for the milling of spiral reamers, cutters, etc., but does not give all possible combinations.
Ratio of head — 1 : 40
Change gear wheels provided — 24, 24, 28, 32, 40, 48, 56, 64, 72, 86, and 100 teeth.

Use any gears as idlers to make up center distance where necessary. Two idlers or no idler gives right-hand helix; one idler gives left-hand helix.

$$\text{Lead obtained} = \frac{B \times D \times 200 \text{ mm}}{A \times C}$$

Lead of table screw — 5 mm

Lead per one revolution expressed in mm	Wheel on the table spindle (A)	Driven (B)	Driving (C)	Wheel on the dividing head (D)
192.0	100	32	24	72
193.5	100	72	64	86
195.36	86	56	48	72
198.42	72	40	56	100
200.0	48	24	28	56
201.6	100	56	40	72
203.5	86	28	32	100
205.72	40	24	28	48
207.4	72	28	24	64
208.38	86	56	40	64
210.0	40	24	32	56
212.62	86	40	28	64
215.0	48	24	40	86
216.0	100	48	32	72
218.74	64	40	32	56
220.42	56	24	28	72
222.22	48	32	24	40
224.0	100	48	24	56
225.0	64	24	24	72
228.58	28	24	24	32
230.4	100	64	40	72
233.34	48	24	24	56
235.14	64	28	32	86
238.1	72	24	28	100
240.0	40	24	24	48
243.06	72	28	32	100
245.0	40	28	32	56
248.06	86	40	24	64
250.0	32	24	24	40
252.0	100	56	32	72
255.16	86	48	28	64
256.0	100	56	28	64
258.0	100	48	32	86
262.5	48	28	32	72
265.78	86	32	28	100
268.74	56	28	32	86
270.0	40	24	32	72
273.0	72	32	28	86
275.2	100	64	40	86
277.78	72	24	24	100
280.0	40	24	24	56
285.72	28	24	24	40
288.0	100	48	24	72
290.72	86	40	32	100
293.02	86	56	32	72
296.3	72	40	24	64
297.68	86	56	28	64
299.0	86	72	56	100
300.0	32	24	24	48
304.76	72	48	28	64
307.14	48	24	28	86
310.08	86	64	48	100
312.5	64	24	24	100
315.0	40	28	32	72
320.0	40	24	24	64
322.5	40	24	32	86
325.6	86	56	40	100
329.12	100	64	28	72
334.44	72	56	40	86
336.0	100	56	24	72
341.24	72	40	28	86
345.5	64	72	56	86
350.0	32	24	24	56
355.56	48	32	24	64
360.0	40	24	24	72
364.58	48	28	32	100
367.34	56	40	28	72
370.38	72	32	24	100
373.26	72	86	64	100
375.0	40	24	32	100
380.96	28	32	24	40
384.0	100	64	24	72
387.0	100	72	32	86
390.7	86	56	24	72
396.8	72	40	28	100
400.0	48	32	24	72
403.2	64	72	48	86
407.0	86	56	32	100
409.6	56	64	48	86
414.8	72	56	24	64
418.0	72	56	32	86
420.0	40	48	32	56
426.6	72	86	56	100
430.0	40	24	24	86
437.6	64	56	40	100
440.8	56	48	28	72
444.4	72	64	40	100
448.0	48	40	32	86
450.0	64	48	24	72
457.2	28	24	24	64
460.8	48	72	56	86
465.2	86	64	32	100
470.4	64	56	32	86
476.2	56	64	48	100
480.0	48	72	40	64
486.2	72	56	32	100
491.4	56	64	40	86
497.6	48	86	72	100
500.0	48	40	24	72
510.2	56	40	28	100
516.0	100	72	24	86
520.8	48	40	32	100
525.0	48	56	32	72
533.4	48	56	28	64
537.6	64	56	28	86
540.0	40	48	32	72
546.0	72	64	28	86
560.0	48	86	64	100
571.4	40	64	56	100
583.4	48	56	48	100
587.8	56	64	28	72
595.2	48	40	28	100
598.0	68	72	28	100
600.0	32	48	28	56
609.6	28	32	24	64
614.4	28	24	24	86
620.2	86	64	24	100
625.0	64	56	28	100
630.0	40	56	32	72
635.0	28	64	72	100
640.0	40	56	28	64
645.0	40	72	48	86
666.6	40	32	24	100
685.8	28	64	48	72
694.4	48	40	24	100
700.0	48	56	24	72
720.0	40	64	32	72
740.8	72	64	24	100
750.0	40	72	48	100
767.8	56	86	40	100
777.8	24	40	24	56
800.0	48	64	24	72
819.2	28	32	24	86
833.4	48	64	32	100
840.0	40	56	24	72
860.0	40	64	32	86
875.0	40	56	32	100
888.8	24	40	24	64
900.0	32	56	28	72
921.4	48	72	28	86
937.6	48	72	32	100
952.4	48	64	28	100
960.0	40	64	24	72
982.8	40	64	28	86
1000.0	40	56	28	100
1028.6	32	64	28	72
1050.8	32	56	24	72
1075.0	32	48	24	86
1105.6	40	72	28	86
1125.0	40	72	32	100
1142.8	40	64	28	100
1166.6	40	56	24	100
1190.0	28	40	24	100
1200.0	32	64	24	72
1254.0	24	56	32	86
1290.0	24	72	48	86
1371.4	24	64	28	72
1433.3	24	64	32	86
1612.0	24	72	32	86
1843.0	24	72	28	86

Figure 10–86 Cutting a helical gear rack with a rack milling attachment *(Courtesy of Cincinnati Milacron)*

Figure 10 87 Milling a helical gear *(Courtesy of Cincinnati Milacron)*

Table 10-9 Rules and Formulas for Helical Gear Calculations *(Courtesy of Cincinnati Milacron)*

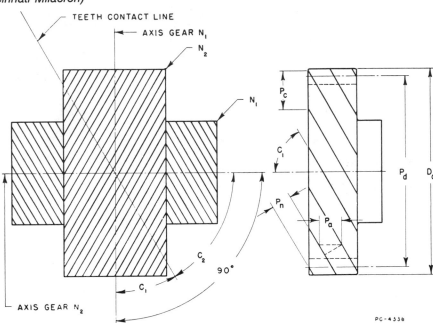

PC-4338

The following symbols are used in conjunction with the formulae for determining the proportions of helical gear teeth.

P_{nd} = Normal diametral pitch (pitch of cutter).

P_c = Circular pitch.

P_a = Axial pitch.

P_n = Normal pitch.

P_d = Pitch diameter.

S = Center distance.

C, C_1, C_2 = Helix angle of the gears.

L = Lead of tooth helix.

T_n = Normal tooth thickness at pitch line.

a = Addendum.

h_t = Whole depth of tooth.

N, N_1, N_2 = Number of teeth in the gears.

D_o = Outside diameter.

N_c = Hypothetical number of teeth for which the gear cutter should be selected.

Table 10–9 Continued

TO FIND	RULE	FORMULA
Normal diametral pitch P_{nd}	Divide the number of teeth by the product of the pitch diameter and the cosine of the helix angle.	$P_{nd} = \dfrac{N}{P_d \cos C_1}$
Circular pitch P_c	Multiply the pitch diameter of the gear by 3.1416, and divide the product by the number of teeth in the gear.	$P_c = \dfrac{3.1416\ P_d}{N}$
Axial pitch P_a	Multiply the circular pitch by the cotangent of the helix angle.	$P_a = P_c \cot C_1$
Normal pitch P_n	Divide 3.1416 by the normal diametral pitch.	$P_n = \dfrac{3.1416}{P_{nd}}$
Pitch diameter P_d	Divide the number of teeth by the product of the normal pitch and the cosine of the helix angle.	$P_d = \dfrac{N}{P_{nd} \cos C_1}$
Center distance S	Divide the sum of the pitch diameters of the mating gears by 2.	$S = \dfrac{P_{d_1} + P_{d_2}}{2}$
Checking Formulae (shafts at right angles)	Multiply the number of teeth in the first gear by the tangent of the tooth angle of that gear, and add the number of teeth in the second gear to the product. The sum should equal twice the product of the center distance multiplied by the normal diametral pitch, multiplied by the sine of the helix angle.	$N_1 + (N_2 \tan C_2) = 2\ S\ P_{nd} \sin C_1$
Lead of tooth helix L	Multiply the pitch diameter by 3.1416 times the cotangent of the helix angle.	$L = 3.1416\ P_d \cot C_1$
Normal circular tooth thickness at pitch line T_n	Divide 1.571 by the normal diametral pitch.	$T_n = \dfrac{1.571}{P_{nd}}$
Addendum a	Divide the normal pitch by 3.1416.	$a = \dfrac{P_n}{3.1416}$
Whole depth of tooth h_t	Divide 2.157 by the normal diametral pitch.	$h_t = \dfrac{2.157}{P_{nd}}$
Outside diameter D_o	Add twice the addendum to the pitch diameter.	$D_o = P_d + 2\ a$
Hypothetical number of teeth for which gear cutter should be selected N_c	Divide the number of teeth in the gear by the cube of the cosine of the helix angle.	$N_c = \dfrac{N_1}{(\cos C_1)^3}$

REVIEW QUESTIONS

1. Define the term helix.
2. What is the basic difference between a helix and a spiral?
3. Compare a helical gear to a spur gear.
4. How do you distinguish between a right-hand helix and a left-hand helix?
5. What is the lead of a helix?
6. In cutting a helix, how is movement of the work table and index plate accomplished?
7. How is the helix angle determined?
8. How is the milling machine table swiveled to cut a left-hand helix?
9. How is the milling machine table swiveled to cut a right-hand helix?
10. What is the formula for calculating change gears for a required lead?

UNIT 10-7

MEASURING GEAR TEETH

OBJECTIVES

After completing this unit, the student will be able to:

- measure gear teeth using a gear tooth vernier.
- describe worm gearing.
- set up and cut a worm and worm gear.
- describe the module system of gearing.
- select the proper cutter for a bevel gear.
- set up and cut an approximate bevel gear.

KEY TERMS

Gear tooth vernier
Chordal (corrected) addendum
Worm gearing
Worm
Worm gear
Lead angle of the worm
Hobbing machine

Hobbing cutter
Module system
Bevel gear
Frustrum
Pitch cone
Pitch cone angle
Miter gears
Pinion

Single involute gear cutter

Set-over or offset method

GEAR TOOTH VERNIER

There is more than one method of measuring the size of gear teeth. A **gear tooth vernier**, Figure 10–88, is used for measuring the chordal thickness of a gear tooth at the pitch line, and the chordal or corrected addendum. This measuring gage is both a vernier caliper and a vernier depth gage.

Corrected Addendum

S is the actual addendum, Figure 10–89, which is the distance from the top of the tooth to the pitch line at the center of the tooth. The gear tooth vernier measures tooth thickness or chordal thickness on the pitch circle. It also measures the chordal or corrected addendum which is H.

$$\text{Chordal addendum} = \text{addendum} + \frac{(\text{arc thickness})^2}{4 \times \text{pitch diameter}}$$

PROBLEM: A pinion has 14 teeth and a diametral pitch of 14. The addendum is 0.0714 inch. The tooth thickness at the pitch line is 0.1122 inch. What is the corrected or chordal addendum?

$$H = 0.0714 + \frac{(0.1122)^2}{(4)\,(14 \div 14)} = 0.0745 \text{ in}$$

WORM GEARING

Worm gearing, Figure 10–90A, is a different type of gearing in that the small gear looks more like an Acme screw thread than a gear. This small gear is called the **worm** and is usually made of hardened steel. Worms can be made in single-start, double-start, triple-start, or quadruple-start threads.

In mesh with the worm is the larger **worm gear**, Figure 10–90B, which is usually made of phosphor bronze, with teeth cut at an angle to the centerline of the gear to match the lead of the worm. The worm gear also has a curved throat to fit more of the worm teeth.

The main purpose of this type of gearing is to reduce speed as in a speed reducer, where an electric motor that may be running at 1725 rpm drives the worm. If the ratio is 50:1, that is, for every 50 turns of the worm the worm gear turns once, the worm gear will actually run at 34.5 rpm, which is quite a reduction. A dividing head with a ratio of 40:1 is another example of the practical use of worm gearing. The index crank will turn 40 times, the spindle will turn only once.

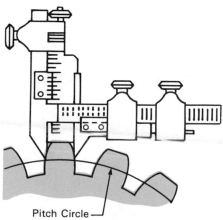

Figure 10–88 Measuring gear teeth

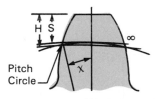

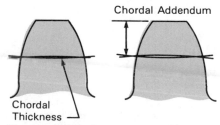

Figure 10–89 Corrected addendum

Cutting Worms and Worm Gears

There is no real trick to cutting a worm gear except for two things. First, the table must be swiveled so that the worm gear matches the lead angle of the worm. Secondly, the gear cutter cannot pass through the gear blank from side to side because of the throat radius. The cutter must be suspended directly over the center of the gear blank, Figure 10–91. The blank is then raised vertically for each cut until the cutter reaches the whole depth of each tooth. The gear cutter should be the same diameter as the worm.

Cutting a worm is like cutting an Acme 29-degree screw thread, but worm teeth are in fact deeper. These can be cut on a lathe. They can also be milled on a milling machine using a helical milling setup and a rack milling attachment, Figure 10–92.

Worm gears are more commonly cut on a hobbing machine. When cutting gears with a single involute gear cutter, each tooth is cut individually. The gear blank is rotated by indexing. In hobbing, Figure 10–93, the gear blank is left free to rotate. A hobbing cutter has a lead angle similar to a worm, and except for the cutting teeth, actually resembles a worm in appearance. As the hobbing cutter touches the gear blank, the lead angle causes the gear blank to rotate. When the hobbing cutter reaches the whole depth of the tooth, all of the teeth are cut. Similarly, in cutting a spur gear, one pass through the gear blank cuts all of the teeth. In both cases, no indexing is necessary. Hobbing cutters are extremely expensive compared to single involute gear cutters.

Tables 10–10A (page 442) and 10–10B (page 444) give the rules and formulas for both worm wheel and worm gear calculations.

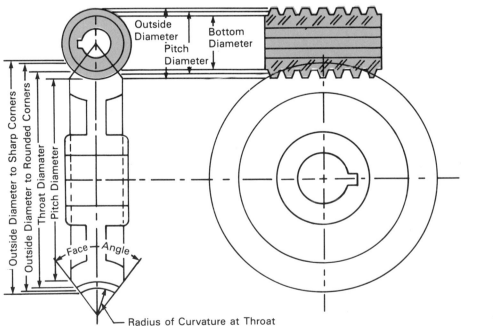

Figure 10–90A Worm gearing

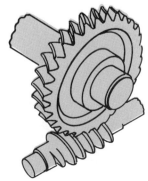

Figure 10–90B Worm and worm gear

THE MODULE SYSTEM OF GEARING

The **module system** is generally used by countries that have adopted the metric system of measurement. Usually, it is understood to mean the pitch diameter in millimeters divided by the number of teeth in the gear. This differs from the diametral pitch system, which is the number of teeth in the gear divided by the pitch diameter.

The module system may also be based on inch measurement, in which case, it is called English module to distinguish it from Metric module.

Module is an actual dimension, whereas diametral pitch is a ratio between the inches of pitch diameter and the number of teeth in the gear. If the pitch diameter of a gear is 40 mm and there are 20 teeth, the module is 2, or there are 2 mm of pitch diameter for each tooth.

The rules for the module system of gearing are given in Table 10-11 (page 446).

Figure 10-91 Gashing the teeth of a worm wheel *(Courtesy of Cincinnati Milacron)*

BEVEL GEARS

Bevel gears are quite different from spur gears. Spur gears are developed from parallel cylinders, and the shafts of mating spur gears are parallel to each other, Figure 10–94 (page 446).

Bevel gears are developed from cones, actually the **frustrum** or a part of a cone. The shafts of two mating bevel gears are at an angle to each other. Shafts can be at 90 degrees to each other or they can be at angles less than 90 degrees, or more than 90 degrees, Figure 10–95A to D (page 447). If two mating bevel gears are the same size, they are called **miter** gears. If they are different sizes, the smaller gear is called a **pinion**.

The following series of sketches shows how a bevel gear is developed from the basic cone diagram, Figure 10–96A (page 448), to the completed sketch of the gear, Figure 10–96B (page 448). The teeth of a bevel gear are thicker and deeper at the large end than they are at the small end, Figure 10–97 (page 448). The parts of the bevel gear are illustrated in Figure 10–98 (page 449). These terms are standard in the industry.

Cutting a Bevel Gear

A correctly formed bevel gear tooth can be cut properly only on a generating type of gear cutting machine such as a hobbing machine.

For many purposes, though, an approximate bevel gear tooth can be cut on a horizontal milling machine using rotary-formed milling cutters, especially where extreme accuracy is not required. Milled bevel gears are not suitable for use in high-speed applications. For milling 14 1/2 degree, pressure angle bevel gears, the standard cutter series is commonly used. Proper bevel gear cutters are thinner than spur gear cutters.

Figure 10–92 Milling a worm *(Courtesy of Cincinnati Milacron)*

Selecting the Cutter

Cutters for bevel gears are selected according to the following formula:

$$N' = \frac{N}{\text{cosine of the pitch cone angle}}$$

N' = the number of teeth for which to select the cutter.

N = the actual number of teeth of the bevel gear.

Milling a bevel gear on a horizontal milling machine is more complicated than milling a spur gear. Figure 10–99 (page 450) illustrates the general setup for bevel gear cutting. The sketched side view, Figure 10–100 (page 450), shows the position of the bevel gear with respect to the gear cutter, with the dividing head tilted at the cutting angle of 41 degrees, 53 minutes.

Using the set-over or offset method, Figure 10–101 (page 451), for milling bevel gear teeth involves a sequence of three operations. The first operation is the rough gashing of the teeth with the gear blank set on center. For the second operation, the gear blank is set off center to the left and the gear blank is rolled to the right until the small end of the tooth touches the cutter. No material is removed from the small end, but the large end is shaved slightly. For the third opera-

Figure 10-93 Hobbing the teeth of a worm wheel *(Courtesy of Cincinnati Milacron)*

Table 10–10A Rules and Formulas for Worm Wheel Calculations *(Courtesy of Cincinnati Milacron)*

(Single and Double Thread—$14\frac{1}{2}°$ Pressure Angle)

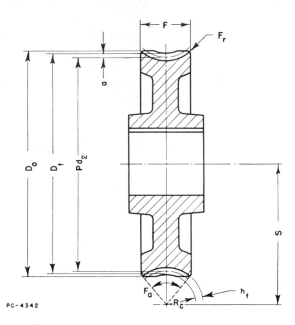

PC-4342

The following symbols are used in conjunction with the formulae for determining the proportions of worm wheel teeth.

P_c = Circular pitch.
P_{d_2} = Pitch diameter.
N = Number of teeth.
D_o = Outside diameter.
D_t = Throat diameter.
R_c = Radius of curvature of worm wheel throat.
D = Diameter to sharp corners.
F_a = Face angle.
F = Face width of rim.
F_r = Radius at edge of face.
a = Addendum.
h_t = Whole depth of tooth.
S = Center distance between worm and worm wheel.
G = Gashing angle.

Table 10–10A Continued

(Single and Double Thread—14½° Pressure Angle)

TO FIND	RULE	FORMULA
Circular pitch P_c	Divide the pitch diameter by the product of 0.3183 and the number of teeth.	$P_c = \dfrac{P_{d_2}}{0.3183\ N}$
Pitch diameter P_{d_2}	Multiply the number of teeth in the worm wheel by the linear pitch of the worm, and divide the product by 3.1416.	$P_{d_2} = \dfrac{NP_L}{3.1416}$
Outside diameter D_o	Multiply the circular pitch by 0.4775 and add the product to the throat diameter.	$D_o = D_t + 0.4775\ P_c$
Throat diameter D_t	Add twice the addendum of the worm tooth to the pitch diameter of the worm wheel.	$D_t = P_{d_2} + 2\ a$
Radius of curvature of worm wheel throat R_c	Subtract twice the addendum of the worm tooth from half the outside diameter of the worm.	$R_c = \dfrac{D_o}{2} - 2\ a$
Diameter to sharp corners D	Multiply the radius of curvature of the worm-wheel throat by the cosine of half the face angle, subtract this quantity from the radius of curvature. Multiply the remainder by 2, and add the product to the throat diameter of the worm wheel.	$D = 2\ (R_c - R_c \times \cos \dfrac{F_a}{2}) + D_t$
Face width of rim F	Multiply the circular pitch by 2.38 and add 0.25 to the product.	$F = 2.38\ P_c + 0.25$
Radius at edge of face F_r	Divide the circular pitch by 4.	$F_r = \dfrac{P_c}{4}$
Addendum a	Multiply the circular pitch by 0.3183.	$a = 0.3183\ P_c$
Whole depth of tooth h_t	Multiply the circular pitch by 0.6866.	$h_t = 0.6866\ P_c$
Center distance between worm and worm wheel S	Add the pitch diameter of the worm to the pitch diameter of the worm wheel and divide the sum by 2.	$S = \dfrac{P_{d_1} + P_{d_2}}{2}$
Gashing angle G	Divide the lead of the worm by the circumference of the pitch circle. The result will be the cotangent of the gashing angle.	$\cot G = \dfrac{L}{3.1416\ d}$

Table 10–10B Rules and Formulas for Worm Gear Calculations *(Courtesy of Cincinnati Milacron)*

(Single and Double Thread—14½° Pressure Angle)

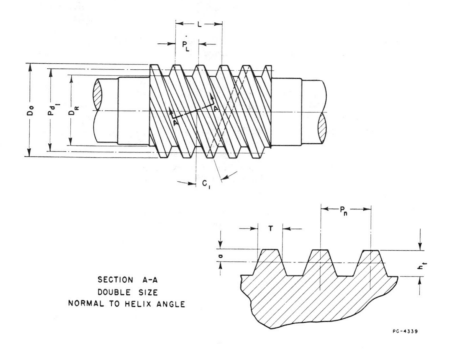

SECTION A-A
DOUBLE SIZE
NORMAL TO HELIX ANGLE

PG-4339

The following symbols are used in conjunction with the formulae for determining the proportions of worm gear teeth.

P_L = Linear pitch.
P_{d_1} = Pitch diameter.
D_o = Outside diameter.
N_w = Number of threads.
D_R = Root diameter.
h_t = Whole depth of tooth.
C_1 = Helix angle.
P_n = Normal pitch.
a = Addendum.
L = Lead.
T = Normal tooth thickness.
t = Width of thread tool at end.

Table 10–10B Continued

(Single and Double Thread—$14\frac{1}{2}°$ Pressure Angle)

TO FIND	RULE	FORMULA
Linear pitch P_L	Divide the lead by the number of threads in the whole worm: i. e., one if single-threaded or four if quadrupled threaded.	$P_L = \dfrac{L}{N_w}$
Pitch diameter P_{d_1}	Subtract twice the addendum from the outside diameter.	$P_{d_1} = D_o - 2\,a$
Outside diameter D_o	Add twice the addendum of the worm to the pitch diameter of the worm wheel.	$D_o = P_{d_1} + 2\,a$
Root diameter D_R	Subtract twice the whole depth of the tooth from the outside diameter.	$D_R = D_o - 2\,h_t$
Whole depth of tooth h_t	Multiply the linear pitch by 0.6866.	$h_t = 0.6866 P_L$
Helix angle C_1	Multiply the pitch diameter of the worm by 3.1416, and divide the product by the lead. The quotient is the cotangent of the helix angle.	$\cot C_1 = \dfrac{3.1416 P_{d_2}}{L}$
Normal pitch P_n	Multiply the linear pitch by the cosine of the helix angle of the worm.	$P_n = P_L \cos C_1$
Addendum a	Multiply the linear pitch by 0.3183.	$a = 0.3183\,P_L$
Lead L	Multiply the linear pitch by the number of threads.	$L = P_L\,N_w$
Normal tooth thickness T	Multiply one-half the linear pitch by the cosine of the helix angle.	$T = \dfrac{P_L}{2} \cos C_1$
Width of thread tool at end t	Multiply the linear pitch by 0.31.	$t = 0.31\,P_L$

tion, the gear blank is first set back on center and then set over to the right. The blank is rolled to the left until the small end of the tooth just touches the cutter and the other side of the large end of the tooth is shaved. In using the set-over method, then, it is necessary to make three separate circuits of the gear blank. Finally, Figure 10–102 (page 451) illustrates the effect of this method on the contour of the tooth.

The rules and formulas for bevel gear calculations are given in Table 10–12 (page 452).

PROCEDURE FOR MILLING A BEVEL GEAR

1. Set the dividing head to the root angle (cutting angle).

2. Center the gear blank to the cutter.

3. Adjust the table until the revolving cutter just touches the gear blank at the outside diameter.

4. Raise the table the whole depth; take a cut; index one tooth; take another cut.

5. Measure the tooth thickness on the pitch line at both the large end and the small end using a gear-tooth vernier caliper.

6. The thickness of the tooth at the small end is about right, but the curve of the tooth is not quite enough.

7. The thickness of the tooth at the large end is too much, but the curve of the tooth is correct.

8. The problem now is to obtain the correct tooth thickness at both ends.

9. Calculate the setover (offset) according to the standard formula given in Table 10–12.

10. Offset the gear, but remember the original centerline setting.

11. Rotate the gear (with the index crank) until the revolving cutter just touches the side of the tooth at the small end.

12. Trim the side of the tooth. This removes metal from the large end but not from the small end.

13. Index each tooth and cut once around the gear blank. This completes one side of the teeth.

Table 10–11   Basic Rules for the Module System of Gearing

TO FIND	FORMULA

1. Metric module	Module = $\dfrac{D}{N}$ D = Pitch diameter (mm)
	        N = Number of teeth
	or = CP × 0.3183
	        CP = Circular pitch (mm)
	or = $\dfrac{OD}{N+2}$ OD = Outside diameter
	        N = Number of teeth

2. English module	Module = $\dfrac{D}{N}$ D = Pitch diameter (inches) N = Number of teeth
3. Metric module which is equivalent to diametral pitch	Equivalent module = $\dfrac{25.4 \text{ (mm/inch)}}{P \text{ (diametral pitch)}}$
4. Diametral pitch which is equivalent to metric module	Equivalent diametral pitch = $\dfrac{25.4 \text{ (mm/inch)}}{\text{Module}}$
5. Pitch diameter	Pitch diameter = N × Module
6. Outside diameter	$OD = (N + 2)$ × Module

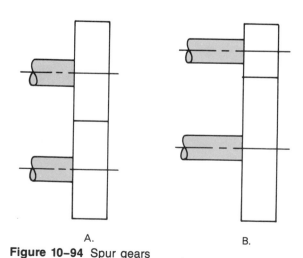

A.   B.

Figure 10–94 Spur gears

14. Return the gear to the original centerline setting.

15. Offset the gear the other way. Rotate the gear again until the revolving cutter touches the side of the tooth at the small end.

16. Trim the side of the tooth. Measure the tooth thickness at both ends of the tooth.

17. Index each tooth and cut once around the gear blank.

ting. The intermediate set of gear cutters, numbered from 1 1/2 to 7 1/2, can be used where greater accuracy is needed. Each of the gear cutters in the set, Figure 10–103 (page 454), is set at the same whole depth. The curve of the cutter is what is different in each case. A number 1 cutter could not be used to cut a 12-tooth gear. The teeth would not mesh properly with other gears in the set because of the wrong curvature of the teeth.

INVOLUTE GEAR CUTTERS

The usual basic set of gear cutters consists of the whole numbered cutters from 1 to 8, Table 10–13 (page 454). Normally, these are all that are necessary for gear cut-

TOOTH SIZES

An approximate comparison of tooth sizes is made for several diametral pitches, Figure 10–104 (page 454). Diametral pitch is more commonly referred to as pitch.

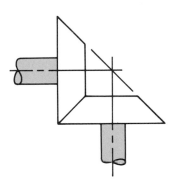

A. Bevel gears at 90°

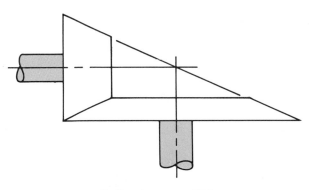

B. Bevel gears at 90°

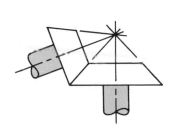

C. Bevel gears at less than 90°

Figure 10–95 Bevel gears

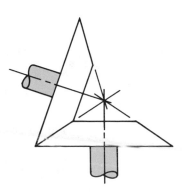

D. Bevel gears at more than 90°

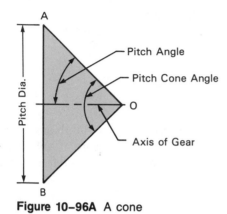

Figure 10–96A A cone

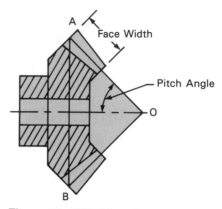

Figure 10–96B A bevel gear

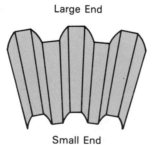

Figure 10–97 Bevel gear teeth

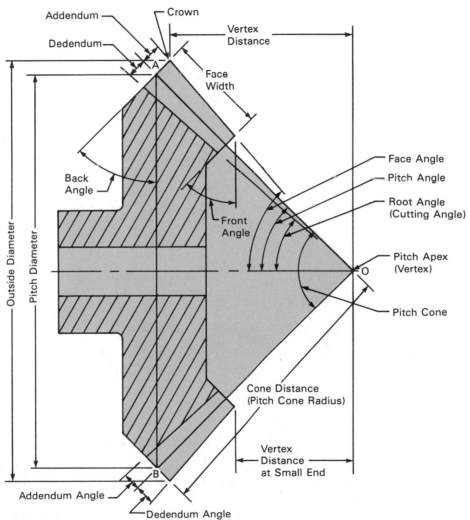

Figure 10-98 The parts of a bevel gear

Figure 10–99 Milling a bevel gear on a horizontal milling machine *(Courtesy of Cincinnati Milacron)*

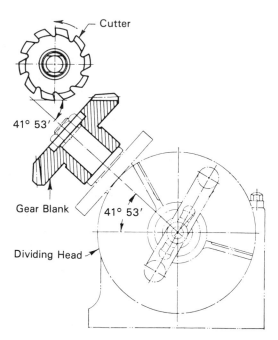

Elevation showing position of gear blank with respect to gear cutter when tilted at the cutting angle of 41° 53′.

Figure 10–100 Position of the gear blank set for cutting *(Courtesy of Cincinnati Milacron)*

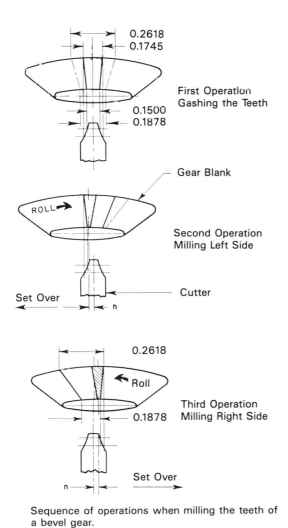

Sequence of operations when milling the teeth of a bevel gear.

Figure 10–101 Using the offset method *(Courtesy of Cincinnati Milacron)*

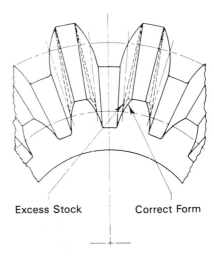

Effect of bevel gear cutter contour on the tooth shape.

Figure 10–102 Corrected tooth shape *(Courtesy of Cincinnati Milacron)*

Table 10–12 Rules and Formulas for Bevel Gear Calculations *(Courtesy of Cincinnati Milacron)*

(Shafts at Right Angles)

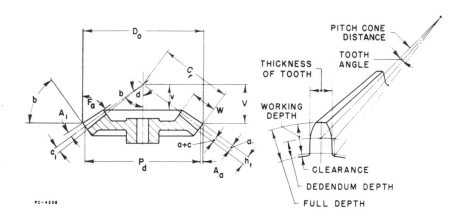

PC-4336

The following symbols are used in conjunction with the formulae for determining the proportions of bevel gear teeth.

P = Diametral pitch.

P_c = Circular pitch.

P_d = Pitch diameter.

b = Pitch angle.

C_r = Pitch cone distance.

a = Addendum.

A_1 = Addendum angle.

A_a = Angular addendum.

D_o = Outside diameter.

c_1 = Dedendum angle.

$a+c$ = Addendum plus clearance.

a_s = Addendum of small end of tooth.

T_L = Thickness of tooth at pitch line.

T_s = Thickness of tooth at pitch line at small end of gear.

F_a = Face angle.

h_t = Whole depth of tooth space.

V = Apex distance at large end of tooth.

v = Apex distance at small end of tooth.

m_g = Gear ratio.

N = Number of teeth.

N_g = Number of teeth in gear.

N_p = Number of teeth in pinion.

d = Root angle.

W = Width of gear tooth face.

N_c = Number of teeth of imaginary spur gear for which cutter is selected.

Table 10-12 Continued
(Shafts at Right Angles)

TO FIND	RULE	FORMULA
Diametral pitch P	Divide the number of teeth by the pitch diameter.	$P = \dfrac{N}{P_d}$
Circular pitch P_c	Divide 3.1416 by the diametral pitch.	$P_c = \dfrac{3.1416}{P}$
Pitch diameter P_d	Divide the number of teeth by the diametral pitch.	$P_d = \dfrac{N}{P}$
Pitch angle of pinion $\tan b_p$	Divide the number of teeth in the pinion by the number of teeth in the gear to obtain the tangent.	$\tan b_p = \dfrac{N_p}{N_g}$
Pitch angle of gear $\tan b_g$	Divide the number of teeth in the gear by the number of teeth in the pinion to obtain the tangent.	$\tan b_g = \dfrac{N_g}{N_p}$
Pitch cone distance C_r	Divide the pitch diameter by twice the sine of the pitch angle.	$C_r = \dfrac{P_d}{2\,(\sin b)}$
Addendum a	Divide 1.0 by the diametral pitch.	$a = \dfrac{1.0}{P}$
Addendum angle $\tan A_1$	Divide the addendum by the pitch cone distance to obtain the tangent.	$\tan A_1 = \dfrac{a}{C_r}$
Angular addendum A_a	Multiply the addendum by the cosine of the pitch angle.	$A_a = a \cos b$
Outside diameter D_o	Add twice the angular addendum to the pitch diameter.	$D_o = P_d + 2A_a$
Dedendum angle $\tan c_1$	Divide the dedendum by the pitch cone distance to obtain the tangent.	$\tan c_1 = \dfrac{a + c}{C_r}$
Addendum of small end of tooth a_s	Subtract the width of face from the pitch cone distance, divide the remainder by the pitch cone distance and multiply by the addendum.	$a_s = a\left(\dfrac{C_r - W}{C_r}\right)$
Thickness of tooth at pitch line T_L	Divide the circular pitch by 2.	$T_L = \dfrac{P_c}{2}$
Thickness of tooth at pitch line at small end of gear T_s	Subtract the width of face from the pitch cone distance, divide the remainder by the pitch cone distance and multiply by the thickness of the tooth at the pitch line.	$T_s = T_L\left(\dfrac{C_r - W}{C_r}\right)$
Face angle F_a	Face cone of blank turned parallel to root cone of mating gear.	$F_a = b + c_1$
Whole depth of tooth space h_t	Divide 2.157 by the diametral pitch.	$h_t = \dfrac{2.157}{P}$
Apex distance at large end of tooth V	Multiply one-half the outside diameter by the tangent of the face angle.	$V = \left(\dfrac{D_o}{2}\right)\tan F_a$
Apex distance at small end of tooth v	Subtract the width of face from the pitch cone distance, divide the remainder by the pitch cone distance and multiply by the apex distance.	$v = V\left(\dfrac{C_r - W}{C_r}\right)$
Gear ratio m_g	Divide the number of teeth in the gear by the number of teeth in the pinion.	$m_g = \dfrac{N_g}{N_p}$
Number of teeth in gear and/or pinion N_g, N_p	Multiply the pitch diameter by the diametral pitch.	$N_g = P_d\,P$ $N_p = P_d\,P$
Cutting angle d	Subtract the addendum plus clearance angle from the pitch angle.	$d = b - c_1$
Number of teeth of imaginary spur gear for which cutter is selected N_o	Divide the number of teeth in actual gear by the cosine of the pitch angle.	$N_c = \dfrac{N}{\cos b}$

Table 10–13 Involute Gear Cutters

CUTTER NUMBER	WILL CUT GEARS	CUTTER NUMBER	WILL CUT GEARS
1	135 teeth to rack	5	21 to 25 teeth
2	55 to 134 teeth	6	17 to 20 teeth
3	35 to 54 teeth	7	14 to 16 teeth
4	26 to 34 teeth	8	12 to 13 teeth
1½	80 to 134 teeth	5½	19 to 20 teeth
2½	42 to 54 teeth	6½	15 to 16 teeth
3½	30 to 34 teeth	7½	13 teeth
4½	23 to 25 teeth		

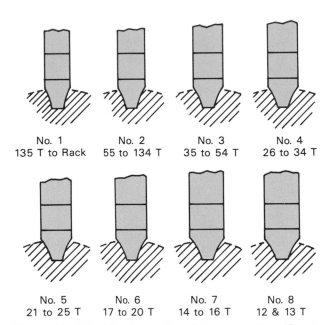

No. 1
135 T to Rack

No. 2
55 to 134 T

No. 3
35 to 54 T

No. 4
26 to 34 T

No. 5
21 to 25 T

No. 6
17 to 20 T

No. 7
14 to 16 T

No. 8
12 & 13 T

Figure 10–103 A basic set of involute gear cutters *(Courtesy of Cincinnati Milacron)*

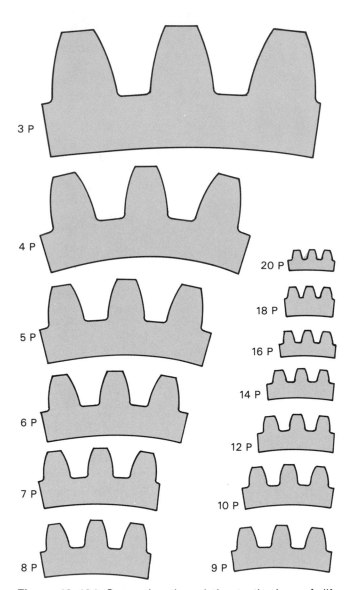

3 P

4 P

5 P

6 P

7 P

8 P

20 P

18 P

16 P

14 P

12 P

10 P

9 P

Figure 10–104 Comparing the relative tooth sizes of different diametral pitches *(Courtesy of Cincinnati Milacron)*

REVIEW QUESTIONS

1. In measuring a gear tooth with a gear tooth vernier, what is actually measured?
2. What is the corrected addendum? What is another name for the corrected addendum?
3. Briefly describe worm gearing.
4. What is the main purpose of worm gearing?
5. List two changes in the procedure for cutting a worm gear as compared to the procedure for cutting a spur gear.
6. Compare the cutting of a worm to the cutting of an Acme screw thread.
7. Compare cutting gears using a single involute gear cutter and a hobbing cutter.
8. What is the module system of gearing?
9. State the formulas for finding the metric module and the English module.
10. How do bevel gears differ from spur gears?
11. Compare the teeth of a bevel gear to the teeth of a spur gear.
12. What type of machine is required to cut a correctly formed bevel gear tooth?
13. Are milled bevel gears suitable for all applications?
14. State the formula used to select the cutter for a bevel gear.
15. To mill a bevel gear, at what angle is the dividing head set?
16. Describe the sequence of operations in using the set-over method for milling gear teeth.

Face milling cast iron on a horizontal boring mill. *Courtesy of Cominco Ltd.*

Milling with a carbide staggered-tooth milling cutter. *Courtesy of Kennametal Inc.*

Milling the driving slots in the flange of an adaptor with an end milling cutter. *Courtesy of Kennametal Inc.*

Milling a steel surface using a KC850 titanium nitride-coated carbide cutter. *Courtesy of Kennametal Inc.*

Milling a steel surface using a heavy-duty face milling cutter. *Courtesy of Kennametal Inc.*

Milling a steel surface with a shear angle face milling cutter. The shear angle cutter is suitable on both ferrous and nonferrous continuous chip materials. An improved chip clearing action is provided by a positive axial rake and a negative radial rake. This facilitates high feed rates and fine finishes. *Courtesy of Kennametal Inc.*

Face milling on a horizontal boring mill using a multitoothed face mill and indexable insert tools. *Courtesy of Cominco Ltd.*

Milling titanium using a helical end milling cutter. *Courtesy of Kennametal Inc.*

CHAPTER ELEVEN

GRINDING

A grinding tool is essentially a finishing tool used to produce fine and accurate surfaces. Equally important, it is also a sharpening tool used for sharpening all of the cutting tools that are used on machine tools.

Grits or abrasives bound together in the form of a disc or wheel make up one type of grinding tool. Another type is the coated abrasive where grits or abrasives are bound together to a cloth or paper belt or disc. These tools are the heart of any grinding operation.

Grinding machines such as bench grinders or hand grinders are used for offhand grinding which is considered to be rough grinding. Surface grinders, cylindrical grinders, centerless grinders, and tool and cutter grinders are precision machines that produce surfaces which are fine and true. Modern manufacturing needs this capability to produce the close tolerances and fine surface finishes that are required for machinery and products of all kinds.

GRINDING SAFETY

In grinding, the chief danger is the moving grinding wheel. The possibility of the work shifting and the wheel exploding are also very real dangers.

1. Permission should be obtained to operate the machine, at the discretion of the instructor.

2. Always roll up loose sleeves and remove ties, rings, watches, and so on, before operating the machine.

3. Always wear specified eye protection. Do not wear contact lenses when operating a grinder.

4. Always use a proper wheel guard to contain the pieces should a wheel burst.

5. Make all adjustments to both the job and the machine with the machine off.

6. If there is a tool rest such as is found on a bench grinder, it should not be more than 1/16 in (1.6 mm) away from the wheel at any time.

7. The standard grinding wheel is designed to cut on its periphery only. No grinding should be done on the sides of the wheel unless it is a proper face wheel or has a metal backing.

8. A wheel should not be used when it is out of round, when it vibrates or chatters, when it is dirty or glazed, or when it is worn excessively. Dress and balance the wheel properly before using it to make it run true and to make it cut properly.

9. Always check the rpm of both the wheel and the machine. Never exceed the recommended wheel speed. If you do not know the recommended speed of a wheel, do not use it.

10. Before installing a wheel, it is very important to test it for cracks and for balance.

11. Check the mounting flanges for equal and correct diameter. Flanges should be at least one-third the diameter of the wheel. Flanges should be recessed and mounting blotters should be used to equalize pressure on the wheel.

12. Stand to one side out of the line of the wheel when first switching on the machine. Allow newly mounted wheels to run at full operating speed for at least one minute before grinding. Warm up the machine thoroughly before doing precise grinding. On ultraprecision work, keep the spindle running until the work is complete.

13. Feed the work slowly and steadily. Do not jam the work into the wheel and never stall a grinding machine.

14. The magnetic chuck on a surface grinder should be thoroughly cleaned. The work should be carefully checked for grit, ragged edges, or bruises before placing it on the chuck.

15. Test the holding power of the chuck before starting the grinding operation.

16. Set all stops and test them before starting the machine.

17. Always turn off the coolant before stopping the wheel to avoid an out-of-balance condition. Allow the wheel to run for at least one minute after shutting off the coolant.

18. Never walk away from a grinder when it is operating.

19. When the work is complete, shut off the power and clean the machine.

UNIT 11-1

THE GRINDING WHEEL

OBJECTIVES

After completing this unit, the student will be able to:

* define the terms offhand grinding and precision grinding.
* select the proper grinding wheel for the material being ground and the type of finish desired.
* explain the purpose of abrasive grits (grains).
* describe the basic categories of natural and manufactured abrasive grains.
* describe the term bond as it relates to a grinding wheel.
* explain what is meant by wheel grade and wheel structure.
* use the Norton Company standard wheel marking system.

KEY TERMS

Grinding
Abrasive grit
Grinding wheel
Coated abrasive
Offhand grinding
Precision grinding
Bench grinding
Bench grinder
Wheel loading
Abrasive grain
Natural abrasive grain
Manufactured
  abrasive grain
Alumina oxide
Zirconia alumina
Silicon carbide
Diamond abrasive
Cubic boron nitride

Bond
Grit size
Coarse grain wheel
Medium grain wheel
Fine grain wheel
Very fine grain wheel
Virtified bond
Resinoid bond
Rubber bond
Shellac bond
Metal bond
Wheel grade
Wheel structure
Norton wheel marking
  system
Mounted points
Mounted wheels

GENERAL

Grinding is the process of removing material from the work by means of **abrasive grits**, which are hard crystals that do the actual cutting. These grits or crystals are **bonded** or held together in the form of discs called **grinding wheels**. They can also be bonded to paper or cloth in the form of sheets, strips, or belts. These are called **coated abrasives**. A grinding wheel or a coated abrasive contains thousands of hard, tough, abrasive grains that move against the work and cut away tiny chips.

There are two general types of grinding: offhand and precision grinding. In **offhand grinding**, the grinding wheel can be applied to the work by hand, as in snagging (cleaning) rough castings, or in weld grinding. The work can also be applied to the grinding wheel, as in bench grinding. The bench grinder, Figure 11–1, can be mounted on a bench or on a separate pedestal. It is used for rough hand grinding or for the hand grinding and sharpening of lathe and hand cutting tools.

Precision grinding includes such operations as cylindrical grinding (between centers or centerless), internal grinding, surface grinding, and tool and cutter grinding. The only difference in these is in the method of grinding.

The Grinding Wheel

The most important rule in machine shop practice is to know your cutting tool thoroughly. Grinding wheels are no exception. A basic relationship between different sized grits and the materials that they cut is illustrated in Figure 11–2. On hard materials, the increased number of cutting points on the face of a moderately fine grit wheel, Figure 11–2A, will remove stock faster than the fewer cutting points of the coarser wheel, Figure 11–2B. The larger abrasive grains cannot penetrate as deeply into the hard material without burning it. On soft, ductile materials, the larger grains penetrate easily and provide the necessary chip clearance to minimize wheel loading, Figure 11–2C. The cylindrical cutting action, Figure 11–3, illustrates the actual chips produced by the cutting action.

ABRASIVE GRITS AND GRAINS

The abrasive grit or grain is the actual cutting tool. The basic requirements of the ideal abrasive grains are resistance to point dulling and the ability of the grain to fracture under normal grinding pressures before serious dulling occurs. When a grain fractures, new, sharp cutting points are exposed.

Abrasive grains fall into two general categories: natural and manufactured. The natural grains include emery, sandstone, corrundum, quartz, and diamond.

The manufactured abrasives or conventional electric furnace abrasives are in two broad categories. One includes aluminum oxide, aluminum oxide/zirconia alloy, and silicon carbide. The other includes diamond and cubic boron nitride.

Aluminum oxide is made from bauxite. Zirconia alumina is made by alloying aluminum oxide and zirconium oxide. Silicon carbide is made from silica sand and carbon in the form of coke.

Aluminum oxide wheels are used for high-tensile strength materials, carbon and alloy steels, malleable cast iron, and wrought iron. Regular Alundum® abrasives (Norton's trade name for aluminum oxide) are widely used for heavy-duty work on a range of steels, but not on the hardest and most heat-sensitive steels. The cooler running Alundum® 32 or 38 is ideal for tool and cutter grinding of tool steels and high-speed steels.

Zirconia alumina abrasive alloys are best suited for high-pressure, high-speed grinding of cast and fabricated steels. Snagging wheels for cleaning castings in a foundry chipping room are one example of this tough type of grinding.

Silicon carbide is extremely brittle and is used for low-tensile strength materials. A black silicon carbide (37 Crystolon®) wheel is used for cast iron, brass, soft bronze, aluminum, copper, and nonmetallic materials. A special, green-colored, silicon carbide wheel (39 Crystolon®) is used for sharpening carbide-tipped tools.

Diamond abrasives, natural or man made, are used mainly for grinding cemented tungsten carbides in both wet and dry applications. The type of diamond abrasive selected depends on how much steel is to be ground in addition to the carbide, or how much brazing material is contacted, and whether or not the grinding operation is wet or dry.

Figure 11–1 Bench grinder *(Courtesy of DoALL Company)*

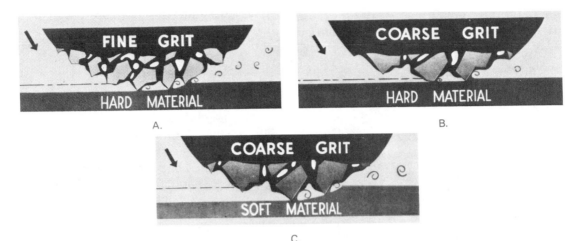

Figure 11-2 The action of grit size on various materials *(Courtesy of Norton Company)*

Cubic boron nitride (CBN), a material manufactured by a process similar to that used to make synthetic diamond abrasive, is used mainly for grinding hard and tough alloy steels and all tool and die steels of Rockwell C50 hardness or better.

Commonly used grit sizes are 10, 12, 14, 16, 20, 24, 30, 36, 46, 54, 60, 70, 80, 90, 100, 120, 150, 180, and 220, with the most commonly used sizes falling in the range of 24 to 80 grits. Occasionally used sizes are 240, 280, 320, 400, 500, and 600 grits.

GRINDING WHEEL COMPONENTS

The heart of any grinding machine, regardless of the application, is the grinding wheel. Like any other machine tool cutter, the efficiency of any grinding operation depends on the use of the proper wheel for the work at hand. Every grinding wheel has two major components, the abrasive grits that do the actual cutting, and the bond that supports and holds the grains together, Figure 11-4.

Grit Size

Grits or grains are sorted according to size by using various sizes of screen to sift them and sort them out. The number designating grit size represents the approximate number of openings per linear inch in the final screen used to size the grain. As an example, 30 grit size is one in which the grains will pass through a screen with 27 openings per linear inch and be held on a screen with 33 openings per linear inch.

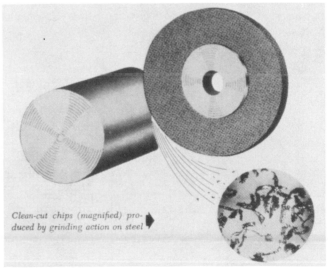

Clean-cut chips (magnified) produced by grinding action on steel

Figure 11-3 Cutting action of a grinding wheel *(Courtesy of Norton Company)*

. . . that every

GRINDING WHEEL

has **2** *components*

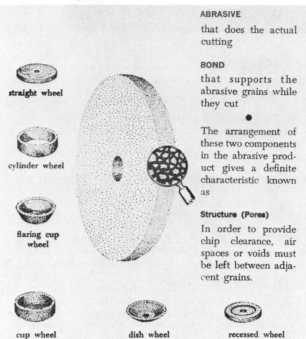

ABRASIVE

that does the actual cutting

BOND

that supports the abrasive grains while they cut •

The arrangement of these two components in the abrasive product gives a definite characteristic known as

Structure (Pores)

In order to provide chip clearance, air spaces or voids must be left between adjacent grains.

straight wheel

cylinder wheel

flaring cup wheel

cup wheel   dish wheel   recessed wheel

Figure 11–4 Components of a grinding wheel *(Courtesy of Norton Company)*

Coarse grain wheels, Figure 11–5A, are sized from 10 to 24, medium grain wheels, Figure 11–5B, from 30 to 60, fine grain wheels, Figure 11–5C, from 70 to 180, and very fine grain wheels from 220 to 600.

Bond

The bond is the material used to hold the abrasive grits together. The basic types of bond are vitrified, resinoid, rubber, shellac, and metal.

A vitrified bond is used for over 75% of the grinding wheels manufactured. It is an extremely hard, glass-like bond that is not affected by water, acid, oils, or ordinary temperature variations.

A resinoid bond is used for high-speed wheels in foundries, welding and billet shops, and is also used in cut-off and thread grinding operations.

A rubber bond is used for most centerless feed wheels, precision ball race grinding wheels, portable grinders where finish is important (stainless steel welds), and cut-off wheels of less than 1/32 in (0.79 mm) in thickness.

A shellac bond is capable of producing high finishes on cam shafts and on steel and paper mill rolls. It is also used for some cut-off wheels.

A metal bond is used on diamond wheels for grinding carbides. The high strength and toughness of the metal bond helps retain costly diamond

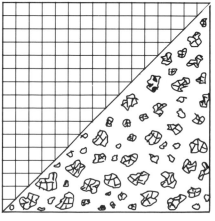

A. Coarse grain

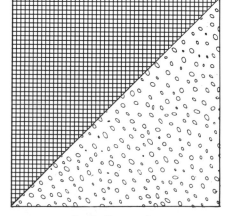

B. Medium grain

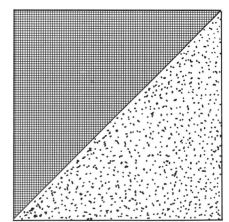

C. Fine grain

Figure 11–5 Grit sizes *(Courtesy of Norton Company)*

abrasives firmly throughout their entire useful life. Metal bonds are used either with diamond or aluminum oxide abrasive. They are also used for electrical discharge grinding and electrochemical grinding where an electrically conductive wheel is needed.

Wheel Grade

Grade indicates the relative strength (holding power) of the bond that holds the abrasive grains in place. With any given bond, it is usually the amount of bond that determines the hardness. Letters of the alphabet are used to grade the wheels, A being the softest wheel and Z the hardest. The light areas in Figure 11–6 are the openings or pores required for chip clearance or temporary storage of chips, dissipation of heat, and application of coolant.

Wheel Structure

Wheel structure refers to the relative spacing of the abrasive grains in the wheel, Figure 11–7. Although structure numbers are not always shown in the standard marking system, the scale used is 0, 1, 2, 3, 4, 5, 6, 7, 8, 9, 10, 11, 12, with zero being the denser structure and twelve being the more open structure.

The Norton Wheel Marking Systems

The Norton wheel marking systems are used as standards in the grinding wheel industry, Figure 11–8.

Wheel Types and Shapes

The most common types of grinding wheels are shown in Figure 11–9 (page 469). There are many more types available.

The wheel shapes illustrated in Figure 11–10 (page 469) are standard in the grinding wheel industry.

Mounted Points or Wheels

Mounted points or wheels, Figure 11–11 (page 470), mounted on their own metal shafts, are normally used in small hand grinders. Grinding with mounted points, Figure 11–12 (page 471), is essentially a hand polishing operation. The motor and flexible shaft is an extremely handy tool for this type of work.

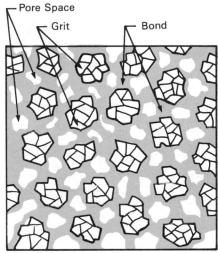

A. Hard bond

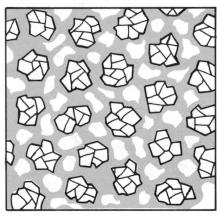

B. Medium bond

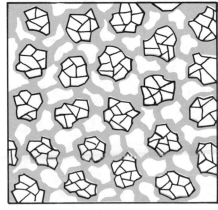

C. Soft bond

Figure 11–6 Wheel grade *(Courtesy of Norton Company)*

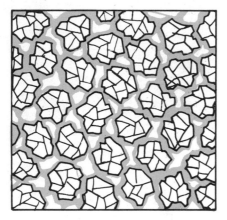

A. Dense structure

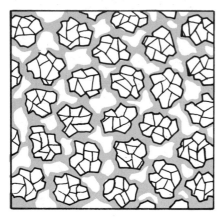

B. Medium structure

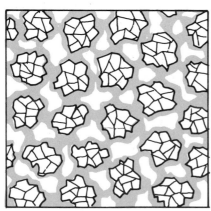

C. Open structure

Figure 11–7 Wheel structure *(Courtesy of Norton Company)*

FACTORS TO BE CONSIDERED IN THE SELECTION OF THE PROPER GRINDING WHEEL

1. The material to be ground and its hardness affect the selection of the abrasive, the grit size, and the grade of the wheel. The abrasive should be aluminum oxide for steel and steel alloys, and silicon carbide for cast iron, nonferrous materials, and nonmetallic materials.

   A fine grit should be selected for hard, brittle materials, and a coarse grit size for soft, ductile materials.

   A hard grade of wheel should be used for soft materials, and a soft grade for hard materials.

2. The amount of stock to be removed and the finish required determine the selection of grit size and type of bond. A coarse grit size should be selected for rapid stock removal and for rough grinding. A high finish would require a fine grit size. Choose a vitrified bond for fast cutting and commercial finish. Choose a resinoid, rubber, or shellac bond for the highest finish.

3. Whether the operation is wet or dry will affect the selection of grade or hardness of a wheel. Wet grinding usually permits the use of wheels one grade harder than for dry grinding, without the danger of burning the work.

4. The speed of the wheel determines the bond you will choose. Standard vitrified wheels can be used with speeds up to (but not over) 6500 surface feet per minute (1950 meters per minute). Resinoid, rubber, and shellac bonds can be used for most applications in the range of 6500 to 16000 sfpm (1950 to 4800 m/min). Never exceed the speed marked on a wheel.

5. Grit size and grade are determined by the area of grinding contact, which is the actual area of the wheel in contact with the work. A coarse grit is best for a large area of contact and a fine grit is best for a small area of contact. The smaller the area of contact, the harder the grade of wheel that should be used.

NORTON GRINDING WHEEL MARKING
32A46-H8VBE

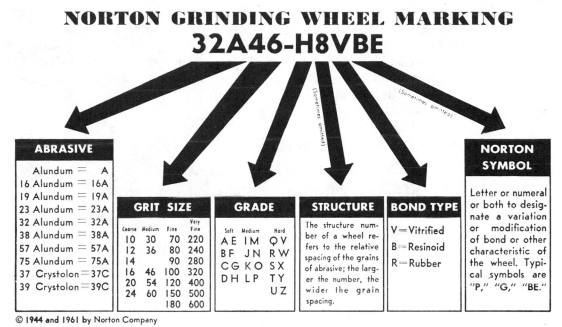

© 1944 and 1961 by Norton Company

Figure 11-8A Standard wheel marking system *(Courtesy of Norton Company)*

Analysis of a typical

Norton diamond wheel markings
ASD100-R100B 56 ⅛ *

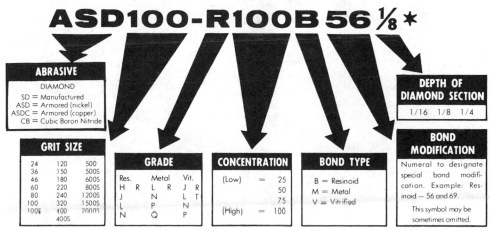

NOTE: No grade is shown for Hand Hones.   *Manufacturer's Identification Symbol

©1961 by Norton Company

Figure 11-8B Diamond wheel marking system *(Courtesy of Norton Company)*

6. The severity of the grinding operation affects the selection of the abrasive. A tough abrasive like 76A or ZS should be chosen for grinding steel and steel alloys under severe conditions. A mild abrasive like 32A or 38A is particularly suitable for grinding all kinds of tool steels including tough vanadium alloy steels. An intermediate abrasive like 23A or 57A is good for work of average severity.

7. The horsepower of the machine is a factor in selecting the grade of the wheel. The harder grades of wheels are used for machines using higher horsepowers.

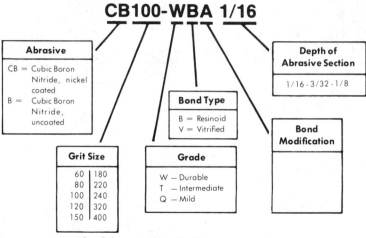

Norton CBN wheel markings

CB100-WBA 1/16

Abrasive
CB = Cubic Boron Nitride, nickel coated
B = Cubic Boron Nitride, uncoated

Grit Size	
60	180
80	220
100	240
120	320
150	400

Bond Type
B = Resinoid
V = Vitrified

Grade
W — Durable
T — Intermediate
Q — Mild

Depth of Abrasive Section
1/16 - 3/32 - 1/8

Bond Modification

Recommended Specifications

For hard to grind M & T type high-speed tool and die steels; and generally steels 50 Rockwell C and harder, CBN wheels are recommended.

	Specification	
Application	**Wet**	**Dry**
Cutter Sharpening Straight Wheel Cup Wheel Dish Wheel		
Surfacing Straight Wheel		
Above specifications recommended for all hardened tool and die steels in "M" and "T" series.		
Internal Mounted Points		
Straight Wheel		
Above internal specifications recommended for essentially all ferrous materials. †Dry grinding not usually recommended on this application. If dry grinding is required, the use of "BD" bond will reduce heat which causes straight wheels to go out of round.		

Figure 11–8C CBN wheel marking system *(Courtesy of Norton Company)*

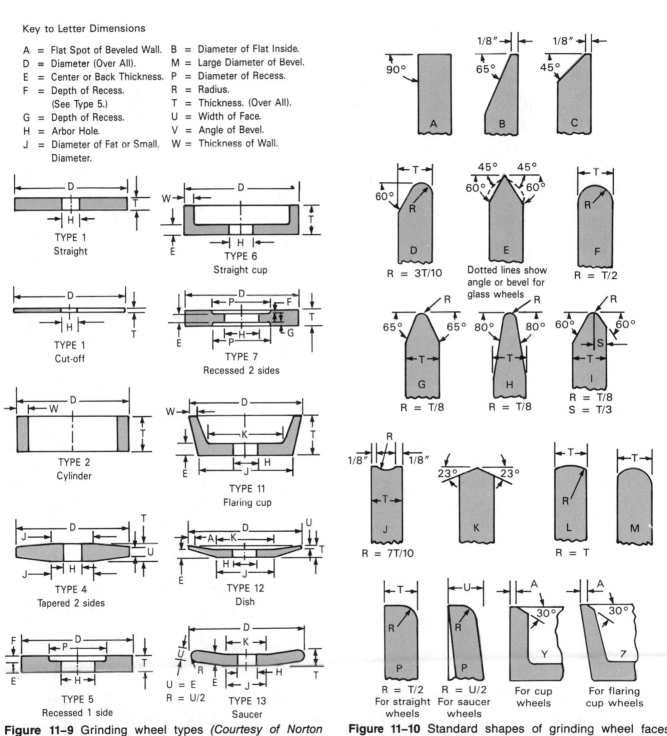

Key to Letter Dimensions

A = Flat Spot of Beveled Wall.
D = Diameter (Over All).
E = Center or Back Thickness.
F = Depth of Recess.
    (See Type 5.)
G = Depth of Recess.
H = Arbor Hole.
J = Diameter of Fat or Small.
    Diameter.

B = Diameter of Flat Inside.
M = Large Diameter of Bevel.
P = Diameter of Recess.
R = Radius.
T = Thickness. (Over All).
U = Width of Face.
V = Angle of Bevel.
W = Thickness of Wall.

TYPE 1
Straight

TYPE 1
Cut-off

TYPE 2
Cylinder

TYPE 4
Tapered 2 sides

TYPE 5
Recessed 1 side

TYPE 6
Straight cup

TYPE 7
Recessed 2 sides

TYPE 11
Flaring cup

TYPE 12
Dish

U = E
R = U/2
TYPE 13
Saucer

Figure 11–9 Grinding wheel types *(Courtesy of Norton Company)*

A   B   C

D
R = 3T/10

E
Dotted lines show angle or bevel for glass wheels

F
R = T/2

G
R = T/8

H
R = T/8

I
R = T/8
S = T/3

J
R = 7T/10

K

L
R = T

M

P
R = T/2
For straight wheels

P
R = U/2
For saucer wheels

Y
For cup wheels

Z
For flaring cup wheels

Figure 11–10 Standard shapes of grinding wheel faces *(Courtesy of Norton Company)*

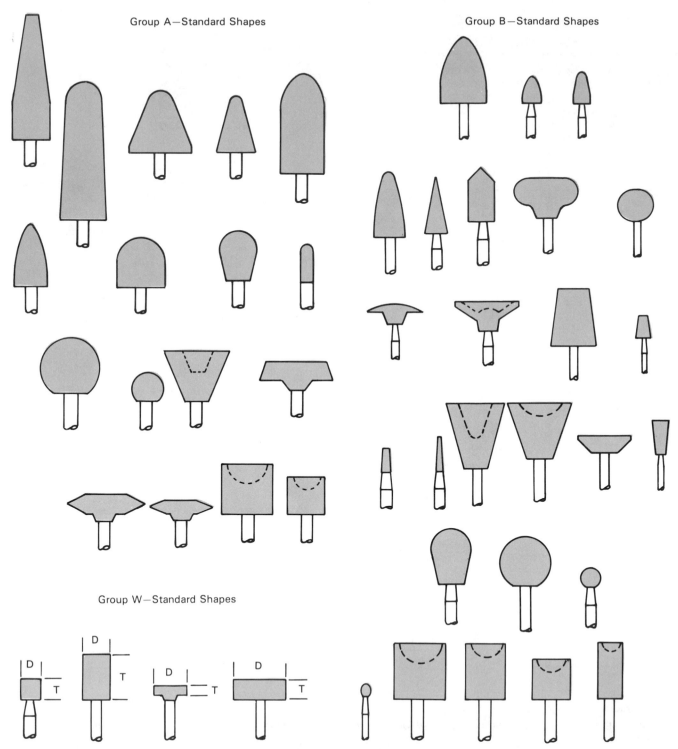

Group A—Standard Shapes

Group B—Standard Shapes

Group W—Standard Shapes

Figure 11–11 A selection of the mounted points or wheels available *(Courtesy of Norton Company)*

GRINDING WHEEL SPECIFICATIONS

To emphasize the importance of having a thorough knowledge of grinding wheels, three examples of grinding work have been selected to show how it should be done according to the Norton numbering system, which is a standard in the grinding industry.

Operation 1

Operation 1 is the use of a tool and cutter grinder to sharpen a high-speed steel milling cutter with dry grinding.

In selecting the abrasive, an aluminum oxide wheel is needed for grinding high-speed steel. High-speed steel is sensitive to grinding heat; i.e., it burns easily, therefore, you should choose a cool cutting abrasive, and one that stays sharper longer, such as a 32 Alundum abrasive. Your specification number would start with:

32A___-_____

A small amount of material is to be removed. Accuracy and finish are important. The wheel must retain its size so that all the teeth around the cutter will be ground to the same height. In this case, a medium-fine grit such as a 60-grit should be used. The specification number then becomes:

32A60-_____

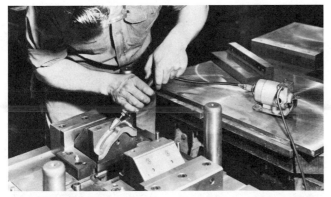

Figure 11–12 Grinding with a mounted point *(Courtesy of Dumore Corporation)*

This is precision work requiring a wheel that will hold its size and have a cool cutting action. A medium-soft grade, about *J*, can be selected. If a very soft grade such as *H* is used, wheel wear may be excessive. If too hard a grade is used, the wheel will dull and burn the cutting edges. The specification number increases to:

32A60-J_____

The structure number is not always shown in a specification number, but from experience, a grain spacing such as a #8 structure, which is relatively open, gives the best performance in tool and cutter grinding when combined with a 60-grit and *BE*-type vitrified bond. The specification number is now:

32A60-J8_____

A vitrified bonded wheel of the Norton *BE* type together with the 32 Alundum abrasive should be used in this application. The *BE* bond was developed to give the maximum benefit of the 32 Alundum abrasive in tool and cutter grinding. The speed of tool and cutter grinders is usually in the range of 4500 to 6000 sfpm (1350 to 1800 m/min), which is within the recommended speed range for vitrified bonded wheels. The full specification number for this operation becomes:

32A60-J8VBE

Operation 2

Operation 2 is weld grinding, the grinding down of heavy welds with a portable grinder. The material to be ground is mild steel weld, and the machine to be used is a portable hand grinder using straight wheels or cup wheels.

Mild steel is ground best with an aluminum oxide abrasive. A 23 Alundum abrasive can be used in this case both for its sharpness and its ability to resist fracture. Mild steel is not particularly heat sensitive to grinding; therefore, a cool cutting abrasive such as 38A or 32A is not required. The specification number starts with:

23A___-_____

The material to be ground is relatively soft and ductile and can be ground readily. A coarse grit size such as 16, 20, or 24 will grind soft, ductile materials faster than the fine sizes. The specification number increases to:

23A16-_____

The grade selection is made according to the amount of work that can be done with the grinding machine. In this case, the portable grinder is limited by the power available. Consequently, a medium-hard grade, about Q is recommended. The specification number increases to:

23A16-Q____

The structure recommended for this work would be a #4 or #5 to indicate a medium density and good chip clearance. The specification number now is:

23A16-Q4___

Since the wheel usually runs at 9500 sfpm (2850 m/min), it is necessary to use an organic bond. Of the various organic bonds, resinoid is the most durable and heat resistant. The operation does not call for the fine finish that would be produced by a rubber- or shellac-bonded wheel. The *BH* modification of resinoid bond would be best for this operation. The full specification number is:

23A16-Q4BH

Operation 3

The third operation selected is the finishing of a heat-treated steel shaft of SAE 1090 steel, using a wet grinding cylindrical grinder.

The steel shaft is best ground with an aluminum oxide abrasive. This is a precision production opera-tion of rough and finish grinding requiring a relatively light stock removal of about 0.010 in (0.25 mm). A 23 Alundum abrasive is chosen because of its uni-formly free cutting action which is especially suited to production grinding. If the operation had required a large amount of stock removal as in rough grinding, a tougher, stronger abrasive such as regular Alundum would have been better. The specification number starts at:

23A____-_____

The material to be ground is moderately hard (Rockwell C42). A good commercial finish, which can be obtained with a medium-fine grit such as 54-grit, is required. The specification number is now:

23A54-_____

Since this is a wet grinding operation and the material is moderately hard, a medium grade of hard-ness such as L or M can be selected. If you are in doubt as to what grade of wheel to use, it is usually wise to choose a softer grade that could do the work rather than a harder grade which you might not be able to use. The specification number becomes:

23A54-L_____

A medium structure like #5 would probably be best for a 54-grit wheel and an L grade of hardness. The specification number expands to:

23A54-L5____

The wheel speed is between 5500 sfpm and 6500 sfpm (1650 m/min and 1950 m/min), which would allow the use of a vitrified bond. Since the initial choice of 23 Alundum abrasive was made, a *BE* type of vitrified bond would be best because it com-plements the cutting action of the abrasive, which would ensure that the most pieces of work for each wheel dressing would be produced. The whole specification number for this operation is:

23A54-L5VBE

SUMMARY

If the proper grinding wheel has been selected for the work being done, the wheel can be considered to be relatively self-sharpening. As the abrasive grains become dull, they tend to fracture and fall out because of the friction between the wheel and the work. As a result, new, sharp abrasive grains are exposed.

If the wheel is not self-sharpening, it must be dressed more often with a wheel dresser such as a rotating hand dresser, or abrasive stick, abrasive wheels, or a type of diamond dresser, and valuable time is lost.

The following points have already been made, but should be emphasized once again.

1. Always check the wheel markings for the correct speed. Never run a wheel faster than the marked speed. If you don't know the wheel speed, don't use it. If these points are ignored, the wheel can explode.
2. Before using any wheel, always check it for cracks. If it is cracked, don't use it. Again, the wheel will explode if this point is ignored.
3. Always use a guard. It will protect you from exploding wheels.
4. Never take chances.

REVIEW QUESTIONS

1. Describe two basic types of grinding.
2. What part of the grinding wheel does the actual cutting?
3. What is the relationship between the grit size and speed of grinding on hard materials?
4. List the two basic requirements of the ideal abrasive grain.
5. List four manufactured abrasive grains and give a use for each.
6. How is grit classified as to size? What is the range of the most commonly used grit sizes?
7. List the grit size range for coarse grain wheels, medium grain wheels, fine grain wheels, and very fine grain wheels.
8. What does the word bond mean in relation to a grinding wheel?
9. Describe five different bonds in grinding wheels and the differences between them.
10. Describe the relationship of grade to a grinding wheel. How is grade designated?
11. What is the relationship of structure to the grinding wheel? What is the rating of the most dense structure?
12. List the basic categories in the Norton standard wheel marking system.
13. Name six different types of grinding wheels.
14. List five factors to be considered in the selection of the proper grinding wheel for a specific job.

UNIT 11-2

PRECISION GRINDING

OBJECTIVES

After completing this unit, the student will be able to:

- perform the ring test to check a grinding wheel for cracks.
- set up the work properly on a surface grinder.
- perform surface grinding.
- mount a wheel properly on a surface grinder.
- dress and true a grinding wheel.
- identify and correct problems in precision surface grinding.

KEY TERMS

Surface grinding	Low spots
Surface grinder	Irregular scratches
Ring test	Patterned scratches
Mounting a wheel	Feed lines
Blotters	Vibration lines
Dressing and trueing	Discolored surface
a grinding wheel	Burnished surface
Diamond dresser	Slide scratches
Wheel glazing	Grinding shift
Traverse lines	scratches
Herringbone finish	

INTRODUCTION

One of the greatest advantages of precision grinding is that cuts as small as 0.0005 in (0.01 mm) or even less can be done repeatedly. Both the grinding wheel and the work are securely positioned and their movements are tightly controlled.

SURFACE GRINDING

The surface grinder, Figure 11–13A, is a very precise machine. The work is held on a table under the grinding wheel and is moved back and forth so that the wheel can remove excess material in a very uniform way.

A surface grinder can do many types of grinding, the most common of which is horizontal surface grinding. The simplest type of work is the grinding of a flat block, Figure 11–13B. Other surface grinding operations include the grinding of a flat block with the center removed, Figure 11–14A, the grinding of a profile, Figure 11–14B (the guard is removed to show the operation), and surface grinding a step block, Figure 11–14C.

The basic types of surface grinding machines available are the horizontal spindle and reciprocating table, Figure 11–15A, the horizontal spindle and rotary table, Figure 11–15B, the vertical spindle and reciprocating table, Figure 11–15C, and the vertical spindle and rotary table, Figure 11–15D.

TESTING WHEELS BEFORE MOUNTING

Before mounting a wheel on any type of grinder at any time, it must first be tested for cracks. This is done by performing the ring test. Figure 11–16 (page 478) explains the method used. The wheel is supported and struck. The handle of a screwdriver or a mallet can be used to produce a ringing sound. If there is a crack in the wheel, you will hear a dead sound rather than a clear ring.

MOUNTING AND DRESSING WHEELS

The proper method of mounting a wheel on a spindle is shown in Figure 11–17 (page 479). The purpose of the **blotters** between the flanges and the wheel is to equalize the pressure exerted on the wheel. The faces of the flanges are also undercut so that only the outer edges of the flanges exert pressure on the grinding wheel.

Make certain that the guard is in place and securely fastened. Never grind without a guard, Figures 11–18A and 11–18B (page 479). A flying piece of grinding wheel is a lethal projectile. A guard will effectively contain an exploding wheel.

DRESSING AND TRUEING THE WHEEL

Before dressing and trueing the wheel, Figure 11–19 (page 480), be very sure that the wheel is tight on the spindle. There is no point in trying to dress and true a loose wheel.

The main work-holding device on a surface grinder is a magnetic table or chuck. The diamond dresser, Figure 11–20 (page 481), must be held securely

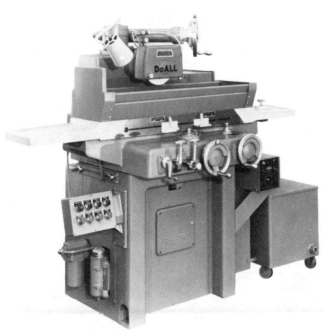

Figure 11–13A Surface grinder *(Courtesy of DoALL Company)*

Figure 11–13B Surface grinding a flat block *(Courtesy of DoALL Company)*

Figure 11–14A Surface grinding a flat block with the center removed *(Courtesy of DoALL Company)*

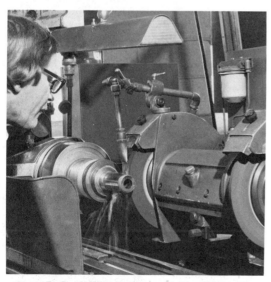

Figure 11–14B Surface grinding a profile. (Note: The guard was removed to show the operation.) *(Courtesy of DoALL Company)*

on a thoroughly clean chuck. The point of the diamond should be set slightly to the left of center, Figure 11–21 (page 481), away from the direction of rotation of the wheel. With this setting, if the diamond moved for any reason, little or no damage would be done to the wheel. The diamond should also be positioned so that dressing starts on the highest point on the wheel periphery. Too deep a cut could cause damage to the wheel or the diamond or both.

A grinding wheel can become dull like any other cutting tool. It can become clogged with dirt or it can become so glazed that it will not cut. Just as a lathe cutter can form a built-up edge, so can the grits of a grinding wheel. Minute pieces of metal cling to the

Figure 11–14C Surface grinding a step block *(Courtesy of DoALL Company)*

cutting points, effectively hindering the cutting action of the wheel. A diamond dresser will rectify this.

Feed the grinding wheel down until it just touches the diamond. Move the diamond across the face of the wheel. Repeat until the wheel is clean, sharp, and square. Feed the wheel down 0.001 in (0.025 mm) for each cut of the diamond.

SETTING THE WORK

1. The table must be absolutely clean before the diamond dresser or the work or anything else is mounted on it. Inspect the work surfaces. Make sure they are clean and free of any sharp projections or ragged edges.

2. If you are grinding dry (without a coolant), it is good practice to place a thin sheet of paper between the work and the table. This will prevent scratching of the table or the work when the completed work is removed. Oil paper strips can be used wet or dry.

3. Mount the work on the table; check to see that the work is secure; set the stops for length of stroke; then feed the wheel down until it almost touches the work.

4. Start the wheel; start the table; feed the wheel down until it is just sparking; move the work across until the wheel has covered the whole surface; then move the work clear of the wheel.

5. Feed the wheel down 0.001 in (0.025 mm) per cut. Repeat until the surface is finished.

6. Redress the wheel as often as necessary.

7. Check with the instructor.

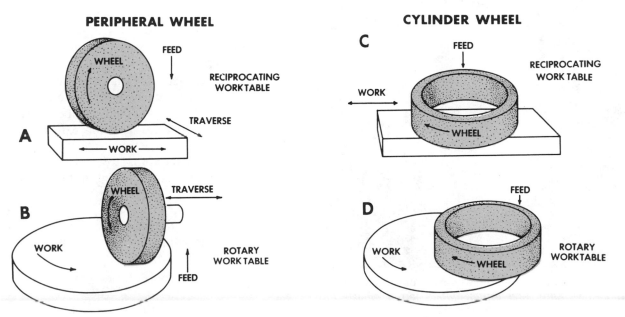

Figure 11–15 Basic types of surface grinding *(Courtesy of DoALL Company)*

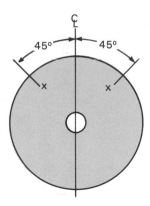

Figure 11–16 Testing grinding wheels. Small wheels can be held in one hand and struck, while larger wheels are suspended as shown. In either case, tap the wheel at the spots indicated by the crosses in the drawing. *(Courtesy of DoALL Company)*

TROUBLESHOOTING PRECISION SURFACE GRINDING

Many problems can occur when you are surface grinding. Following are some of the most common problems and the probable solutions.

1. The work surface is not flat.
   - (a.) Excessive heat can expand the work causing surface low spots to develop. Check the coolant system for dirt, clogged filters, or clogged lines.
   - (b.) Too slow a table speed can cause too much heat buildup in the center of the work. Warping and low spots can occur.
   - (c.) The chuck must be flat to within 0.0001 in (0.0025 mm). If it is not, it must be reground with a relatively coarse wheel.
   - (d.) Too much oil on the ways can cause the table to lift at the center of each reciprocating stroke. The lubricator must be adjusted.

2. Thin work can present problems.
   - (a.) The chuck can exert too much holding power causing the work to distort or warp. It may be possible to reduce the holding power or it may be possible to use blocks to hold the work in place.
   - (b.) Excessive heat can cause work distortion. An improper wheel, poor coolant, incorrect table speed, or too deep a cut are some of the causes.
   - (c.) Work distortion can be due to residue left on the magnetic chuck. Coolant, grit, or burs may be between the chuck and the work.

3. The hydraulic table traverse can sometimes be slow.
   - (a.) The hydraulic system could be cold. Allow some time for the system to warm

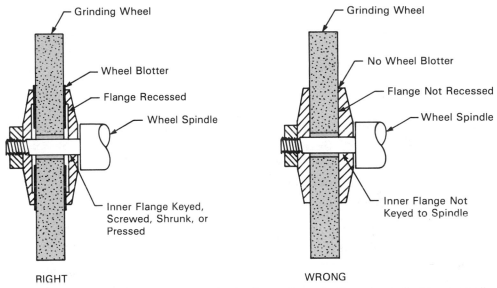

Grinding Wheel

Wheel Blotter

Flange Recessed

Wheel Spindle

Inner Flange Keyed, Screwed, Shrunk, or Pressed

RIGHT

Grinding Wheel

No Wheel Blotter

Flange Not Recessed

Wheel Spindle

Inner Flange Not Keyed to Spindle

WRONG

Figure 11–17 Proper and improper methods of mounting grinding wheels that have small holes *(Courtesy of DoALL Company)*

Figure 11–18A Always use a guard; in this case, the guard is in place.

Figure 11–18B In this case, the guard has been removed to show why a guard is necessary

up. The table should be started gradually.

(b.) The table ways may not be lubricated properly. Check the lubrication system to make sure that the oil is coming through properly.

(c.) Check for a clogged hydraulic system filter.

4. The table traverse may be erratic..

(a.) There may be air in the system or the hydraulic oil level may be low.

(b.) Do not use the wrong hydraulic oil. Check the oil specifications with your local oil agent. Do not use a poor quality oil.

(c.) Check the ways for dirt or poor lubrication.

5. The downfeed handwheel sometimes sticks and becomes difficult to turn.

(a.) Check the grease and oil lubrication on the elevating unit.

(b.) Check for foreign matter or burs between the gear teeth.

(c.) Check the metal shrouds on the elevating unit column to make certain that they are telescoping properly.

6. During the actual grinding operation, the surface formed by the grinding wheel can tell you many things.

(a.) Widely spaced spots are usually caused by glazed spots on the wheel. Proper wheel dressing will rectify this.

(b.) Fine, even spiral lines are caused by a wheel that has been dressed too quickly. Diamond marks on the wheel are transferred to the work surface.

(c.) Uneven, fine lines and isolated deep marks are a result of faulty wheel dressing. To correct this, first make sure that the wheel is tight on the spindle and then redress it properly.

(d.) Wavy traverse lines happen when faulty wheel dressing leaves ragged wheel edges.

(e.) Deep irregular marks are usually caused by a loose wheel.

(f.) A herringbone finish is a result of wheel wobble. Check the wheel mounting and pressure washers for grit or burs. The sides of the wheel may also have to be trued.

(g.) Low spots can be caused by a loaded or glazed wheel that would not cut prop-

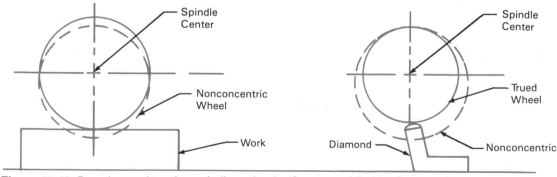

Figure 11–19 Dressing and trueing grinding wheels *(Courtesy of DoALL Company)*

erly, causing overheating and expansion of the work. The wheel may need redressing, or a harder or coarser wheel may be needed. The table speed may be too slow, or there may not be enough coolant, which would cause excessive heating of the work.

(h.) Irregular scratches are formed, but there is no regular pattern. These can be caused by loose particles of abrasive which can fall from the wheel guard or can be carried to the work surface in the coolant. Both the guard and the coolant tank must be cleaned. If the grinding wheel is too soft, the abrasive grains might be released too easily. Too coarse a wheel can also cause irregular scratches.

(i.) Patterned scratches may be due to improper wheel dressing, possibly resulting from a defective or dull diamond dresser. The diamond could be cracked or loose in its mounting. In setting up the diamond dresser, it should sit at a slope of about 15 degrees in the direction of wheel rotation. Dur-

ing the dressing operation, it should pass across the wheel face as slowly and evenly as possible.

(j.) Feed lines can be caused by too deep a cut or too fast a cross feed or a combination of both.

(k.) Vibration marks are caused by a slipping action between the grinding wheel and the work. One reason for this is that one section of the wheel becomes glazed or loaded and stops cutting, slipping on the work surface instead of cutting. Make sure that the wheel is tight on the spindle, and redress the wheel. Other causes can be poor wheel bearings, a defective traverse system, or a wheel out of balance because it is soaked on one side with coolant, a situation that occurs when a wheel is not allowed to spin dry after wet grinding. Outside vibrations from other machinery can

Figure 11–20 Diamond wheel dresser *(Courtesy of DoALL Company)*

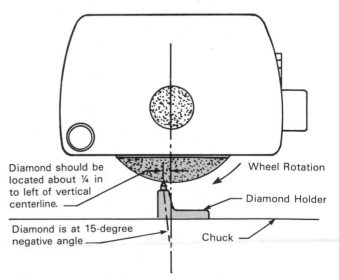

Diamond should be located about ¼ in to left of vertical centerline.

Wheel Rotation

Diamond Holder

Diamond is at 15-degree negative angle

Chuck

Figure 11–21 Diamond dresser set for wheel dressing *(Courtesy of DoALL Company)*

easily result in vibration marks or a lack of rigidity or cushioning in the foundations of the grinder itself.

(l.) A discolored surface is usually a result of overheating the work, even when using sufficient coolant. The solution is to use a softer cutting wheel or to make the wheel that you have act like a softer wheel. This can be done by speeding up the table movement, or by dressing the wheel coarser, or by taking lighter cuts. A generous use of coolant is also important.

(m.) A burnished surface is made smooth and shiny by rubbing. Burnishing can be caused by a glazed grinding wheel, or too fine a wheel, or too hard a wheel, and often happens on broad surfaces.

(n.) Slide scratches can be caused when sliding the work from the chuck after the grinding operation. If you are dry grinding, one solution is to use a thin piece of good-quality paper between the work and the chuck. If you are wet grinding, oil paper strips are often used. Although it is difficult, try to pick the work up starting from the back, with a hinging motion rather than a sliding motion.

(o.) Grinding shift scratches result when the work slides on the chuck during grinding each time a pass is made over the work. The work must be held securely. Block the work, use a vise, or use a combination of these in addition to the holding power of the magnetic chuck. Make certain that the work will not move.

REVIEW QUESTIONS

1. Explain how a grinding wheel is tested before mounting.
2. Why are blotters used when a grinding wheel is mounted between the flanges?
3. How important is the flange on each side of a grinding wheel?
4. Describe how a diamond dresser is mounted for the proper dressing of a grinding wheel.
5. For what reason is a grinding wheel dressed?
6. What feed is recommended for each pass of the diamond dresser?
7. List six precautions to be observed to ensure a safe grinding operation.
8. List three causes of work distortion due to grinding.
9. Irregular scratches are formed on the work. List possible causes for the scratches and the corrective action to be taken.
10. What can cause patterned scratches in the work? What corrective action can be taken?
11. List three causes of vibration marks on the work when grinding. How can these problems be corrected?
12. When a discolored work surface results from grinding, how can the problem be overcome?

UNIT 11-3

OTHER GRINDING OPERATIONS

OBJECTIVES

After completing this unit, the student will be able to:

- describe the operation of cylindrical grinding.
- describe the operation of centerless grinding.
- state the normal wheel speeds used for cylindrical and centerless grinding.
- set up and perform a simple tool and cutter grinding operation.

KEY TERMS

Cylindrical grinding
Plunge grind
Traverse grind
Centerless grinding
Through feed
  grinding machine

Infeed grinding
  machine
Tool and cutter
  grinding

CYLINDRICAL GRINDING

In **cylindrical grinding**, often called O.D. or outside diameter grinding, the work is secured on both ends, between centers, and rotates during the grinding operation. The grinding wheel of a cylindrical grinder, located behind the work, Figure 11–22, rotates and is fed into and away from the work. Either the grinding wheel or the work may be traversed with respect to the other so that grinding may be done continuously from one end of the part to the other. Plunge grinding and traverse grinding are illustrated in Figure 11–23. One method of cylindrical grinding is to use a series of plunge grinds followed by a final traverse grind to bring the work to the required diameter and finish.

Plain cylindrical grinders are used primarily to produce straight cylindrical-shaped parts. With special modifications, however, shapes like cones, shafts with collars or sharp inside or outside corners

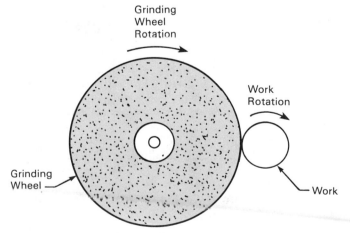

Figure 11–22 Cylindrical grinding *(Courtesy of Norton Company)*

can also be ground. Crankshaft grinding is a modified form of cylindrical grinding. Some of the newer grinding machines feature wheel speeds up to 8500 sfpm (2550 m/min), although conventional grinders usually operate at wheel speeds up to 6500 sfpm (1950 m/min). Work speed should be about 100 sfpm (30 m/min).

Centerless grinding is a type of precision grinding used to produce long, highly accurate, cylindrical shapes quickly and easily. Many types of centerless grinding machines are in use. These include through-feed machines for grinding bar stock and tubing, and infeed machines for grinding work with shoulders or multiple diameters.

A typical grinding operation, Figure 11–24, illustrates an end view showing the relative positions of the grinding wheel at the back of the machine, the regulating wheel, the work, and the support blade. Blade materials used are nonferrous cast alloys, cemented carbide, high-speed steel, cast iron, and meehanite, which is an alloy of cast iron, steel, and other metals. An angular-top support blade is used for the majority of centerless grinding operations.

Most grinding wheels used in centerless grinding machines are vitrified bonded, used at speeds up to 6500 sfpm (1950 m/min). Some of the newer grinders use wheel speeds up to 8500 sfpm (2550 m/min). The average starting speed of a regulating wheel is from 22 to 39 rpm.

TOOL AND CUTTER GRINDING

Tool and cutter grinding is a precision grinding operation that is very highly specialized. The successful sharpening of cutting tools requires a knowledge of the use of the correct grinding wheel, tool setup, and method of grinding for each kind of tool, whether it is a carbide, high-speed steel, or cast alloy tool.

Tool and cutter grinding is used widely to recondition and sharpen cutting tools of all kinds, such as milling cutters, reamers, single-point tooling, saw blades, drills, and shears. This type of grinding utilizes nearly every kind of grinding machine. Suc-

VIEW FROM ABOVE

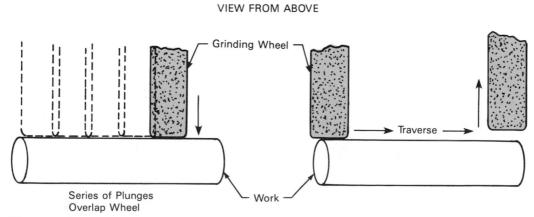

Figure 11–23 Plunge grinding and traverse grinding *(Courtesy of Norton Company)*

cessful machining depends on sharp, well-maintained cutting tools. Correct relief angles and a sharp edge will produce a tool that will maintain its cutting ability for the maximum length of use.

Tool grinding, particularly the offhand grinding of tool bits and drills, has already been discussed in this textbook. Clearance and sharp cutting edges can be ground on a plain milling cutter on a precision tool and cutter grinder as illustrated by the end view in Figure 11–25, using either a straight grinding wheel or a cup wheel. Each of the teeth of the multi-tooth milling cutter must be identical. Precision tool and cutter grinding is the best way to accomplish this. Table 11–1 contains the relief values for sharpening high-speed steel milling cutters. Table 11–2 shows grinding wheel speeds.

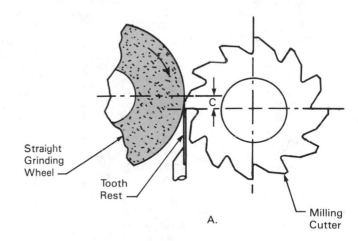

A.

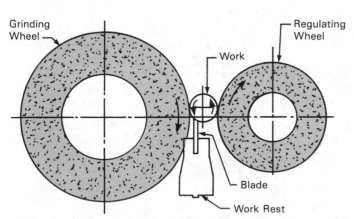

Figure 11–24 A typical centerless grinding setup *(Courtesy of Norton Company)*

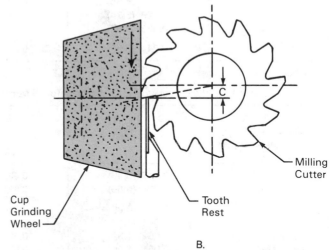

B.

Figure 11–25 Grinding clearance on a multitooth plain milling cutter, using a straight grinding wheel (A.) or a cup wheel (B.) *(Courtesy of Norton Company)*

Table 11–1 Relief or Clearance Tables for Grinding High-Speed Steel Cutting Tools *(Courtesy of Norton Company)*

I—Using Straight Wheels

C = Distance in inches to set center of cutter and tip of tooth rest below (or above) center of wheel when grinding with a straight wheel.

WHEEL DIAMETER (inches)	C FOR 4° RELIEF ANGLE inch (mm)	C FOR 5° RELIEF ANGLE inch (mm)	C FOR 6° RELIEF ANGLE inch (mm)	C FOR 7° RELIEF ANGLE inch (mm)
3	0.104 (2.64)	0.131 (3.33)	0.157 (3.99)	0.183 (4.65)
3¼	0.113 (2.87)	0.141 (3.58)	0.170 (4.32)	0.198 (5.03)
3½	0.122 (3.10)	0.152 (3.86)	0.183 (4.65)	0.213 (5.41)
3¾	0.131 (3.33)	0.163 (4.14)	0.196 (4.98)	0.227 (5.76)
4	0.139 (3.53)	0.174 (4.42)	0.209 (5.31)	0.242 (6.15)
4¼	0.150 (3.81)	0.185 (4.70)	0.222 (5.64)	0.259 (6.58)
4½	0.157 (3.99)	0.195 (4.95)	0.235 (5.97)	0.274 (6.96)
4¾	0.165 (4.19)	0.207 (5.26)	0.248 (6.30)	0.289 (7.34)
5	0.174 (4.42)	0.218 (5.54)	0.261 (6.63)	0.305 (7.75)
5¼	0.183 (4.65)	0.228 (5.79)	0.274 (6.96)	0.319 (8.10)
5½	0.191 (4.85)	0.239 (6.07)	0.287 (7.29)	0.335 (8.51)
5¾	0.200 (5.08)	0.250 (6.35)	0.300 (7.62)	0.350 (8.89)
6	0.209 (5.31)	0.261 (6.63)	0.313 (7.95)	0.365 (9.27)
6¼	0.218 (5.54)	0.272 (6.91)	0.326 (8.28)	0.381 (9.68)
6½	0.226 (5.74)	0.283 (7.19)	0.339 (8.61)	0.396 (10.06)
6¾	0.235 (5.97)	0.294 (7.47)	0.352 (8.94)	0.411 (10.44)
7	0.244 (6.20)	0.305 (7.75)	0.365 (9.27)	0.426 (10.82)

II—Using Cup Wheels

C = Distance in inches to set tip of tooth rest below or above center of cutter when grinding the peripheral teeth of cutters with a cup wheel.

CUTTER DIAMETER (inches)	C FOR 4° RELIEF ANGLE inch (mm)	C FOR 5° RELIEF ANGLE inch (mm)	C FOR 6° RELIEF ANGLE inch (mm)	C FOR 7° RELIEF ANGLE inch (mm)
½	0.017 (0.43)	0.022 (0.56)	0.026 (0.66)	0.031 (0.79)
¾	0.026 (0.66)	0.033 (0.84)	0.040 (1.02)	0.046 (1.17)
1	0.035 (0.89)	0.044 (1.12)	0.053 (1.35)	0.061 (1.55)
1¼	0.044 (1.12)	0.055 (1.40)	0.066 (1.68)	0.077 (1.95)
1½	0.052 (1.32)	0.066 (1.68)	0.079 (2.01)	0.092 (2.34)
1¾	0.061 (1.55)	0.076 (1.93)	0.092 (2.34)	0.108 (2.74)
2	0.070 (1.78)	0.087 (2.21)	0.105 (2.67)	0.123 (3.12)
2½	0.087 (2.21)	0.109 (2.77)	0.131 (3.33)	0.153 (3.89)
2¾	0.096 (2.44)	0.120 (3.05)	0.144 (3.66)	0.168 (4.27)
3	0.104 (2.64)	0.131 (3.33)	0.158 (4.01)	0.184 (4.67)
3½	0.122 (3.10)	0.153 (3.89)	0.184 (4.67)	0.215 (5.46)
4	0.139 (3.53)	0.174 (4.42)	0.210 (5.33)	0.245 (6.22)
4½	0.157 (3.99)	0.197 (5.00)	0.237 (6.02)	0.276 (7.01)
5	0.174 (4.42)	0.219 (5.56)	0.263 (6.68)	0.307 (7.80)
5½	0.192 (4.88)	0.241 (6.12)	0.289 (7.34)	0.338 (8.58)
6	0.207 (5.26)	0.262 (6.65)	0.315 (8.00)	0.368 (9.35)

44 GRINDING **487**

Table 11-2 Grinding Wheel Speeds *(Courtesy of Norton Company)*

To find the number of revolutions of the wheel spindle, having been given the surface or peripheral speed and the diameter of the wheel, divide the surface speed in feet per minute by the circumference (diameter x 3.14) in feet.

To find the surface speed of a wheel in feet per minute, multiply the circumference in feet by the revolutions per minute.

Diam. of Wheel in Inches	mm (Approx.)	Peripheral Speed in Feet per Minute					
		4000 Ft 1200 m	4500 Ft 1350 m	5000 Ft 1500 m	5500 Ft 1650 m	6000 Ft 1800 m	6500 Ft 1950 m
		Revolutions per Minute					
1/4	6	61,116	68,756	76,392	84,032	91,672	99,212
3/8	9	40,744	46,594	50,928	56,021	61,115	66,141
1/2	13	30,558	34,378	38,196	42,016	45,836	49,656
5/8	16	24,446	27,502	30,557	33,615	36,669	39,685
3/4	19	20,372	22,918	25,464	28,011	30,557	33,071
7/8	22	17,462	21,826	21,826	24,009	26,192	28,346
1	25	15,279	17,189	19,098	21,008	22,918	24,828
2	50	7,639	8,594	9,549	10,504	11,459	12,414
3	75	5,093	5,729	6,366	7,003	7,639	8,276
4	100	3,820	4,297	4,775	5,252	5,729	6,207
5	125	3,056	3,438	3,820	4,202	4,584	4,966
6	150	2,546	2,865	3,183	3,501	3,820	4,138
7	175	2,183	2,455	2,728	3,001	3,274	3,547
8	200	1,910	2,148	2,387	2,626	2,865	3,103
10	250	1,528	1,719	1,910	2,101	2,292	2,483
12	305	1,273	1,432	1,591	1,751	1,910	2,069
14	355	1,091	1,228	1,364	1,500	1,637	1,773
16	405	955	1,074	1,194	1,313	1,432	1,552
18	455	849	955	1,061	1,167	1,273	1,379
20	505	764	859	955	1,050	1,146	1,241
22	560	694	781	868	955	1,042	1,128
24	610	637	716	796	875	955	1,034
26	660	588	661	734	808	881	955
28	710	546	614	682	750	818	887
30	760	509	573	637	700	764	828
32	810	477	537	597	656	716	776
34	860	449	505	562	618	674	730
36	910	424	477	530	583	637	690

LASER MACHINING ON A METAL LATHE

The use of a laser beam as a cutting tool is a several-year-old technology. Laser machining operations such as slotting and drilling have been done successfully, although some extra attention usually must be given to surface finish. In fairness, though, finishing is an operation that also follows many conventional cutting operations.

Machine tool vibration, cutting tool wear, and pressures exerted by the cutting tool are not factors in laser beam machining as they are in conventional machining. A laser beam is a beam of light, not a conventional cutting tool. Therefore, no pressures are exerted either on the work or on the cutting tool. There is no machine vibration or chatter from the cutting action and there is no cutting tool wear.

One of the limitations with laser machining is that the cutting speed is not fast, especially when compared to some conventional tooling such as carbide and coated carbide cutting tools. At the present time, conventional carbide and coated carbide cutting tools simply remove more metal faster. At one time, it was difficult to produce a three-dimensional shape with a laser beam. This limitation is being overcome as further development of the process continues.

Laser machining on a lathe is a developing technology. One possible method of using a laser beam as the cutting tool is to split the beam into two beams. One beam, known as the axial beam, is directed parallel to the centerline of the work. The second beam, or the radial beam, is directed perpendicular to the centerline of the work. The radial beam cuts toward the centerline from the outside diameter to the required diameter of the workpiece as the job is revolving. The axial beam cuts parallel to the centerline from the end of the workpiece toward the headstock to intersect the radial beam. When the two cuts meet, a solid section of material can be removed. The beams are set for a second cut, and again a solid section of material is removed. The operation is continued until all of the necessary material is removed from the workpiece. This operation demonstrates that it is feasible to use a laser as a cutting tool to turn a shoulder. Figure 1 illustrates the turning of a shoulder using both conventional turning and laser beam turning.

This same method can be considered for turning tapers, as outlined in Figure 2. At this stage in the development of laser turning it appears that it is possible to turn any of the materials listed in the article, Laser Cutting and Machining. Chip removal and removal of the solid rings of material are problems that must be addressed. Additional development should readily solve these details.

If turning can be done by removing a solid section of material, then cutting off stock on a

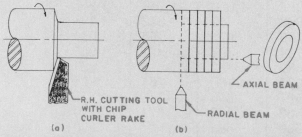

Figure 1 Turning a shoulder by conventional turning and laser beam turning

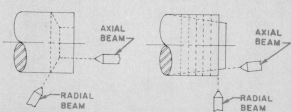

Figure 2 Taper turning with a laser beam

lathe should be possible using a single laser beam, as shown in Figure 3. Limitations may exist for the maximum diameter that can be cut off. Smooth cuts in material 3/4 inch and more in diameter are obtainable now, at least in the cutting of sheet material, using a CO_2 gas assisted laser beam. At this time, experimental laser beam turning has used a 1200-W CO_2 laser beam as the cutting tool.

Another very important point in laser turning is that the support of a tailstock is not required, since there is no pressure exerted on the work by the cutting instrument. The lack of pressure alone eliminates many of the problems encountered now by students in the conventional cutting off of materials on a lathe.

The laser lathe is a new concept in the turning of materials. At the present time, many of the operations that a conventional lathe can do with ease, such as the boring of deep holes and the cutting of screw threads, are difficult to do on a laser lathe. Since this is a thermal machining process in that heat does the actual cutting, the resulting surface finish will require extra attention. In fact, for some of the super finishes that are required today, extra finishing operations and setups will be needed.

As stated earlier, one of the major limitations of the laser lathe is that of cutting speed. Conventional cutting methods are much faster and much less expensive. There are, however, definite advantages to laser turning. Since the laser light beam is so easy to control and it can be directed virtually anywhere in the shop, any number of laser machine tools can be supplied with power from a single source as shown in Figure 4. Individual machine tools will not require their own power source. The actual lathes will be much lighter in weight and simpler in construction mainly because no pressure will be exerted by the cutting instrument and cutting tool vibration or chatter will not exist. It is certain that the laser lathe will become an important manufacturing tool in the near future.

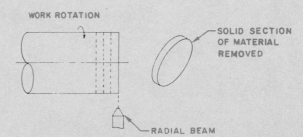

Figure 3 Cutting off stock on a lathe with a laser beam

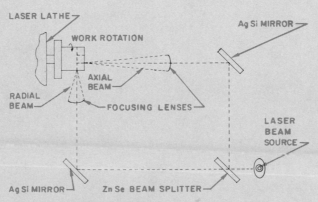

Figure 4 Supplying power to a laser lathe

REVIEW QUESTIONS

1. Briefly describe cylindrical grinding.
2. Explain plunge grinding and traverse grinding.
3. What is centerless grinding?
4. What wheel speeds are common in cylindrical and centerless grinding?
5. What information must be known if tool and cutter grinding is to be performed correctly?

Hand dressing a cup wheel with a dressing stick on a surface grinder. *Courtesy of Norton Company*

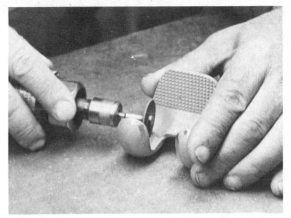

Hand grinding a small casting. *Courtesy of Norton Company*

Hand grinding a small casting. *Courtesy of Norton Company*

Hand grinding with a mounted point. *Courtesy of Norton Company*

Hand grinding a fillet weld with a mounted point. *Courtesy of Norton Company*

Hand grinding castings in a foundry chipping room. *Courtesy of Norton Company*

Hand grinding with a disc grinder. *Courtesy of Norton Company*

Cutting off sprues in a foundry. *Courtesy of Norton Company*

Surface grinding on a rotary surface grinder. *Courtesy of Norton Company*

Internal grinding. *Courtesy of Norton Company*

Squaring the end of a piece of steel on a face wheel. *Courtesy of Norton Company*

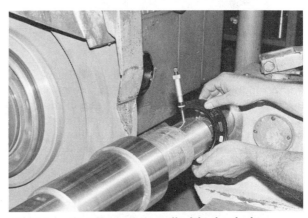

Checking diameters on a cylindrical grinder. *Courtesy of Norton Company*

Cylindrical grinding a straight shaft. *Courtesy of Norton Company*

Cylindrical grinding a stepped shaft. *Courtesy of Norton Company*

Drill sharpening on an automatic drill sharpening machine. As the drill rotates, the grinding face wheel oscillates to give each cutting lip the required clearance. This machine can be set for two, three, or four flute drills. *Courtesy of Cominco Ltd.*

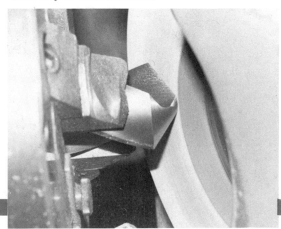

Grinding the flutes of a threading tap on a tool and cutter grinder. *Courtesy of Norton Company*

Setting up to grind the cutting lips of an end milling cutter on a tool and cutter grinder. *Courtesy of Norton Company*

Indexing a two-flute end milling cutter for sharpening both cutting lips. *Courtesy of Cominco Ltd.*

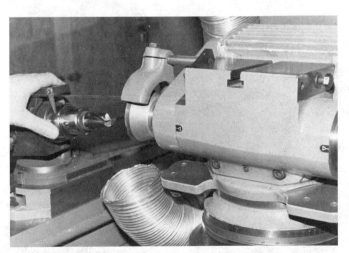

Sharpening the cutting lips on a two-flute end milling cutter, showing the indexing mechanism. *Courtesy of Cominco Ltd.*

Sharpening the cutting lips of a two-flute end milling cutter. *Courtesy of Cominco Ltd.*

Cutting off the old center of a reground end milling cutter on a tool and cutter grinder. *Courtesy of Norton Company*

Dipping the reconditioned end milling cutters in liquid oil-treated plastic. The liquid plastic firms up quickly to protect the milling cutters from damage. It is easily peeled off when the cutters are needed. The oil protects the cutters from rusting. *Courtesy of Cominco Ltd.*

CHAPTER TWELVE

NUMERICAL CONTROL (N/C), COMPUTER NUMERICAL CONTROL (CNC), AND ROBOTICS

Each of the processes described in this chapter has played an important role in the development of modern machining technology. They continue to contribute to the evolution of manufacturing. The original numerical control (N/C) process brought to industry the ability to control the movements of machine tools by means of programmed numerical data that was electronically converted to machine movements. The addition of a computer to numerical control (CNC) added much more flexibility and speed to the process. CNC provided industry with the means of producing all manner of parts automatically and with extreme accuracy.

Robots are also playing an increasing role in manufacturing. In the early stages of development, they were used to load work into a machine tool for machining and then to unload the finished part. In modern manufacturing applications, that role has been extended to include transporting work automatically to a machine tool, removing work from a machine tool, and then transporting it to a storage area. Robots are also used to transport tooling for machine tools. New functions for robots will be added as further development occurs in the years to come in manufacturing technology.

UNIT 12-1

NUMERICAL CONTROL (N/C)

OBJECTIVES

After completing this unit, the student will be able to:

- define the numerical control process, including listing the major components in a numerical control system.
- explain how an EIA or ASCII tape is used to deliver commands to a machine tool to cause it to perform specified operations.
- define absolute programming and incremental programming.
- state the machine axis nomenclature with appropriate direction signs.
- list advantages of N/C and CNC machine tools in manufacturing.
- describe the typical machining practices for N/C and CNC lathes.

KEY TERMS

Numerical control
  (N/C)
Standard code
Electronic Industry
  Association (EIA)
American Standard
  Code for Information
  Interchange (ASCII)
N/C control unit
Track or channel
Command memory
  circuit
Block of instructions
Zero position
End of block (EB)
Flexowriter

Binary code
Eight-channel tape
End of line (EL)
Odd parity
Even parity
Preparatory code
Absolute programming
Incremental
  programming
Axis orientation
X axis
Y axis
Z axis
Word address letter
Setup point or
  home position

INTRODUCTION

Numerical control or **N/C**, as it is more commonly called, is one of the most important developments in the machining industry and dates from about 1950. It is a method of automatically producing machine parts with extreme accuracy and with a minimum of lost time or lost motion. An N/C machine is just as accurate at the end of a work shift as it was at the beginning of the shift. There is little lost time or lost motion during the various machining operations as would be inevitable with manual equipment. The operator becomes more tired and loses concentration on manual equipment. On N/C equipment, the machine consistently repeats its cycle. It is faster, more consistently accurate, and more predictable. Many people think of N/C in terms of large quantity production only. N/C is, in fact, sometimes even better suited to small lot production, one significant factor being that of reduced setup time.

As with any machining operation, N/C starts with a drawing of the required part. A program is then made up in the form of a chart. The program must note each machine movement, each distance and direction covered by the cutter or cutters, and each of the required feeds and speeds necessary to make the required part. This information is then transferred to a tape that is punched with holes according to a standard code developed by the Electronics Industry Association (EIA). This code is one of the main standards in N/C machining. American Standard Code for Information Interchange (ASCII) is another standard.

The workpiece is set up; the required tooling is set up and checked; and then the punched tape is placed into an N/C control unit, usually called the **control**, Figure 12–1, which reads the tape and converts the tape information into machine instructions to perform the various machine operations. The control is the heart of the N/C operations.

TAPES AND CODES

N/C machines usually receive their signals from holes punched into a tape. A standard N/C tape, Figure 12–2, is 1.000 in plus or minus 0.003 in (25.4 mm plus or minus 0.077 mm) wide and is usually 0.004 in plus or minus 0.0003 in (0.10 mm plus or minus 0.007 mm) thick. There are eight tracks or channels numbered one to eight, plus a row of sprocket holes between channels three and four, which moves the tape and feeds it through the reader in the control. A light shines down onto the tape passing first through a round prism. The prism causes the light beam to focus into a thin straight line across the width of the tape. Wherever there is a hole punched in one or more of the channels, the light passes through the hole activating a diode or switch below the tape. It would be the same as closing the switch. This causes an electrical impulse that is decoded and stored in a command memory circuit in blocks of information much like a calculator would collect a series of numbers for addition. Once all the numbers are collected, the calculator gives a total.

Similarly, in the command memory circuit, once a block of instructions is received it can be acted upon. These instructions are then passed to the various motors on the turning center (lathe), or to the machining center, or to the N/C milling machine, depending on the particular machining operation being used.

A block of instructions may tell a cutter to start from a zero position, turn a shoulder of a certain diameter and length at a certain speed, and return to the zero position. At this point, an **end of block (EB)** would be signalled and the next block of instructions

Figure 12–1 A numerical control, usually referred to as the control *(Courtesy of General Electric Company)*

would be read and acted upon. According to the Electronics Industry Association (EIA) Code, an EB is a single hole punched in channel eight. At the end of a completed program, after all the blocks of information have been acted upon, the control could command the tape to be rewound, and if no more parts were to be made, there would be no further tool movement.

On a flexowriter automatic writing machine, coded instructions, e.g., N001 G04 X02 S38 T11 M03, indicating a sample block of instructions, would be typed out. As this is being typed out, the machine is punching out the tape by punching holes in the various columns as needed. N001 is a sequence number that identifies Block 1 (first block of instructions). G04 is a preparatory code for dwell (pause or delay of the movement). X02 is the amount of dwell specified by the X word, in this case, 2 seconds. S38 is the code for the spindle speed. M03 turns the spindle clockwise. T11 indicates the turret, turret position, and tool offset assignment.

Tapes are usually punched using a **binary code**, *binary* meaning two. If a signal is sent to an electric circuit, there are only two possibilities, the circuit is either "on" or "off." In N/C, it is understood that "zero" is an "off" circuit and "one" is an "on" circuit. In a tape, if a hole is punched, a "one" is understood; if no hole is punched, a "zero" is understood. In another sense, a punched hole can designate a "minus" sign and absence of a hole can designate a "plus" sign.

In a standard eight-hole or eight-channel tape, six holes are used for information or data. The seventh hole is used as a check where coding for a character or symbol would normally contain an even number of coded holes. An extra hole is punched in the check row (usually channel five) to make the total number of code holes an odd number as is required in the EIA code. This permits the detection of improper code combinations and reduces the possibility of an error in the system. The eighth hole is used for only one purpose, either an "end of block" (EB) or an "end of the line" (EL).

It is standard that an EIA (Electronics Industry

Association) tape always contains an odd number of holes (odd parity) punched across eight channels. An ASCII (American Standard Code for Information Interchange) tape always contains an even number of holes (even parity) punched across the eight channels. EIA coding was established before ASCII codes and was comprised of 63 different characters including numbers, letters, and symbols. The letters were all lower case. The computer industry needed more letters and numbers so the ASCII code was developed and was comprised of 128 different characters, lower and upper case letters and numbers. Both codes are used today in N/C and CNC machining. The EIA code is used on older programming and controls. Machine tools today usually can be switched to read either code. Code lettering is all upper case today.

PROGRAMMING: ABSOLUTE AND INCREMENTAL

When a machine is equipped with the absolute or incremental option, preparatory commands (G codes) are used to select the mode of operation; G90 will select absolute programming and G91 will select incremental programming.

In **absolute programming**, Figure 12–3, the control receives the following instructions: Start at the origin and move z + 3; refer to the origin and move z + 5; refer to the origin and move z +7. In absolute programming, the origin is used as the dimensional reference point for each move.

A G90 code sets the control up to accept data that is programmed in the absolute mode. The zero or origin is used as the dimensional reference point for each move. On some machines, the zero can be selected anywhere on the work. Most common is a zero at the left and/or right end of the work and the work centerline. If the zero is selected at the left end of the work, all commands will be plus (+). If the zero is at the right end, all commands will be minus (–).

When a tool is in its index position for tool change, we have to know the exact location of the tool edge with respect to zero. The distances are deter-

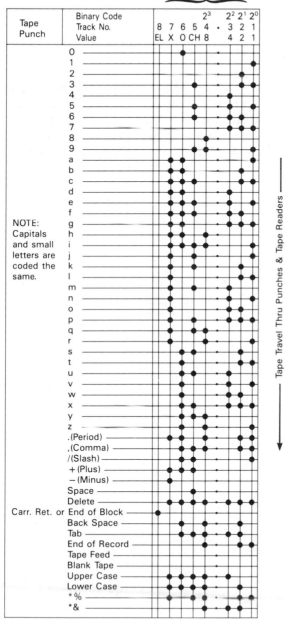

*Not EIA Codes

Figure 12–2 A numerical control tape *(Courtesy of General Electric Company)*

mined by the turret's home position and the distance *GLX* and *GLZ*, Figure 12–4. *GLX* is the distance from the centerline of the turret to the nose of the cutting tool, measured along the x-axis. *GLZ* is the distance from the center of the turret to the nose of the cutting tool, measured along the z-axis.

In absolute programming, these distances are used to preset the x and z position registers. This is done by programming a G92 code with the x distance as the x command and the z distance as the z command. G92 sets up the control to accept data that is absolute programmed. The control knows then exactly where the tool edge is located in respect to the zero. All commands then are referenced from zero.

In **incremental measuring**, Figure 12–5, the dimensional reference point for each move is the last previous position. The instruction would be: Start at the origin and move z + 3, from that point move z + 2, and from that point move another z + 2.

A G91 sets up the control to accept data that is incrementally programmed. In incremental programming, the dimensional reference point for each move is the last previous position. When programming is incremental, it is advantageous to keep an absolute column on the programming sheet. This column will tell where the tool point is located at all times with respect to the zero.

A program can be seen as several tapes in series with each other. Each tool has its own section of the tape. To find out if any errors have been made in the program, add up all plus or minus commands in the x and z columns. The total in x or z has to be zero.

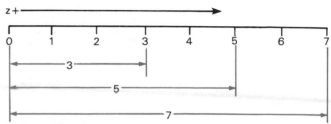

Figure 12–3 Absolute measuring where zero (the origin) is used as the reference point for each measurement

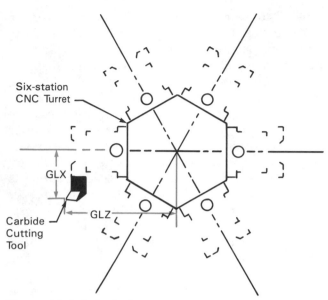

Figure 12-4 A vertical turret

AXIS ORIENTATION

In a two-axis turning center, aside from its rotation, the workpiece does not move in any way. The tool moves in two ways, toward or away from the centerline of the work (along the x-axis) and from side to side between headstock and tailstock along the z-axis, Figure 12-6. The z-axis is specified as running along the center of the spindle on any machine. Since the machine has two axes of control, a word address letter must be assigned to each axis to ensure that dimensional information is assigned to the correct slide.

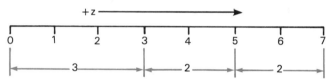

Figure 12-5 Incremental measuring where the reference point for the next step is the last previous position

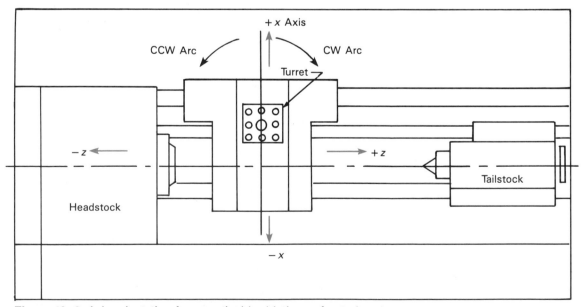

Figure 12-6 Axis orientation for a vertical bed lathe — front elevation

The machine axis nomenclature and direction signs have been assigned as follows: the longitudinal motion is the z axis, and the cross slide is the x axis. Longitudinal motion toward the headstock is the z – (minus) direction, and the motion toward the right is the z + (plus) direction. Similarly, cross-slide motion toward the machine centerline is the x + (plus) direction, and the opposite is x – (minus).

The axis orientation for a horizontal bed lathe is given in Figure 12–7.

The z axis on any machine is the centerline of the spindle. This is standard. The zero point or origin where the x, y, and z axes meet is the point from which all coordinate dimensions are measured, as on a vertical milling machine, Figure 12–8. The setup point may be the intersection of two machined edges or the center of a previously machined hole. This point is generally referred to as the **home** position.

MACHINING WITH N/C

N/C and CNC machine tools are meant for heavy metal removal, precision finishing, and fast cycling of the work from start to finish. The turning center (lathe),

Figure 12–9, is a modern N/C machine tool. The guards are removed, showing the chuck and a six-tool vertical turret. The cutter cuts on top of the work rather than on the side as with a conventional lathe. Chips fall into the chip pan for easy cleanout either manually or with an automatic chip conveyor. Another factor is that the chips do not have a chance to scratch the work surface of the work because they fall so naturally into the chip pan. Behind the door of the turning center, Figure 12–10A, can be seen an eight-tool, upper, outside diameter (OD) turret and a specially designed four-tool shaft turning turret for work between centers for rough and finish turning, grooving, threading, and related operations. Figure 12–10B (page 506) shows the same lathe with the same eight-tool upper turret, but with a lower inside diameter (ID) turret substituted for the shaft turning turret. The two turrets (upper and lower) are mounted on a common slide. Tools in both turrets mount directly and simply.

In a heavy-duty turning operation, Figure 12–11 (page 506), it is easy to see the deep cuts, the shape of the chips which resemble a number nine, and the natural fall of the chips into the chip pan. The boring operation, Figure 12–12 (page 506), shows the cutter

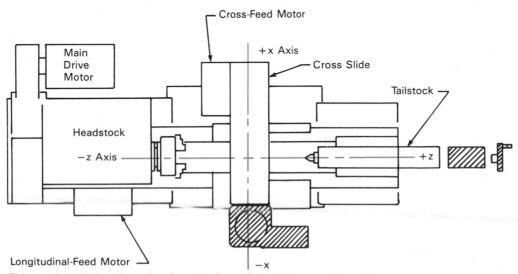

Figure 12–7 Axis orientation for a horizontal bed lathe — plan view

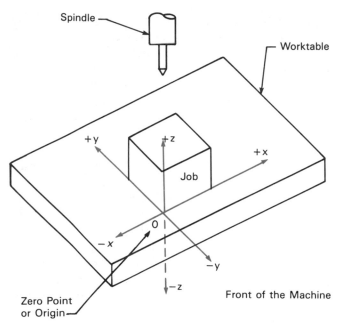

Figure 12-8 X, Y, and Z axes for a vertical milling machine

at the top of the bore. Chips fall away leaving an unobstructed view of the point of the cut. Chip jams that can mar fine finishes are virtually eliminated. A cushioned stock stop, Figure 12–13, is used with a push-through bar system to position the stock in the chuck. Drilling bar stock, Figure 12–14, is a typical drilling operation. The upper OD turret, Figure 12–15, is used for parting or cutting off stock. In the threading operation, Figure 12–16 (page 508), note how rigidly the threading cutter is being held by the tool holder. Also note how the chips would fall naturally into the chip pan and how it is impossible for them to jam in the thread being cut.

Machining Practices

Following are typical machining practices for N/C and CNC lathes:

    1. Use positive rake tooling on aluminum, on

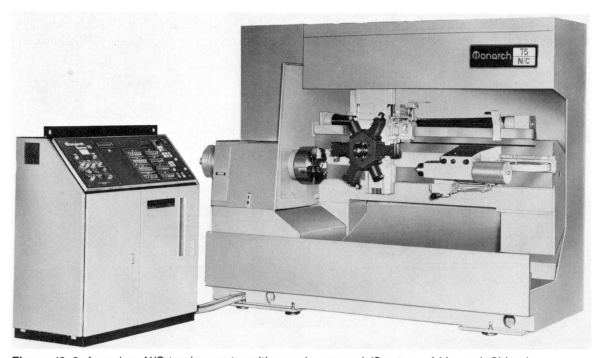

Figure 12-9 A modern N/C turning center with guards removed *(Courtesy of Monarch Sidney)*

very soft steels, and on heat-treated steels.

2. Use a coolant wherever possible.
3. Be cautious when using 80-degree tools on contouring because the insert may move in the tool holder. Eighty-degree tools are for turning and facing operations.
4. It is preferable to use 5-degree, negative-rake tools. A 7-degree negative rake produces too much chip load.
5. Use 1/32-in (0.787-mm) tool nose radius on most applications.
6. Break all corners on the work. It is best to lead out of material on an angle.
7. On boring bars, keep the bar diameter to length ratio to within a 1.5 to 5.0 ratio. Allow

for coolant and chip removal.

8. For profiling, use 3 degrees negative lead and allow 0.005-in (0.127-mm) stock on shoulders.
9. Use a maximum allowable flat on single-point threading tools. Reduce the in-feed (29 or 30 degrees on a V-thread) on each pass, in proportion as the contact area increases.
10. On forgings and castings, use a lighter cut on the first pass due to run-out and stock variation.
11. Use indexable inserts wherever possible. Make sure the insert is held securely and use a clamp on the insert where possible.
12. Keep tools in order of use in the turret.
13. Back up drills, boring bars, and so on with

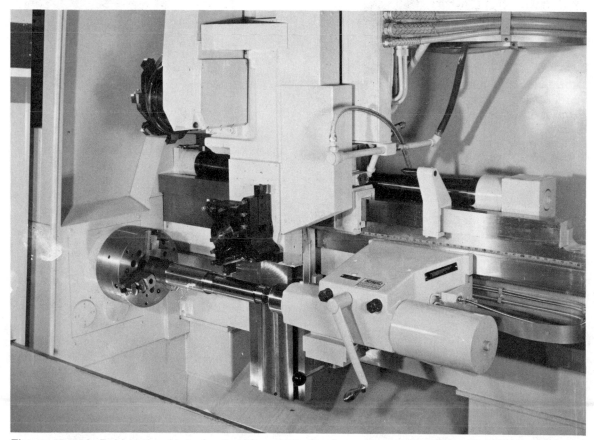

Figure 12–10A Behind the door of a turning center showing a turning operation *(Courtesy of Monarch Sidney)*

Figure 12–10B A lower ID turret set for boring. The upper OD turret is not being used at the moment. *(Courtesy of Monarch Sidney)*

Figure 12–12 ID (inside diameter) cutting (boring) *(Courtesy of Monarch Sidney)*

Figure 12–11 A heavy-duty turning operation *(Courtesy of Cincinnati Milacron)*

Figure 12–13 Positioning stock against a cushioned stock stop *(Courtesy of Monarch Sidney)*

Figure 12–14 Drilling bar stock *(Courtesy of Monarch Sidney)*

plugs. Keep flange bushings back against the face of the holder.

14. Use proper shank length (qualified tooling preferred) to that the tool contacts the sides in the holder. Be sure that all the hold-down bolts are used. **Qualified tooling** is tooling made almost exactly to size within close tolerances. Unqualified tooling is not made to close tolerances. The width and length of a turning tool made to within plus or minus 0.005 inch (0.13 mm) would be a qualified tool.

15. Use spade drills where practical. They are better for both chip removal and coolant use in the hole.

Figure 12–15A Using the upper OD turret for grooving or parting (cutting off) stock *(Courtesy of Monarch Sidney)*

Figure 12–15B Another high-speed grooving operation *(Courtesy of Monarch Sidney)*

16. Determine a method of chucking so that the work is able to be chucked a second time.
17. With twist drills, do not drill deeper than the flute length. At a depth of three times the diameter of the drill, withdraw the drill to clean the flutes. Then withdraw the drill for every one and one-half times the diameter of the drill in depth.
18. Check the tooling in the turret after the machining of every part on the initial setup.
19. Run titanium-coated inserts one and one-half times faster on roughing and finishing speeds over normal carbide grades.
20. Hone carbide insert edges for greater shock applications. On titanium-coated inserts, hone the edge behind the insert radius but not the tool radius.
21. Shaft diameter to length ratios of one to twelve or less should be used for standard turning. Do not keep the tailstock quill extended any farther than necessary.

Figure 12–16 A threading operation *(Courtesy of Monarch Sidney)*

REVIEW QUESTIONS

1. What is N/C? List the major steps in setting up for N/C.
2. List three advantages of N/C machining as compared to machining with manual equipment.
3. What is the control and what are its functions?
4. In N/C a punched tape delivers the information required to perform the machining operations. Briefly describe how the information is transmitted to the machine.
5. What is a block of instructions?
6. Describe EIA coding. How does it differ from the ASCII code?
7. What is absolute programming?
8. What is incremental programming?
9. State the assigned machine axis nomenclature and direction signs.
10. What is the standard Z axis on any machine?
11. What is a two-axis turn center?
12. List six common machining practices for N/C and CNC lathes.
13. What is meant by qualified tooling?

UNIT 12-2

COMPUTER NUMERICAL CONTROL (CNC)

OBJECTIVES

After completing this unit, the student will be able to:

- define the term computer numerical control.
- describe the differences between N/C and CNC.
- list the advantages of computer numerical control.
- read a monitor and run through a program following the computer prompts.
- define the term computer graphics.
- describe applications of computer graphics.

KEY TERMS

Computer numerical control (CNC)
Computer
Computer memory
Program
Cassette tapes
Computer diskettes
Computer keyboard

Programmer
Preprogrammed routine
Computer terminal
Computer prompts
Computer graphics
Word abbreviations

INTRODUCTION

It is difficult to buy an N/C machine today in the mid-1980s. They are no longer on the market as such. Computers have been added to them and the machining centers are now called **Computer Numerical Control** or **CNC**.

A **computer** is an electronic machine that collects and stores data in memory banks. It can sort out the data very quickly and transmit that data in the form of electrical impulses to electric switches and motors to switch machines on and off and activate motors to move cutters. The machine may be a lathe or a machining center.

The major difference between N/C and CNC is that in N/C the tape controls the entire program. The N/C machine must follow the dictates of the tape. In CNC, the tape places all of its information in the computer memory and the computer then controls the program. The tape can be removed and stored once the computer has been programmed. Cassette tapes and computer diskettes are also used for storing information. Changes in the program can be made directly on the computer keyboard. It must be remembered, however, that any career in CNC requires a solid understanding of both the fundamentals of N/C and the basics of machining.

A computer can do many things. It can handle vast amounts of facts and figures and it can solve complicated problems at incredibly high speeds, but a computer can't think for itself. An operator or programmer has to instruct the computer in exactly what to do with the data it receives. These instructions are called a **program**.

We have already studied the basic N/C control unit, which is actually a type of computer in that it will collect data from a standard tape that is fed through the control. That data is then transmitted to the machining center itself. The tape, in this case, can be programmed manually. If the data is calculated manually it can take hours of work. The computer is capable of cutting the time required by the programmer by about two-thirds, thus reducing costs to less than half.

Throughout the United States (e.g., Cleveland, Ohio), there are computer program centers that contain host or central computers. These centers are actually like libraries containing data with prepro-

grammed routines such as simple rough and finish turning, facing, boring, tapering, threading, and cutting curves. What may be missing from a particular preprogrammed routine are the dimensions of a particular part that is being made at a particular factory. These necessary dimensions to make the part can be transmitted to the central computer by telecommunications directly from the machine shop office of the factory. Very quickly, the programmed data would be returned and fed into a computer terminal that would make the control tape. The control tape would then be placed in the N/C control for the transmission of data to the machining center itself.

A computer saves the programmer many hours of calculating, planning, and fitting dimensions to machining routines. An early refinement of the system was to have a computer, Figure 12–17, mounted directly

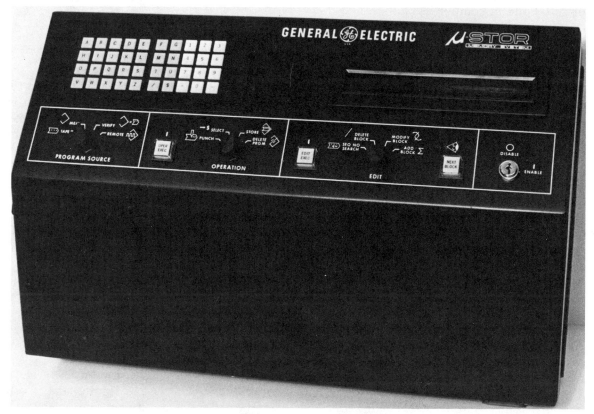

Figure 12–17 An N/C computer *(Courtesy of General Electric Company)*

on the N/C control and wired into the system so that, in effect, it became a computer control. The computer accepts the programmed data, transfers that data to the control, and feeds the information to the machining center. More modern computer controls, Figure 12–18, are much more compact and can do much more than the early models. They also include a monitor or screen which earlier computers did not have.

Today, controls like the Cinmill, Figure 12–19A, shown installed on a vertical milling center, are designed to make the programming of machining centers faster and easier. This enables the operator to concentrate on the part being made rather than on the programming. An expert machine operator is not necessarily an expert programmer. Figure 12–19B shows the Cinmill control with the keyboard in the center and the cathode-ray tube (screen) above and to the left.

To program the making of a part, the operator needs to have an understanding of blueprints and the basic information included on the blueprint of the part. The control uses prompts to guide the operator, step by step, through initial setup, material definition, feeds and speeds, clamping, and machining operations. Once the operator defines the operation and where it is to take place, the control automatically determines the tools to be used, feeds and speeds of each tool, and the machining sequence for that operation. If a situation arises where standard machining practices don't apply, the control allows changes to be made in the program to rectify the situation. The necessary information is typed in on the keyboard.

COMPUTER GRAPHICS

Computer **graphics** plays a prominent role in CNC machining. Graphics is a way of producing a picture or chart directly on the screen of the CRT (cathode-ray tube). The computer converts the written data to a picture. The different steps of a program such as initial setup, definition of materials, clamping the work to the worktable, machining sequences, and so on can each be shown graphically, using just the screen of the Cinmill control to illustrate them.

Definition of materials, Figure 12–20A, is one application of graphics as used in a machining program. From the blueprint the operator selects the material for the part (an SAE 1040 steel). This is a carbon steel containing 40-point or 0.4% carbon, which is shown on the screen as part of a chart. The appropriate feeds, speeds, and coolant to be used for each tool type are automatically determined by the computer and are also included in the chart.

Clamping the work to the worktable, Figure 12–20B (page 514), shows another example of graphics. The picture illustrates the best position for the clamps used in the setup. Clamping is a necessary part of any machining center operation. The control provides for a definition of up to nine clamps and their locations. As each clamp is defined, it is displayed in a window separate from the overall view of the part, making it very clear to the operator which clamp he is working with, Figure 12–20C (page 514). Once the clamp is defined, the control will ensure that during the actual machining, no cutting tool will enter an area occupied by a clamp.

Figure 12–18 A modern computer control *(Courtesy of General Electric Company)*

Using the information provided in the preceding groups, the control can determine optimum cutting tool motions for all types of machining operations. These operations are also presented to the operator in logical sequence, providing for the machining of pockets, Figure 12–20D, grid hole patterns, Figure 12–20E, the end mills used, Figure 12–20F (page 516), and area of mill path, Figure 12–20G (page 516), again showing the window feature displaying one specific area.

For every operation the information required is clearly indicated on the CRT. The window feature enables the operator to concentrate only on that portion of the part being programmed. Progression from

one operation to the next is prompted in plain English. To review a previously programmed step, the operator can work backward through the program, one step at a time.

Each machine on the shop floor can be controlled by its own built-in computer, or the computer can be set up in the shop office. Once the program is created, it can be sent electronically from the office to one machine or several machines, depending on the situation, Figure 12–21 (page 517). In addition, a program created at a machine can be sent to the office for storage on a disk, or it can be sent to another machine in the shop.

Word abbreviations are commonly used in pro-

Figure 12–19A A Cinmill control installed on a vertical milling center *(Courtesy of Cincinnati Milacron)*

gramming. The geometry of the part might make use of abbreviations such as LIN for line, CIR for circle, and so on. Terms for setting up might be SET for setup and CLA for clamp. Machining operations might be expressed as TAP for tap, BOR for bore, and REA for ream. Capital letters are used for the abbreviations.

For the operator at the machine, talk of an intermediate language and EIA codes might only be confusing. These do not even exist for the operator when using graphics. Graphics provides the fastest and easiest means for programming the making of a part.

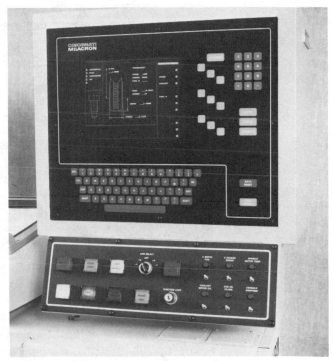

Figure 12–19B A Cinmill control *(Courtesy of Cincinnati Milacron)*

UNITS:	IPR	RPM	1040Stel	
Tool Type	Feed	Speed	Coolant	Feed Speed Coolant
Face Mill	.005	500	Pulse	
End Mill	.003	400	Pulse	
Plunge	.003	400	Mist	
Edge	.002	300	Mist	
Tap		100	Pulse	
Drill	.010	250	Pulse	
Bore	.006	300	Pulse	
Ream	0.003	75	Mist	
MATERIAL TECHNOLOGY				

Figure 12–20A Definition of materials *(Courtesy of Cincinnati Milacron)*

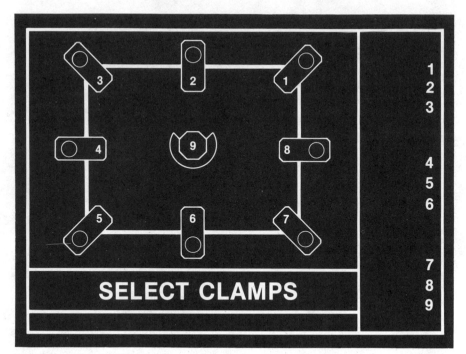

Figure 12–20B Selecting clamps *(Courtesy of Cincinnati Milacron)*

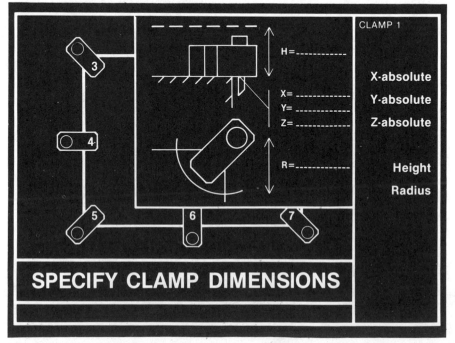

Figure 12–20C Specifying clamp dimensions *(Courtesy of Cincinnati Milacron)*

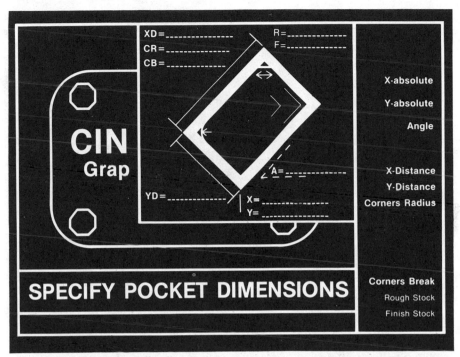

Figure 12–20D Pocket dimensions for the machining of pockets *(Courtesy of Cincinnati Milacron)*

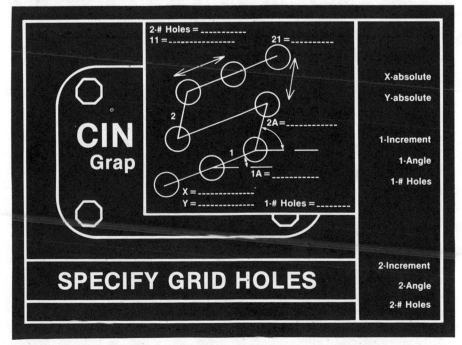

Figure 12–20E Grid hole patterns *(Courtesy of Cincinnati Milacron)*

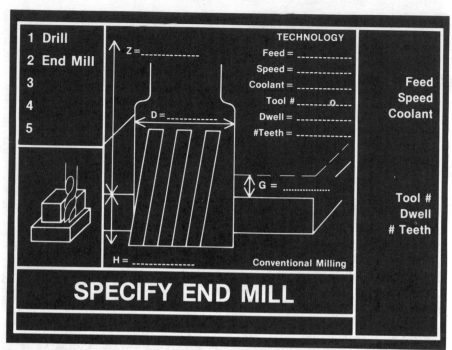

Figure 12–20F End milling cutters *(Courtesy of Cincinnati Milacron)*

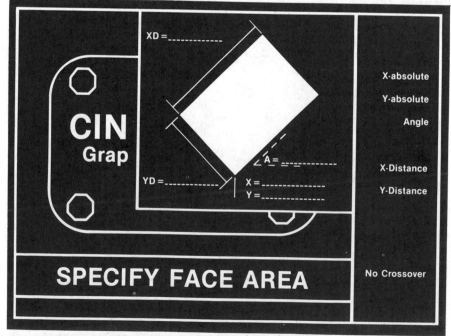

Figure 12–20G Specifying face area *(Courtesy of Cincinnati Milacron)*

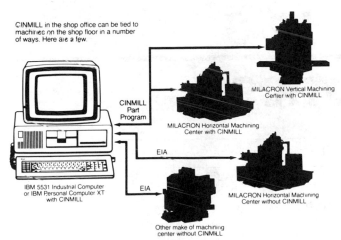

CINMILL in the shop office can be tied to machines on the shop floor in a number of ways. Here are a few.

CINMILL Part Program

MILACRON Vertical Machining Center with CINMILL

MILACRON Horizontal Machining Center with CINMILL

IBM 5531 Industrial Computer or IBM Personal Computer XT with CINMILL

EIA

EIA

MILACRON Horizontal Machining Center without CINMILL

Other make of machining center without CINMILL

Figure 12–21 Control possibilities with the Cinmill control *(Courtesy of Cincinnati Milacron)*

REVIEW QUESTIONS

1. What is a computer?
2. What is the basic difference between N/C and CNC?
3. What is a central host computer?
4. What is meant by the term preprogrammed routine and how does such a routine assist the programmer?
5. How does a computer help the programmer?
6. How is a computer added to the N/C system?
7. Describe the steps by which the operator programs information for machining operations. What information is then provided by the computer control?
8. What is computer graphics?
9. How does computer graphics help the machine operator?
10. List three typical applications of computer graphics.
11. What are word abbreviations? Give three examples.
12. What are the advantages of computer graphics in programming the making of a part?

UNIT 12-3

ROBOTICS

OBJECTIVES

After completing this unit, the student will be able to:

- state the advantages of using robots in manufacturing.
- define the term industrial robot.
- list five applications of industrial robots.
- define the term tool center point (TCP).
- describe the control system of a robot, including its primary functions.
- list and define the three main control algorithms for programming a robot.
- explain the two modes of operation of a robot system.

KEY TERMS

Robot

Industrial robot

Taught (programmed)
  sequences

Handling application

Processor application

Inspection and
  gaging

Gripper jaws

Tool center point

Jointed arm system

Jointed arm working
  volume

Teach pendant

Algorithm (concept)

Point-to-point algorithm

Continuous path
  algorithm

Controlled path
  algorithm

Teach mode

Automatic mode

Tracking function

Interface

Feedback

Artificial vision

Automatic Speech
  Recognition (ASR)

THE ADVENT OF THE ROBOT

Robots have joined the work force. In every plant throughout North America and Europe there are jobs that are boring, repetitive, back breaking, dirty, noisy, and dangerous. Industrial robots are capable of doing these jobs day in and day out with astonishing dependability, boosting productivity and improving quality while freeing people from unpleasant, even hostile environments.

The word robot was suggested by the Czechoslovakian word robota meaning work. It has been used for years in science fiction to mean mechanical man, but the modern day **industrial robot**, in simple terms, is a mechanical device that does some of the work of human beings. It is basically a single-arm type of device that can manipulate parts or tools through previously "taught" (programmed) sequences of motions and operations. These sequences and operations are not necessarily repetitive, as some robots have the ability to make logical decisions. It is a device that can be applied to many different operations in industry and that is capable of being removed from one application and easily taught to do another.

Industrial robots are capable of performing many different applications which are classed as either handling or processor applications. Handling applications include, but are not limited to, loading and unloading machine tools such as lathes, grinding machines, and even inspection or gaging stations. It also includes materials handling such as parts retrieval from storage areas or conveyor systems. Processor applications include such operations as spot welding, seam welding, paint spraying, metallizing (metal spraying), and cleaning; in fact, any operation in which the robot can manipulate a tool to carry out a manufacturing process.

DESIGN OF AN INDUSTRIAL ROBOT

A robot system is a means of moving a pair of grippers or a tool and directing it through predetermined sequences of motions. At certain points in the sequences, the system will direct some functions to take place; e.g., "close grippers" or "start machine tool cycle." Realistically, the user is concerned mostly with the position of a given point on the end of the arm, for instance, the center of the gripper jaws. This point is called the **tool center point (TCP)**.

The mechanical system of a robot is essentially the means of moving the TCP in space. Various types of movement are available. They range from simple linear (straight line) movement to linear and rotary combinations of movement to the all rotary or jointed-arm system, Figure 12–22. The robot arm has a six-axis, rotary-jointed configuration, Figure 12–23. The device is hydraulic on some robots, five of which are hydraulic rotary actuators. The elbow uses a hydraulic cylinder. The newest robots are all electrical drive units. The arm takes up less than six square feet (approximately two square meters), yet it can move the TCP within a volume in excess of 1000 cu ft (approximately 28 cu meters), Figure 12–24 (page 522).

The Control System

The purpose of a control system for an industrial robot is basically two-fold. It provides the operator with a means to teach or program the robot in what it should accomplish. It also directs the machine through its movements and sequences of operations during normal automatic running.

The control system uses a minicomputer in its system, an adaptation of a CNC system for machine tools, that is responsible for the performance of many duties. When the robot operates automatically, the path between points should be predictable; e.g., when the path between consecutive points is in a straight line traversed at constant velocity with built-in acceleration and deceleration spans, Figure 12–25 (page 523). There should be communication between the robot and the machine it serves so that machining cycles are complete before the robot moves. Malfunctions in the system should be provided for. During the teaching routine, the operator should be able to guide the TCP through a familiar movement system as explained in Figure 12–22. There should be communication between the operator and the control to allow changes to be made in the program and also to allow the control to indicate whether or not the changes can be met.

The control unit, Figure 12–26 (page 523), contains a minicomputer, a cathode-ray tube (CRT) and keyboard, a control panel, a tape punch and reader, axis servocontrol electronics, a servotest panel, a teach pendant, and terminals for communication with other equipment. Figure 12–27 (page 524) shows how the computer fits into the overall control system, and also shows the flow of information through the system.

Control Algorithms

There are three main control **algorithms (concepts)** available for programming a machine such as a robot. The first two are well-known, and the third was developed to satisfy many of the desirable requirements of a robot system.

In the **point-to-point algorithm**, only the coordinates of the programmed points are stored in the control memory. The path taken by the TCP between the programmed points is run at maximum or limited rate and is not easily predicted. Teaching is done by moving each axis individually until a combination of axis positions yields the required position of the TCP.

In the **continuous-path algorithm**, the TCP is physically grasped and led by the teacher or operator through the required path. The position of each axis is recorded by the control on a constant time base during the teach motion. Every motion that the TCP makes during teaching, whether intentional or not, will be reproduced in the automatic running of the robot.

In the **controlled-path algorithm**, the taught positions of the TCP (position and orientation) are entered into the control memory together with the velocity at which the TCP is required to approach those positions. In the automatic mode, all the axes move smoothly and proportionally between consecutive points, providing predictable controlled-path motion. The computer in the system is needed to solve real-time transformations, relating the rectangular coordinates of the taught positions of the TCP to the rotary coordinate system of the machine. In the teach mode, the operator has the choice of moving the TCP in either rectilinear or cylindrical coordinated motions.

The three control algorithms represent a range from the simplest in terms of control system complexity (point-to-point) to the simplest from the point of view of teaching and predictability of motions (controlled path).

GEOMETRIC CONFIGURATIONS

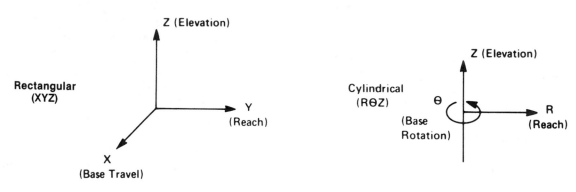

Figure 12–22 Geometric configurations for the types of movement available from robots *(Courtesy of Cincinnati Milacron)*

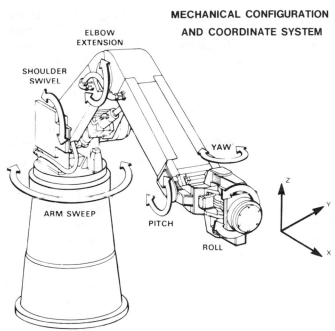

MECHANICAL CONFIGURATION
AND COORDINATE SYSTEM

ELBOW EXTENSION

SHOULDER SWIVEL

YAW

ARM SWEEP

PITCH

ROLL

Z

Y

X

Figure 12–23 Axis motions — modular system *(Courtesy of Cincinnati Milacron)*

MODES OF OPERATION

The robot system has two modes of operation. The **teach mode** enables the operator to teach an application program to the robot. The **"auto" (automatic) mode** enables the robot to automatically execute the taught program.

A robot program consists of one or more taught sequences of "points." Each point consists of positional, functional, and speed data. To teach the robot the operator uses the teach pendant, Figure 12–28 (page 525), the cathode-ray tube (CRT), and the keyboard, Figure 12–26. When the operator is satisfied that he has the correct data for each point, the program button on the teach pendant is pressed and the data is stored in the computer. The operator can also replay the taught program and can add or delete points at will. The computer will direct the CRT display to indicate to the operator whether or not a particular position, speed, or function can be used.

In the automatic mode of operation, the robot is guided by the control system through the taught sequences of points and performs the taught functions associated with each point. The TCP moves between taught points along straight line paths at the programmed speeds, with built-in acceleration and deceleration. Exceptions occur in two cases: If a point is programmed with a **continue** function, the TCP will not decelerate but will pass through the point at its taught velocity. When the **tracking** function is in use, the position of the TCP is updated in real time to correspond with the position of the moving target. The velocity of the TCP will also be modified by an amount equal to the velocity of the moving target.

During automatic operation, the computer monitors the robot system for malfunctions. If one occurs, an appropriate message is displayed on the CRT screen and, if necessary, the control directs the robot into an appropriate shut-down mode.

During automatic running, the taught program resides completely in the computer memory. However, when the robot is required to carry out a different application, the tape punch may be directed to punch out a perforated tape of the program. Later, that same program may be read back into computer memory by means of the tape reader.

PRACTICAL APPLICATIONS FOR THE ROBOT

Figure 12–29 (page 526) shows two grinding machines being serviced by one robot. Twin grippers on the end of the arm unload a finished part and immediately load an unfinished part, reducing wasted motions and overall cycle time. The robot is interfaced (coordinated) with both grinding machines and the loading racks. The wheelheads of the grinding machines are retracted before unloading and loading, and the work is held securely and located between centers prior to moving the robot away and starting the grinding cycle. If one loading rack is empty, the robot will continue to service the other grinding machine, utilizing expensive machinery to a maximum.

Figure 12–30A (page 526) shows an overhead view of two back-to-back CNC turn centers with rear loading; i.e., loading from the back of the lathe by an industrial robot. The robot is conveniently located in the center between the two lathes.

The double grippers on the robot, Figure 12–30B (page 526), enable the robot to exchange finished and rough shaft-type workpieces in a machine chuck. In this case, loading is being done from the back of the turning center.

Two Cinturn CNC turning centers, Figure 12–30C (page 527), are being loaded from the front of the lathe by an industrial robot. A laser gage in the foreground checks the final part size. An out-of-tolerance dimension will initiate feedback to the control, which automatically triggers tool offsets to correct the condition.

To summarize, a six-axis, computer-controlled, robot system, Figure 12–31 (page 528), can satisfy many requirements. It can carry high loads at fast speeds through a large working volume in predictable paths. It can easily interface with peripheral equipment and has the ability to make logical decisions. Its very simple mechanical configuration and its computer control, which is based on a proven CNC system, result in maximum reliability. Its modular configuration allows it to be placed in restricted areas.

This type of industrial robot system can interface (be coordinated with) higher level computers and thus has the potential of contributing greatly to computer-aided manufacturing systems of the future. The word *interface* also implies communication and understanding between robots and machines.

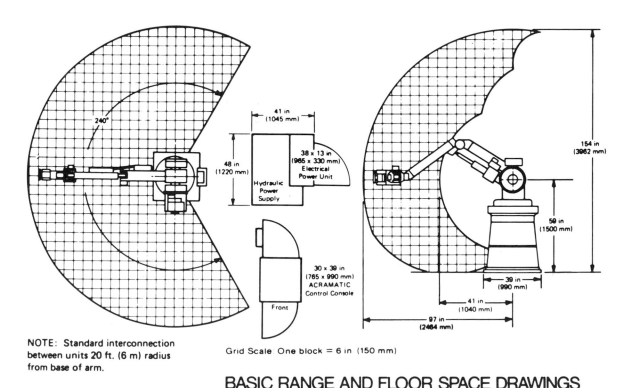

NOTE: Standard interconnection between units 20 ft. (6 m) radius from base of arm.

Grid Scale: One block = 6 in (150 mm)

BASIC RANGE AND FLOOR SPACE DRAWINGS

Figure 12–24 Working volume of the jointed arm of a robot *(Courtesy of Cincinnati Milacron)*

NEW DEVELOPMENTS IN ROBOTS

Artificial vision systems have been developed by combining the use of television cameras with computers and computer logic in the form of a digital code. Television cameras can feed images to the computer which converts these images to a digital code and compares this digital representation with images previously photographed and stored in its memory. The robot can then recognize a single part by its shape or silhouette and can make the decision and the move to pick up that part from a number of other random parts. This is two-dimensional information; i.e., recognition and comparison.

Researchers are working to develop a vision system with a third dimension, whereby a camera system can provide data so that a computer can determine distance and depth. Three-dimensional information can give a robot depth perception so that it can be used to perform complicated assembly work.

From the viewpoint of safety, a robot could be programmed to recognize a human. If that human happened to be standing in the robot's way, the robot would recognize the human and all movement would automatically stop.

Researchers are also working to give robots a sense of touch. For example, the robot gripper could be covered with a thin plastic skin such as material con-

CONTROLLED PATH - STRAIGHT LINE

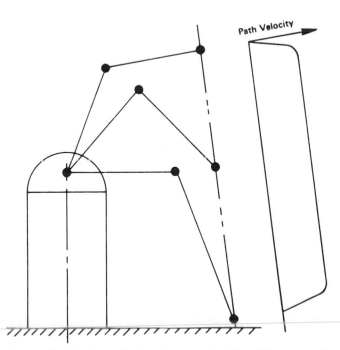

Figure 12–25 Controlled path — straight line *(Courtesy of Cincinnati Milacron)*

Figure 12–26 Robot control console *(Courtesy of Cincinnati Milacron)*

taining thousands of tiny electrical sensors. Pressing this material would generate small electric currents that would be applied to a microprocessor. The information would be analyzed and transmitted back to the gripper. Such a system could tell the gripper how tight to grip an object or could even direct a robot to feel its way to a position that could not be seen. This would almost parallel our own human nervous system, with electrical nerve impulses operating on a system between the brain (robot computer) and the fingers (robot grippers).

Researchers are also working on a system called Automatic Speech Recognition (ASR), where a computer can accept data from a human voice. Speech is the easiest and most natural method of communication. To be able to transmit instructions to a robot in this manner would greatly simplify programming, making it quicker, less costly, and more accurate.

CONTROL SYSTEM DIAGRAM

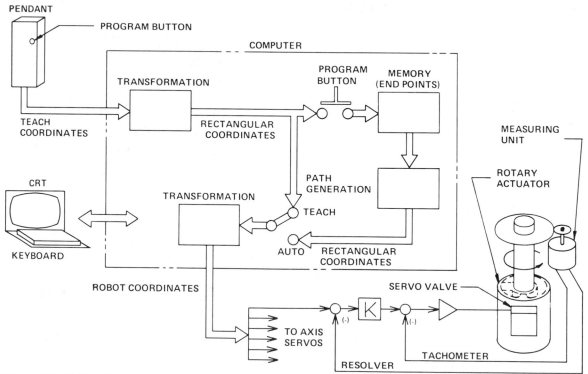

Figure 12–27 Functions of the computer in the control system *(Courtesy of Cincinnati Milacron)*

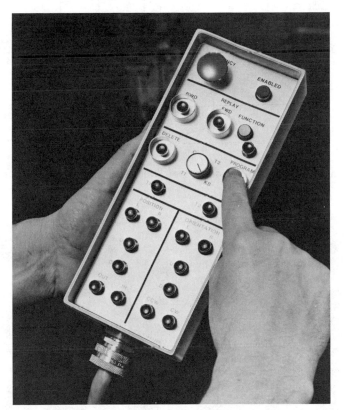

Figure 12-28 The teach pendant *(Courtesy of Cincinnati Milacron)*

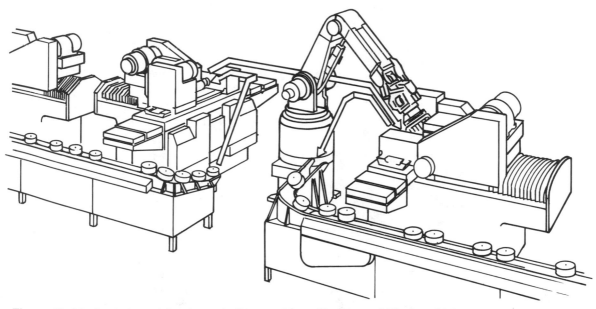

Figure 12-29 A robot servicing two grinding machines *(Courtesy of Cincinnati Milacron)*

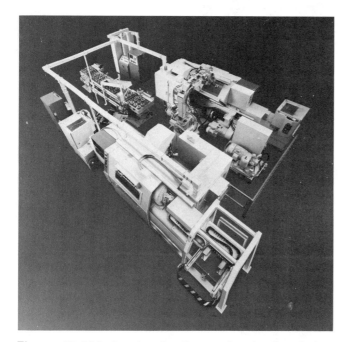

Figure 12-30A A robot loading and unloading lathes *(Courtesy of Cincinnati Milacron)*

Figure 12-30B A robot changing a shaft *(Courtesy of Cincinnati Milacron)*

Figure 12–30C A robot servicing two lathes from the front *(Courtesy of Cincinnati Milacron)*

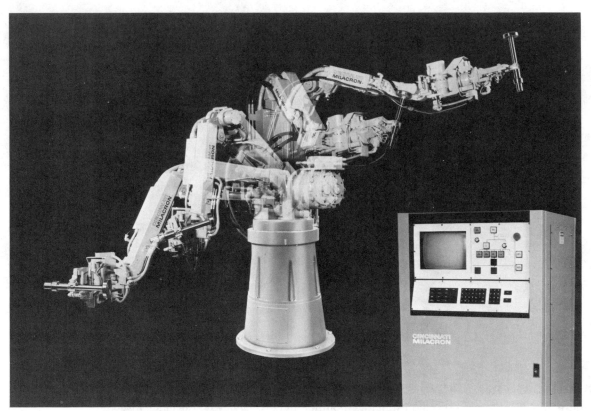

Figure 12–31 Computer-controlled industrial robot *(Courtesy of Cincinnati Milacron)*

REVIEW QUESTIONS

1. What is an industrial robot?
2. What types of work can an industrial robot do?
3. Define the term tool center point and describe the operation of the robot in relation to the TCP.
4. Name the four basic geometric configurations for robot movements.
5. Name the six axis motions available on many robots.
6. What type of drive system is used on an industrial robot?
7. What is the purpose of the control system?
8. What is the role of the computer in the control system?
9. What is meant by teaching the robot?
10. What is the point-to-point concept?
11. What is the continuous-path concept?
12. What is the controlled-path concept?
13. What is automatic operation?
14. Explain loading and unloading of machines.
15. What is a teach pendant?
16. In the automatic mode of operation what is the purpose of the tracking function?
17. In automatic operation what happens if there is a malfunction in the robot system?
18. List four features of a computer-controlled robot system.
19. How can vision be given to a robot?
20. How can a sense of touch be given to a robot?

GALLERY TWELVE

A single-arm robot unloading a CNC turning center. The finished part is being unloaded from the rear door. The unfinished or "green" parts are at the bottom of the photo, waiting to be loaded. *Courtesy of Cincinnati Milacron*

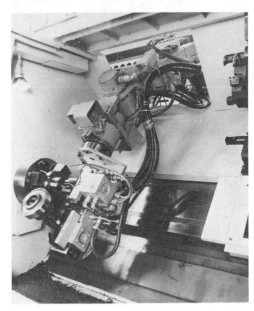

A single-arm fixed robot unloads a finished part from a chuck, working through the rear trap door of a CNC turning center. *Courtesy of Cincinnati Milacron*

Double turning in which two diameters are turned at the same time. *Courtesy of Cincinnati Milacron*

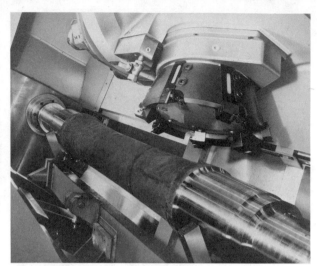

Turning different diameters on a large shaft.
Courtesy of Cincinnati Milacron

Forming curves and tapers with a profiling insert tool on a CNC turning center. *Courtesy of Monarch Sydney*

Cutting screw threads. *Courtesy of Cincinnati Milacron*

Continuous seam welding using a single-arm robot. The robot has been "taught" or programmed to follow the path to be welded. *Courtesy of Cincinnati Milacron*

A robot is spot welding around a door opening of an automobile body on an assembly line, as viewed from the inside. *Courtesy of Cincinnati Milacron*

Welding on an automotive assembly line. A robot is spot welding around a door opening of an automobile body on an assembly line, as viewed from the outside. *Courtesy of Cincinnati Milacron*

COMPUTED TOMOGRAPHY – TESTING FINISHED PARTS IN THE MACHINE SHOP

Finished parts can be tested for quality in different ways. Test samples can be made to certain sizes and specifications. The samples are then subjected to a series of stress and strain tests. The samples are deliberately extended beyond the breaking point by pushing, pulling, bending, or striking. As an example, a test piece is turned to a diameter for a certain length and to close tolerances. The test piece is then placed in a holding device. It is pulled apart, or bent, or struck hard enough to break it. The pounds of force required to do the pulling, or bending, or striking to the breaking point is measured and noted. This type of procedure is called destructive testing.

However, it is not always desirable to destroy a part to test the quality. Another type of testing known as nondestructive testing X-rays the actual parts to determine if there are any internal flaws. A common flaw in a casting is a blowhole. Only the X-ray technique, or one similar to it, will detect the blowhole since it is not obvious on the surface of the casting. A blowhole will definitely weaken a casting and may cause the casting to fail under pressure.

Another type of flaw appears in machined parts that are heat treated. Heat treating may cause inner cracks. The only way that these cracks can be detected is by the use of X-rays.

Even X-rays may not expose the flaw. Originally developed for medical applications, X-ray techniques used in nondestructive testing of machined parts are limited in that the part is shown as two dimensional when actually it is three dimensional. All flaws may not be visible in the two-dimensional image.

A newer X-ray scanning technique employing the use of a computer was developed first for medical applications. This system is known as Computed Tomography (CT) and now is an advanced technique for testing the quality of machined parts. Computed tomography is the reconstruction by computer of a tomographic plane (or slice) of an object. In this technique, a collimated X-ray beam (one that is adjusted parallel and is very accurate) is passed through the machined part as if it were slicing the part. The beam is picked up and measured by an array

Figure 1 The principle of computed tomography. In computed tomography, the X-ray beam slices through the part to be tested. Detectors on the other side of the test piece measure the part and transmit the data to the computer. The computer analyzes the data, constructs the image from the data, and displays the image on the monitor. *(Courtesy of Automatix Inc.)*

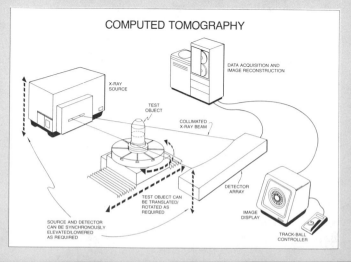

of detectors on the other side of the part. Figure 1 shows the principle of computed tomography which makes use of a translate-rotate scanning method. This means that the object is scanned and then rotated slightly and scanned or X-rayed again. This procedure is repeated until the part has been rotated a total of 180 degrees. A tomographic system exposes only a small cross-sectional slice of the test object during each scan. The detectors pass the data to the computer which then recreates a very accurate cross-sectional image. Figure 2 illustrates a typical tomographic section through a test object. The X-ray beam and detector array can be raised or lowered so that any part of the object can be scanned. Any number of images can be created and displayed in this way. A picture of the entire object from top to bottom can actually be revealed. Because the information is processed by a computer, image enhancement techniques can be used to emphasize various parts of the image. Since this image is a cross-sectional view, much more information about the part is revealed than if a normal X-ray picture were viewed. Figure 3 shows a computed tomography system with the X-ray source on the left, the raising and lowering tower, and the solid-state detector array on the right. The test object

rests on the rotating table in the center.

Computed tomography is an exceptional nondestructive method for testing and evaluating the quality of a machine part or even assembled parts. It is possible to measure dimensions or detect defects, something that a plain X-ray cannot necessarily do. Remember that the part is three dimensional and that an X-ray actually looks at an object and portrays it in two dimensions. In simplest terms, this is basically why computed tomography reveals so much more than conventional X-rays.

The first CT system was developed to establish this technique as an economically viable nondestructive evaluation (NDE) tool for scanning missile and solid rocket motor hardware components. One advanced Air Force CT system can scan objects up to 39 inches (990 mm) in diameter. Another Air Force CT system, the world's largest, is capable of inspecting objects up to 8 ft (2438 mm) in diameter by 17 ft (5181 mm) long, and weighing 110,000 pounds. The fabrication of these first industrial CT systems represents a major advancement in nondestructive evaluation and analysis techniques. Wherever critical industrial inspection needs exist, computed tomography will prove its worth.

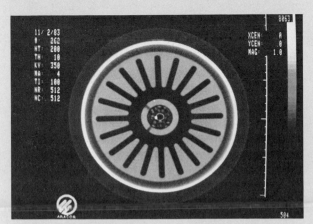

Figure 2 A typical tomographic section through a test object *(Courtesy of Automatix Inc.)*

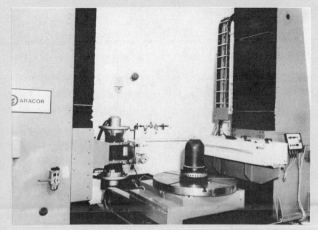

Figure 3 A computed tomography system *(Courtesy of Automatix Inc.)*

CHAPTER THIRTEEN

SPECIAL-PURPOSE PROCESSES

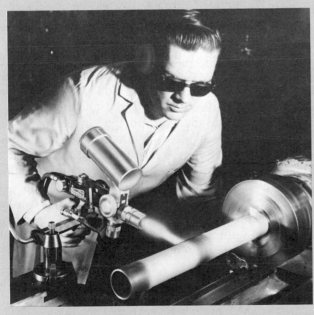

Each of the special-purpose processes described in this chapter has a particular role to play in machining technology and manufacturing. Powder metallurgy is used to produce the thousands of different carbide cutting tips that are needed in machining. In addition, many small parts such as gears can be produced economically and easily by this method.

Electrical discharge machining, both conventional and wire cut, is also playing an increasing role in manufacturing. The addition of wire cut EDM and modifications to conventional EDM, including CNC capabilities for both, make this an indispensable and very flexible type of machining, especially in tool and die applications.

Thermal spraying techniques were initially regarded as a repair method. In fact, this is still a major role of thermal spraying. However, the ability to spray ceramic powders and mixed metal and ceramic (cermet) powders makes thermal spraying important in electronics manufacturing. The aerospace industry also has increasing uses for thermal spraying of protective ceramic coatings. CNC capabilities and robotics also play a part in thermal spraying applications in many fields.

These three processes will continue to play important roles in industry in the years to come.

UNIT 13-1

POWDER METALLURGY

OBJECTIVES

After completing this unit, the student will be able to:

- explain the general process of powder metallurgy.
- list the steps required to make carbide tips.
- define the terms cemented carbide and sintering.

KEY TERMS

Powder metallurgy

Powder selection and
   blending

Filling the dies

Pressing or impacting

Ejecting the blank

Sintering

Controlled-atmosphere
   furnace

Pressure sintering

INTRODUCTION

Powder metallurgy is the production of various articles from powders, mainly metallic, but also incorporating nonmetals, metallic compounds, or chemical additions in various proportions. A major part of powder metallurgy today involves the production of carbide cutters in hundreds of different shapes and styles. Many small machine parts such as gears are also produced by this method.

Basically, the process, Figure 13–1, involves the thorough mixing of the various powders, the pressing or impacting of the powders within a die, the ejecting of the blank from within the die, and finally the sintering of the blank in an oxygen-free furnace (called a controlled atmosphere) at a temperature below the melting point of the metal or alloy.

Pure tungsten (W) powder is thoroughly mixed with carbon (C) powder. They are heated to produce tungsten carbide. Tungsten carbide powder is then mixed thoroughly with powdered cobalt (Co). The cobalt acts as a binder to bind the powder particles together or cement them together, hence the name cemented carbide. These powders are poured into a die and compressed under tons of pressure, after which the molded blanks are ejected from the die. At this stage, the blanks are "green" or soft and can be easily broken. The molded blanks are then placed in a controlled atmosphere furnace and heated so that the particles are cemented together.

DETAILS IN THE PROCESS OF POWDER METALLURGY

Powder Production

All the materials that go into the dies initially must be in the form of powder. The production of powders is normally done by mechanical or chemical means.

Powder Selection and Blending

Powder selection and blending are perhaps the most important parts of the powder process. Kennametal, for example, uses tungsten, tantalum, columbium, titanium, and cobalt in more than 40 different compositions that have been developed for specific applications. These powders must be thoroughly blended. This can take place over a number of hours or days. It can be done in air, or in a controlled atmosphere, or under liquid to minimize oxidation. Once the powders are blended, they can be transferred directly to the press hopper for immediate pressing, or they can be stored under air or liquid, or in a vacuum in sealed containers.

Pressing

The filling of the dies with powders can be done by hand or it can be done automatically from a hopper to the press. Vibration can be used during filling to create packing. Individual blanks can be compacted, or it can be done automatically and continuously as in mass production. Mechanical or hydraulic presses exert pressures up to 100 tons per square inch. Compacting can also be done by sheet rolling, hot pressing, continuous compaction, extrusion, hydrostatic pressing, and centrifugal pressing.

Ejection of the blank after pressing can be done by hand or it can be done automatically.

Sintering

Sintering *is the bonding or cementing together of particles by heating. There are many different types of furnaces used, but a controlled atmosphere inside the furnace is essential to ensure proper sintering without oxidation taking place.*

As a rule, the temperatures used are well below the melting point of the metal or the major portion of the alloy. The sintered product is generally porous, with the degree of porosity varying from zero to 80°. Normally, there is no pressure exerted during sintering, but hot pressing or pressure sintering can be done. Temperatures can vary from a few hundred degrees to approximately 5400 degrees F (nearly 3000 degrees C), depending on the melting point of the metal or alloy involved. The length of time that temperatures are held can vary from minutes to a number of hours.

Additional Processes

It may be necessary to alter shape, form, density, surface character, chemical composition, or physical characteristics. Straightening, rolling, forging, extruding, welding, coining, or machining might be necessary to obtain a final shape or size of the blank in question.

Certain mechanical or electrical properties might be required. For example, the pores of the blank might be impregnated with graphite to form a motor bearing that would need no lubrication. The pores of an elec-

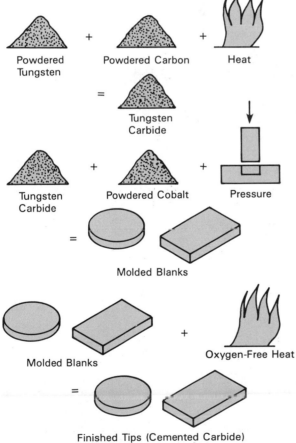

Figure 13–1 The powder metallurgy process

trical contact may need a particular hardness, or a better conductivity may be required. There are many possibilities.

REVIEW QUESTIONS

1. What is powder metallurgy?
2. For what types of parts is powder metallurgy used primarily?
3. List the steps in the basic powder metallurgy process.
4. Explain the origin of the term cemented carbide.
5. Explain how the blending of the powders is so critical to the success of the powder metallurgy process.
6. List six methods by which the powder mixture can be compacted in the dies.
7. Explain sintering in powder metallurgy.
8. What does controlled atmosphere mean?
9. What is the temperature of sintering compared to the melting point of the primary metal component of the powder?
10. What characteristics of the powder may be altered to obtain a desired blank?

UNIT 13-2

ELECTRICAL DISCHARGE MACHINING (EDM)

OBJECTIVES

After completing this unit, the student will be able to:

- explain the process of conventional electrical discharge machining (EDM).
- list the components of the conventional EDM system.
- explain the process of wire-cut electrical discharge machining (EDM).
- list the components of the wire-cut EDM system.
- state the function of the dielectric in EDM.
- explain how surface finish is affected by the current and frequency of discharge in conventional EDM.
- list uses of both types of electrical discharge machining.

KEY TERMS

Electrical discharge machining (EDM)

Discharge
Deionization

Electric spark
Spark erosion
Dielectric fluid
Kerosene
Conventional EDM

Frequency of discharge
Surface finish
Heat-affected zone
Overcut

Wire-cut EDM
Electrode
Spark gap
Direct current
Rectifier
Dielectric constant
Dielectric circulation
  system

Electrode wear
Rotary impulse
  generator (RIG)
Controlled pulse
  circuit (CPC)

INTRODUCTION

Electrical discharge machining (EDM) is a method of cutting the hardest and toughest of metals, including carbides, whether or not they are heat treated. The cutter is an electric spark that actually removes a chip of material from the surface of the work. This is sometimes called **spark erosion**. The cutting takes place under the surface of a dielectric fluid like kerosene.

There are essentially two methods used in EDM, conventional and wire cut. In conventional EDM, an electrode moves vertically down onto the work surface. Electric sparks jump the gap between the end of the electrode and the work surface, removing a chip of metal. The resulting hole is the size and shape of the electrode and progresses either part way or all the way through the work.

In wire-cut EDM, a thin copper wire, which is the electrode, passes through the work and either the wire or the work moves horizontally, acting like a bandsaw. The sparks jump the gap horizontally and many intricate shapes can be cut out of the metal.

Both methods are adaptable to N/C and CNC. The conventional electrical discharge machine, Figure 13–2, is just one example of the machinery used in EDM.

EDM has gained wide acceptance in industry today and is used in many of the more sophisticated machine shops, particularly in tool and die making. The first EDM machines, in the 1950s, were built to perform simple drilling operations by utilizing the phenomenon of electric spark discharge. Today they play an indispensable role in machining technology.

CONVENTIONAL EDM

In conventional EDM, the machine consists of a worktable to which the work can be clamped, and then both are submerged in a bath of kerosene. There is also a holding device to position the electrode which must be able to move accurately up and down without backlash (looseness or play). The electrode, which is the exact shape of the hole or depression required, is placed directly over the work. It is moved down until it touches the work, and then is withdrawn slightly to create a slight gap. The power source causes the electricity to flow through the electrode, jumping the gap as a spark that melts and blasts a particle of metal off the surface of the work. Since there are thousands of sparks, the tiny craters merge into each other form-

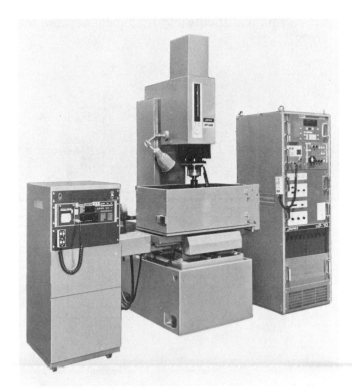

Figure 13–2 Electrical discharge machine, conventional type, with N/C control *(Courtesy of Japax, Inc.)*

ing a new surface as the cutting progresses. This forms either a cavity to a certain depth, or a hole that goes right through the work. Figure 13–3 illustrates the cutting action below the electrode when using conventional EDM.

The sparks must move in one direction only — toward the work. This is done by making the electrode negative and the work positive. This also means that the machine must put out direct current; therefore, a rectifier must be included in the circuit to change alternating current to direct current.

Dielectrics

The **dielectric** acts as an insulator between the tool and the work, provides a path for the spark, acts as a coolant because the sparks create a great deal of heat, and also flushes away the chips.

Sparks can occur in air (like lightning), but in practice this is difficult and inefficient because the dielectric constant is too low. The **dielectric constant** is the amount of energy that each spark can carry. The dielectric constant of air is approximately 1; kerosene is more than 2; alcohol is 25; water is 80. The voltage in the water will cause the water to dissociate into hydrogen and oxygen, an explosive pair. Alcohol is expensive and its vapors are explosive. Kerosene is the most popular choice of industry (there are others) because it is cheap, universally available, not too flammable, and does not cause rust.

Methods of Circulating Dielectrics. All machines have some form of dielectric circulation system. There must be a reservoir (usually in the base of the machine or sometimes in an auxiliary reservoir), a pressure pump to circulate the dielectric, a filter to filter the debris from the dielectric, a workpan (which allows the dielectric to cover the worktable, work, and electrode), valves, gages, and the necessary plumbing.

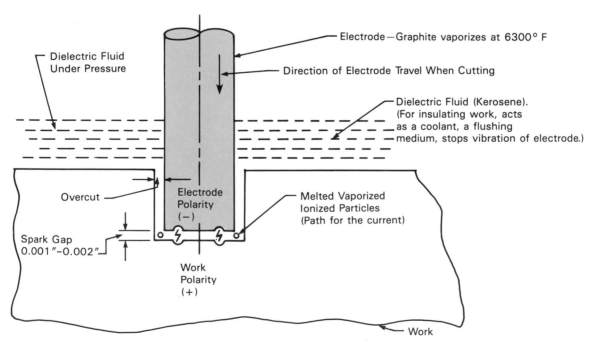

Figure 13–3 Electrode action during cutting in conventional EDM

Electrodes

Electrodes can be made from soft, easily machinable materials. Yellow brass is used with pulse-type circuits; copper is often used when higher voltages are used; and high-density, high-purity carbon or graphite is also used.

Discharge

The amount of metal removed, or the size of the crater cut, depends on the size of the spark or the energy that fires that spark at the metal surface. This energy is determined by the gap voltage during the discharge, the discharge current, and the length of time the current flows. A low voltage and a high amperage is put through the electrode.

Deionization

The electrical charge ionizes the dielectric, thus making it a good conductor for the electric sparks.

When discharge is completed, the gap voltage is held to a low value until the dielectric fluid is deionized and will no longer conduct the electric sparks. If this is not done, current will immediately start to flow through the gap.

Effect of Frequency on Surface Finish

Figure 13–4 shows the effect of current and frequency of discharge on the surface finish and on metal removal rate. Increased discharge frequency can improve surface finish. Figure 13–4A illustrates the effect of one spark per unit of time. Figure 13–4B shows two sparks per unit of time, each making a cut with less depth than one spark per unit of time but producing a smoother surface. Figure 13–4C shows four sparks per unit of time, each making a cut with less depth than two sparks per unit of time but producing a still smoother surface.

Heat-affected Zone

Heat generated by the sparks has a hardening effect on the layer of metal on the sides and bottom of the hole cut by the electrode, Figure 13–5.

Maintaining the Gap

The spark gap between the electrode and the work must be constantly maintained. The electrode must be slowly fed down into the work maintaining a spark

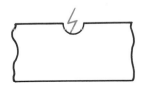

A. 5 amperes; 1 spark per unit of time

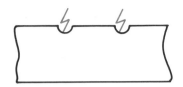

B. 5 amperes; 2 sparks per unit of time

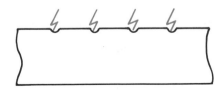

C. 10 amperes; 4 sparks per unit of time

Figure 13–4 Effect of frequency on surface finish

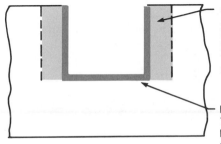

Annealed Layer; 2–5 points below the hardness of the parent metal; 0.002″ finish cut; 0.008″ rough cut

Hardened layer; 68 To 70 Rockwell "C" hardness 0.0001″ finish cut; 0.0005″ rough cut

Figure 13–5 The heat-affected zone

gap of 0.001 in to 0.002 in (approximately 0.02 mm). If the gap is too large, cutting does not take place, and if the gap becomes too small, the electrode and work could be fused together. A servomechanism (an electrohydraulic control) automatically advances the electrode into the work at the proper rate, or in the event the gap becomes blocked in any way by particles from the work, it will retract the electrode until the cutting area is clear. The electrode is then returned to the cutting position and the cutting is resumed.

Overcut

Overcut means that the machined hole in the work is slightly larger than the size of the electrode. This is caused by the initial voltage and discharge energy. A higher current would increase the discharge energy which would increase the size of overcut.

Electrode Wear

As the sparks jump the gap, a crater is produced in the electrode as well as in the work surface. The size of the crater produced depends on the electrode material and the discharge energy. This means that the electrode wears as it cuts into the work. One of the principal materials used for electrodes is graphite. Graphite does not melt, it vaporizes.

Rotary Impulse Generators (RIG)

With the circuit for a rotary impulse generator (RIG), Figure 13–6, a standard, high-frequency (HF), alternating-current (AC) generator can be used to produce pulses that move in one direction.

RIG is capable of removing metal quickly but results in a rough surface finish because the operating frequency is low and is not adjustable.

Controlled Pulse Circuits (CPC)

Controlled pulse circuits, Figure 13–7, provide a fast, positive means to electronically stop current flow. Transistors are used as switching devices. The result is faster metal removal, better surface finish, and improved electrode wear.

WIRE-CUT EDM

Since 1975, the use of wire-cut EDM machines has spread rapidly. In conventional EDM, Figure 13–8, the sparks move vertically. The wire-cut machine, Figure 13–9, uses a very thin wire, 0.0008 to 0.012 in (0.02 to 0.3 mm) in diameter, as an electrode that machines the work with an electrical discharge, acting much like a bandsaw by moving either the work or the wire. Erosion of the metal by spark discharge is the same as in conventional EDM, except that the sparks move horizontally. The prominent feature of a moving wire, Figure 13–10, is that a complicated cutout can be easily machined without using a forming electrode.

A wire-cut EDM machine basically consists of a machine proper, composed of the work contour movement control unit (Numerical Control unit or copying unit), a worktable for mounting the work, and a wire drive section for accurately moving the wire at constant tension, a machining power supply that applies

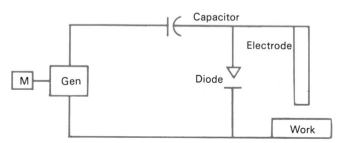

Figure 13–6 A basic circuit for a rotary impulse generator

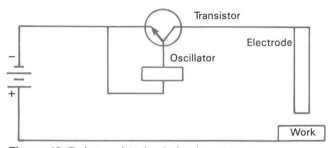

Figure 13–7 A transistorized circuit

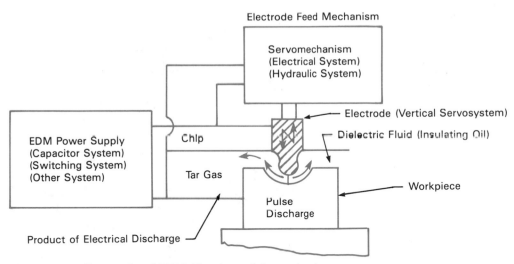

Figure 13–8 Conventional EDM *(Courtesy of Japax Inc.)*

electrical energy to the wire electrode, and a unit that supplies a dielectric fluid (distilled water) with constant specific resistance.

The wire most frequently used is 0.008-in (0.2-mm) copper wire. One spool of such wire can generally machine for 50 to 60 hours. Since the wire electrode is fed constantly during machining, its wear can be practically ignored.

Since a new portion of wire electrode is constantly supplied at a speed of about 0.39 to 1.17 in per second (10 to 30 mm per sec), machining can be continued without any accumulation of chips and gases. Since the electrode is very thin, an extremely small amount of discharge energy suffices for one spark. Cut surfaces are uniform and smooth. Machined dies can be used for production purposes without any polishing.

Machining a gear profile, Figure 13–11, is one example of the possibilities of wire-cut EDM. Free stamping dies and aluminum extrusion dies, Figures 13–12 and 13–13, are two more examples of the intricate cuts that can be made with wire-cut EDM.

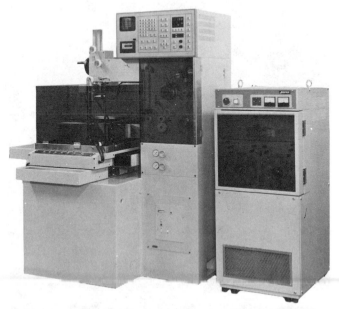

Figure 13–9 Japax wire-cut EDM with JAPT-3F N/C control unit *(Courtesy of Japax Inc.)*

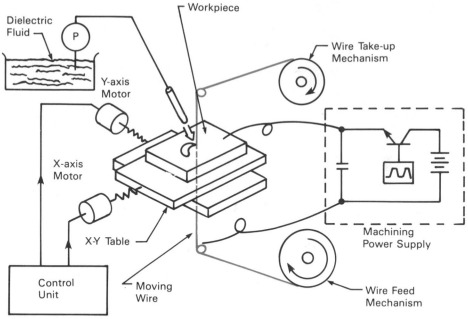

Figure 13–10 Wire-cut EDM *(Courtesy of Japax Inc.)*

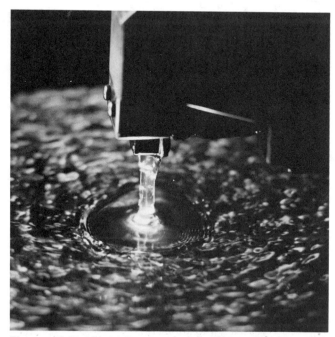

Figure 13–11 Machining an involute gear profile *(Courtesy of Japax Inc.)*

Figure 13–12 EDM wire-cut free stamping dies *(Courtesy of Japax Inc.)*

Figure 13–13 EDM aluminum extrusion dies, wire-cut (*Courtesy of Japax Inc.*)

REVIEW QUESTIONS

1. What is electrical discharge machining?
2. Describe the basic process of conventional electrical discharge machining.
3. In which direction does the spark cross the gap between the electrode and the work in conventional electrical discharge machining?
4. What is the purpose of a dielectric?
5. Define the term dielectric constant.
6. Why is kerosene a popular choice as a dielectric?
7. What materials are commonly used to make electrodes?
8. Name the factors that determine the size of the crater in conventional electrical discharge machining.
9. What is meant by deionization?
10. Describe the effect of current and frequency of discharge on the surface finish and on the rate of metal removal.
11. How does the heat of the process affect the work?
12. What is the recommended gap space between the electrode and the work?
13. What is the effect if the gap is too large? Too small?
14. What is overcut?
15. Compare wire-cut EDM to conventional EDM.
16. List three advantages of the use of wire in the wire-cut EDM process.

UNIT 13-3

THERMAL SPRAY PROCESSES

OBJECTIVES

After completing this unit, the student will be able to:

- state the basic function of the thermal spray process.
- describe each of the spray processes: electric arc spraying, plasma spraying, powder flame spraying, and wire flame spraying.
- explain how surface preparation affects the success of the metallizing process.
- list four factors to be considered in the machining of metal sprayed surfaces.
- list the types of coating materials used in the thermospray process.
- prepare a worn shaft for spraying.
- spray a worn shaft.

KEY TERMS

Thermal spray process	Metallizing
Thermal spray coating	Sprayed molten metal
Hardfacing alloy	Metallizing gun
Wire flame spraying	Undercutting
Powder flame spraying	Stepped undercutting
Electric arc spraying	Shoulder
Plasma spraying	Thermospray gun
Surface preparation	Powder orifice
Platelets	Self-fluxing spray
Arc gun	Oxidation-resistant
Plasma-forming gas	alloy
Convergent/divergent	Self-bonding coating
nozzle	Grit blasting

INTRODUCTION

The thermal spray processes apply thermal spray coatings of hardfacing alloys, metals and alloys, self-bonding and one-step materials, carbides, ceramics, cermets (ceramics and metals), plastic-base powders, and plastic-bond coats. Thermal spray coatings are produced from materials in either wire or powder form. The material is melted in a heat source and projected onto a surface or substrate to form a coating. As the molten particles strike the surface, they flatten out and form thin platelets that adhere to one another and the surface to form a dense, functional, protective coating. The four processes used are wire flame spraying, powder flame spraying, electric arc spraying, and plasma spraying. Thermal spray coatings are used to rebuild worn machine parts, to salvage mismachined parts, and to remanufacture used equipment. Other coating functions include spraying for wear resistance, heat and oxidation resistance, atmospheric and immersion corrosion resistance, and electrical conductivity or resistivity.

Thermal spray systems can range from a simple hand-held gun, to a fully automated system with a heavy-duty gun and an electronic control unit, to a computer-controlled system (CNC) with a heavy-duty gun and a six-axis, articulated arm robot and computer-controlled tilting turntable. These systems can coat virtually any size or configuration of parts at virtually any desired production rate.

ELECTRIC ARC SPRAYING

Electric arc spraying applies coatings of selected metals in wire form. Two electrically charged wires are fed through the arc gun to contact tips at the gun head. An arc that melts the wires at temperatures higher than 7200 degrees F (3982 degrees C) is created. Compressed air atomizes the molten metal and projects it onto a prepared surface. This process is excellent for applications that require a heavy coating build-up, or where wide, large surfaces must be sprayed.

PLASMA SPRAYING

Plasma spraying uses a plasma-forming gas (usually argon or nitrogen) as both the heat source and the propelling agent for the coating. A high-voltage arc struck between an anode and cathode within a specially designed spray gun excites the plasma gas by ionization. The excited gas is forced through a convergent/divergent nozzle. As the gas exits from the nozzle, it returns to its natural state, liberating extreme heat. Powdered spray material is injected into the hot plasma stream in which it is melted and projected at high velocities onto a prepared surface. Wherever parts wear or corrode, there is a potential plasma application.

METALLIZING

Metallizing is the process of spraying molten metal onto a surface to form a coating. Pure or alloyed metal is melted in a flame (e.g., oxy-hydrogen is one gas combination used) and atomized by a blast of compressed air into a fine spray. This spray builds up onto a previously prepared surface to form a solid metal coating. Because the molten metal is accompanied by a large blast of air, the object being sprayed does not heat up very much. Metallizing is known, therefore, as a "cold" process of building up metal.

Sprayed metal is a metallurgical material with entirely different physical properties than the original metal. Sprayed metal is generally harder, more brittle and porous than the original metal, and has excellent bearing characteristics due to oil retention in the pores of the metal.

Sprayed metal is most commonly used for building up worn parts such as spraying stainless steel on worn bronze pump runners, or salvaging mismachined parts. It is particularly good for this type of work because the process does not heat up the parts and cause warpage. Sprayed metal is also used extensively for corrosion resistance, e.g., aluminum and zinc sprayed on iron or steel. There are many more applications such as electrical shielding, electrical conductive elements for radiant heaters, and soldered connections for carbon resistors and brushes.

There are two types of metallizing or metal spraying: the wire process and the powder process. In the wire process, a metallizing gun, Figure 13–14, is used to spray on the metal. Wire can be seen being fed through the gun. Figure 13–15 shows the complete metallizing installation for the wire process.

Figure 13–14 A large shaft being built up by metal spraying. The financial savings on one job of this size could more than pay for the cost of the spraying gun. *(Courtesy of Metco Inc.)*

FLAME SPRAYING, WIRE PROCESS

Surface Preparations

The reason for grooving the surface to be sprayed, Figures 13–16 and 13–17, is explained as follows. Sprayed metal has a laminated structure and is made up of many fine, flat particles. The strength of sprayed metal is much greater parallel to these flat particles and is weaker in a direction perpendicular to the laminations. The sprayed metal splits parallel to the laminations more easily than it can be broken in the other direction. If the laminations of sprayed metal are folded up and down over relatively large grooves or bumps, the strength of the sprayed metal is improved materially and it cannot split. The folded laminations eliminate shear planes. The grooves restrain any tendency that the metal has to shrink and loosen the bond.

Undercutting

Stepping the undercut on heavily worn shafts, Figure 13–18, is done to avoid extra spraying and expense. The corners should always be rounded under the sprayed metal.

Sprabond wire is very hard. Tapering off the coating thickness eliminates hard rings, reduces stresses at the ends of the buildup, and requires a minimum of machining or grinding. It also results in a nearly invisible line where sprayed metal and base metal meet at the shoulder. Figure 13–19 illustrates regular undercutting and Sprabond wire undercutting.

Treatment of Shoulders

Deep, square, shoulders should be avoided where possible. Stepping down the undercuts or tapering the undercuts is preferred practice, Figure 13–20.

Shoulders should usually be cut square with a 0.015-in (0.38-mm) to 0.020-in (0.51-mm) radius in the corner. A fairly common practice with some machinists is dovetailing the shoulder about 15 to 20 degrees. This is not ordinarily recommended because it leaves a weak spot at the top of the dovetail.

Leave a shoulder at the end of the undercut section wherever possible, Figure 13–21. When a shoulder cannot be left at the end of the undercut section, undercut the end of the shaft, Figure 13–22. The round corner should be cut on the sprayed metal when finishing.

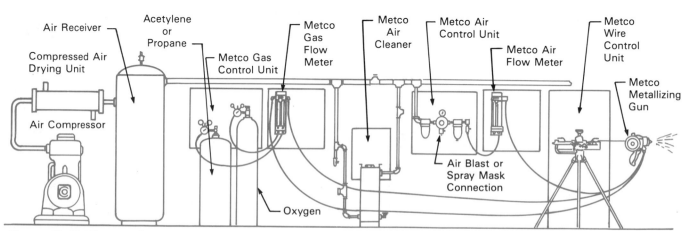

Figure 13–15 The metallizing installation for the wire process *(Courtesy of Metco Inc.)*

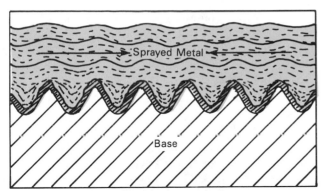

Figure 13-16 Spraying over grooves. Shrinkage is restrained by the grooves. *(Courtesy of Metco Inc.)*

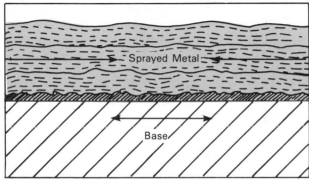

Figure 13-17 Spraying on a smooth surface. Shear stress on the bond is due to shrinkage. *(Courtesy of Metco Inc.)*

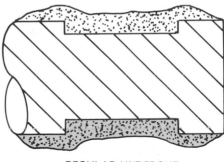

REGULAR UNDERCUT

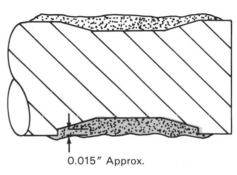

0.015″ Approx.

SPRABOND WIRE UNDERCUT

Figure 13-19 Regular undercutting and Sprabond wire undercutting *(Courtesy of Metco Inc.)*

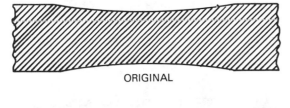

ORIGINAL

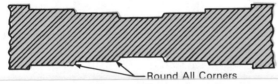

Round All Corners

STEPPED UNDERCUT

Figure 13-18 Stepped undercutting *(Courtesy of Metco Inc.)*

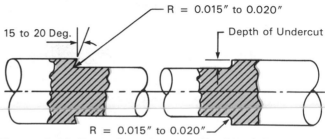

15 to 20 Deg.

R = 0.015″ to 0.020″

Depth of Undercut

R = 0.015″ to 0.020″

Figure 13-20 Dovetails and square shoulders *(Courtesy of Metco Inc.)*

Shaft Preparation

The shaft to be metallized should be mounted in a lathe and the worn section turned undersize to clear up the worn section. Groove the shaft with a tool bit ground as shown in Figure 13–23. A cut-off tool blade makes a good cutter for this work. Grind it to a thickness of 0.045 in (1.15 mm) to 0.050 in (1.28 mm), and round the end. Rake and clearance should be as in standard practice for the material of the shaft, to give a good cut. The radius of the corners of the tool should not be over 0.020 in (0.51 mm), as the sides of the groove should be straight at the surface of the shaft.

The grooves should be cut 0.025 in (0.64 mm) deep and may be cut as a continuous thread of 16 threads per inch (1.5-mm pitch) or by cutting a number of separate grooves. In the case of cast iron, the lathe should be set to cut 14 threads per inch (1.75-mm pitch). When cutting separate grooves, the ridges should be 0.015 in (0.38 mm) wide except for cast iron which calls for ridges 0.025 in (0.64 mm) wide, Figure 13–24.

Next, apply the Metco rotary-shaft preparing tool. The tool is mounted in the lathe tool post and is run back and forth over the shaft to roughen the surface on the top of ridges. Cutting speed should be 200 to 300 ft per min (60 to 90 m per min), and a feed of about 0.020 in (0.51 mm) per revolution should be used. About three passes should be made with this tool.

One of the oldest methods of shaft preparation is the rough thread method. Cut an ordinary V-thread of 60 degrees. The cutter should have negative back rake and be set below center so as to tear as much as possible. Usually 20 to 32 threads per inch (1.25-mm to 0.75-mm pitch) is used.

Spraying

In the actual spraying operation, the distance between the gun and the work will vary from 4 to 10 in (100 to 250 mm), depending on the equipment used. The angle at which the sprayed metal strikes the surface of the work should be 90 degrees wherever possible and never less than 45 degrees, Figure 13–25. The flame setting of the gun should be neutral, not oxidizing or carburizing. Figure 13–26 illustrates a cross-section of the sprayed surfaces before final finishing takes place.

Machining Metal Sprayed Surfaces

After a shaft has been properly prepared and sprayed, there is usually a ragged edge of sprayed metal at the ends of the undercut section. This ragged edge may cause the start of a crack that could penetrate into the main section of the coating. It is good practice to remove this ragged edge immediately after spraying by machining or grinding it off before the main finishing is begun. Blunt-nose tools are recom-

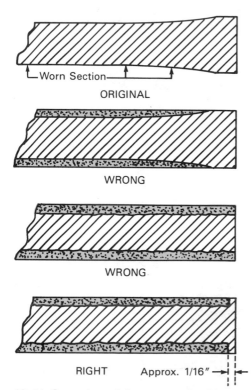

Figure 13–21 Correct and incorrect shoulder treatment *(Courtesy of Metco Inc.)*

mended for machining the undercut section before finishing takes place, Figure 13–27.

One problem frequently encountered is the difficulty of machining the hard ring that might form adjacent to the ends of the undercuts. This can be minimized by concentrating the spray into the corners when first starting to spray so as to partially fill and round the corners. It is advisable to finish these hard sections first with the blunt-nose tool which should

be fed into the raised sections and set laterally into the "flash" or curled up edges.

The machinable metals may be machined quite readily with high-speed cutters, if desired. The usual procedure is to machine within 0.010 in to 0.015 in (0.256 mm to 0.38 mm) of the finished dimension in the first cut, taking the balance on the finishing cut, allowing for filing and polishing, Figure 13–28.

Carbide cutters, Figure 13–29, should always be

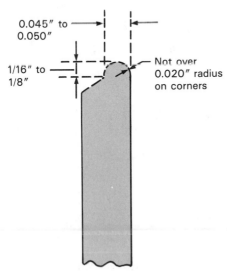

Figure 13–22 Treating the end of a shaft *(Courtesy of Metco Inc.)*

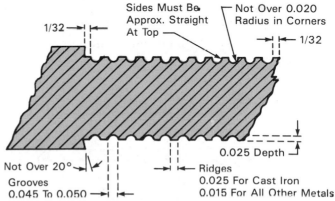

Figure 13–24 Standard grooving *(Courtesy of Metco Inc.)*

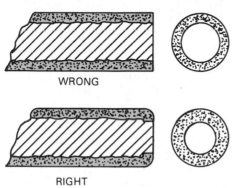

Figure 13–23 Top view of a grooving tool

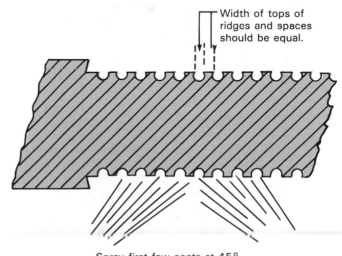

Figure 13–25 Metal spraying *(Courtesy of Metco Inc.)*

used if possible. They are the best for all machining operations on sprayed metal. Sprayed metal, including the softer sprayed metals, wears tools more rapidly than ordinary metal.

Carbide cutters simplify the machining of sprayed metals and give better finishes. Machining speeds are faster, and some of the harder metals, which otherwise require grinding, may be readily machined.

Due to the particular structure of sprayed metals and to the oxide and carbide inclusions, the nose of the cutting tool is subjected to a great deal of wear and abrasion. Carbide tools reduce this difficulty to a minimum because of their extreme hardness and wear resistance.

It is advisable to use cutting oils (mineral oil and kerosene, 50–50) on some sprayed metals. Sprabronze C, Sprasteel L.S., and Sprasteel 10 and 25 are those metals that would benefit, but metals such as Metcoloy #1, Nickel, Monel, Sprabronze A, and Sprabronze P would not. These are Metco brand names. The cutting oil is brushed on and allowed to stand for 20 to 30 minutes before machining.

FLAME SPRAYING, THERMOSPRAY OR POWDER PROCESS

Coating applications with the wire-type of flame spray equipment are limited to those materials that can be formed into wire or rod. The thermospray processes permit the use of an almost infinite variety of metals, alloys, ceramics, and cermets (ceramic and metal) that are available as powders, Figure 13–30.

A typical thermospray operation is illustrated in Figure 13–31. First, the worn shaft is undercut, Figure 13–31A. This is followed by grooving the surface and preparing the shoulders, Figure 13–31B. Next, a bond-

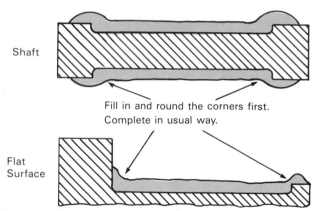

Figure 13–26 A metal-sprayed surface before machining (*Courtesy of Metco Inc.*)

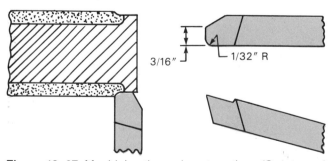

Figure 13–27 Machining the undercut sections (*Courtesy of Metco Inc.*)

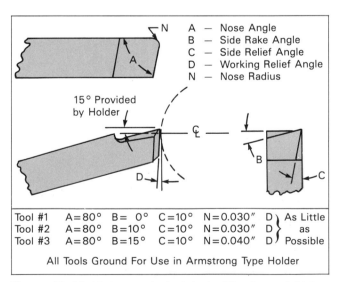

Tool #1	A=80°	B= 0°	C=10°	N=0.030″	D) As Little
Tool #2	A=80°	B=10°	C=10°	N=0.030″	D } as
Tool #3	A=80°	B=15°	C=10°	N=0.040″	D) Possible

A — Nose Angle
B — Side Rake Angle
C — Side Relief Angle
D — Working Relief Angle
N — Nose Radius

All Tools Ground For Use in Armstrong Type Holder

Figure 13–28 High-speed steel tools (*Courtesy of Metco Inc.*)

ing undercoat is applied, Figure 13–31C, to ensure the bonding of the build-up coat, Figure 13–31D. The built-up shaft is finally finished by turning, Figure 13–31E.

The Thermospray Gun

The thermospray gun, Figure 13–32, is designed for maximum versatility in the application of self-bonding alloys, ceramics, cermets, and certain oxidation-resistant alloys. High thermal efficiency permits the application of high melting point materials and at the same time affords top speeds in spraying metals and alloys.

Ordinarily, the gun requires no air and only two light hoses are used to supply oxygen and fuel gas (acetylene or hydrogen, as a rule). The powder is fed from a reservoir attached directly to the gun. A small reservoir for hand use, a larger reservoir for lathe-mounted guns or for large-scale production work is used, as these powders are quite heavy, Figure 13–33.

By changing powder orifices, any required powder feed rate can be obtained. This permits the spraying of the entire range of metals, alloys, ceramics, and cermets which can be applied by the thermospray processes. The only other gun change needed is the nozzle. There are several nozzle types for different purposes.

Types of Coating Material

There are four basic types of coating materials used in the powder process. These are alloys for fused coatings, oxidation-resistant metals and alloys generally used in the "as sprayed" state, ceramics, and self-bonding alloys.

Alloys for fused coatings are often called self-fluxing alloys. They are highly resistant to corrosion

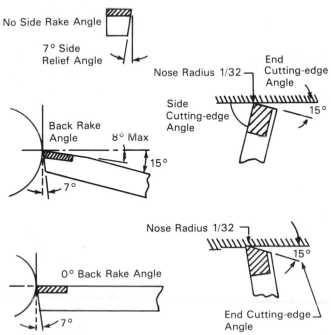

Figure 13–29 Tool angles for cemented carbide tools (*Courtesy of Metco Inc.*)

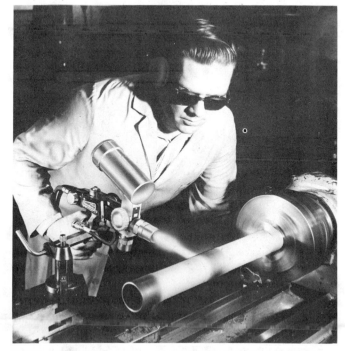

Figure 13–30 Thermospray coating being applied (*Courtesy of Metco Inc.*)

and oxidation, generally superior to premium stainless steels, and range in hardness from Rc 30 to Rc 62. They are highly resistant to wear and abrasion and will often outwear steel of equivalent hardness by as much as 20 to 1. Some of the common powders used are nickel-chromium-boron alloys, cobalt-base, hard facing alloys, tungsten carbide blended with a self-fluxing alloy, and tungsten carbide in a nickel-chromium matrix.

Oxidation-resistant alloys are used in the "as sprayed" state, which means that no additional finishing is required after spraying. These are selected primarily on the basis of oxidation resistance. These powders are premium stainless steel, high-chromium stainless steel, high-purity, spherical, aluminum powder, and spherically shaped, high-purity, copper powder.

The ceramic materials most commonly used are alumina and zirconia. Zirconia is used principally as a thermal barrier for high-temperature service, and alumina is used as heat barrier and for wear resistance. Alumina has also been used as an undercoat for plastics such as Teflon.

A self-bonding coating is both bond and buildup. All coatings in this category require only the one step. No separate bond coating is needed and less undercutting is required. Metallurgical bonds have strengths of up to 6000 pounds per square inch and operator training is minimal.

A Undercutting the worn shaft

B Grooving the surface and preparing the shoulders

C Applying bonding undercoat (Metco 450)

D Applying the build-up coat

E Finish turning the built-up shaft

Figure 13–31 Flame spraying powder process *(Courtesy of Metco Inc.)*

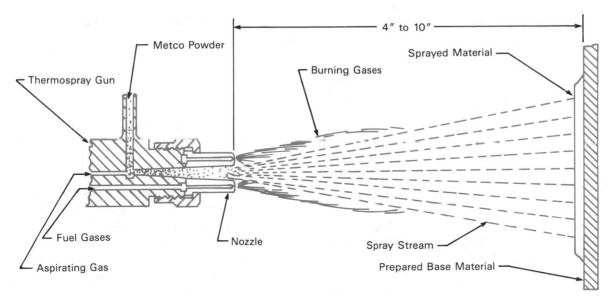

Figure 13-32 A Metco thermospray gun nozzle *(Courtesy of Metco Inc.)*

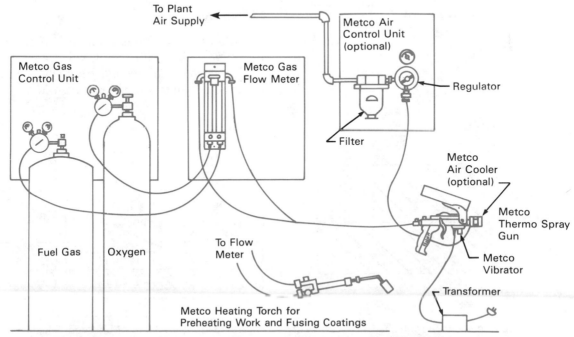

Figure 13-33 A thermospray installation for fused coatings *(Courtesy of Metco Inc.)*

Surface Preparation

The importance of proper surface preparation cannot be overemphasized. It is often the most critical part of the total project. The surface to be coated and also the adjacent surfaces should be scrupulously clean. Oil, grease, paint, or other foreign matter should be removed.

Cylindrical work such as a shaft section or a pump plunger usually requires undercutting of the area to be coated, Figure 13–34. Depth of the undercut is usually determined by service considerations. If maximum wear allowance is 0.020 in (0.51 mm) on the radius, the work should be undercut 0.025 in to 0.030 in (0.64 mm to 0.76 mm) on the radius in order to leave a continuous coating after maximum wear has taken place.

Shoulders at the ends of the undercut section should be opened at about a 40 to 45 degree angle to the axis of the work, Figure 13–35.

One of the many methods of surface preparation before metal spraying is grit blasting. Grit or abrasive blasting is the most satisfactory and most versatile of several methods of surface preparation. Many types of grits such as angular chilled iron, aluminum oxide, silica sand, Joplin flint, and crushed clay are used.

Spraying Techniques

The surface of the work should be properly prepared and preheated to 200 degrees F (93 degrees C) or higher. On cylindrical work rotating on a lathe, surface speed should be about 20 to 100 ft per min (6 to 30 m per min), with a traverse of 0.250 in to 0.500 in (6.35 mm to 12.7 mm) per revolution.

This will apply a coating of 0.004 in to 0.006 in (0.103 mm to 0.154 mm) per pass. For hand-held guns on stationary work, the work should be traversed at a rate of 12 in (305 mm) in 2 to 3 seconds, laying down "stripes" of about 1/2 in (12.7 mm) wide at each pass.

On work that is undercut, the spray should be directed into the corners of the undercut at an angle of about 45 degrees. This tends to fill in the area next to the shoulders and prevents excess dust and oxidation at these points.

Finishing

Most fused coatings of the self-fluxing alloys must be finished by grinding. Oxidation-resistant alloy coatings can usually be machined by carbide cutters. Ceramic coatings are generally ground.

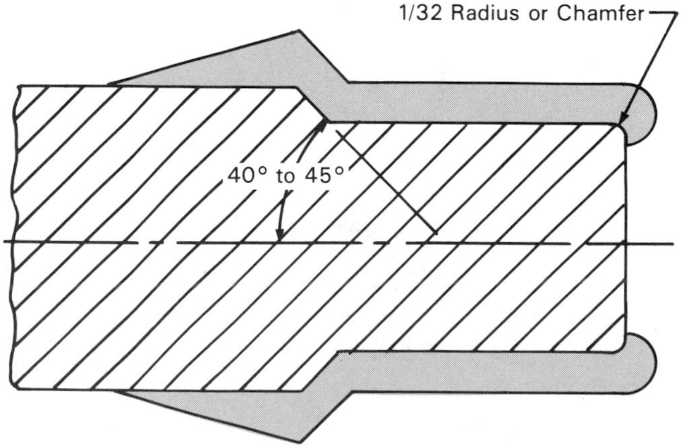

Figure 13–34 Undercut shoulder and end chamfer *(Courtesy of Metco Inc.)*

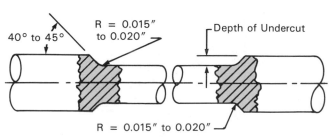

Figure 13–35 Shoulders of undercut sections *(Courtesy of Metco Inc.)*

REVIEW QUESTIONS

1. What are the basic functions of the thermal spray processes?
2. What are the basic components of an electric arc spraying system?
3. Name the preferred applications for electric arc spraying.
4. What gases are commonly used in plasma spraying as both heat source and propellant for the coating material?
5. What is metallizing?
6. Define the term sprayed metal and list four common applications.
7. Briefly describe the wire process of flame spraying.
8. What surface preparation is necessary before spraying takes place?
9. Why is surface preparation important to the success of the wire process?
10. What is undercutting?
11. How does Sprabond undercutting differ from regular undercutting?
12. What shoulder preparation is necessary before wire spraying?
13. What shaft preparation is required before wire spraying?
14. What is the average distance between the gun and the work in the wire spray process?
15. What type of flame is required for the process?
16. Why is machining of metal sprayed surfaces required?
17. What is the procedure to minimize the formation of a hard ring adjacent to the ends of the undercuts?
18. Name the type of cutter preferred for machining of sprayed metals.
19. What are some limitations of the wire-type method of spraying?
20. What are the steps in the basic thermospray (powder) process?
21. Briefly describe the operation of the gun in the thermospray process.
22. In thermospray, what are four basic types of coating materials?
23. List three characteristics of self-fluxing alloys.
24. What alloys require no finishing after spraying?
25. What are two common ceramic materials used for spraying? Give an application of each.
26. List two advantages of the use of bonding coatings.
27. What shoulder preparation is necessary before thermospraying?
28. What is meant by grit blasting?

ELECTRICAL DISCHARGE MACHINING (EDM) – NEW TRENDS

Conventional EDM is discussed in this text. Now, exciting innovations are being added to EDM. Advanced power supplies developed for EDM permit accurate discharging with minimum waste, enabling the workpiece to be machined at high speed with minimum electrode wear. The ability of the power supply to put out an increased number of discharge pulses, plus controls that can adjust the discharge energy more accurately, make it possible to produce better and more evenly finished surfaces.

In the single-axis control in conventional EDM, the spark jumps vertically to the work and cuts out a piece of material. In other words, conventional EDM is movement and electrical discharge along the Z axis. However, newer EDM machines permit electrical discharging to be done horizontally. This is called lateral machining and the applications possibilities are endless. The ability to discharge electrically on the X and Y axes as well as on the Z axis means that some EDM machines are capable of two-axis machining, that is, that can cut any two of the axes. Further, two-axis control has been refined into three-axis control.

Another innovation in EDM is the development of the Automatic Rotary Indexing Chuck. A schematic drawing of the rotary head mechanism is shown in Figure 1. This chuck enables the electrode to rotate as it is cutting. The process is called C-axis cutting. Adding these new capabilities to CNC programming and more accurate and sensitive servo systems means that

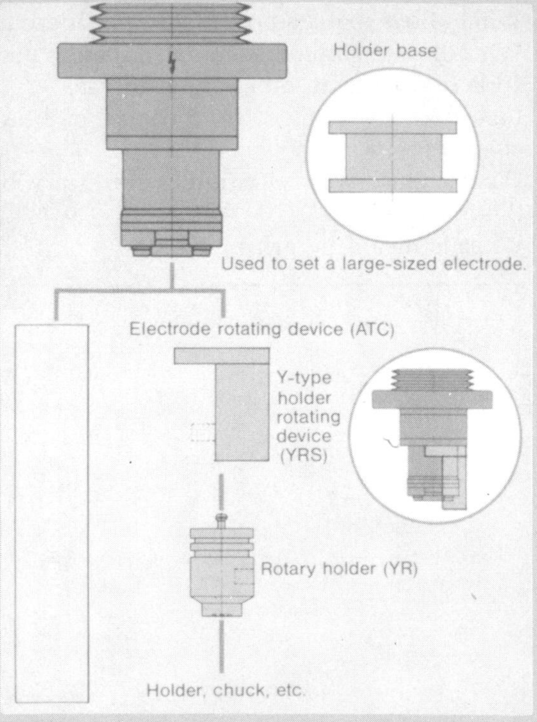

Figure 1 Schematic of an automatic rotary chuck *(Courtesy of LeBlond Makino)*

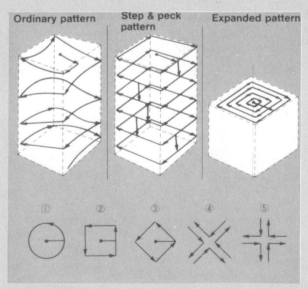

Figure 2 Pattern machining *(Courtesy of LeBlond Makino)*

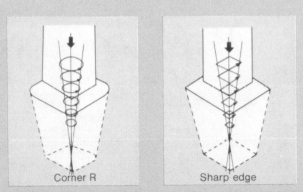

Figure 3 Taper machining *(Courtesy of LeBlond Makino)*

the new generation of EDM has the flexibility to do pattern machining, taper machining, thread cutting, and more. Pattern machining, Figure 2, makes use of one of the axes (X, Y, or Z) and follows such pattern modes as Ordinary Pattern, Step and Peck Pattern, or Expanded Pattern.

In taper machining, a nontapered electrode moves in a gradually reduced pattern radius along with vertical machining to produce a tapered shape, Figure 3. Threads of any pitch can be machined when digitally set for depth (Z axis)

and rotating direction (C axis). It is possible to cut threads in cemented carbides and in steels after heat treatment has been done. The process may be applied to the machining of helical gears, worm shafts, and screws. Finely finished lateral surfaces are produced.

Wire-cut EDM is relatively new in machining technology, dating to about the mid 1970s. The ability of an EDM machine to cut horizontally rather than just vertically was a key in its development. This process brought a new dimension to EDM machining. The development of new and better control mechanisms, CNC controls and programming, improved machining speeds, and better surface finish all contribute to the new flexibility for EDM machining.

The schematic drawing, Figure 4, illustrates a typical setup showing the wire as it is automatically threaded through the EDM machine. The workpiece is shown in a sectional view (cross-hatched). Figure 5 shows intricate cuts by

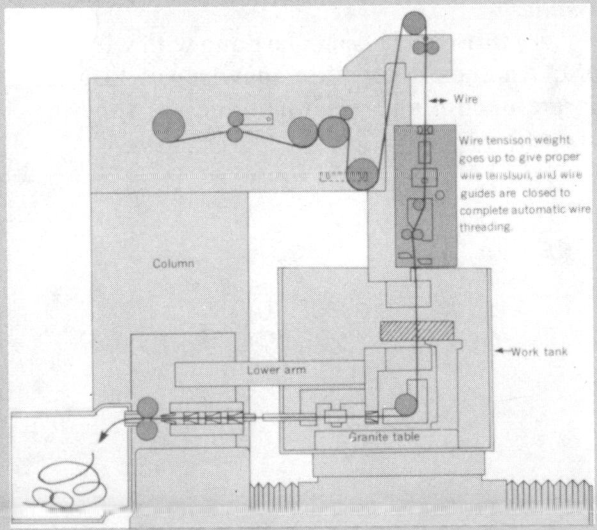

Figure 4 Schematic of wire-cut EDM *(Courtesy of LeBlond Makino)*

Figure 5 Intricate cuts by wire-cut EDM *(Courtesy of Japax Inc.)*

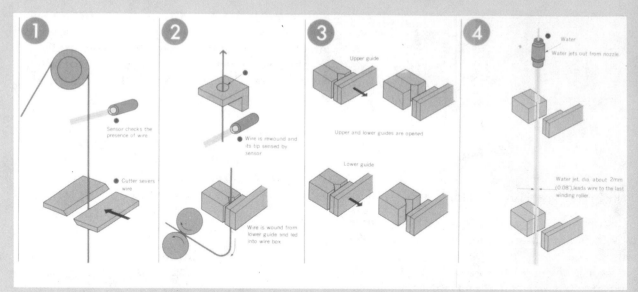

Figure 6 Schematic of automatic wire threading *(Courtesy of LeBlond Makino)*

wire-cut EDM. Once a hole is cut, an automatic wire threading function (new in wire-cut EDM), will cut the wire when the hole is completed. The work will be moved into position for the next hole and new wire will automatically be threaded through for the next cut. The wire threading sequence is shown in Figure 6. This sequence is programmed into the computer. The program also includes data on selecting machining conditions, changing dielectric quantities and other functions. The use of the automatic wire threader program means both rough machining and finish machining can be carried out. Changing the machining conditions, dielectric amount, and other variables by means of the computer programming, makes unmanned machining possible.

Very intricate cuts can be done with wire-cut EDM. Another innovation in wire-cut EDM is micron machining which can achieve superior finishes.

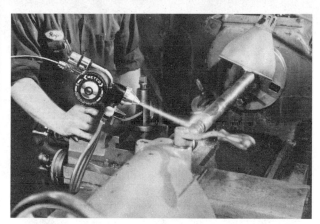

A typical metallizing job using a wire-type spray gun. The metal wire can be seen feeding through the back of the gun. Both eye and ear protection are required for this type of job. In this operation, compressed air is mixed with the gas combination, resulting in a very loud hissing noise. Hearing damage can result if the proper protective equipment is not used. *Courtesy of Metco Inc.*

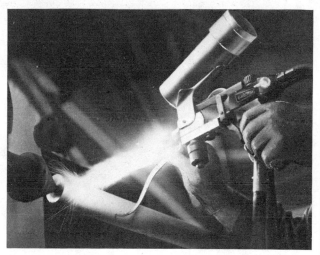

A pump part is being sprayed with a powder spray gun. The powder container can be seen on the top of the gun. *Courtesy of Metco Inc.*

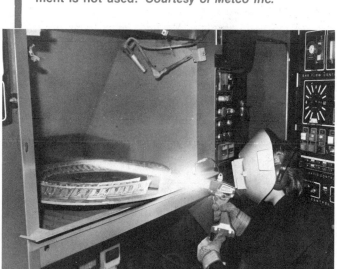

Applying a wear-resistant coating to a jet engine seal surface using a hand-held plasma gun. *Courtesy of Metco Inc.*

In metallizing, a different metal can be sprayed on a part. For example, stainless steel (chromium/nickel) is being sprayed on bronze pump runners so that a pump can be quickly returned to service and will last longer than the original. *Courtesy of Metco Inc.*

Resurfacing a paper mill drier roll with a heavy-duty, high-energy plasma gun. *Courtesy of Metco Inc.*

Using the plasma spray process to apply a premium material coating over lightweight or low-cost substrates. *Courtesy of Metco Inc.*

Applying a protective coating with the plasma gun held by an automated, computer-controlled robot. *Courtesy of Metco Inc.*

Applying a protective coating against wear and heat oxidation to an aircraft turbine engine part. *Courtesy of Metco Inc.*

CHAPTER FOURTEEN

METALLURGY

A machinist must learn to cut various materials. It is also necessary for the machinist to know a great deal about the composition of those materials. Metallurgy, or the study of metals, can be divided into two broad categories: ferrous metallurgy dealing with those metals that contain iron, and nonferrous metallurgy which is concerned with those metals that do not contain iron. The machinist will work with metals in both categories; therefore, a thorough knowledge of these metals is a must.

In manufacturing processes, the first step is to produce the ferrous metals. Second, the ferrous metals must be shaped. Third, many ferrous metal parts must be heat treated and tested for hardness. The machinist is not normally involved in the first step, but is usually involved directly in the second and third steps.

Similarly, the machinist must also have a thorough knowledge of nonferrous metals if they are to be shaped properly to form metal parts. This knowledge is also critical to the proper design and use of cutting tools.

UNIT 14-1

THE METALLURGY AND MACHINING OF FERROUS METALS

OBJECTIVES

After completing this unit, the student will be able to:

- describe how the blast furnace converts raw materials to pig iron used to make cast iron or steel.
- list the primary components of cast iron.
- define the terms graphite, ferrite, cementite, and pearlite.
- describe each of the following, including composition and characteristics: gray cast iron, white cast iron, malleable cast iron, and nodular cast iron.
- state the effect of each of the following on cast iron: silicon, sulfur, phosphorus, and manganese.
- state the effect of each of the following on cast iron: copper, aluminum, tin, vanadium, and titanium.
- describe the machining characteristics of white cast iron, gray cast iron, malleable cast iron, and nodular cast iron.

KEY TERMS

Metallurgy	Tap off
Ferrous metal	Pig iron
Cast iron	Cupola furnace
Steel	Electric furnace
Iron ore	Graphite
Limestone	Ferrite
Flux	Cementite
Bituminous coal	Pearlite
Coke	Slow cooling
Blast furnace	Rapid cooling
Iron-carbon compound	Gray cast iron
Slag	White cast iron
Malleable cast iron	Ferromanganese

Nodular cast iron	Sand pocket
Spiegeleisen	Diamond

INTRODUCTION

Two of the most common ferrous metals machined in a machine shop are cast iron and steel, each in one form or another. Neither metal resembles the other either physically or in cutting action, yet both metals have a common beginning. The raw materials needed for producing both of these metals are iron ore, limestone, and bituminous coal in the form of coke.

Iron, which is found in the form of an oxide, must be separated from the oxygen in order to be usable. Other elements such as phosphorus, sulfur, silicon, and manganese, along with many impurities are also found in the iron ore.

Limestone, which is mostly calcium carbonate, is used as a flux or cleanser. It combines with the impurities in the iron ore and forms a slag that floats on top of the molten iron in the blast furnace and is tapped off separately from the iron.

A bituminous coal is made into coke which is mostly carbon. This gives heat in the blast furnace, but more important, it supplies the carbon that is needed in this process.

THE BLAST FURNACE

The three raw materials are charged into the top of the blast furnace, Figure 14–1, sometimes in layers and sometimes mixed together. The main objective in the blast furnace is to separate the iron from the oxygen. Carbon combines with the oxygen in the ore and separates it from the iron. At the same time, carbon combines with the iron and forms an iron-carbon compound. The limestone combines with many of the impurities in the iron ore to form a slag that floats on top of the molten iron and is tapped off from the furnace. The iron is tapped off next to form a product called "pig iron." Pig iron is made up mainly of the following elements:

Iron	Carbon	Silicon	Manganese	Phosphorus	Sulfur
94.25%	3.5%	1.25%	0.9%	0.06%	0.04%
to	to		to	to	
88–96%	4.25%		2.5%	3.0%	

The slag, which was once considered a waste product, is used in the manufacture of Portland cement, which is one constituent of concrete. It is also used in the place of crushed rock in the making of concrete and as a filler for railroad beds.

Pig iron in the molten state or in the solid state can go in either of two directions. It can go to the iron foundry to be refined into cast iron, or it can go to the steel plant to be refined into steel. The carbon content of cast iron is about 3.5%, while that of steel varies from about 0.05% to a maximum of 1.6% for practical purposes. Pig iron has also been used directly from the blast furnace to make cast-iron castings. Sometimes a mixture from several blast furnaces is used to secure the proper proportioning of chemical elements.

A sketch of an early blast furnace is shown in Figure 14–2. Figure 14–3 illustrates how the term *pig iron* reputedly arose. Molten iron from the blast furnace flowed from the main runner into smaller molds resembling suckling pigs. Whether or not this story is true, the term *pig* is applied to many different metals in this form.

CAST IRON

Cast iron is usually made from pig iron in a cupola furnace, Figure 14–4, after which it is cast into molds

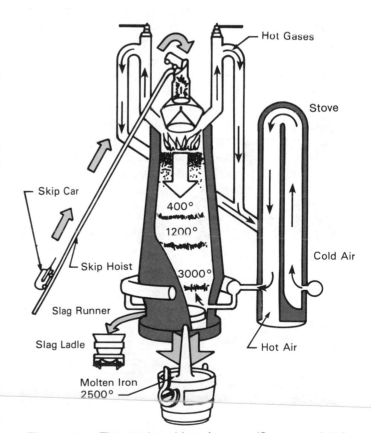

Figure 14–1 The modern blast furnace *(Courtesy of U.S. Steel Corporation)*

or is stored in the form of pigs. As much as two-thirds of scrap cast iron is used in the charge along with fuel (coke) and flux (limestone). The cupola furnace, which is somewhat like a small blast furnace, is the simplest and cheapest method of making cast iron. The materials are charged in alternate layers, slag is tapped off, cast iron is tapped off, and the furnace is emptied.

Electric furnaces are sometimes used to make cast iron, but they are more expensive in initial cost and in operation. Control is easy and a superior cast iron is made, but the bulk of cast iron made is still cupola iron.

The second most important element in cast iron (next to iron) is carbon. It may be present as free carbon (graphite), or as combined carbon and iron to form cementite, or as cementite combined with free iron (ferrite) to form pearlite. Cast iron, then, may be composed of ferrite, graphite, cementite, and pearlite.

Ferrite is free iron that is soft and weak, yet tough. **Graphite** is free carbon, a flaky, greasy, weak material. **Cementite** is combined carbon and iron composed of one part carbon and fifteen parts iron. It has great strength, is very brittle, and is harder than the hardest steel. Pearlite is six parts ferrite and one part cementite. It is more ductile, is not as strong nor as hard as cementite, but is stronger and harder than ferrite. The proportions of these four materials may vary considerably, with some cast iron containing graphite, and others containing practically no cementite.

Slow cooling in the solidifying of cast iron allows the carbon to separate out as graphite. If aluminum and silicon are present, slow cooling aids in the separation of carbon and iron during solidification. Slow cooling tends to produce gray cast iron.

In the **rapid cooling** of cast iron, the carbon tends to stay combined with the iron because it does not have the time to separate out, and white cast iron is produced. Chromium and manganese, when present, help to keep the carbon in the combined form.

Carbon has more effect on cast iron than any other element except the iron itself. Carbon content usually varies from 3% to 4%. The properties of cast iron depend to a large extent on the amount of carbon present and also on the form, i.e., free or combined.

Classes of Cast Iron

In gray cast iron, graphite content varies from 2% to 4% and cementite is less than 1.5%. It is a mixture, then, of ferrite, graphite, and cementite. Gray cast iron is softer, tougher, weaker, and less brittle than white cast iron. It is dark gray in color and has a greasy feel (due to the graphite flakes).

Figure 14–2 An early blast furnace (Courtesy of U.S. Steel Corporation)

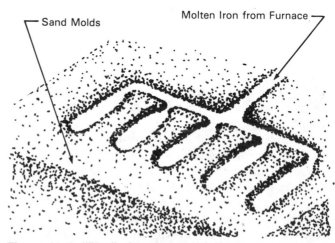

Sand Molds

Molten Iron from Furnace

Figure 14–3 "Pigs" (Courtesy of U.S. Steel Corporation)

White cast iron is essentially cementite and pearlite. It is harder than the hardest steel and more brittle than gray cast iron. Ductility is practically zero. Most white cast iron is used in the making of malleable cast iron.

Malleable cast iron is annealed white cast iron in which the carbon has been separated from the iron without forming flakes of graphite as in gray cast iron. Malleable cast iron is quite tough and is much more able to withstand shocks and blows than the cast iron from which it is made. Malleable cast iron has all the advantages of gray cast iron with respect to casting into various shapes, plus toughness, ductility, and strength nearly equal to that of some steels.

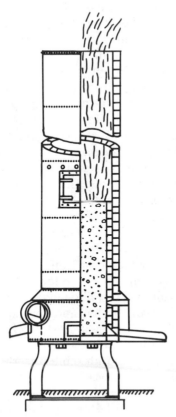

Figure 14-4 A cupola furnace *(Courtesy of U.S. Steel Corporation)*

In **nodular cast iron**, graphite is present in a ball-like form rather than flake form. The addition of small amounts of magnesium or cerium-bearing alloys together with special processing produces this spheroidal graphite structure and results in a casting of high strength and appreciable ductility. It is machinable, but it is much harder than gray cast iron.

The Effect of Silicon, Sulfur, Phosphorus, and Manganese on Cast Iron

In addition to iron and carbon, cast iron contains small amounts of silicon, sulfur, phosphorus, and manganese. The amount of silicon in cast iron usually varies from 0.5% to 4.0%. From 0.8% to 1.8% of silicon makes the iron soft and tough. A greater or a lesser amount makes the iron brittle and hard. About 3.0% makes the carbon separate out in flake or graphite form. Iron with silicon is good in foundry work because it tends to increase the fluidity, eliminate blow holes, and decrease shrinkage when used properly. It reduces the chill in castings (reduces the tendency to become surface hardened due to rapid cooling).

Good cast iron usually contains more than 0.15% of sulfur. This element helps to keep the carbon in the combined form and tends to make the iron hard, brittle, and weak. It also causes "red shortness," i.e., makes the iron very brittle in a red heat. Such iron is not good for steel manufacture. In foundry work, sulfur reduces the fluidity and increases the tendency to chill.

The fusibility and fluidity of the iron is increased by the addition of from 2% to 5% of phosphorus, thus helping in the making of fine castings in the molds. From 1.0% to 1.5% is often used for fluidity and softness, but more than 1.5% tends to make the iron brittle and hard. For best strength results, not over 0.55% of phosphorus should be used. Phosphorus in iron to be used for steel making should not be more than 0.07%.

The amount of manganese present may vary from 0% to 80%, but rarely exceeds 2% in ordinary castings. Less than 1.0% of manganese has practically no effect on the iron, while about 1.5% makes the iron fine grained and hard to machine. Foundry iron usually contains less than 1.0% of manganese. This element increases the shrinkage, decreases the magnetism, and increases the solubility of carbon in the iron. **Spiegeleisen** (sometimes called spiegel) is iron containing from 10% to 50% of manganese. It is capable of taking a high polish and is very hard, resisting cutting by hard cast alloy tools. If the iron contains more than 50% manganese, it is called **ferromanganese**.

The Effect of Other Chemical Elements on Cast Iron

Many other elements are often found in cast iron, usually in very small quantities. These elements may have some effect on the properties of the iron.

Varying amounts of copper, from 0.1% to 1.0%, close the grain of cast iron but do not cause much brittleness. Copper does, however, make the iron unsuitable for making malleable iron.

When aluminum, from 0.2% to 1.0%, is added to the ladle in the form of an iron-aluminum alloy, it increases the softness and strength of white cast iron. When added to gray cast iron, it softens and weakens it. About 0.1% of aluminum has the same effect as 1.5% of silicon. Too much aluminum is undesirable.

Tin increases hardness and fusibility and decreases malleability as well as making the iron unfit for conversion into malleable cast iron.

Vanadium, in very small quantities, increases softness and ductility. As much as 0.15% added to the ladle in the form of a ground iron vanadium alloy greatly increases the strength of cast iron. Vanadium also acts as a deoxidizer and alloying material.

Titanium increases strength when added in small amounts, such as a 2% or 3% titanium-iron alloy containing about 10% titanium.

Machining the Cast Irons

Of the four cast irons described, white cast iron is generally nonmachinable except with some of the newer carbide cutting tools. Gray, malleable, and nodular cast irons are all machinable, but they each machine differently and none of them cut at all like steel. Cast irons produce discontinuous chips while steels produce continuous chips.

Gray cast iron can be soft, tough, weak, and brittle. The casting can have sand impregnated into the outer surfaces depending on how the cast iron and casting were made. Sand pockets can be found within the casting. These are formed in the foundry from loose sand in the sand mold. Hard spots can be found anywhere in the casting. These are partly due to uneven cooling. When either of these situations occurs, the cut must be stopped. The sand pocket must be removed with a cold chisel. The hard spot, often called a "diamond," must be ground out with a hand grinder. To overcome sand on the outer surfaces, it is best to make a deep cut so that the cutter gets under the sand.

A good gray iron casting can be machined almost as fast as cold rolled steel, but generally speaking, the cutting speed for gray cast iron is about 60 to 80 sfpm (18 to 24 m/min) and that for cold rolled steel is about 80 sfpm (24 m/min) when using an HSS cutter. Carbide cutting tools can be used at speeds about three times faster, but again, the cutting speed for gray cast iron is less than that for cold rolled steel.

Malleable cast iron is a much finer-grained material than gray cast iron. It is also tougher, stronger, and more ductile. It is easily machinable with HSS or carbide cutting tools. Cutting speeds are slightly less than those for gray cast iron.

Nodular cast iron is harder than gray cast iron and is better machined with a carbide cutting tool rather than an HSS tool bit. Rpm should be about one-third slower than for gray cast iron.

Other cutting tools used for cutting cast irons are ceramic-oxide and cast-alloy tools. Ceramic-oxide cutting tools have the highest red hardness of the cutting tool materials, but they are extremely brittle and are highly susceptible to chipping and breakage. They are used generally as indexable inserts. Because the cutters are relatively weak, negative rakes, sometimes as high as 25 degrees, are always used. Edges are honed to minimize failure by chipping.

Cast-alloy cutting tools are made from cobalt, chromium, and tungsten. They can be used at twice the cutting speed of HSS tool bits. They can be used for form turning and for interrupted cutting where carbide tools might chip or where the hardness of the work might cause HSS tool bits to fail.

Tables 14–1 and 14–2 give recommended machining data for cast irons. Figure 14–5 illustrates the clearances and rakes for carbide tools. In addition to the clearance and rake data, a side cutting-edge angle (lead angle) of 15 to 45 degrees should be used. This is to prevent metal breakout at the exit edge of the work. A 45 degree angle, while giving the least amount of breakout, also gives the greatest radial pressure, and in some work could cause too much deflection.

Table 14–1 Clearance and Rake For High-Speed Steel Tool Bits

HIGH-SPEED STEEL CUTTERS				
CAST IRON	FRONT CLEARANCE	SIDE CLEARANCE	BACK RAKE	SIDE RAKE
Gray	5 to 7	4 to 6	0 to 2	0 to 4
Malleable	4 to 6	4 to 6	0 to 8	0 to 4
Nodular	Not used as a rule — too hard.			

Table 14–2 Clearance and Rake For Carbide Cutting Tools

CARBIDE CUTTERS				
CAST IRON	FRONT CLEARANCE	SIDE CLEARANCE	BACK RAKE	SIDE RAKE
Gray	P 2 to 4 S 8 to 12	P 2 to 4 S 8 to 12	0 0	– 2 to 4
Malleable	P 2 to 4 S 8 to 12	P 2 to 4 S 8 to 12	0	– 2 to 4
Nodular	P 2 to 4 S 8 to 12	P 2 to 4 S 8 to 12	– 2 to 0	– 2 to 0
	Land 1/32″ is best 1/16″ max.		Better	Better

P is primary clearance.
S is secondary clearance.

Round inserts allow higher feed rates and maintain good surface finish, but both this and the steeper lead angle cause too much pressure and more heat is generated.

Nose radius should be as large as possible for good surface finish. Normally, nose radius is 1/64 in to 1/8 in, but where no shoulder or shoulder radius is required, larger radii can be used.

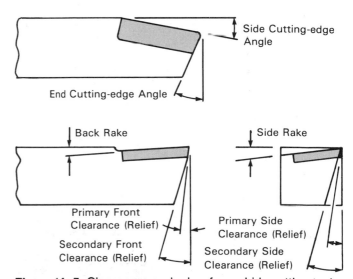

Figure 14–5 Clearances and rakes for carbide cutting tools

REVIEW QUESTIONS

1. What raw materials are required to produce pig iron? What is the function of each of these?
2. Briefly describe what happens to these raw materials in the blast furnace.
3. What happens to pig iron when it leaves the blast furnace?
4. In what type of furnace is cast iron made? Briefly describe the process.
5. What are the primary components of cast iron?
6. Describe the composition and characteristics of the following: graphite, ferrite, cementite, and pearlite.

7. What is the effect of slow cooling and what type of cast iron is produced?
8. What is the effect of rapid cooling and what type of cast iron is produced?
9. How much carbon is there in cast iron?
10. What factors determine the properties of cast iron?
11. Give the composition and basic physical characteristics of gray cast iron, white cast iron, malleable cast iron, and nodular cast iron.
12. How is malleable cast iron produced?
13. What effects do each of the following elements have on cast iron: silicon, sulfur, phosphorus, and manganese?
14. Describe the iron compounds spiegeleisen and ferromanganese.
15. What effect does copper have on cast iron?
16. When aluminum is added to white cast iron, what is its effect? When added to gray cast iron?
17. List the effects of the addition of tin to cast iron.
18. What are the effects of vanadium in cast iron?
19. What amount of vanadium added to cast iron increases its strength?
20. Which cast irons are readily machinable?
21. When machining gray cast iron, what procedure is followed when a hard spot or diamond is found?
22. What are the preferred cutting speeds and tools for gray cast iron, malleable cast iron, and nodular cast iron?
23. List two other types of cutting tools used for cast irons.
24. To achieve a good surface finish when cutting cast iron, what is the recommended nose radius?

UNIT 14-2

HOW STEEL IS MADE

OBJECTIVES

After completing this unit, the student will be able to:

- list the steps in the four main methods of making steel: the Bessemer process, the basic oxygen process, the open hearth process, and the electric furnace process.
- name the basic elements in steel, including the typical percentage range for each.
- state the primary function of each of the basic elements present in steel.
- describe the two main classifications of steel, including how one is distinguished from the other.
- use the iron-carbon equilibrium diagram to determine the composition of carbon steels at different temperatures.
- describe how the rake and clearance angles of cutting tools are varied to machine different types of steel.
- machine steels properly.
- machine Monel and nickel alloys properly.

KEY TERMS

Steel	Alloy steels
Bessemer process	High-speed steel
Basic oxygen process	Alloying elements
Open hearth process	Iron-carbon
Electric furnace	equilibrium diagram
process	Nickel alloy
Iron and carbon alloy	Monel metal
Carbon steels	Work harden

INTRODUCTION

The flow chart, Figure 14–6, shows how steel is made from the original iron ore to some of the finished products. Pig iron from the blast furnace, either in the molten state or in the solid state, and scrap steel (as much as two-thirds of the quantity) is moved to the steel plant to be refined into steel. The main methods used to make steel are the Bessemer process, the basic oxygen process, the open hearth process, and the electric furnace process. All are similar in that the impurities are burned out of the pig iron and the scrap steel, and proper amounts of manganese and other desirable elements are added to give the steel the desired chemical composition.

The Bessemer process was the first widely used process for the quantity production of steel. The first inventor was William Kelly of Kentucky in 1854, Figure 14–7. Henry Bessemer of England experimented with a similar process in the same year. Neither man really got beyond the experimental stages, but by one of those queer quirks of fate, Bessemer's process became better known and was consequently named after him. It was Robert Mushet, an English metallurgist, who first made steel by the process.

The principle of the Bessemer process, Figure 14–8, is the oxidation of carbon and some of the other impurities by blowing a blast of cold air through a bath of molten pig iron in a converter. After the impurities have been burned out, the steel is poured into a ladle, then the necessary amounts of manganese and other desirable elements are added to give the steel the desired chemical composition.

The **basic oxygen converter**, Figure 14–9, substitutes high-purity oxygen (98% or better) for air and blows it at supersonic speed onto the top of the molten metal bath, rather than blowing it through the bath of molten metal. This process is now widely used in steel making because it makes high quality steel in quantity and at high-speed. As much as half of the total steel production in the United States is made by this method.

The open hearth process, Figure 14–10, (page 577), used to be the workhorse of the steel industry. At one time it was used to make nine out of every ten tons of steel made in America. The principle of this process is the oxidation of the impurities in the metal under the direct action of an oxidizing flame of gas and air.

The electric process, Figure 14–11 (page 577), uses an electric arc instead of gas and air to produce heat. Operation is expensive, but control is extremely accurate and there is no contamination in the furnace atmosphere. These furnaces are used especially for special alloy steels used in a wide variety of industries.

METALLURGY OF STEEL

Steel can be called an alloy of iron and carbon. Carbon is not found in the original iron ore. Some of the other elements found in steel are phosphorus, sulfur, silicon, and manganese. Iron makes up the main bulk of steel, but carbon has the greatest physical effect on steel.

Elements in Steel

Iron makes up 65% to 98% of steel, forming the main bulk of the elements present in steel. The quality of the steel depends on the quality of the original iron.

Carbon is added and makes up 0.05% to 1.6%. With up to 0.9% carbon, the hardness, tensile strength, and yield strength increases, but ductility is reduced. Carbon is an embrittling element.

Phosphorus is present from 0.02% to 0.05% and improves machining properties. It also prevents sheet

Flowchart of Steelmaking

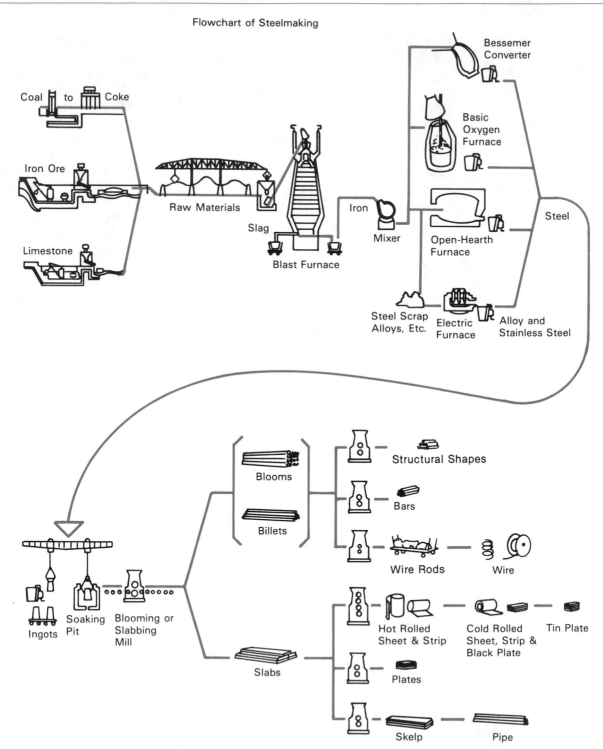

Figure 14-6 How steel is made *(Courtesy of U.S. Steel Corporation)*

and bar steel from sticking to rolls during rolling processes.

Sulfur makes up from 0.01% to 0.05% and improves machining properties.

Silicon is added in amounts of from 0.01% to 0.30% and acts chiefly as a deoxidizer and purifier in steels. A larger percentage of silicon puts steel in the alloy class of steels.

Manganese content generally ranges from 0.2% to 0.85%. Manganese acts as a deoxidizer and purifier in steels. A larger percentage of manganese would form an alloy steel.

CLASSIFICATION OF STEELS

Steel is usually graded into two main groups. They are carbon steels and alloy steels. Carbon steels contain all of the preceding elements in varying percentages. Carbon steels account for over 90% of total steel production.

Alloy steels contain all of the preceding elements plus one or more extra elements such as nickel, chromium, manganese, vanadium, molybdenum, tungsten, cobalt, and so on. Some of these elements have a toughening effect, and some act as hardeners. Some elements may act to prevent oxidation, as in stainless steels, and some, such as tungsten, will give

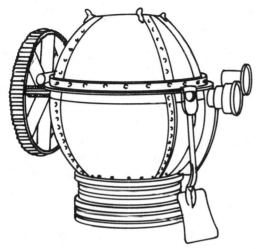

Figure 14–7 Converter used by William Kelly of Kentucky in 1854 *(Courtesy of U.S. Steel Corporation)*

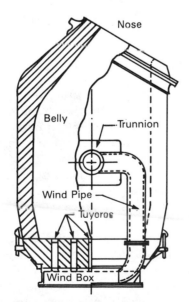

Figure 14–8 Bessemer converter *(Courtesy of U.S. Steel Corporation)*

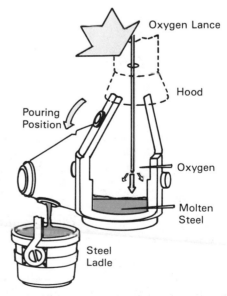

Figure 14–9 Basic oxygen furnace *(Courtesy of U.S. Steel Corporation)*

steel a heat-resistant quality. Alloy steels are usually named after the main alloying element, as in nickel steel, manganese steel, or silicon steel. The family of alloy steels broadly called stainless steels are steels with varying percentages of chromium and nickel. They are often called chrome-nickel steels.

Summary of the Classification of Steels

The main elements in steels are as follows:

Carbon Steels
Iron up to 98%
Carbon 0.1 to 1.6%
Phosphorus
Sulfur
Silicon
Manganese

Alloy Steels

Iron	Plus one or more	Molybdenum
Carbon	of these other	Tungsten
Phosphorus	elements	Cobalt
Sulfur		Chromium
Silicon		Nickel
Manganese		Vanadium
		Silicon
		Manganese

High-speed steel cutting tools are made of alloy steels. A typical analysis of just one type of high-speed steel used as a lathe tool bit would be: carbon 1.6%, manganese 0.30%, silicon 0.30%, chromium 4.75%, tungsten 12.50%, vanadium 5.00%, and cobalt 5.50%. This makes a total of approximately 30%. The balance of 70% would be made up largely of iron.

The Relationship of Carbon Content to the Uses of Steels

To give some indication of the effect of the presence of carbon in steels, a list of the uses for which different carbon steels can be used follows:

Carbon %	Uses
0.1	wire, nails, rivets, etc.
0.2	hot and cold rolled steel, parts for case hardening
0.4	crank pins, axles, etc.

The following are carbon tool steels:

0.5	parts subjected to heavy shock and reversals of stress
0.8	cold chisels, screwdrivers, wrenches, pickaxes, etc.
1.2 to 1.3	files
1.6	saws for cutting steels, dies for wire drawing, etc.

Steels below 0.3% carbon will not harden to any great extent when they are heated to a hardening temperature and quenched suddenly to cold. These steels must be case hardened by carburizing, nitriding, or cyaniding. Steels of 0.5% carbon and higher, can be hardened and tempered by heating and quenching. The iron-carbide diagram, Figure 14–12, shows the effect of various temperatures on carbon steels and the compositions formed in the carbon steels at different temperatures.

Alloying Elements in Steel

Chromium is important in many grades of alloy steels and is necessary in all stainless and heat-resisting steels. Small amounts improve hardening qualities. Larger amounts improve resistance to corrosion.

Nickel is one of the most important alloying elements. It increases toughness, strength, and ductility. In large amounts, it increases resistance to heat and acids.

Cobalt is used mainly as an alloying element in magnet steels and permanent magnet alloys, in special alloys for severe high-temperature service, and for tool steels that hold a cutting edge at high temperatures.

Manganese is another of the more important alloy-

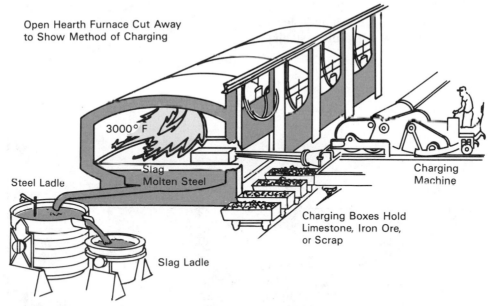

Open Hearth Furnace Cut Away to Show Method of Charging

3000° F

Steel Ladle

Slag
Molten Steel

Slag Ladle

Charging Machine

Charging Boxes Hold Limestone, Iron Ore, or Scrap

Figure 14–10 Open hearth furnace *(Courtesy of U.S. Steel Corporation)*

ing elements. In small quantities, 1% to 2%, it increases strength and toughness. In large quantities, 12% to 14%, it greatly increases toughness and resistance to wear and abrasion.

Molybdenum improves hardening qualities and resistance to shock. It is used to increase high-temperature strength in many alloys designed for high-temperature service.

Silicon is most important as an alloying element in steels used for electrical equipment. It improves the magnetic characteristics of steel.

Tungsten is added to steel to increase hardness and toughness at high temperatures. It is used in many tool steels for metal-cutting tools, particularly in the high-speed steels. It is also used in alloys for severe high-temperature service.

Vanadium is added to alloy steels to improve their heat-treating characteristics, toughness, and mechanical properties.

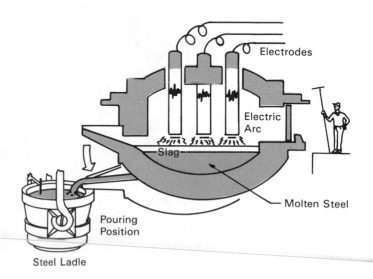

Electrodes

Electric Arc

Slag

Molten Steel

Pouring Position

Steel Ladle

Figure 14–11 Electric furnace *(Courtesy of U.S. Steel Corporation)*

All steels usually take the name of their principal alloying element, e.g., carbon steels, chromium steels, nickel steels, chrome-nickel steels (stainless steels), and so on.

MACHINING STEELS

Up to this point, the metallurgy of iron and steel has consisted of very brief descriptions of the first of three steps in the preparation of iron and steel; i.e., obtain-

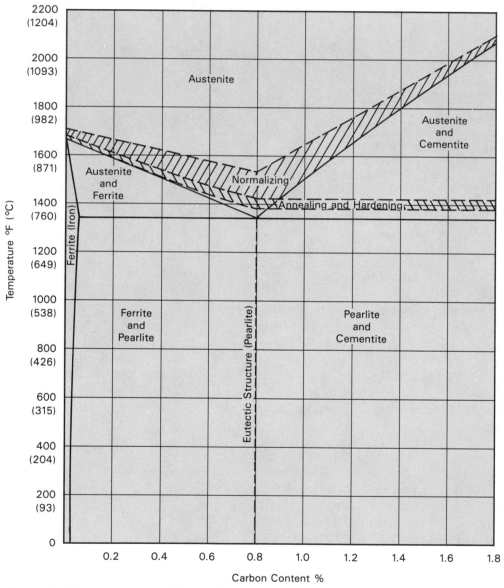

Figure 14–12 Iron-carbon equilibrium diagram

ing the precise chemical composition required. Of more importance to the machinist are the final two steps. They are machining, which is the shaping of steel into its desired form, and finally, for many purposes, the heat treatment of steels.

Types of side rake on a tool bit, Figures 14–13, 14–14, and 14–15, are reviewed here to give a fresh understanding of the HSS cutting tool geometry needed for machining iron and steel.

The following tool bits, Figures 14–16 to 14–25, are

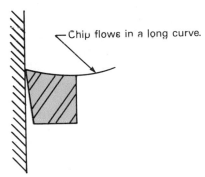

Figure 14–13 Standard side rake

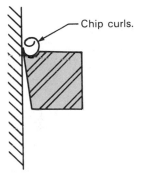

Figure 14–14 Chip curler

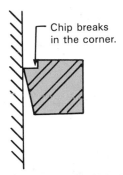

Figure 14–15 Chip breaker

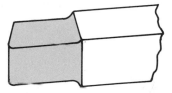

Figure 14–16 Tool bit with no lead angle

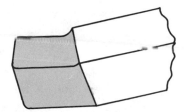

Figure 14–17 Tool bit with lead angle

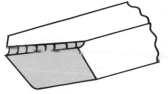

Figure 14–18 Chip curler tool bit with lead angle

effective with different types of steels, using either left- or right-hand cutting tools in all of the basic cuts. The changes required for the different types of steel are the amounts of back and side rake that are presented to the work. With tougher and harder steel, the cutting edge needs more support. Rake and clearance angles are ground less than those used for the machining of softer steels.

When using a combination back and side rake, the chip slides off the face at an angle toward the side and also toward the back of the tool bit, Figures 14–16 and 14–17. The problem with this cutting tool is that chips tend to pile up and tangle around the tool bit and tool holder, presenting a real danger to the operator. Chip disposal is poor. One answer to this problem is to increase the feed, but this takes more power. A more efficient way of removing this metal is with a deep cut and a finer feed, using a chip curler groove.

A right-hand cutter with chip curler, Figures 14–18 and 14–19, is used for removing a lot of metal quickly. A deep cut with a depth at least equal to the width of the cutter, and a reasonably fine feed, will remove the most metal in the shortest time. The chip curler groove causes chips to curl tightly and snap off either against the side of the tool bit or against the tool holder. There is no danger to the operator and chip disposal is easy.

The chip curling groove can also be adapted to a modified-nose cutting tool, Figure 14–20A. With this tool, it is possible to get in closer to the chuck without the carriage or compound being struck by the chuck. This setup, Figure 14–20B, must be watched to make sure that pressure does not force the tool holder to revolve, undercutting the surface of the work.

A chip curler groove on a full-nose cutting tool, Figure 14–21A, is used by many machinists. Again, this setup, Figure 14–21B, should be watched to make

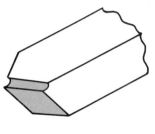

A. Tool bit

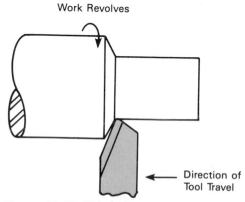

Figure 14–19 Chip curler tool bit

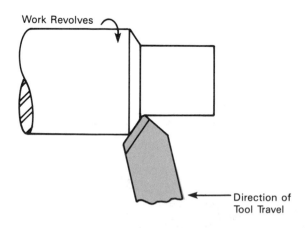

B. Setup

Figure 14–20 Modified-nose tool bit with chip curling groove

sure that pressure does not force the tool holder to revolve, undercutting the surface of the work. A slight side rake on the curler groove will cause the curled chip to move slightly sideways to fall into the chip pan.

The purpose of chip breakers in cutting tools is to prevent the formation of long continuous chips. Another problem is the disposal of such long chips. Chip breakers break the chips into small pieces against a sharp shoulder, presenting less danger to the operator and making chip disposal much easier. The carbide cutters in Figure 14–22 are just a few of the types of chip breakers in general use. Lead angles can be zero to as much as 45 degrees on carbide cutters. In the angular shoulder type, Figure 14–22A, the angle

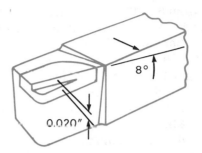

A. Angular shoulder

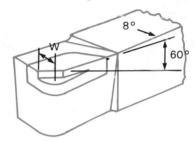

B. Double angular shoulder

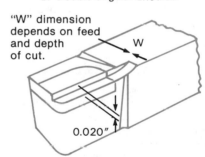

C. Parallel shoulder

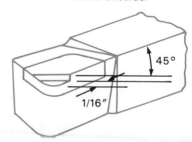

D. Angular shoulder

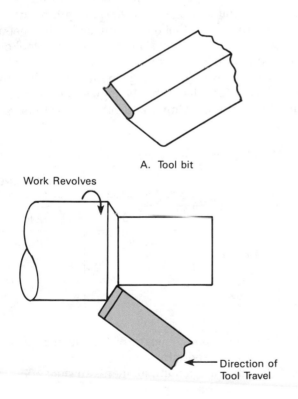

A. Tool bit

B. Setup

Figure 14–21 Full-nose tool bit with chip curling groove

Figure 14–22 Chip breakers *(Courtesy of Kennametal Inc.)*

can vary from 6 to 15 degrees. In the parallel shoulder type, Figure 14–22C, chips come off in short curled sections. There are differences of opinion regarding the merits of both this type and a groove type with a reinforcing flat at the cutting edge, but both types have proven to be satisfactory. Another method of reinforcing a carbide cutting edge (used especially in Europe) employs a 45-degree negative bevel along the cutting edge, but this requires much more power. The angular shoulder type, Figure 14–22D, is sometimes used for finishing cuts.

It is important in all chip breakers that the chip-bearing surfaces have a fine finish. This can be obtained by honing with a silicon-carbide pocket hone, which greatly increases the life of the cutter.

MACHINING MONEL AND NICKEL ALLOYS

One of the tougher metals that the machinist is required to cut is Monel metal, a nickel alloy. One of the tougher Monel metals is "K" Monel, which contains the following:

- Nickel — 66%
- Copper — 29%
- Aluminum — 2.75%
- Iron — 0.9%
- Manganese — 0.4%
- Carbon — 0.15%
- Silicon — 0.25%
- Sulfur — 0.005%

This is classed as a nonferrous alloy since it contains very little iron, less than 1%. Both high-speed steel cutters and carbide cutters (standard cemented carbides and indexable tips) are used successfully in machining nickel alloys. Sulfurized oil, mineral-lard oil, or soluble oil are all recommended for rough and finish turning. For heavy-duty cutting, sulfurized oils are best.

Nickel alloys have a high tendency to work harden. This can be minimized by using adequate relief angles and positive rake angles, but above all, by keeping the cutting edges sharp. Depth of cut and feed should be such that the cutting tool will penetrate the work without rubbing. Keep the cutting tool cutting. Don't allow it to rub on the work. This is the major cause of work hardening. To minimize chipping, hone the cutting edge. Table 14–3 gives suggested rake and clearances for Monel and other nickel alloys.

Table 14–3 Recommended Rakes and Clearances for Cutting Monel and Other Nickel Alloys

CUTTER	CUTTING SPEED	BACK RAKE	SIDE RAKE	SIDE CLEARANCE	FRONT (END) CLEARANCE
High-speed steel	40–70 sfpm (12–21 m/min)	5 to 7	10 to 15	8 to 10	12 to 15
Carbide	100–150 sfpm (30–45 m/min)	Chip curler (Kenloc N/P insert)	Chip curler (Kenloc N/P insert)	4 to 6	6 to 8

REVIEW QUESTIONS

1. What is the principle of the Bessemer process in steel production?
2. How does the basic oxygen furnace differ from the Bessemer furnace?
3. What is the open hearth process?
4. What are two advantages of the electric furnace process in producing steel?
5. What are the six main elements in steel and give the percentage range for each.
6. Which element has the greatest physical effect on steel?
7. For each of the elements listed in answer to #5, state the primary function in steel.

8. What is the difference between carbon steel and alloy steel?
9. How are alloy steels named?
10. The various stainless steels are also known as what type of steels?
11. What is the percentage range of carbon in carbon tool steels?
12. What forms of heat treatment are used for steels with a carbon content of 0.5% or more?
13. For what purpose is the iron-carbon equilibrium diagram used?
14. For each of the following alloying elements, state its primary function in steel: chromium, nickel, cobalt, manganese, molybdenum, silicon, tungsten, and vanadium.
15. What changes in cutting tool geometry are required for different types of steel?
16. What is the advantage of using a cutting tool with a chip curler groove when machining steels?
17. Name the types of cutting tools preferred for machining monel and nickel alloys.
18. When machining nickel alloys, how can work hardening be minimized?

UNIT 14-3

HEAT TREATMENT OF STEELS

OBJECTIVES

After completing this unit, the student will be able to:

- list the types of heat treatment which involve no change in the chemical composition of the steel.
- state the purpose of each of the heat treatments in the first group.
- list the types of heat treatment which involve a change in the surface composition of the steel.
- state the purpose of each of the treatments in the second group.
- briefly describe each heat treatment process.
- state how the heat treatment of tool steels differs from that of other steels.
- harden and temper a tool steel item such as a cold chisel.
- describe the Brinell and Rockwell hardness tests.
- perform the Rockwell hardness test on heat treated steel parts.

KEY TERMS

Heat treatment
Normalizing
Transformation range
Annealing

Pack hardening
Pack carburizing
Liquid bath
   carburizing

Quenching
Hardening
Tempering (drawing)
Temper rolling
Case hardening
  (surface hardening)
Carburizing

Hardness testing
Brinell hardness test
Rockwell hardness
   test
Penetrating ball
Brale penetrator
Major load

Nitriding
Cyaniding
Liquid bath
Letting-down process

Minor load
"C" scale
"B" scale

FORMS OF HEAT TREATMENT

Heat treatment is the final step in the preparation of steel for many purposes. This treatment falls into two main divisions. The first division involves no change in the chemical composition of the steel. The steel is heated to specific temperatures and cooled at various speeds in various cooling mediums. Normalizing, annealing, hardening, and tempering (drawing) are in this division. The second division involves a change in the surface of the metal through the absorption into that surface or surfaces of carbon, or nitrogen, or carbon and nitrogen together. Case hardening by carburizing, nitriding, or cyaniding are in this division.

Normalizing

In the first division, normalizing is used to restore the structure of steel back to "normal" after cold working or hot working. It may be applied to castings or forgings, as well as to various rolled products. It is done by heating the steel to a temperature above the transformation range and cooling it in still air at room temperature. The transformation range when heating steels is the range of temperatures during which austenite forms. During the cooling of steels, it is the range of temperatures during which austenite disappears.

Annealing

In annealing, the chief purpose is to soften the steel by relieving the internal stresses caused by rolling and wire drawing. Annealing also tends to refine the grain structure of the steel. Another use is to soften some steels so that they may be more easily machined. The annealing operation consists of heating the steel to a temperature above the transformation range and then after holding the steel for a proper time at this temperature, cooling it slowly over a number of hours to atmospheric temperature.

Hardening

When hardening, steel of 0.5% carbon or better is heated to a suitable temperature and cooled or quenched very rapidly. The most rapid cooling develops maximum hardness and the slowest cooling develops the greatest softness. Quenching is done by plunging the red-hot steel into a liquidlike water, brine or oil, each of which cools steel at different rates of speed. Subjecting the steel to a cold blast of air is another method of quenching.

Tempering

Tempering or **drawing** is a form of heat treatment applied to hardened steel immediately after hardening. Its primary purpose is to regulate the hardness of steel, to toughen it, or to free it from internal stresses that could cause hardening cracks. Tempering is an extremely important step after hardening, especially if the steel has been hardened by very rapid quenching. In such cases, if the tempering is delayed too long, stresses may be set up in the steel causing cracks. Tempering immediately after hardening prevents the development of harmful effects from the stresses. Actually, the hardness is being tempered or some of the hardness is being drawn out. The hardened steel is heated to a low temperature until the metal reaches a certain color (light straw, purple, or blue, depending on the type of steel). Then it is rapidly cooled by quenching.

The terms *temper* and *tempering* are also employed in connection with cold rolling. Cold rolling of steel makes it hard. Hardness developed in this manner is referred to as temper, and cold rolling for this purpose is referred to as temper rolling. Mills employed to peform this operation are known as temper mills.

Case Hardening

In the second division of heat treatment, the purpose of case hardening is to develop a "case" or hardened surface on the steel while still retaining the same interior as before. This is done by adding carbon or nitrogen or both carbon and nitrogen to the surface of the steel. Cold rolled steel, for example, with 0.2% carbon in it, will not harden when heated and quenched. By adding enough carbon to the surface to bring it into the carbon tool steel range of 0.5% or better, the surface of the steel will then harden when heated and quenched. A case-hardened steel does not require tempering after quenching.

Carburizing

Carburizing means that carbon is absorbed into the surface of the steel, increasing the carbon content of the surface to 0.5% or better. The steel is heated in direct contact with carbon-bearing materials such as charcoal, charred bone, charred leather, bituminous coal, or coke. The steel may also be heated in carbon-bearing gases or liquids.

Nitriding

Nitriding is the process by which the surface of the steel is made to absorb nitrogen. Nitrogen in steel is like seasoning in food. The right amount in the right places can be most desirable, but too much in the wrong place is undesirable. Nitrogen causes low-carbon steel to become brittle. One of the most common methods of adding nitrogen is to heat the steel in a closed container filled with a gas containing active nitrogen. The container is slowly heated and kept at a proper temperature until the desired amount of nitrogen has been absorbed into the surface of the steel. Nitrided steels have greater surface hardness and they can withstand wear better than any other steel with the exception of manganese steel under certain conditions.

Cyaniding

Cyaniding is the absorption of both carbon and nitrogen into the surface of the steel. This is done by heating the steel in a molten bath containing sodium cyanide, calcium cyanide, or potassium cyanide. This is followed by quenching.

Heat Treatment of Tool Steels

In all of the heat treatments just described, steels should be heated slowly and uniformly to the various temperatures involved. Tool steels, that is, those steels containing between 0.5% and 1.6% carbon, require different hardening and tempering temperatures. Tool steels are first hardened and then immediately tempered. Modern heat treatment is a very precise and scientific process, especially in temperature control.

Heating for hardening is carried out by furnaces with oxidizing atmospheres or controlled atmospheres (oxygen free). Work can be covered with protective coatings or can be heated in molten liquid baths to prevent decarburization. In all quenching operations, the work usually goes directly from the furnace to the proper quenching medium; either water, oil, or cold air blast. All temperatures and times involved in the heat treatment process are rigidly controlled.

Tools and dies should be tempered or drawn immediately after the hardening quench, to relieve hardening strains brought about by the sudden change from soft steel to hardened steel. Temperatures for tempering are 300 degrees F (149 degrees C) or more. Accurately controlled circulating air furnaces and liquid baths are used for heating. As a safety precaution when using liquid baths, the work should be preheated to drive off any moisture before immersion in the bath. One drop of moisture on the work could cause a salt bath to explode.

Tool steels hardened below 1700 degrees F (925 degrees C) are usually charged directly to the furnace. Sometimes, if a tool or die is large or complex in design, it may require preheating at 1200 to 1300 degrees F (650 to 705 degrees C) to minimize warping or distortion during the heat treatment procedure. With tool steels hardened between 1700 and 2000 degrees F (925 to 1095 degrees C), the preferred practice is to preheat at 1250 to 1350 degrees F (675 to 32 degrees C) unless tools or dies are small or are relatively simple in design. Tool steels hardened above 2000 degrees F (1095 degrees C) should always be preheated at 1500 to 1600 degrees F (815 to 871 degrees C).

An approximate method of hardening and tempering can be done on such tools as cold chisels, center punches, drift punches, pry bars, screwdriver blades, and so on. These are tool steels containing about 0.8% of carbon. This is the so-called letting-down process as used by blacksmiths. This can be done in either one or two heats. A blowtorch or welding torch and a quenching medium of oil or water is needed. Use of oil or water would depend on whether the tool steel was an oil-hardening steel or a water-hardening steel.

USING THE LETTING-DOWN PROCESS ON A COLD CHISEL

To use this method on a cold chisel, for example, heat half the blade length to a bright cherry red color, about 1425 to 1500 degrees F (775 to 815 degrees C). At this temperature, the heated portion is nonmagnetic and can be quickly tested with a magnet. Plunge the red hot steel straight into the quenching medium, moving it in a figure-eight pattern. This will ensure uniform cooling and a blending of the hardened portion of the blade into the softer portion of the blade. Temper the hardness immediately. Polish the blade so that colors can be seen. Reheat the top end of the blade until it turns a dark blue, about 580 degrees F (300 degrees C). The heat will spread both ways on the chisel but will move to the sharp end of the blade quicker because it is thinner. Slowly the end of the blade will turn to a straw color, about 430 degrees F (220 degrees C), then to a brown, about 480 degrees F (250 degrees C), then to a light purple, at about 500 degrees F (260 degrees C). Immediately quench in clear, cold water. If the end of the blade had been allowed to turn a dark blue, it would probably have become too soft and would have required rehardening and tempering. This is the major reason why such care has to be taken when you are regrinding carbon tool steel cutting tools. If they turn blue from the heat of grinding, the hardness has been lost and they require rehardening and tempering. This is not so with high-speed steel cutting tools. They can be red hot and will still keep cutting. As they cool, they will reharden themselves. But care must be taken in quenching them while hot as they tend to crack.

The process of case hardening the lower carbon steels is one of the oldest processes known to metallurgists. The objective is to produce a hardened surface over a tough, ductile core. A very common method that is still used in some shops is pack hardening or pack carburizing. Steel parts and catalysts such as sodium or barium carbonate that speed up the action are packed in a steel case containing a charcoal or coke mixture. The pack is heated to temperatures of 1500 to 1675 degrees F (815 to 912 degrees C) over a period of 30 minutes or more. Under these conditions, extra carbon is absorbed into all of the steel surfaces in contact with the charcoal or coke mixture to a depth of about 0.015 in (0.38 mm). The surface carbon content is increased from 0.2% to 0.5% or better. The inner core of the steel parts remains at 0.2% carbon. The steel parts are then removed from the pack and quenched in clear, cold water, effectively leaving the parts with a hardened surface and a soft inner core. This is carburizing, which is the absorption of carbon into a steel surface to increase the carbon content. Another method is to use a hydrocarbon gas in which CO (carbon monoxide) is present.

Liquid bath carburizing makes use of neutral carburizing salts with a cyanide content. The result is a true carbon case that can penetrate up to 0.160 in (about 4 mm). This is different from the more brittle high-nitride, low-carbon case produced from straight cyanide hardening. The time involved is much less than with pack carburizing. The parts are heated within the molten liquid bath more rapidly and more evenly, with a minimum of distortion and no scaling from oxidation. Adequate ventilation, however, is a necessity because of the presence of cyanide (CN).

HARDNESS TESTING

Hardness of a material can be defined as a resistance to being changed or deformed. In the hardness testing of metals, the most common types of testing will measure the resistance of the metal to penetration by a hard object such as a hardened steel ball of known diameter or a conical-shaped diamond penetrator. The two most widely used tests are the Brinell hardness test and the Rockwell hardness test, although there are other hardness tests such as the Vickers or diamond pyramid hardness test, the Shore scleroscope test, the microhardness test, and the Knoop test for special applications.

In the Brinell hardness test, a hardened steel ball is forced into the material or specimen under a definite static load. The load applied to the penetrating ball (in kilograms) is divided by the area of the indentation in square millimeters to produce a Brinell number usually designated as BHN.

The Rockwell hardness test is a simple, accurate, versatile test, and is used more than any other hardness test. In this test, a 1/16-in hardened steel ball is used to penetrate brass, bronze, cast iron, and soft steel. About one-third of the ball is exposed for penetration into the specimen. For softer materials, 1/8-in, 1/4-in, and 1/2-in ball penetrators can be used. A sphero-conical diamond penetrator ("Brale") is used to penetrate hardened tool steel, case-hardened steel, and heat-treated spring steel. The depth of penetration by a specific penetrator into the specimen surface under a specific load is measured.

A Rockwell hardness tester is illustrated in Figure 14–23A. The principle of the Rockwell hardness test is presented in Figure 14–23B & C.

The Rockwell hardness tester is a precision depth measuring instrument. To eliminate errors due to surface imperfections or distortions around the periphery of the indentations, a minor and major load are applied, making two superimposed indentations.

The minor load of 3 or 10 kgf (kilograms force) is applied first, creating a reference point. Without removing the minor load, the major load of 15, 30, 45, 60, 100, or 150 kgf is applied.

Without moving the test piece, the major load is removed and the hardness number is read directly on the dial or the digital readout.

Loads of 15, 30, 45, 60, 100, or 150 kilograms can be used in testing. A 100-kilogram load is usually used with the 1/16-in ball for testing soft metals. The 150-kilogram load is usually used with the "Brale" diamond penetrator for testing hardened steel. The Rockwell dial gage is an accurate, direct-reading, micrometer gage with a reversed scale. The deeper the indentation, the lower the dial reading. The harder the material is, the higher will be the hardness number. The regular gage has two scales: one black, one red. The black scale (the "C" scale) is used with the diamond penetrator, Figure 14–23D (page 590), and the red scale (the "B" scale) with the ball penetrator, Figure 14–23E (page 590). One revolution of the regular gage equals a depth of 0.008 in (0.2 mm). Figures 14–23F and G (page 591) show a tool bit being tested for hardness.

To give some examples of hardness, a high-speed steel tool bit should have a hardness of about Rc 64 on an average reading. A case-hardened specimen of cold rolled steel should read about Rc 25 to Rc 35 for good hardness.

In a theoretical situation, suppose a piece of unknown steel were to be case hardened and your hardness reading was Rc 64. It would be easy to guess that the unknown steel has a high carbon content rather than a low carbon content. This would mean that this piece of steel is in the totally hard condition and would need immediate tempering to prevent hardening cracks from forming. The hardness conversion table, Table 14–4 (page 595), gives a comparison of some of the hardness numbers in use today.

Dial-type hardness testers, Figure 14–24 (pages

596–97), are still manufactured and used, but digital readouts and microprocessor technology has been added, Figure 14–25 (page 598). This makes these measuring devices much easier to operate. The instrument performs most of the test cycle automatically and will also prevent many costly operator errors, actually inhibiting the test from proceeding until the mistake is corrected, or it will display clear instructions for the operator. It is also possible to connect these hardness testers with computers, monitors, and printers so that complete statistical analyses of test results can be supplied at any time.

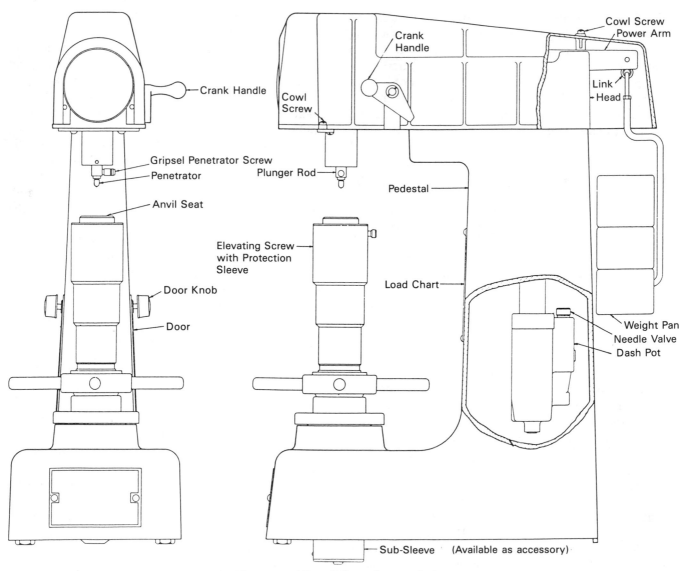

Figure 14–23A Rockwell hardness tester *(Courtesy of Page-Wilson Corporation)*

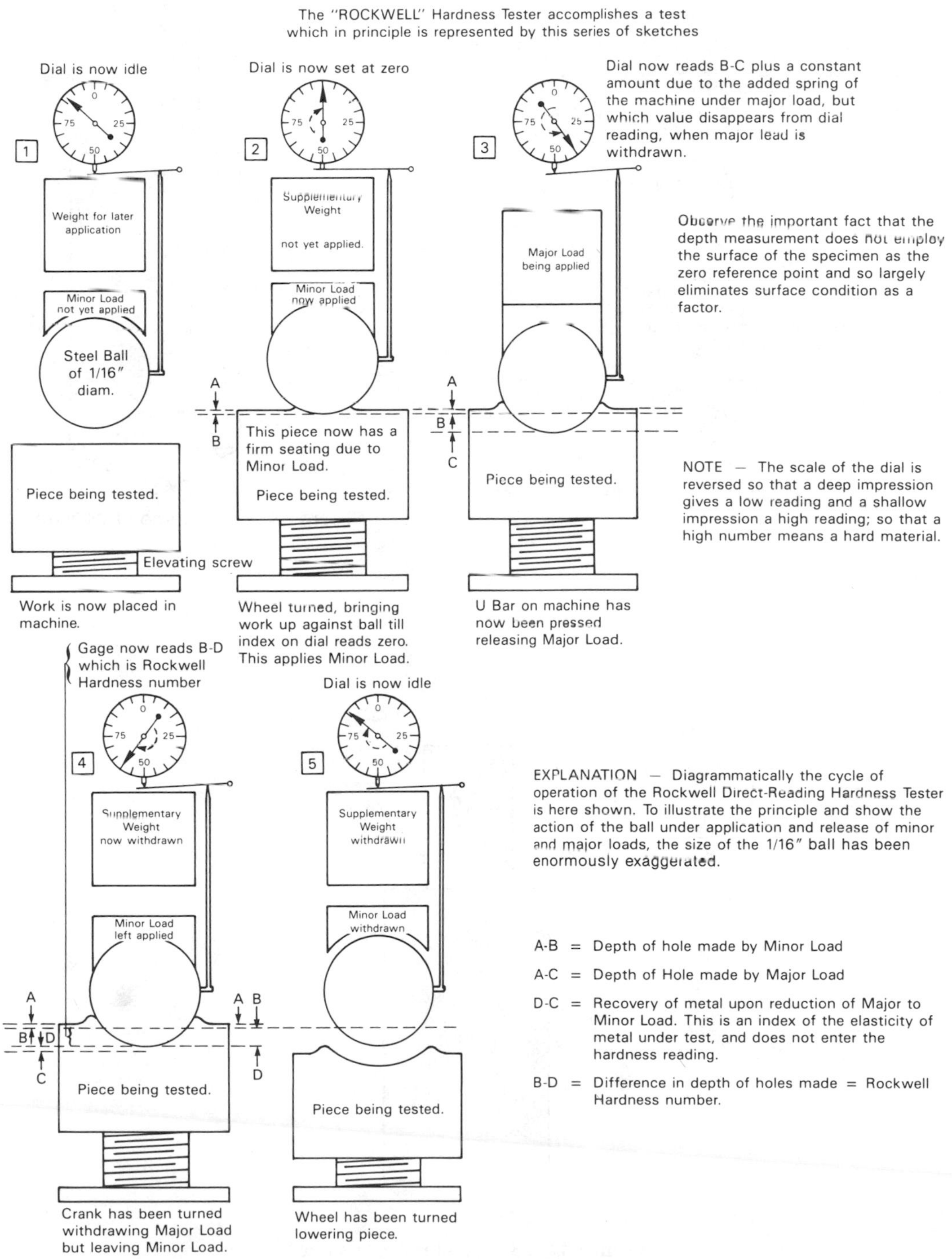

The "ROCKWELL" Hardness Tester accomplishes a test
which in principle is represented by this series of sketches

1 Dial is now idle

Weight for later
application

Minor Load
not yet applied

Steel Ball
of 1/16″
diam.

Piece being tested.

Elevating screw

Work is now placed in
machine.

2 Dial is now set at zero

Supplementary
Weight

not yet applied.

Minor Load
now applied

A
B

This piece now has a
firm seating due to
Minor Load.

Piece being tested.

Wheel turned, bringing
work up against ball till
index on dial reads zero.
This applies Minor Load.

3 Dial now reads B-C plus a constant
amount due to the added spring of
the machine under major load, but
which value disappears from dial
reading, when major load is
withdrawn.

Major Load
being applied

A
B
C

Piece being tested.

U Bar on machine has
now been pressed
releasing Major Load.

Observe the important fact that the
depth measurement does not employ
the surface of the specimen as the
zero reference point and so largely
eliminates surface condition as a
factor.

NOTE — The scale of the dial is
reversed so that a deep impression
gives a low reading and a shallow
impression a high reading; so that a
high number means a hard material.

4 Gage now reads B-D
which is Rockwell
Hardness number

Supplementary
Weight
now withdrawn

Minor Load
left applied

A
B D
C

Piece being tested.

Crank has been turned
withdrawing Major Load
but leaving Minor Load.

5 Dial is now idle

Supplementary
Weight
withdrawn

Minor Load
withdrawn

A B

D

Piece being tested.

Wheel has been turned
lowering piece.

EXPLANATION — Diagrammatically the cycle of
operation of the Rockwell Direct-Reading Hardness Tester
is here shown. To illustrate the principle and show the
action of the ball under application and release of minor
and major loads, the size of the 1/16″ ball has been
enormously exaggerated.

A-B = Depth of hole made by Minor Load

A-C = Depth of Hole made by Major Load

D-C = Recovery of metal upon reduction of Major to
Minor Load. This is an index of the elasticity of
metal under test, and does not enter the
hardness reading.

B-D = Difference in depth of holes made = Rockwell
Hardness number.

Figure 14–23B The principle of hardness testing *(Courtesy of Page-Wilson Corporation)*

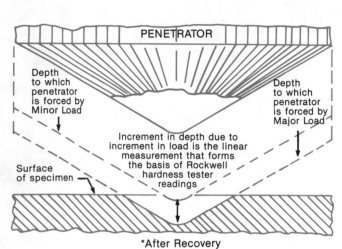

PENETRATOR

Depth
to which
penetrator
is forced by
Minor Load

Depth
to which
penetrator
is forced by
Major Load

Increment in depth due to
increment in load is the linear
measurement that forms
the basis of Rockwell
hardness tester
readings

Surface
of specimen

*After Recovery

Figure 14–23C The principle of the Rockwell hardness test
(Courtesy of Page-Wilson Corporation)

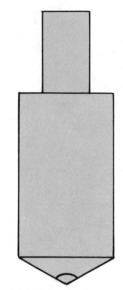

Figure 14–23D Diamond penetrator (Brale)

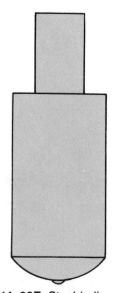

Figure 14–23E Steel ball penetrator

Figure 14–23F Testing the hardness of a tool bit — applying the minor load

Figure 14–23G Testing the hardness of a tool bit — final turning of the crank to withdraw the major load but leave the minor load

ARTIFICIAL MACHINE VISION

Artificial machine vision is a reality in modern machining technology. Applications of machine vision include inspection of machined parts, identification of parts, counting of parts, gaging for accuracy, part sorting and positioning, and orientation of parts. In addition, information provided by the vision system is used to reject parts due to defects or missing components, and to control manufacturing processes such as material handling, welding or assembly.

The vision system camera can view up to 1200 parts per minute. During viewing, the parts can be randomly oriented or they can be moving. As each part is viewed, it can be measured and classified according to preprogrammed characteristics. Such characteristics include area, perimeter, number of holes, center of gravity, gray scale statistics, and so on. The information compiled from this viewing is then used by the system to determine part completeness and adherence to specifications. Decisions are made by the system as to whether or not the part is acceptable. Other decisions are also made such as the adjustment of process procedures, robot device controls, tool adjustments, and other similar functions.

Figure 1 shows five artificial vision camera inspection stations inside a typical assembly line. All of the stations are controlled by two computers. A photo of the same assembly line is shown in Figure 2. This system is an example of noncontact inspection. The part to be inspected in the assembly line shown was a tie rod, part of an automobile steering linkage. Changes in manufacturing design for different models of the tie rod are accomplished by reprogramming the system.

To illustrate a sample inspection, the system is to evaluate the swage of the tie rod. The swage is that part of the housing that is formed or bent over the ball to hold the ball end of the tie rod in the housing. Insufficient swaging allows the tie rod to pull free of the housing. Too much swaging will not permit the degree of movement required by the tie rod. A computer program

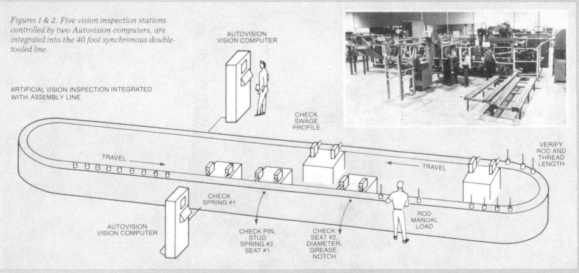

Figures 1 & 2. Five vision inspection stations controlled by two Autovision computers, are integrated into the 40 foot synchronous double-tooled line.

ARTIFICIAL VISION INSPECTION INTEGRATED WITH ASSEMBLY LINE

AUTOVISION VISION COMPUTER

CHECK SWAGE PROFILE

TRAVEL

VERIFY ROD AND THREAD LENGTH

TRAVEL

AUTOVISION VISION COMPUTER

CHECK SPRING #1

CHECK PIN, STUD SPRING #2, SEAT #1

CHECK SEAT #2, DIAMETER, GREASE, NOTCH

ROD MANUAL LOAD

Figure 1 Sketch of a typical assembly line inspection system *(Courtesy of Automatix Inc.)*

Figure 2 A typical assembly line inspection system *(Courtesy of Automatix Inc.)*

directs the inspection system to follow a series of step-by-step instructions. The visual inspection of the swaged housing is done by actually taking a picture. The vision computer can then make a comparison to stored data and a decision as to whether or not the swaged housing is satisfactory.

Swage inspection requires testing of the major axis angle of the top of the housing. Part length, width, area, roundness, number of holes, and so on, are also parameters evaluated during the inspection program.

In another example, a robot equipped with special cameras can locate the actual position of a part and its orientation, or how it is lying in that position. This type of system can also be a stand-alone inspection system for the quality control of parts. Figure 3 shows an integrated vision system for robots. Figure 4 shows a vision system that is used to guide robots for precision assembly. In Figure 5, a welding robot in the center with its torch raised waits between the two solid-state cameras. The cameras will find the part (which is below, out of sight) for the robot. Although the tolerances in car body assembly may be as much as two inches, the cameras ensure that the welding torch is spotted exactly.

In another welding situation, also involving a car body on an assembly line, the location, width, and depth of a welding seam are variable. The cameras will locate the seam and correct the motion of the robot. As the seam is visually gaged, the correct welding parameters, such as

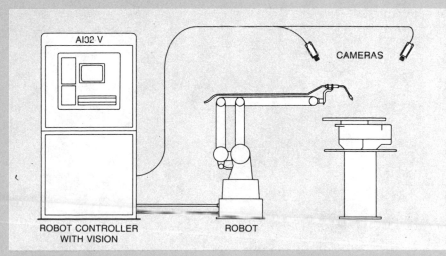

Figure 3 An integrated vision system *(Courtesy of Automatix Inc.)*

speed, voltage, and weave, are automatically selected. Programming for welding parameters for day-to-day operations is done by the "teach and show" method. In other words, the hand-held "teach" module is used. The "teach" module prompts the user with messages written in English, but a program can also be made on the main keyboard. During production runs, if changes must be made, the vision system will calculate the new positions. Adaptation to the new program will be done automatically. The robot "offsets" the nominal program in location and orientation so that it corresponds to the new part location. The system continuously adapts to varied part location.

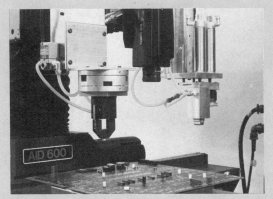

Figure 4 A vision system for assembly robots *(Courtesy of Automatix Inc.)*

At this time, artificial vision is still an emerging technology, although it is in practical use on the factory floor. As the development of more rugged camera hardware, faster vision computers and more software progresses, artificial vision systems will become an indispensable tool in all phases of manufacturing.

Figure 5 A welding robot with vision *(Courtesy of Automatix Inc.)*

Table 14-4 Hardness Conversions *(Courtesy of Page-Wilson Corp.)*

APPROXIMATE RELATIONS BETWEEN BRINELL, ROCKWELL, SHORE, VICKERS AND FIRTH HARDNESS AND THE TENSILE STRENGTHS OF S.A.E. CARBON AND ALLOY CONSTRUCTIONAL STEELS

BRINELL		VICKERS or Firth	ROCKWELL			
Dia. (mm) 3,000 kg 10 mm Carbide Ball	Hardness Number	Diameter Pyramid (50 kg Brale)	C. Scale (150 kg Brale)	B. Scale 100 kg 1/16" Ball	Shore	Tensile Strength (x 1000 psi)
—	—	940	68	—	97	—
2.30	712	860	66	—	92	—
2.35	682	800	64	—	88	—
2.40	653	737	62	—	85	—
2.50	601	697	60	—	81	—
2.55	578	677	59	—	80	328
2.60	555	640	57	—	77	309
2.65	534	591	55	120	73	285
2.70	514	579	54	119	71	279
2.75	495	547	52	119	70	263
2.80	477	528	51	117	68	253
2.85	461	508	50	117	67	247
2.90	444	494	49	116	66	237
2.95	429	472	47	115	63	225
3.00	415	455	46	115	61	217
3.05	401	440	45	114	59	212
3.10	388	425	43	113	58	200
3.15	375	410	42	112	56	196
3.20	363	396	40	112	54	186
3.25	352	383	39	110	52	181
3.30	341	372	38	110	51	177
3.35	331	360	37	109	50	174
3.40	321	350	36	109	48	168
3.45	311	339	34	108	47	158
3.50	302	328	33	108	46	154
3.55	293	319	32	107	45	150
3.60	285	309	31	106	43	146
3.65	277	301	30	105	42	142
3.70	269	292	29	104	41	138
3.75	262	284	28	104	40	135
3.80	255	276	27	103	39	131
3.85	248	269	25	102	38	125
3.90	241	261	24	101	37	121
3.95	235	253	23	100	36	119
4.00	229	247	22	99	35	117
4.05	223	241	21	98	34	113
4.10	217	234	19	97	33	110
4.15	212	228	18	96	33	107
4.20	207	222	16	95	32	102
4.25	202	218	15	95	32	100
4.30	197	212	14	94	31	98
4.35	192	207	13	93	30	96
4.40	187	202	12	92	29	94
4.45	183	196	10	91	—	90
4.50	179	192	9	90	28	89
4.55	174	188	8	89	27	87
4.60	170	182	6	88	—	84
4.65	166	178	5	87	26	82
4.70	163	175	4	86	—	80
4.75	159	171	3	85	25	78
4.80	156	167	2	84	—	77
4.85	153	163	1	83	24	76
4.90	149	160	—	82	—	75
4.95	146	156	—	81	23	74
5.00	143	153	—	80	—	72
5.05	140	150	—	79	22	71
5.10	137	147	—	78	—	70
5.15	134	143	—	76	21	67
5.20	131	140	—	75	—	66
5.25	128	137	—	74	—	65
5.30	126	134	—	73	—	64
5.35	124	132	—	72	20	63
5.40	121	129	—	71	—	62
5.45	118	127	—	70	19	60
5.50	116	124	—	69	—	59
5.55	114	122	—	68	18	58
5.60	111	119	—	67	—	57
5.65	109	117	—	66	15	56
5.70	107	—	—	65	—	—
5.75	105	—	—	64	—	—
5.80	103	—	—	62	—	—
				61		

Figure 14–24A A modern, dial-type Rockwell hardness tester *(Courtesy of Page-Wilson Corporation)*

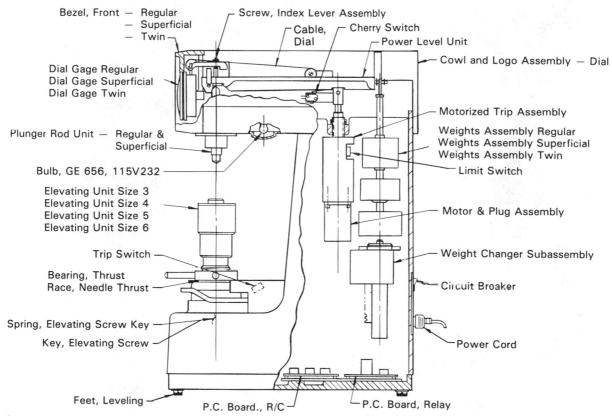

Bezel, Front — Regular
— Superficial
— Twin

Screw, Index Lever Assembly

Cable, Dial

Cherry Switch

Power Level Unit

Cowl and Logo Assembly — Dial

Dial Gage Regular
Dial Gage Superficial
Dial Gage Twin

Motorized Trip Assembly

Weights Assembly Regular
Weights Assembly Superficial
Weights Assembly Twin

Plunger Rod Unit — Regular &
Superficial

Limit Switch

Bulb, GE 656, 115V232

Elevating Unit Size 3
Elevating Unit Size 4
Elevating Unit Size 5
Elevating Unit Size 6

Motor & Plug Assembly

Trip Switch

Weight Changer Subassembly

Bearing, Thrust
Race, Needle Thrust

Circuit Broaker

Spring, Elevating Screw Key
Key, Elevating Screw

Power Cord

Feet, Leveling

P.C. Board., R/C

P.C. Board, Relay

Figure 14–24B Parts of a modern, dial-type Rockwell hardness tester *(Courtesy of Page-Wilson Corporation)*

Figure 14–25A A modern, digital-type hardness tester *(Courtesy of Page-Wilson Corporation)*

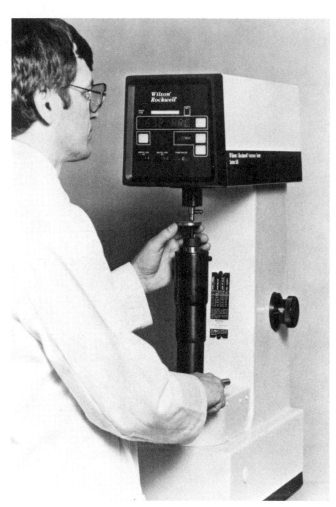

Figure 14–25B Hardness testing a specimen on a modern, digital-type hardness tester *(Courtesy of Page-Wilson Corporation)*

REVIEW QUESTIONS

1. Describe the two main divisions in heat treatment.
2. Describe the process of normalizing steel. What is its primary purpose?
3. What are three functions of the annealing process?
4. How is annealing of steel accomplished?
5. Describe the process of hardening of steel.
6. Following hardening, how is the steel quenched?
7. Once steel has been hardened, why is it tempered?
8. Describe the process of tempering.
9. How is steel case hardened? What is the purpose of case hardening?
10. Define the term carburizing and briefly describe the process.
11. How is the surface of steel made to absorb nitrogen in the process of nitriding?
12. Describe the process of cyaniding.
13. How does the heat treatment of tool steels differ from that of other steels?
14. What happens to a carbon steel cutting tool if it heats to a dark blue color during regrinding?
15. Name three methods of case hardening lower carbon steels.
16. Define the term hardness.
17. List the two most common methods of testing the hardness of metals.
18. List the main steps in the hardness testing of metals using the most common test.
19. What type of penetrator is used to test the hardness of hardened tool steel?
20. What is an average hardness value for a high speed steel tool bit in the Rockwell scale?

UNIT 14-4

THE METALLURGY AND MACHINING OF NONFERROUS METALS

OBJECTIVES

After completing this unit, the student will be able to:

- state the basic properties of aluminum and aluminum alloys.
- define the terms wrought aluminum alloys and casting aluminum alloys.
- list three considerations in the machining of high-silicon aluminum alloys.
- list four factors in the machining of cast aluminum alloys.
- state the characteristics of the copper alloys brass, and bronze.
- list four recommendations each for the machining of copper, brass, and bronze.
- describe the cutting fluids recommended for machining copper alloys.
- list four recommendations for the machining of babbitts.

KEY TERMS

Aluminum
Aluminum alloys
Wrought aluminum alloys
Casting aluminum alloys
Copper
Brass
Bronze

Free machining
Naval brass
Tobin bronze
Bearing bronze
Phosphor bronze
Manganese bronze
Bearing metals
Babbitt

ALUMINUM AND ITS ALLOYS

There is a wide variety of uses for aluminum and its alloys, and each use is a result of the properties of these metals. The principal properties are light weight, corrosion resistance, and high electrical conductivity.

Many metals and combinations of metals can be alloyed with aluminum. The wide range of alloys presents an equally wide variety of properties such as mechanical strength, ductility, electrical conductibility, and corrosion resistance. A few of the alloying elements used with aluminum are silicon, magnesium, copper, nickel, and chromium. There are many others as well.

By special refining, aluminum higher than 99.85% purity can be obtained, but even with no alloying elements added, the small percentages of metallic elements present as impurities have to be carefully controlled. The resultant aluminum can be considered to be an alloy in spite of the high purity.

Aluminum alloys are divided into two main groups: wrought alloys and casting alloys. **Wrought alloys** are those in which the cast metal is rolled, drawn, extruded, or forged to shape, while **casting alloys** are cast to their final shape in a mold. Both groups present different problems in machining.

Machining Aluminum

Aluminum alloys have better machining characteristics than pure aluminums. The softer alloys and, to a lesser extent, some of the harder alloys tend to form a built-up edge on the cutting tool, which means that particles of aluminum tend to become welded to the face of the tool at the cutting edge because they were melted by the heat generated during the cutting action. The built-up edge can be minimized by the use of cutting fluids and by the honing and polishing of cutting edges and of the cutting tool face.

Alloys with more than 10% silicon are the most abrasive and difficult to machine. The hard particles of free silicon cause rapid tool wear. Alloys with more than 5% silicon will not machine to a bright surface as will other high-strength alloys. Chips are torn rather than sheared from the work so that special precautions must be taken. These include the use of cutting fluids containing lubricants, polishing and honing the cutting edge and cutter face to minimize the built-up edge, use of larger rake angles, and use of lower speeds and feeds for economical machining. High-silicon alloys can be cut successfully with high-speed steel tool bits, but cemented carbide cutting tools will wear better. Use clearances of 12 to 14 degrees, or back rakes of 0 to 15 degrees, and side rakes of 8 to 30 degrees with HSS tool bits. A high-speed steel tool bit with standard side rake, Figure 14–26A, and a high-speed steel tool bit with chip curler, Figure 14–26B, work well with aluminum.

Cast alloys containing copper, magnesium, or zinc as the main alloying elements present few machining problems such as burring on the work or built-up cutting edges. Smaller rake angles can be used than in cutting aluminum-silicon alloys. Most wrought alloys in this category have excellent machining characteristics. Chips are usually continuous, but not always. Generally speaking, larger relief and rake angles are used than in cutting steels. These larger angles direct the chips away from the work quickly and prevent surface scratches caused by work-hardened chips rubbing the surface. Best results are obtained by using cutting fluids. For many purposes, soluble oils are good. Turning with high-speed steel tool bits requires clearances of 14 to 16 degrees, back rakes of 5 to 20 degrees, and side rakes of 15 to 35 degrees. Chip curlers work well. Cut-off tools and necking tools should have end clearances of 8 to 12 degrees and back rakes of 12 to 20 degrees.

When threading aluminum with 60-degree threads, some machinists prefer large back and side rakes, but this requires a modification in the included angle of the cutter to produce the correct thread contour. With zero rake angles, the cutter is ground to the included angle of the thread. If this is followed by grinding a positive back rake, the included angle of the cutter becomes smaller. Grinding side rake is permissible.

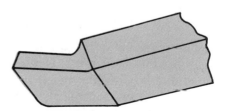

A. Standard side rake

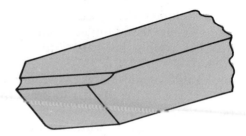

B. Chip curler

Figure 14–26 High-speed steel (HSS) tool bit

A chip curler on the leading edge also works well. In all cases, keep the cutter sharp, honed, and polished.

When threading with taps, standard hand and machine taps can be used, but spiral-fluted ground taps are superior.

When milling aluminum, standard milling cutters can be used successfully, but best results are obtained with coarse tooth, large-helix angle cutters, with clearance angles up to 10 to 12 degrees recommended.

When end milling a profile using peripheral teeth on the end mill, climb milling will generally produce a better finish than conventional milling. Climb milling should not be used unless the machine is equipped with a backlash eliminator that will prevent the cutter from "grabbing" or pulling the job into and under the cutter. Face milling cutters should have a large axial rake angle.

When drilling aluminum, standard twist drills can be used without difficulty, but drills with high-helix angles and wide, polished flutes are better equipped to help clear the chips more easily. Carbide-tipped twist drills wear better in long runs of production work.

COPPER AND ITS ALLOYS

Copper is the parent metal of this group. It is quite a heavy metal and reddish in color. It is relatively soft, very malleable, ductile, flexible, and it is also tough and strong. It does not machine easily because of its softness and because of the sticky way in which it resists the cutting action and sticks to the cutter.

Brass, in its many forms, is essentially copper and zinc. It is heavy, and golden in color, with a distinct greenish tinge when viewed under a light. It is much more brittle than copper, but it is extremely easy to machine.

Bronze, in its many forms, is essentially copper and tin. It is heavy and is similar to brass in color except for a distinct reddish tinge when viewed under a light. It is also much more brittle than copper, but it is stringier, tougher, and harder than brass. It is also easy to machine.

Machining Copper and Its Alloys

Copper. In the drilling of copper, chips tend to cling to the drill. Because of this, the drill must be removed from the hole and the flutes must be cleared frequently. In tapping, the material fills the flutes of the tap quickly. The metal also deforms or bulges very easily, so for best results a slightly oversized hole should be drilled for tapping. Another problem that occurs is that copper heats very quickly and holds its heat. Cutting fluids are recommended. Ample clearances should be used and rakes should be generous. Cutting speeds of 200 sfpm (60 m/min) are recommended for cutting copper with high-speed steel cutters. Chip curler grooves work well when machining copper.

There is a danger. Copper is poisonous and will cause infection. Cuts should be carefully and quickly attended to.

Brass. Brass is nonmagnetic and is very easily machined, provided you do not use steel cutting tools with positive rakes for most brasses. Rakes, both side and back, should generally be zero, Figure 14–27. High-speed steel cutters are used successfully at cutting speeds to 200 sfpm (60 m/min). Chips are discontinuous, coming off in small chips.

Standard twist drills may be used, provided the cutting lips are flattened to a zero rake, Figure 14–28, to prevent the drill from "grabbing" the work. Straight-fluted drills are made especially for brasses. The exceptions to using zero rakes would be when cutting Naval brass and Tobin bronze.

Brass containing 90% copper is frequently referred to as bronze, but that term is usually reserved for copper-tin alloys. Its color is reddish, like bronze.

Free machining, yellow brass rod is composed of 60% to 63% copper, 2.5% to 3.7% lead, a maximum of 0.35% iron, a maximum of 0.5% of other elements, and the remaining, approximately 36% of zinc. This alloy is used for small screw machine parts, pins, nuts, screws, plugs, etc. The addition of lead increases machinability.

Brass wire, made of approximately 70% copper and 30% zinc, is used for making springs. Red brass, with 80% to 85% copper, is used for plumbing fixtures and costume jewelry among other things.

Naval brass, with approximately 62% copper, 37% zinc, and 1% tin, is stronger, tougher, and more corrosion resistant than yellow brass rod. It is used for brass castings, handwheels, and water pump and pro-

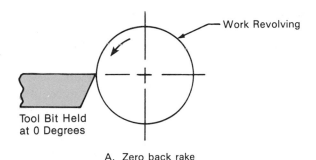

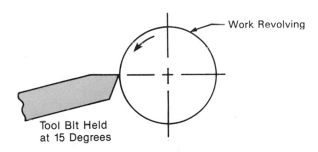

Figure 14–27 Zero back rake, tool bit held at 0° (A.) and at 15° (B.)

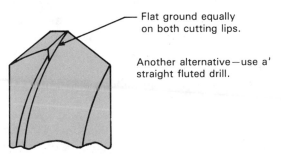

Figure 14–28 Drill rake

peller shafts. It is moderately machinable, and is best machined with positive rakes.

Tobin bronze is actually a 60–40 brass with approximately 60% copper, 1 1/2% tin, and the balance zinc. It is strong, extremely tough, and highly noncorrosive. It is used for sleeves, bearings, propellers, and propeller shafts. Similar to Naval brass, it is also tough to machine, requiring positive rakes.

Bronze. Bronze is tougher, stringier, and harder than brass. The outer surface of castings has a hard glass-like structure that will very quickly knock the point from a cutter. The cutter must go deep to get under this skin. High-speed steel cutters and carbide cutters can be used successfully. Chips are generally of the discontinuous type. Clearances should be generous. Rakes should generally be zero. Standard twist drills should have the cutting edges flattened to a zero rake. The exceptions to a zero rake would be metals like phosphor-bronze and manganese-bronze.

Bearing bronze has about 70% copper, 26% lead, and 4% tin. This metal is used for high-speed bearings where moderate pressure is a requirement. It is easily machined, producing small, discontinuous chips.

Phosphor-bronze contains approximately 78% to 81% copper, 9% to 11% tin, 9% to 11% lead, and 0.05% to 0.25% phosphorus. This metal has good corrosion resistance. One use is in bearings for heavy-duty service. In sheet and strip, it can be used for springs. It is tough to machine, producing a long, tough, stringy chip.

One type of manganese-bronze has 60% to 68% copper, 2% to 4% iron, 3% to 6% aluminum, 2.5% to 5% manganese, a maximum of 0.5% tin, 0.2% lead, 0.55% nickel, and some zinc. It is stronger and tougher than phosphor-bronze and is used mainly for propellers, shafts, and gears. It is tough to machine, the cutters generally require positive rakes, and the chip is long, tough, and stringy.

Cutting Tools for Brasses and Bronzes

The copper family of brasses and bronzes can present a variety of machining problems. For example, the tool bit used for free machining yellow brass cannot be used successfully to machine manganese-bronze. A standard twist drill cannot successfully drill yellow brass unless the rake is decreased. Table 14–5 gives rakes and clearances for free machining alloys such as yellow brass, leaded yellow brass, bearing bronze, leaded bronze, and tin bronze. For moderately machinable alloys such as red brass, Tobin bronze, and Naval brass, use rakes and clearances recommended in Table 14–6. Table 14–7 suggests rakes and clearances for difficult to machine alloys such as phosphor-bronze and manganese-bronze.

Table 14–5 Rakes and Clearances for Free Machining Alloys

Free machining alloys — yellow brass, leaded yellow brass, bearing bronze, leaded bronze, tin bronze

CUTTER	BACK RAKE	SIDE RAKE	FRONT CLEARANCE	SIDE CLEARANCE
High-speed steel	0	0	10 to 12	8 to 10
Carbide	0	0	5 to 7	5 to 7

Table 14–6 Rakes and Clearances for Moderately Machinable Alloys

Moderately machinable alloys — red Brass, Tobin bronze, Naval brass

CUTTER	BACK RAKE	SIDE RAKE	FRONT CLEARANCE	SIDE CLEARANCE
High-speed steel	5 to 10	5 to 10	10 to 12	8 to 10
Carbide	0 to 5	4 to 8	5 to 7	5 to 10

Table 14–7 Rakes and Clearances for Difficult to Machine Alloys

Difficult to machine alloys — phosphor-bronze, manganese-bronze

CUTTER	BACK RAKE	SIDE RAKE	FRONT CLEARANCE	SIDE CLEARANCE
High-speed steel	10 to 15	15 to 20	10 to 12	10 to 12
Carbide	4 to 8	10 to 15	7 to 10	7 to 10

Cutting Fluids for Copper Alloys

Some machining of copper alloys is done dry. Sometimes yellow brass has the annoying habit of a small piece of brass welding itself to the chisel edge of a drill point. This effectively stops the drill from cutting and must be removed. A 10:1 soluble oil is good for most purposes, allowing an increase in speed and feed and an increase in productivity. It also improves surface finish and accuracy, and lengthens tool life. The more stringy the chip, the heavier the solution of cutting oil should be. For the difficult to machine alloys, a mineral-lard oil is good. Sulfurized or chlorinated cutting fluids cause staining and should be avoided. If nothing else is available and staining does occur, the work can be cleaned by immersing it for twenty minutes in a solution of 10% sodium cyanide in water.

BABBITT BEARINGS

Babbitt is a nonferrous alloy composed essentially of lead, tin, antimony, and copper. It is a soft metal used for support bearings in steel shafts. There is a wide range of applications for this type of bearing in the automotive industry, industrial plants, marine engineering, and ship building.

A tin-based babbitt is composed of 87.5% tin, 6.75% antimony, and 5.75% copper. These have steel, brass, or bronze backs. Uses include main bearings and motor bushings. The metal is soft, has fair fatigue resistance and good corrosion resistance.

A lead-based babbitt contains 75% lead, 15% antimony, and 10% tin. These have a steel, brass, or bronze back, or are poured into a bearing housing of cast iron. Uses include main and connecting rod bearings. The metal is soft, moderately fatigue resistant, with good corrosion resistance.

A copper-lead based babbitt is made from 76% copper and 24% lead. These have steel backs. Uses include main and connecting rod bearings for hard shafts. The metal is moderately hard, has good fatigue resistance, and is somewhat subject to oil corrosion.

A copper-lead-tin based babbitt is composed of

67% copper, 28% lead, and 5% tin. These have steel backs. Uses include main and connecting rod bearings for hardened or cast shafts. The metal is moderately hard, has fairly good fatigue resistance, and improved corrosion resistance.

Machining Babbitts

Babbitt is usually cut dry. Bushings are sometimes difficult to bore dry because the metal has a tendency to roll around the cutting tool in a hard ball. Kerosene and lard oil or kerosene and mineral oil of about a 50:50 blend work well. Back rake should be 0 to 20 degrees, side rake slightly more, end clearance 12 to 15 degrees, and side clearance 6 to 10 degrees. Tin-based and lead-based babbitts deform easily because of their softness, and sometimes cutting tools need extra clearances. Copper-lead based and copper-lead-tin based babbitts are harder because of the high percentage of copper. Cutting fluids improve cutting action. Besides the two cutting fluids mentioned previously, a light soluble oil mix, for example, a 20:1 mix, works well.

REVIEW QUESTIONS

1. Name the two main groups of aluminum alloys.
2. What are the main properties of aluminum?
3. What are three considerations in the machining of aluminum alloys?
4. What are four recommended practices for machining high-silicon aluminum alloys?
5. For the general category of cast aluminum alloys, list five recommended machining practices.
6. When end milling aluminum, is climb milling or conventional milling recommended? Why?
7. What is brass?
8. What is bronze?
9. To what metal family do brass and bronze belong?
10. Are all brasses the same?
11. Are all bronzes the same?
12. List four considerations in the machining of copper.
13. What cutting tool rake values are generally recommended for machining brass?
14. What are the major uses of (a) Naval brass and (b) Tobin bronze?
15. List five considerations in the machining of bronze.
16. What cutting fluids are recommended for machining copper alloys?
17. Describe the composition of babbitt.
18. What is the primary use of babbitt?
19. What are the clearances recommended for machining babbitt?
20. Name the cutting fluids recommended for the machining of babbitt.

CHAPTER FIFTEEN

PLASTICS

More and more plastic materials are being used in industry for machine parts or for parts of machine parts. The reasons for this increased use include the fact that some plastic materials are lighter and stronger than comparable metal parts in some applications. In addition, some plastic materials have been found to make good bearing materials.

One outcome of this increased use is the necessity for machinists to learn more about plastic materials and how to machine them properly and safely. There are some dangers in machining plastics, as described in this chapter. For the most part, plastic materials machine easily, but the proper cutting tool must be used and must be keenly sharp.

UNIT 15-1

TYPES OF PLASTICS

OBJECTIVES

After completing this unit, the student will be able to:

- explain the differences between thermosets and thermoplastics.
- list five plastics included in the thermoset category.
- list precautions to be taken when machining thermosets.
- list six plastics included in the thermoplastic category.

KEY TERMS

Thermosets	Thermoplastics
Reinforcing materials	Acrylic plastic
Amino plastic	Cellulose plastic
Casein plastic	Fluorocarbon plastic
Epoxy plastic	Polyamide plastic
Phenolic plastic	Polyolefin plastic
Polyester plastic	Styrene plastic
Urethane plastic	Vinyl plastic
Fiberglass –	
glass dust	

Plastics are divided into two broad families. These are the thermosets and the thermoplastics. All of these can be machined, and most of them present little difficulty in the machining process. Machinists are coming into contact more and more with these materials and a knowledge of how to deal with them is very important.

THERMOSETS

Thermosets are plastics that are set or cured to their permanent shape by heat. They cannot be remelted and returned to their original state. These plastics are combined with such reinforcing materials as glass fibers, cotton fibers, paper, or asbestos. Resins are used to produce different grades of plastics like melamine, phenolic, epoxy, and silicone. Plastic groups with such names as amino plastics include such items as melmac dishware and Formica. Casein plastic, a component in water resistant adhesives, was the first bond used to make waterproof plywood. Epoxy plastics are used for metal-to-metal adhesives and ceramic gluing. Because they possess a foaming ability, they are also used as a core in lightweight sandwich construction. Phenolic plastics like bakelite, were the first plastics used in consumer goods and are still used heavily. The polyester plastics are used in such common items as enamels, lacquers, Dacron fabrics, Mylar film, and fiberglass. Urethane plastic is a cushioning material used for lining winter clothing. It is also used as foam liners in crash helmets, and is a good insulator.

Special precautions should be taken when machining thermosets, particularly fiberglass. This can be dusty work. A vacuum system should be used to collect dust and chips. The operator should wear a prescribed respirator, plus safety glasses that completely enclose the eyes. This is to prevent glass dust from getting into the lungs and eyes.

THERMOPLASTICS

Thermoplastics become soft when exposed to heat and harden when cooled, no matter how often the process is repeated. These plastics can be reshaped many times. Plexiglas and Lucite are examples of acrylic plastics. Cellulose plastics like celluloid come from cellulose fibers in wood or plants. A very common item made from this plastic is movie film. Fluorocarbon plastic is used as wire insulation, lining for pipe systems, bearings, and for Teflon. Nylon is a polyamide plastic. Polyethylene and polypropylene are both polyolefin plastics. Electrical insulators are made from styrene plastics, and vinyl plastics are used for such common items as floor tile and the inner layer in safety glass.

As diverse as these plastics are, machinists will sooner or later come into contact with them and will have to machine them in one way or another. A thorough knowledge of the cutter geometry is necessary.

REVIEW QUESTIONS

1. What are the two major groups of plastics?
2. Describe how thermosets are formed and give the primary characteristic of this group of plastics.
3. List five major subgroups of thermoset plastics.
4. What precautions must be taken when machining thermosets?
5. Describe the primary characteristics of thermoplastics.
6. List five major subgroups of thermoplastics.

UNIT 15-2

MACHINING PLASTICS

OBJECTIVES

After completing this unit, the student will be able to:
- describe how the machining characteristics differ in each of the plastic groups.
- set the proper cutting speeds for machining plastics.
- describe the annealing procedures for selected plastics: polycarbonate, acetal, nylon, and acrylic.
- state the proper procedure for safely machining acetal plastic.

KEY TERMS

Abrasive

Deform

Annealing

Acetal plastic –
    formaldehyde gas

MACHINING RECOMMENDATIONS

Most plastics can be machined easily and economically using standard machine shop equipment such as twist drills, lathe cutters, and milling cutters. In general turning, boring, and drilling practice, rakes and clearances vary with the different plastics, but speeds and feeds are similar to those used in brass, Tables 15–1 and 15–2.

The thermoset group of plastics, which contains reinforcing materials like paper, cotton, fiberglass, and asbestos, can be very abrasive on cutting tools. In the thermoplastic group, plastics such as acrylics can be very brittle. Some of this group can shatter and crack if proper cutters are not used.

In lathework, cutting speeds of 250 to 500 sfpm (75 to 150 m/min) are common with high-speed steel tool bits, and cutting speeds of 500 to 1500 sfpm (150 to 450 m/min) when using carbide cutters. Feeds can range from 0.002 to 0.016 in/rev (0.05 to 0.41 mm/rev). Drilling speeds can be from 150 to 350 sfpm (45 to 105 m/min) with feeds of 0.007 to 0.015 in/rev (0.179 to 0.38 mm/rev). Clearances should be increased, varying from 10 degrees to as much as 30 degrees in cutting some plastics. Some plastics are deformed easily, that is, they tend to bulge slightly when cut,

so that the cutting edge needs more clearance than usual to prevent the "bulge" from rubbing.

Cutting tools must be kept sharp at all times. Smooth, honed cutting edges are excellent. Rakes on some plastics are at least zero and sometimes even slightly negative, but there are exceptions. The cutting action on some plastics should be scraping rather than slicing. This is especially true when cutting the more brittle plastics like Lucite. Again, there are exceptions. Heat buildup can melt some plastics and gum up the cutter. Most plastics are cut dry, but a cold air jet is commonly used if cooling becomes necessary.

ANNEALING

Although it is not required for most work, the annealing process helps to relieve stresses built up during machining. Some plastics such as polycarbonates should be annealed following the machining opera-

Table 15–1 Recommended Machining Data for Thermosets

	PAPER, COTTON REINFORCED	GLASS, GRAPHITE REINFORCING	ASBESTOS REINFORCING
Cutting Speed Lathe — sfpm — m/min	250 to 500 75 to 150	250 to 500 75 to 150	250 to 500 75 to 150
Cutting Speed Drill — sfpm — m/min	400 to 500 120 to 150	250 to 300 75 to 90	250 to 300 75 to 90
Feed — Lathe in/rev mm/rev	0.004 to 0.012 0.10 to 0.30	0.004 to 0.012 0.10 to 0.30	0.004 to 0.012 0.10 to 0.30
Feed — Drill in/rev mm/rev	0.002 to 0.004 0.05 to 0.10	0.002 to 0.006 0.05 to 0.15	0.002 to 0.010 0.05 to 0.25
Clearance — Lathe	20° to 30°	20° to 30°	20° to 30°
Clearance — Drill	10° to 20°	15° to 20°	15° to 30°
Rake — Lathe	0° to −10°	0° to −10°	0° to −10°
Rake — Drill	0° to −10°	0° to −10°	0° to −10°

*N.B.: Cutting speeds can be as much as 1500 sfpm (450 m/min) when using carbide cutters, which is up to three times faster than when using high-speed steel cutters.

Machining data is a suggested guide only.

tion, and in some cases, prior to machining also. Annealing procedures are as follows:

- Polycarbonates — Uniformly raise the temperature to 250 degrees F (121 degrees C). Do not exceed this temperature. Maintain the temperature for 30 to 40 minutes.
- Acetal — Immerse in agitated vegetable oil or refined mineral oil, etc. Anneal at 315 degrees F (157 degrees C) for 15 minutes per 1/4 in (6.35 mm) thickness. Anneal sections more than 1 in (25.4 mm) thick for 1 hour per 1 in (25.4 mm) thickness at 300 degrees F (149 degrees C). Avoid drafts when cooling.
- Nylon — Immerse in oil at a maximum temperature of 300 degrees F (149 degrees C) for 15 minutes per 1-1/2 in (38 mm) thickness. Sections 1 in (25.4 mm) thick require 4 hours; each

additional 1 in (25.4 mm) of thickness requires an extra 2 hours. Cool slowly. Avoid drafts.

- Acrylic — Anneal at 175 degrees F (79 degrees C) only after completing fabrication and polishing. Anneal 10 hours for thicknesses up to 0.150 in (3.8 mm). Each additional 1/4 in (6.35 mm) thickness up to 1-1/2 in (38 mm) requires another 1/2 hour. Allow 1 hour per 1/4 in (6.35 mm) for thicker sections. Air is the annealing medium. Support parts and cool slowly.

SPECIAL NOTES

There is a danger when machining acetal plastic. Short broken acetal chips burn with no visible flame and produce a dangerous formaldehyde gas. When disposing of acetal chips, care should be taken to

Table 15–2 Recommended Machining Data for Thermoplastics

	ACETATE	ACRYLICS	FLOUROCARBONS	NYLON	POLYOLEFINS
Cutting Speed Lathe — sfpm — m/min	450 to 600 135 to 180	300 to 600 90 to 180	400 to 500 120 to 150	500 to 700 150 to 210	300 to 450 90 to 135
Cutting Speed Drill — sfpm — m/min	300 to 600 90 to 180	200 to 400 60 to 120	200 to 500 60 to 150	180 to 450 54 to 135	200 to 600 60 to 180
Feed — Lathe in/rev mm/rev	0.004 to 0.010 0.10 to 0.25	0.003 to 0.008 0.07 to 0.20	0.004 to 0.008 0.10 to 0.20	0.002 to 0.016 0.05 to 0.41	0.0015 to 0.004 0.04 to 0.10
Feed — Drill in/rev mm/rev	0.004 to 0.010 0.10 to 0.25	0.002 to 0.004 0.05 to 0.10	0.002 to 0.010 0.05 to 0.25	0.003 to 0.012 0.01 to 0.30	0.004 to 0.020 0.10 to 0.50
Clearance — Lathe	10° to 25°	10° to 20°	15° to 30°	5° to 10°	15° to 25°
Clearance — Drill	10° to 25°	12° to 15°	20°	10° to 15°	10° to 20°
Rake — Lathe	0° to 15°	0° to −10°	3° to 20°	5° to 10°	0° to 15°
Rake — Drill	0° to 10°	0° to 10°	0° to −10°	0° to 10°	0° to −5°

N.B.. Thermoplastics are not nearly as abrasive as Thermosets.

In turning, it is best to use high speeds, thin cuts (low feeds). Cuttings that curl away from the cutting edge in long, thin spirals similar to aluminum cuttings indicate proper speed, feed, and depth of cut. Long, continuous, thin, spiral chips from a drill indicate the proper drilling technique.

Machining data is a suggested guide only.

segregate them when possible. Proper ventilation of the work area is required.

Checking close tolerance dimensions on fluorocarbon parts requires special treatment. Fluorocarbons may undergo a dimensional change. Because of that, stock should be kept at temperatures above 75 degrees F (24 degrees C) prior to machining. It is best if the measuring instrument does not exert pressure on the work.

REVIEW QUESTIONS

1. Of the two main divisions of plastics, which one is generally abrasive to cutting tools?
2. Which of the two divisions contains plastics that are very brittle?
3. In lathework, what cutting speeds are commonly used to machine plastics?
4. What drilling speeds are generally recommended for plastics?
5. What does it mean when some plastics are deformed and what can be done to rectify this?
6. List four recommended practices in cutting plastics.
7. What can be done to offset heat buildup when machining plastics?
8. What is the purpose of annealing plastics?
9. What precautions must be taken when machining acetal plastic?
10. What treatment is required to prevent dimensional changes in fluorocarbons?

MANUFACTURING AUTOMATION PROTOCOL (MAP)

Consumers around the world are demanding more high quality goods at reasonable prices. To keep pace with world competition and cheap production costs, North American factories must replan their manufacturing strategies from the ground up to enable them to produce better quality goods at reasonable prices. In a sense, this is a new revolution, as was the Industrial Revolution which changed an agricultural civilization to an industrial civilization beginning about the middle of the 18th century.

Now industry is trying to change an entrenched industrial society into a flexible "high tech" society and to adjust to the resulting disruption that accompanies a major change. Other, more extensive changes will be required in the entrenched type of thinking and long term planning (or lack of it) that has characterized manufacturing until recent years.

Like any new process, automation has developed fitfully and slowly. As a result, small islands of automation have been created within the present factory system. These islands are capable of communicating within themselves, but are unable to communicate with other islands. Two powerful computer-based trends are spreading in today's manufacturing environment. One is a communication tool. It is called Manufacturing Automation Protocol (MAP). It is a specification for computer and control equipment intended to allow multivendor (multiseller) communications that can eventually lead to a totally computer-controlled factory. The other trend is a control tool which uses a host or master computer to monitor and operate one or many machines.

MAP represents a formalized, broad-based approach that eventually will allow communication between all local computer networks. It is hoped that in time many or even all industries will adopt the MAP approach. If robots, for example, are MAP compatible, they will be able to "talk" to other equipment through the MAP network of communications. In the robot industry, the concept of a host computer for data acquisition and storage, program storage and retrieval, and multiple machine control, is not yet a reality. It is hoped that uncomplicated programs can be created to allow procedures like the following to take place:

- Send a program to the robot from the host computer
- Read the directory of programs in the robot memory
- Transfer a program from the robot to the host computer
- Start a robot program
- Stop a robot program
- Read the contents of the registers in the robot
- Read the value of digital inputs
- Turn on digital outputs
- Read spontaneous messages from the robot
- Examine robot status
- Read robot axis positions
- Move the robot to a specific position

What is planned ultimately is that host computers will be able to communicate with each other through the MAP network of communications. A number of major companies, notably General Motors and Boeing, sponsored a demonstration project using the concept of MAP and also Technical and Office Protocol (TOP). The MAP systems designed for the plant floor and the TOP systems intended for the engineering and office environment were connected to allow for

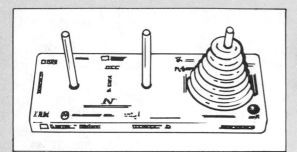

Figure 1 The Towers of Hanoi project *(Courtesy of ASEA Robotics Inc.)*

easy data exchange. GM is the catalyst for the adoption of MAP, while Boeing has the same support role for the adoption of TOP.

The MAP demonstration project was an ancient game called the Towers of Hanoi. It consisted of a base with three holes, pegs inserted into the holes, and seven concentric rings stacked on a peg, Figure 1. Available colors were red, white, and blue. Manufacture was carried out on two parallel assembly lines, Line A and Line B.

A customer ordered a custom made product by specifying the color of the base and pegs. Order-entry information was delivered directly to the job scheduler/dispatcher systems. At that point, job orders were ranked, raw material inventory was tabulated, and orders were sent to the production line and programmable controllers which guided the production machinery. All functions were monitored. A robot, centered between the beginning of production Lines A and B, picked bases from a parts bin to feed both lines. The first operation was a simulated drill station (the base holes had been predrilled to avoid chips during the demonstration). The holes were checked for accuracy by a vision system. Once the holes were drilled and inspected, the bases were sent to the peg insertion station where the colored pegs were inserted in the base (fed by an overhead hopper). Next, the presence of the pegs and color correctness were verified by a vision

system, and the finished jobs were moved to an unload station where they were approved or rejected. Finally, the accepted products were manually inspected and defects were relayed to the system via a voice activated input/output module.

Figure 2 is a synopsis of what MAP is all about. It is certain that the operation/communication movement, which will mean factory-wide communications and true cooperation between members of the MAP communication system, will have a far-reaching effect on the world industrial marketplace.

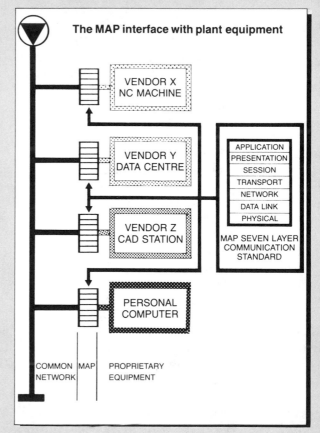

Figure 2 A synopsis of MAP *(Courtesy of ASEA Robotics Inc.)*

Drilling a fiberglass flange. When machining fiberglass, safety requires the use of a vacuum system plus a respirator for the operator. *Courtesy of Cominco Ltd.*

Facing a fiberglass flange with a brazed-tip carbide tool (viewed from the side). *Courtesy of Cominco Ltd.*

Facing a fiberglass flange with a brazed-tip carbide tool (looking toward the chuck). When machining fiberglass, safety requires the use of a vacuum system plus a respirator for the operator. *Courtesy of Cominco Ltd.*

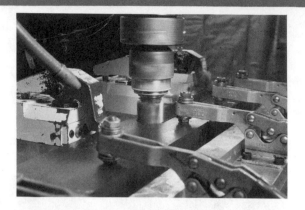

Slot milling thermoplastic. When machining thermoplastics, sharp tools are required, preferably honed. Clearances should be generous and rakes should generally be zero. High-speed steel tools can be used for short runs, but carbide tooling is preferred for long production runs. Cutting speeds with carbide tooling can be from 600 sfpm to 900 sfpm (190 m/min to 286 m/min). *Courtesy of The Polymer Corporation*

Automatic screw machining of thermoplastic parts. Thermoplastics such as nylons, nylatrons, acetal, fluorosint, and so on, are all ideal materials for automatic screw machine work. Turning, plunge-forming, knurling, and cutting off are examples of operations performed. In turning, the cutting tool should be exactly on center. Rakes should be zero, although carbide inserts at a 5° positive rake are satisfactory. *Courtesy of The Polymer Corporation*

Cutting a thermoplastic gear with a gear hobbing cutter. When using a hobbing cutter, the gear blank is left free to revolve. The cutter is set to full depth. The lead angle of the cutter causes the gear blank to revolve, cutting all of the gear teeth in one pass through the gear blank. *Courtesy of The Polymer Corporation*

Cutting lucite on a table saw. *Courtesy of Cominco Ltd.*

Turning a seal ring from fluorosint tubular bar stock. Fluorosint is a reasonably firm, abrasive thermoplastic. It should be machined exclusively with carbide inserts or carbide-tipped tools. For turning, rakes should generally be zero, with low clearances. A suggested cutting speed is 300 sfpm (95 m/min) with feeds of 0.006 to 0.010 in/rev (0.15 to 0.25 mm/rev). For roughing, a feed of 0.002 in/rev (0.05 mm/rev) or less and a higher cutting speed are recommended for finishing. A slight nose radius is recommended. *Courtesy of The Polymer Corporation*

Drilling lucite. *Courtesy of Cominco Ltd.*

Sawing engineering thermoplastics on a band saw. *Courtesy of Cadillac Plastic Co.*

Routing engineering thermoplastics on a vertical milling machine. *Courtesy of Cadillac Plastic Co.*

Facing nylatron bar stock. The recommended cutting speeds for nylatron are 600 sfpm to 900 sfpm (190 m/min to 286 m/min). Feeds are determined by the finish required: up to 0.015 in/rev for roughing, and 0.003 to 0.007 in/rev for finishing. Depths of cut recommended for finishing are 1/16" to 1/8" (1.59 mm to 3.18 mm) and for roughing up to 3/8" (9.53 mm). *Courtesy of The Polymer Corporation*

Turning nylatron. The recommended cutting speeds, feeds, and depth of cut for turning are the same as those given for facing (refer to no. 12). *Courtesy of The Polymer Corporation*

APPENDIX 1

Table 1. English (U.S. Customary) to Metric Conversions *(Courtesy of The L.S. Starrett Company)*

Decimals to Millimeters

Decimal	mm	Decimal	mm
0.001	0.0254	0.500	12.7000
0.002	0.0508	0.510	12.9540
0.003	0.0762	0.520	13.2080
0.004	0.1016	0.530	13.4620
0.005	0.1270	0.540	13.7160
0.006	0.1524	0.550	13.9700
0.007	0.1778	0.560	14.2240
0.008	0.2032	0.570	14.4780
0.009	0.2286	0.580	14.7320
0.010	0.2540	0.590	14.9860
0.020	0.5080	0.600	15.2400
0.030	0.7620	0.610	15.4940
0.040	1.0160	0.620	15.7480
0.050	1.2700	0.630	16.0020
0.060	1.5240	0.640	16.2560
0.070	1.7780	0.650	16.5100
0.080	2.0320	0.660	16.7640
0.090	2.2860	0.670	17.0180
0.100	2.5400	0.680	17.2720
0.110	2.7940	0.690	17.5260
0.120	3.0480	0.700	17.7800
0.130	3.3020	0.710	18.0340
0.140	3.5560	0.720	18.2880
0.150	3.8100	0.730	18.5420
0.160	4.0640	0.740	18.7960
0.170	4.3180	0.750	19.0500
0.180	4.5720	0.760	19.3040
0.190	4.8260	0.770	19.5580
0.200	5.0800	0.780	19.8120
0.210	5.3340	0.790	20.0660
0.220	5.5880	0.800	20.3200
0.230	5.8420	0.810	20.5740
0.240	6.0960	0.820	20.8280
0.250	6.3500	0.830	21.0820
0.260	6.6040	0.840	21.3360
0.270	6.8580	0.850	21.5900
0.280	7.1120	0.860	21.8440
0.290	7.3660	0.870	22.0980
0.300	7.6200	0.880	22.3520
0.310	7.8740	0.890	22.6060
0.320	8.1280	0.900	22.8600
0.330	8.3820	0.910	23.1140
0.340	8.6360	0.920	23.3680
0.350	8.8900	0.930	23.6220
0.360	9.1440	0.940	23.8760
0.370	9.3980	0.950	24.1300
0.380	9.6520	0.960	24.3840
0.390	9.9060	0.970	24.6380
0.400	10.1600	0.980	24.8920
0.410	10.4140	0.990	25.1460
0.420	10.6680	1.000	25.4000
0.430	10.9220		
0.440	11.1760		
0.450	11.4300		
0.460	11.6840		
0.470	11.9380		
0.480	12.1920		
0.490	12.4460		

Fractions to Decimals to Millimeters

Fraction	Decimal	mm	Fraction	Decimal	mm
1/64	0.0156	0.3969	33/64	0.5156	13.0969
1/32	0.0312	0.7938	17/32	0.5312	13.4938
3/64	0.0469	1.1906	35/64	0.5469	13.8906
1/16	0.0625	1.5875	9/16	0.5625	14.2875
5/64	0.0781	1.9844	37/64	0.5781	14.6844
3/32	0.0938	2.3812	19/32	0.5938	15.0812
7/64	0.1094	2.7781	39/64	0.6094	15.4781
1/8	0.1250	3.1750	5/8	0.6250	15.8750
9/64	0.1406	3.5719	41/64	0.6406	16.2719
5/32	0.1562	3.9688	21/32	0.6562	16.6688
11/64	0.1719	4.3656	43/64	0.6719	17.0656
3/16	0.1875	4.7625	11/16	0.6875	17.4625
13/64	0.2031	5.1594	45/64	0.7031	17.8594
7/32	0.2188	5.5562	23/32	0.7188	18.2562
15/64	0.2344	5.9531	47/64	0.7344	18.6531
1/4	0.2500	6.3500	3/4	0.7500	19.0500
17/64	0.2656	6.7469	49/64	0.7656	19.4469
9/32	0.2812	7.1438	25/32	0.7812	19.8438
19/64	0.2969	7.5406	51/64	0.7969	20.2406
5/16	0.3125	7.9375	13/16	0.8125	20.6375
21/64	0.3281	8.3344	53/64	0.8281	21.0344
11/32	0.3438	8.7312	27/32	0.8438	21.4312
23/64	0.3594	9.1281	55/64	0.8594	21.8281
3/8	0.3750	9.5250	7/8	0.8750	22.2250
25/64	0.3906	9.9219	57/64	0.8906	22.6219
13/32	0.4062	10.3188	29/32	0.9062	23.0188
27/64	0.4219	10.7156	59/64	0.9219	23.4156
7/16	0.4375	11.1125	15/16	0.9375	23.8125
29/64	0.4531	11.5094	61/64	0.9531	24.2094
15/32	0.4688	11.9062	31/32	0.9688	24.6062
31/64	0.4844	12.3031	63/64	0.9844	25.0031
1/2	0.5000	12.7000	1	1.0000	25.4000

Table 2. Metric to English (U.S. Customary) Conversions *(Courtesy of The L.S. Starrett Company)*

mm	Decimal	mm	Decimal	mm	Decimal	mm	Decimal	mm	Decimal
0.01	.00039	0.41	.01614	0.81	.03189	21	.82677	61	2.40157
0.02	.00079	0.42	.01654	0.82	.03228	22	.86614	62	2.44094
0.03	.00118	0.43	.01693	0.83	.03268	23	.90551	63	2.48031
0.04	.00157	0.44	.01732	0.84	.03307	24	.94488	64	2.51969
0.05	.00197	0.45	.01772	0.85	.03346	25	.98425	65	2.55906
0.06	.00236	0.46	.01811	0.86	.03386	26	1.02362	66	2.59843
0.07	.00276	0.47	.01850	0.87	.03425	27	1.06299	67	2.63780
0.08	.00315	0.48	.01890	0.88	.03465	28	1.10236	68	2.67717
0.09	.00354	0.49	.01929	0.89	.03504	29	1.14173	69	2.71654
0.10	.00394	0.50	.01969	0.90	.03543	30	1.18110	70	2.75591
0.11	.00433	0.51	.02008	0.91	.03583	31	1.22047	71	2.79528
0.12	.00472	0.52	.02047	0.92	.03622	32	1.25984	72	2.83465
0.13	.00512	0.53	.02087	0.93	.03661	33	1.29921	73	2.87402
0.14	.00551	0.54	.02126	0.94	.03701	34	1.33858	74	2.91339
0.15	.00591	0.55	.02165	0.95	.03740	35	1.37795	75	2.95276
0.16	.00630	0.56	.02205	0.96	.03780	36	1.41732	76	2.99213
0.17	.00669	0.57	.02244	0.97	.03819	37	1.45669	77	3.03150
0.18	.00709	0.58	.02283	0.98	.03858	38	1.49606	78	3.07087
0.19	.00748	0.59	.02323	0.99	.03898	39	1.53543	79	3.11024
0.20	.00787	0.60	.02362	1.00	.03937	40	1.57480	80	3.14961
0.21	.00827	0.61	.02402	1	.03937	41	1.61417	81	3.18898
0.22	.00866	0.62	.02441	2	.07874	42	1.65354	82	3.22835
0.23	.00906	0.63	.02480	3	.11811	43	1.69291	83	3.26772
0.24	.00945	0.64	.02520	4	.15748	44	1.73228	84	3.30709
0.25	.00984	0.65	.02559	5	.19685	45	1.77165	85	3.34646
0.26	.01024	0.66	.02598	6	.23622	46	1.81102	86	3.38583
0.27	.01063	0.67	.02638	7	.27559	47	1.85039	87	3.42520
0.28	.01102	0.68	.02677	8	.31496	48	1.88976	88	3.46457
0.29	.01142	0.69	.02717	9	.35433	49	1.92913	89	3.50394
0.30	.01181	0.70	.02756	10	.39370	50	1.96850	90	3.54331
0.31	.01220	0.71	.02795	11	.43307	51	2.00787	91	3.58268
0.32	.01260	0.72	.02835	12	.47244	52	2.04724	92	3.62205
0.33	.01299	0.73	.02874	13	.51181	53	2.08661	93	3.66142
0.34	.01339	0.74	.02913	14	.55118	54	2.12598	94	3.70079
0.35	.01378	0.75	.02953	15	.59055	55	2.16535	95	3.74016
0.36	.01417	0.76	.02992	16	.62992	56	2.20472	96	3.77953
0.37	.01457	0.77	.03032	17	.66929	57	2.24409	97	3.81890
0.38	.01496	0.78	.03071	18	.70866	58	2.28346	98	3.85827
0.39	.01535	0.79	.03110	19	.74803	59	2.32283	99	3.89764
0.40	.01575	0.80	.03150	20	.78740	60	2.36220	100	3.93701

Table 3. Metric and English (U.S. Customary) Conversion Units *(Courtesy of The L.S. Starrett Company)*

Measures of Length

1 millimeter (mm) = 0.03937 inch
1 centimeter (cm) = 0.39370 inch
1 meter (m) = 39.37008 inches
        = 3.2808 feet
        = 1.0936 yards
1 kilometer (km) = 0.6214 mile
1 inch = 25.4 millimeters (mm)
        = 2.54 centimeters (cm)
1 foot = 304.8 millimeters (mm)
        = 0.3048 meter (m)
1 yard = 0.9144 meter (m)
1 mile = 1.609 kilometers (km)

Measures of Area

1 square millimeter = 0.00155 square inch
1 square centimeter = 0.155 square inch
1 square meter = 10.764 square feet
        = 1.196 square yards
1 square kilometer = 0.3861 square mile
1 square inch = 645.2 square millimeters
        = 6.452 square centimeters
1 square foot = 929 square centimeters
        = 0.0929 square meter
1 square yard = 0.836 square meter
1 square mile = 2.5899 square kilometers

Measures of Capacity (Dry)

1 cubic centimeter (cm³) = 0.061 cubic inch
1 liter = 0.0353 cubic foot
        = 61.023 cubic inches
1 cubic meter (m³) = 35.315 cubic feet
        = 1.308 cubic yards
1 cubic inch = 16.38706 cubic centimeters (cm³)
1 cubic foot = 0.02832 cubic meter (m³)
        = 28.317 liters
1 cubic yard = 0.7646 cubic meter (m³)

Measures of Capacity (Liquid)

1 liter = 1.0567 U.S. quarts
        = 0.2642 U.S. gallon
        = 0.2200 Imperial gallon
1 cubic meter (m³) = 264.2 U.S. gallons
        = 219.969 Imperial gallons
1 U.S. quart = 0.946 liter
1 Imperial quart = 1.136 liters
1 U.S. gallon = 3.785 liters
1 Imperial gallon = 4.546 liters

Measures of Weight

1 gram (g) = 15.432 grains
        = 0.03215 ounce troy
        = 0.03527 ounce avoirdupois
1 kilogram (kg) = 35.274 ounces avoirdupois
        = 2.2046 pounds
1000 kilograms (kg) = 1 metric ton (t)
        = 1.1023 tons of 2000 pounds
        = 0.9842 ton of 2240 pounds
1 ounce avoirdupois = 28.35 grams (g)
1 ounce troy = 31.103 grams (g)
1 pound = 453.6 grams
        = 0.4536 kilogram (kg)
1 ton of 2240 pounds = 1016 kilograms (kg)
        = 1.016 metric tons
1 grain = 0.0648 gram (g)
1 metric ton = 0.9842 ton of 2240 pounds
        = 2204.6 pounds

Table 4. Geometry and Circles *(Courtesy of The L.S. Starrett Company)*

To Find Circumference—
Multiply diameter by 3.1416 . . . . Or divide diameter by 0.3183

To Find Diameter—
Multiply circumference by 0.3183 . . . . Or divide circumference by 3.1416

To Find Radius—
Multiply circumference by 0.15915 . . . . Or divide circumference by 6.28318

To Find Side of an Inscribed Square—
Multiply diameter by 0.7071
Or multiply circumference by 0.2251 . . . . Or divide circumference by 4.4428

To Find Side of an Equal Square—
Multiply diameter by 0.8862 . . . . Or divide diameter by 1.1284
Or multiply circumference by 0.2821 . . . . Or divide circumference by 3.545

Square—
A side multiplied by 1.4142 equals diameter of its circumscribing circle.
A side multiplied by 4.443 equals circumference of its circumscribing circle.
A side multiplied by 1.128 equals diameter of an equal circle.
A side multiplied by 3.547 equals circumference of an equal circle.

To Find the Area of a Circle—
Multiply circumference by one-quarter of the diameter.
Or multiply the square of diameter by 0.7854
Or multiply the square of circumference by .07958
Or multiply the square of ½ diameter by 3.1416

To Find the Surface of a Sphere or Globe—
Multiply the diameter by the circumference.
Or multiply the square of diameter by 3.1416
Or multiply four times the square of radius by 3.1416

Table 5. Basic Trigonometric Functions

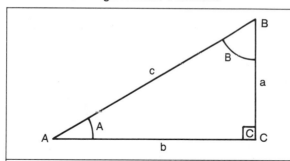

THE SIX BASIC TRIGONOMETRIC FUNCTIONS

1. $\text{SIN A} = \dfrac{\text{Opposite side}}{\text{Hypotenuse}} = \dfrac{a}{c} = \dfrac{\text{Opp}}{\text{Hyp}}$

2. $\text{COS A} = \dfrac{\text{Adjacent side}}{\text{Hypotenuse}} = \dfrac{b}{c} = \dfrac{\text{Adj}}{\text{Hyp}}$

3. $\text{TAN A} = \dfrac{\text{Opposite side}}{\text{Adjacent side}} = \dfrac{a}{b} = \dfrac{\text{Opp}}{\text{Adj}}$

4. $\text{COSEC A} = \dfrac{\text{Hypotenuse}}{\text{Opposite side}} = \dfrac{c}{a} = \dfrac{\text{Hyp}}{\text{Opp}}$

5. $\text{SEC A} = \dfrac{\text{Hypotenuse}}{\text{Adjacent side}} = \dfrac{c}{b} = \dfrac{\text{Hyp}}{\text{Adj}}$

6. $\text{COTAN A} = \dfrac{\text{Adjacent side}}{\text{Opposite side}} = \dfrac{b}{a} = \dfrac{\text{Adj}}{\text{Opp}}$

Sine is written sin.
Cosine is written cos.
Tangent is written tan.
Cosecant is written cosec.
Secant is written sec.
Cotangent is written cotan.

Table 6. A Comparison of Morse Tapers and Browne and Sharpe Tapers

MORSE TAPERS			
No. of Taper	TAPER		
	in/ft	mm/mm	mm/100mm
0	0.625	1 : 19	5.21
1	0.599	1 : 20	4.99
2	0.599	1 : 20	4.99
3	0.602	1 : 20	5.02
4	0.623	1 : 19	5:19
5	0.631	1 : 19	5.26
6	0.626	1 : 19	5.22
7	0.624	1 : 19	5.20

BROWN AND SHARPE TAPERS			
No. of Taper	TAPER		
	in/ft	mm/mm	mm/100mm
1	0.502	1 : 24	4.18
2	0.502	1 : 24	4.18
3	0.502	1 : 24	4.18
4	0.502	1 : 24	4.18
5	0.502	1 : 24	4.18
6	0.503	1 : 24	4.19
7	0.501	1 : 24	4.18
8	0.501	1 : 24	4.18
9	0.501	1 : 24	4.18
10	0.516	1 : 23	4.30
11	0.501	1 : 24	4.18
12	0.500	1 : 24	4.17
13	0.500	1 : 24	4.17
14	0.500	1 : 24	4.17
15	0.500	1 : 24	4.17
16	0.500	1 : 24	4.17
17	0.500	1 : 24	4.17
18	0.500	1 : 24	4.17

Table 7. Tapers and Angles *(Courtesy of The L.S. Starrett Company)*

Taper per Foot	Included Angle			Angle With Center Line			Taper per Inch	Taper per Inch from Center Line
	Deg.	Min.	Sec.	Deg.	Min.	Sec.		
⅛	0	35	47	0	17	54	.010416	.005208
3⁄16	0	53	44	0	26	52	.015625	.007812
¼	1	11	38	0	35	49	.020833	.010416
5⁄16	1	29	31	0	44	46	.026042	.013021
⅜	1	47	25	0	53	42	.031250	.015625
7⁄16	2	5	18	1	2	39	.036458	.018229
½	2	23	12	1	11	36	.041667	.020833
9⁄16	2	41	7	1	20	34	.046875	.023438
⅝	2	59	3	1	29	31	.052084	.026042
11⁄16	3	16	56	1	38	28	.057292	028646
¾	3	34	48	1	47	24	.062500	.031250
13⁄16	3	52	42	1	56	21	.067708	.033854
⅞	4	10	32	2	5	16	.072917	.036456
15⁄16	4	28	26	2	14	13	.078125	.039063
1	4	46	19	2	23	10	.083330	.041667
1¼	5	57	45	2	58	53	.104166	.052084
1½	7	9	10	3	34	35	.125000	.062500
1¾	8	20	28	4	10	14	.145833	.072917
2	9	31	37	4	45	49	.166666	.083332
2½	11	53	38	5	56	49	.208333	.104166
3	14	2	0	7	1	0	.250000	.125000
3½	16	35	39	8	17	49	.291666	.145833
4	18	55	31	9	27	44	.333333	.166666
4½	21	11	20	10	37	10	.375000	.187500
5	23	32	12	11	46	6	.416666	.208333
6	28	4	20	14	2	10	.500000	.250000

Table 8. Double Depth of Screw Threads *(Courtesy of The L.S. Starrett Company)*

$$D.D. = \frac{1.732}{N} \text{ For V Thread}$$

$$D.D. = \frac{1.299}{N} \text{ For American Nat. Form, U. S. Std.}$$

$$D.D. = \frac{1.28}{N} \text{ For Whitworth Standard}$$

Threads per Inch N	V Threads D.D.	Am. Nat. Form D.D. U.S. Std.	Whitworth Standard D.D.	Threads per Inch N	V Threads D.D.	Am. Nat. Form D.D. U.S. Std.	Whitworth Standard D.D.
2	.86600	.64950	.64000	28	.06185	.04639	.04571
2¼	.76978	.57733	.56888	30	.05773	.04330	.04266
2⅜	.72926	.54694	.53894	32	.05412	.04059	.04000
2½	.69280	.51960	.51200	34	.05094	.03820	.03764
2⅝	.65981	.49485	.48761	36	.04811	.03608	.03555
2¾	.62982	.47236	.46545	38	.04558	.03418	.03368
2⅞	.60243	.45182	.44521	40	.04330	.03247	.03200
3	.57733	.43300	.42666	42	.04124	.03093	.03047
3¼	.53292	.39966	.39384	44	.03936	.02952	.02909
3½	.49485	.37114	.36571	46	.03765	.02823	.02782
4	.43300	.32475	.32000	48	.03608	.02706	.02666
4½	.38488	.28869	.28444	50	.03464	.02598	.02560
5	.34640	.25980	.25600	52	.03331	.02498	.02461
5½	.31490	.23618	.23272	54	.03207	.02405	.02370
6	.28866	.21650	.21333	56	.03093	.02319	.02285
7	.24742	.18557	.18285	58	.02986	.02239	.02206
8	.21650	.16237	.16000	60	.02887	.02165	.02133
9	.19244	.14433	.14222	62	.02794	.02095	.02064
10	.17320	.12990	.12800	64	.02706	.02029	.02000
11	.15745	.11809	.11636	66	.02624	.01968	.01939
11½	.15061	.11295	.11130	68	.02547	.01910	.01882
12	.14433	.10825	.10666	70	.02474	.01855	.01829
13	.13323	.09992	.09846	72	.02406	.01804	.01778
14	.12371	.09278	.09142	74	.02341	.01752	.01729
15	.11547	.08660	.08533	76	.02279	.01714	.01684
16	.10825	.08118	.08000	78	.02221	.01665	.01641
18	.09622	.07216	.07111	80	.02165	.01623	.01600
20	.08660	.06495	.06400	82	.02112	.01584	.01560
22	.07872	.05904	.05818	84	.02062	.01546	.01523
24	.07216	.05412	.05333	86	.02014	.01510	.01488
26	.06661	.04996	.04923	88	.01968	.01476	.01454
27	.06415	.04811	.04740	90	.01924	.01443	.01422

Table 9. Diagonal Measurements of Hexagons and Squares

ACROSS FLATS	ACROSS CORNERS		ACROSS FLATS	ACROSS CORNERS	
	Hexa-gon	Squares		Hexa-gon	Squares
1/16	0.072	0.088	1 1/16	1.226	1.502
1/8	0.144	0.177	1 1/8	1.299	1.591
3/16	0.216	0.265			
			1 3/16	1.371	1.679
1/4	0.288	0.353	1 1/4	1.443	1.767
5/16	0.360	0.441	1 5/16	1.515	1.856
3/8	0.432	0.530			
			1 3/8	1.587	1.944
7/16	0.505	0.618	1 7/16	1.659	2.032
1/2	0.577	0.707	1 1/2	1.732	2.121
9/16	0.649	0.795			
			1 9/16	1.804	2.209
5/8	0.721	0.883	1 5/8	1.876	2.298
11/16	0.793	0.972	1 11/16	1.948	2.386
3/4	0.865	1.060			
			1 3/4	2.020	2.470
13/16	0.938	1.149	1 13/16	2.092	2.563
7/8	1.010	1.237	1 7/8	2.165	2.651
15/16	1.082	1.325			
			1 15/16	2.237	2.740
1	1.155	1.414	2	2.309	2.828

Table 10. Cutting Speeds and Rpm *(Courtesy of The L.S. Starrett Company)*

Drill Diam., Inches	FEET PER MINUTE										
	30	40	50	60	70	80	90	100	150	200	250
	REVOLUTIONS PER MINUTE										
1/16	1833	2445	3056	3667	4278	4889	5500	6112	9167	12223	15279
1/8	917	1222	1528	1833	2139	2445	2750	3056	4584	6112	7639
3/16	611	815	1019	1222	1426	1630	1833	2037	3056	4074	5093
1/4	458	611	764	917	1070	1222	1375	1528	2292	3056	3820
5/16	367	489	611	733	856	978	1100	1222	1833	2445	3056
3/8	306	407	509	611	713	815	917	1019	1528	2037	2546
7/16	262	349	437	524	611	698	786	873	1310	1746	2183
1/2	229	306	382	458	535	611	688	764	1146	1528	1910
5/8	183	244	306	367	428	489	550	611	917	1222	1528
3/4	153	204	255	306	357	407	458	509	764	1019	1273
7/8	131	175	218	262	306	349	393	473	655	873	1091
1	115	153	191	229	267	306	344	382	573	764	955
1 1/8	102	136	170	204	238	272	306	340	509	679	849
1 1/4	92	122	153	183	214	244	275	306	458	611	764
1 3/8	83	111	139	167	194	222	250	278	417	556	694
1 1/2	76	102	127	153	178	204	229	255	382	509	637
1 5/8	71	94	118	141	165	188	212	235	353	470	588
1 3/4	66	87	109	131	153	175	196	218	327	437	546
1 7/8	61	82	102	122	143	163	183	204	306	407	509
2	57	76	96	115	134	153	172	191	287	382	477
2 1/4	51	68	85	102	119	136	153	170	255	340	424
2 1/2	46	61	76	92	107	122	138	153	229	306	382
2 3/4	42	56	70	83	97	111	125	139	208	278	347
3	38	51	64	76	89	102	115	127	191	255	318
3 1/4	35	47	59	71	82	94	106	118	176	235	294
3 1/2	33	44	55	66	76	87	98	109	164	218	273
3 3/4	31	41	51	61	71	81	92	102	153	204	255
4	29	38	48	57	67	76	86	96	143	191	239
4 1/2	26	34	42	51	59	68	76	85	127	170	212
5	23	31	38	46	54	61	69	76	115	153	191
5 1/2	21	28	35	42	49	56	63	70	104	139	174
6	19	26	32	38	45	51	57	64	96	127	159
6 1/2	18	24	29	35	41	47	53	59	88	118	147
7	16	22	27	33	38	44	49	55	82	109	136
7 1/2	15	20	26	31	36	41	46	51	76	102	127
8	14	19	24	29	33	38	43	48	72	96	119

English (U.S. Customary)

$$Rpm = \frac{4 \times CS \text{(sfpm)}}{D \text{ (inches)}}$$

Metric

$$Rpm = \frac{CS \text{(m/min)}}{D \text{ (meters)}}$$

Table 11. Cutting Speeds and Rpm of Drills (Fractional)
(Courtesy of SKF & Dormer Tools)

PERIPHERAL SPEED						
Ft./min.	20	30	50	70	80	100
m/min.	6	9	15	21	24	30
Diam. in.	REVOLUTIONS PER MINUTE					
1/64	4,897	7,346	12,243	17,140	19,588	24,485
1/32	2,449	3,673	6,121	8,570	9,794	12,243
3/64	1,629	2,443	4,072	5,701	6,515	8,144
1/16	1,222	1,833	3,056	4,278	4,889	6,112
5/64	978	1,467	2,445	3,424	3,913	4,891
3/32	814	1,222	2,036	2,851	3,258	4,072
7/64	698	1,047	1,746	2,444	2,793	3,492
1/8	611	917	1,528	2,139	2,445	3,056
5/32	489	734	1,223	1,712	1,956	2,445
3/16	407	611	1,019	1,426	1,630	2,037
7/32	349	524	873	1,222	1,397	1,746
1/4	306	458	764	1,070	1,222	1,528
5/16	244	367	611	856	978	1,222
3/8	204	306	509	713	815	1,019
7/16	175	262	437	611	698	873
1/2	153	229	382	535	611	764
9/16	136	204	340	475	543	679
5/8	122	183	306	428	489	611
11/16	111	167	278	389	444	556
3/4	102	153	255	356	407	509
13/16	94	141	235	329	376	470
7/8	87	131	218	306	349	437
15/16	81	122	204	285	326	407
1	76	115	191	267	306	382
1, 1/8	68	102	170	238	272	340
1, 1/4	61	92	153	214	244	306
1, 3/8	56	83	139	194	222	278
1, 1/2	51	76	127	178	204	255
1, 5/8	47	71	118	165	188	235
1, 3/4	44	65	109	153	175	218
1, 7/8	41	61	102	143	163	204
2	38	57	95	134	153	191
2, 1/4	34	51	85	119	136	170
2, 1/2	31	46	76	107	122	153
2, 3/4	28	42	69	97	111	139
3	25	38	64	89	102	127
3, 1/2	22	33	55	76	87	109
4	19	29	48	67	76	95
4, 1/2	17	25	42	59	68	85
5	15	23	38	53	61	76
5, 1/2	14	21	35	49	56	69
0	13	19	32	45	51	64
6, 1/2	12	18	29	41	47	59
7	11	16	27	38	44	55
7, 1/2	10	15	25	36	41	51
8	9	14	24	33	38	48

R.P.M. for Peripheral Speeds not given, can be obtained by simple addition or subtraction, e.g.: 150 ft./min. 100 + 50 = 1,146 R.P.M. (for 1/2'' dia.) 60 ft./min. = 80 — 20 = 4,886 R.P.M. (for 3/64'' dia.)

Table 12. Cutting Speeds and Rpm of Drills (Metric)
(Courtesy of SKF & Dormer Tools)

PERIPHERAL SPEED						
Ft./min.	20	30	50	70	80	100
m/min.	6	9	15	21	24	30
Diam. mm.	REVOLUTIONS PER MINUTE					
0.5	3,878	5,817	9,695	13,573	15,512	19,389
1.0	1,939	2,908	4,847	6,786	7,756	9,695
1.5	1,293	1,939	3,232	4,524	5,171	6,463
2.0	971	1,456	2,427	3,397	3,883	4,854
2.5	776	1,165	1,941	2,717	3,105	3,882
3.0	647	970	1,617	2,264	2,587	3,234
3.5	554	832	1,386	1,940	2,218	2,772
4.0	485	728	1,213	1,698	1,940	2,425
4.5	431	647	1,078	1,509	1,724	2,156
5.0	388	582	970	1,359	1,552	1,940
6.0	323	485	809	1,132	1,294	1,617
7.0	277	416	693	970	1,109	1,386
8.0	243	364	606	849	970	1,213
9.0	216	323	539	755	862	1,078
10.0	194	291	485	679	776	970
11.0	176	265	441	617	706	882
12.0	162	243	404	566	647	809
13.0	149	224	373	522	597	746
14.0	139	208	346	485	554	693
15.0	129	194	323	453	517	647
16.0	121	182	303	424	485	606
17.0	114	171	285	399	457	571
18.0	108	162	269	377	431	539
19.0	102	153	255	357	409	511
20.0	97	146	243	340	388	485
22.0	88	132	221	309	353	441
24.0	81	121	202	283	323	404
26.0	75	112	187	261	299	373
28.0	69	104	173	243	277	346
30.0	65	97	162	226	259	323
35.0	55	83	139	194	222	277
40.0	49	73	121	170	194	243
45.0	43	65	108	151	172	216
50.0	39	58	97	136	155	194
60.0	32	49	81	113	129	162
70.0	28	42	69	97	111	139
80.0	24	36	61	85	97	121
90.0	22	32	54	75	86	108
100.0	19	29	49	68	78	97
115.0	17	25	42	58	66	83
130.0	15	22	38	51	59	73
145.0	13	20	33	46	53	66
160.0	12	18	30	42	48	60
180.0	11	16	27	37	42	53
200.0	10	14	24	33	38	40

R.P.M. for Peripheral Speeds not given, can be obtained by simple addition or subtraction, e.g.: 45 m/min. = 15 + 30 = 1,455 R.P.M. (for 10 mm dia.) 18 m/min. = 24 — 6 = 3.878 R.P.M. (for 1.5 mm dia.)

Table 13. Grinding Wheel Speeds *(Courtesy of Norton Company)*

To find the number of revolutions of the wheel spindle, having been given the surface or peripheral speed and the diameter of the wheel, divide the surface speed in feet per minute by the circumference (diameter x 3.14) in feet.

To find the surface speed of a wheel in feet per minute, multiply the circumference in feet by the revolutions per minute.

Diam. of Wheel in Inches	mm (Approx.)	Peripheral Speed in Feet per Minute					
		4000 Ft / 1200 m	*4500 Ft* / 1350 m	*5000 Ft* / 1500 m	*5500 Ft* / 1650 m	*6000 Ft* / 1800 m	*6500 Ft* / 1950 m
		Revolutions per Minute					
1/4	6	61,116	68,756	76,392	84,032	91,672	99,212
3/8	9	40,744	46,594	50,928	56,021	61,115	66,141
1/2	13	30,558	34,378	38,196	42,016	45,836	49,656
5/8	16	24,446	27,502	30,557	33,615	36,669	39,685
3/4	19	20,372	22,918	25,464	28,011	30,557	33,071
7/8	22	17,462	21,826	21,826	24,009	26,192	28,346
1	25	15,279	17,189	19,098	21,008	22,918	24,828
2	50	7,639	8,594	9,549	10,504	11,459	12,414
3	75	5,093	5,729	6,366	7,003	7,639	8,276
4	100	3,820	4,297	4,775	5,252	5,729	6,207
5	125	3,056	3,438	3,820	4,202	4,584	4,966
6	150	2,546	2,865	3,183	3,501	3,820	4,138
7	175	2,183	2,455	2,728	3,001	3,274	3,547
8	200	1,910	2,148	2,387	2,626	2,865	3,103
10	250	1,528	1,719	1,910	2,101	2,292	2,483
12	305	1,273	1,432	1,591	1,751	1,910	2,069
14	355	1,091	1,228	1,364	1,500	1,637	1,773
16	405	955	1,074	1,194	1,313	1,432	1,552
18	455	849	955	1,061	1,167	1,273	1,379
20	505	764	859	955	1,050	1,146	1,241
22	560	694	781	868	955	1,042	1,128
24	610	637	716	796	875	955	1,034
26	660	588	661	734	808	881	955
28	710	546	614	682	750	818	887
30	760	509	573	637	700	764	828
32	810	477	537	597	656	716	776
34	860	449	505	562	618	674	730
36	910	424	477	530	583	637	690

Table 14. Fractional, Wire or Number, Letter and Metric Drill Sizes *(Courtesy of The Cleveland Twist Drill Company)*

Table 14. Fractional, Wire or Number, Letter and Metric Drill Sizes *(Courtesy of The Cleveland Twist Drill Company)*

Decimal	Fract. Wire Letter	mm.	Decimal	Fract. Wire Letter	mm.	Decimal	Fract. Wire Letter	mm.	Decimal	Fract. Wire Letter	mm.
.0059	07		.0413		1.05	.1065	36		.1960	9	
.0063	96		.0420	58		.1083		2.75	.1969		5.
.0067	95		.0430	57		.1094	**7/64"**		.1990	8	
.0071	94		.0433		1.1	.1100	35		.2008		5.1
.0075	93		.0453		1.15	.1102		2.8	.2010	7	
.0079	92	.2	.0465	56		.1110	34		.2031	**13/64"**	
.0083	91		.0469	**3/64"**		.1130	33		.2040	6	
.0087	90	.22	.0472		1.2	.1142		2.9	.2047		5.2
.0091	89		.0492		1.25	.1160	32		.2055	5	
.0095	88		.0512		1.3	.1181		3.	.2067		5.25
.0098		.25	.0520	55		.1200	31		.2087		5.3
.0100	87		.0531		1.35	.1220		3.1	.2090	4	
.0105	86		.0550	54		.1250	**1/8"**		.2126		5.4
.0110	85	.28	.0551		1.4	.1260		3.2	.2130	3	
.0115	84		.0571		1.45	.1280		3.25	.2165		5.5
.0118		.3	.0591		1.5	.1285	30		.2188	**7/32"**	
.0120	83		.0595	53		.1299		3.3	.2205		5.6
.0125	82		.0610		1.55	.1339		3.4	.2210	2	
.0126		.32	.0625	**1/16"**		.1360	29		.2244		5.7
.0130	81		.0630		1.6	.1378		3.5	.2264		5.75
.0135	80		.0635	52		.1405	28		.2280	1	
.0138		.35	.0650		1.65	.1406	**9/64"**		.2283		5.8
.0145	79		.0669		1.7	.1417		3.6	.2323		5.9
.0156	**1/64"**		.0670	51		.1440	27		.2340	A	
.0157		.4	.0689		1.75	.1457		3.7	.2344	**15/64"**	
.0160	78		.0700	50		.1470	26		.2362		6.
.0177		.45	.0709		1.8	.1476		3.75	.2380	B	
.0180	77		.0728		1.85	.1495	25		.2402		6.1
.0197		.5	.0730	49		.1496		3.8	.2420	C	
.0200	76		.0748		1.9	.1520	24		.2441		6.2
.0210	75		.0760	48		.1535		3.9	.2460	D	
.0217		.55	.0768		1.95	.1540	23		.2461		6.25
.0225	74		.0781	**5/64"**		.1562	**5/32"**		.2480		6.3
.0236		.6	.0785	47		.1570	22		.2500	**1/4"** E	
.0240	73		.0787		2.	.1575		4.	.2520		6.4
.0250	72		.0807		2.05	.1590	21		.2559		6.5
.0256		.65	.0810	46		.1610	20		.2570	F	
.0260	71		.0820	45		.1614		4.1	.2598		6.6
.0276		.7	.0827		2.1	.1654		4.2	.2610	G	
.0280	70		.0846		2.15	.1660	19		.2638		6.7
.0292	69		.0860	44		.1673		4.25	.2656	**17/64"**	
.0295		.75	.0866		2.2	.1693		4.3	.2657		6.75
.0310	68		.0886		2.25	.1695	18		.2660	H	
.0312	**1/32"**		.0890	43		.1719	**11/64"**		.2677		6.8
.0315		.8	.0906		2.3	.1730	17		.2717		6.9
.0320	67		.0925		2.35	.1732		4.4	.2720	I	
.0330	66		.0935	42		.1770	16		.2756		7.
.0335		.85	.0938	**3/32"**		.1772		4.5	.2770	J	
.0350	65		.0945		2.4	.1800	15		.2795		7.1
.0354		.9	.0960	41		.1811		4.6	.2810	K	
.0360	64		.0965		2.45	.1820	14		.2812	**9/32"**	
.0370	63		.0980	40		.1850	13	4.7	.2835		7.2
.0374		.95	.0984		2.5	.1870		4.75	.2854		7.25
.0380	62		.0995	39		.1875	**3/16"**		.2874		7.3
.0390	61		.1015	38		.1890	12	4.8	.2900	L	
.0394		1.	.1024		2.6	.1910	11		.2913		7.4
.0400	60		.1040	37		.1929		4.9			
.0410	59		.1063		2.7	.1935	10				

Table 14. Continued

Decimal	Fract. Wire Letter	mm.	Decimal	Fract. Wire Letter	mm.	Decimal	Fract. Wire Letter	mm.	Decimal	Fract. Wire Letter	mm.
.2950	M		.4688	15/32"		.9843		25.	1.4961		38.
.2953		7.5	.4724		12.	.9844	63/64"		1.5000	1-1/2"	
.2969	19/64"		.4844	31/64"		1.0000	1"		1.5156	1-33/64"	
.2992		7.6	.4921		12.5	1.0039		25.5	1.5157		38.5
.3020	N		.5000	1/2"		1.0156	1-1/64"		1.5312	1-17/32"	
.3031		7.7	.5118		13.	1.0236		26.	1.5354		39.
.3051		7.75	.5156	33/64"		1.0312	1-1/32"		1.5469	1-35/64"	
.3071		7.8	.5312	17/32"		1.0433		26.5	1.5551		39.5
.3110		7.9	.5315		13.5	1.0469	1-3/64"		1.5625	1-9/16"	
.3125	5/16"		.5469	35/64"		1.0625	1-1/16"		1.5748		40.
.3150		8.	.5512		14.	1.0630		27.	1.5781	1-37/64"	
.3160	O		.5625	9/16"		1.0781	1-5/64"		1.5938	1-19/32"	
.3189		8.1	.5709		14.5	1.0827		27.5	1.5945		40.5
.3228		8.2	.5781	37/64"		1.0938	1-3/32"		1.6094	1-39/64"	
.3230	P		.5906		15.	1.1024		28.	1.6142		41.
.3248		8.25	.5938	19/32"		1.1094	1-7/64"		1.6250	1-5/8"	
.3268		8.3	.6094	39/64"		1.1220		28.5	1.6339		41.5
.3281	21/64"		.6102		15.5	1.1250	1-1/8"		1.6406	1-41/64"	
.3307		8.4	.6250	5/8"		1.1406	1-9/64"		1.6535		42.
.3320	Q		.6299		16.	1.1417		29.	1.6562	1-21/32"	
.3346		8.5	.6406	41/64"		1.1562	1-5/32"		1.6719	1-43/64"	
.3386		8.6	.6496		16.5	1.1614		29.5	1.6732		42.5
.3390	R		.6562	21/32"		1.1719	1-11/64"		1.6875	1-11/16"	
.3425		8.7	.6693		17.	1.1811		30.	1.6929		43.
.3438	11/32"		.6719	43/64"		1.1875	1-3/16"		1.7031	1-45/64"	
.3445		8.75	.6875	11/16"		1.2008		30.5	1.7126		43.5
.3465		8.8	.6890		17.5	1.2031	1-13/64"		1.7188	1-23/32"	
.3480	S		.7031	45/64"		1.2188	1-7/32"		1.7323		44.
.3504		8.9	.7087		18.	1.2205		31.	1.7344	1-47/64"	
.3543		9.	.7188	23/32"		1.2344	1-15/64"		1.7500	1-3/4"	
.3580	T		.7283		18.5	1.2402		31.5	1.7520		44.5
.3583		9.1	.7344	47/64"		1.2500	1-1/4"		1.7656	1-49/64"	
.3594	23/64"		.7480		19.	1.2598		32.	1.7717		45.
.3622		9.2	.7500	3/4"		1.2656	1-17/64"		1.7812	1-25/32"	
.3642		9.25	.7656	49/64"		1.2795		32.5	1.7913		45.5
.3661		9.3	.7677		19.5	1.2812	1-9/32"		1.7969	1-51/64"	
.3680	U		.7812	25/32"		1.2969	1-19/64"		1.8110		46.
.3701		9.4	.7874		20.	1.2992		33.	1.8125	1-13/16"	
.3740		9.5	.7969	51/64"		1.3125	1-5/16"		1.8281	1-53/64"	
.3750	3/8"		.8071		20.5	1.3189		33.5	1.8307		46.5
.3770	V		.8125	13/16"		1.3281	1-21/64"		1.8438	1-27/32"	
.3780		9.6	.8268		21.	1.3386		34.	1.8504		47.
.3819		9.7	.8281	53/64"		1.3438	1-11/32"		1.8594	1-55/64"	
.3839		9.75	.8438	27/32"		1.3583		34.5	1.8701		47.5
.3858		9.8	.8465		21.5	1.3594	1-23/64"		1.8750	1-7/8"	
.3860	W		.8594	55/64"		1.3750	1-3/8"		1.8898		48.
.3898		9.9	.8661		22.	1.3780		35.	1.8906	1-57/64"	
.3906	25/64"		.8750	7/8"		1.3906	1-25/64"		1.9062	1-29/32"	
.3937		10	.8858		22.5	1.3976		35.5	1.9094		48.5
.3970	X		.8906	57/64"		1.4062	1-13/32"		1.9219	1-59/64"	
.4040	Y		.9055		23.	1.4173		36.	1.9291		49.
.4062	13/32"		.9062	29/32"		1.4219	1-27/64"		1.9375	1-15/16"	
.4130	Z		.9219	59/64"		1.4370		36.5	1.9488		49.5
.4134		10.5	.9252		23.5	1.4375	1-7/16"		1.9531	1-61/64"	
.4219	27/64"		.9375	15/16"		1.4531	1-29/64"		1.9685		50.
.4331		11.	.9449		24.	1.4567		37.	1.9688	1-31/32"	
.4375	7/16"		.9531	61/64"		1.4688	1-15/32"		1.9844	1-63/64"	
.4528		11.5	.9646		24.5	1.4764		37.5	1.9882		50.5
.4531	29/64"		.9688	31/32"		1.4844	1-31/64"		2.0000	2"	

Table 15. Tap Drill Sizes, English (U.S. Customary)

TAP SIZE	PITCH	TAP DRILL	ALT. TAP DRILL mm	TAP SIZE	PITCH	TAP DRILL	ALT. TAP DRILL mm	TAP SIZE	PITCH	TAP DRILL	ALT. TAP DRILL mm
0	00 UNF	3/64	1.25	5/16	18 UNC	F	6.50	1 1/4	7 UNC	1 7/64	28.00
1	64 UNC	53	1.55		24 UNF	1	6.90		12 UNF	1 11/64	29.50
	72 UNF	53	1.55	3/8	16 UNC	5/16	8.00	1 3/8	6 UNC	1 7/32	30.75
2	56 UNC	50	1.85		24 UNF	Q	8.50		12 UNF	1 19/64	32.75
	64 UNF	50	1.90	7/16	14 UNC	U	9.40	1 1/2	6 UNC	1 11/32	34.00
3	48 UNC	47	2.10		20 UNF	25/64	9.90		12 UNF	1 27/64	36.00
	56 UNF	45	2.15	1/2	13 UNC	27/64	10.80	1 3/4	5 UNC	1 9/16	39.50
4	40 UNC	43	2.35		20 UNF	29/64	11.50	2	4 UNC	1 25/32	45.00
	48 UNF	42	2.40	9/16	12 UNC	31/64	12.20	TAPER PIPE TAPS NPT			
5	40 UNC	38	2.65		18 UNF	33/64	12.90	1/16	27 NPT	D	6.30
	44 UNF	37	2.70	5/8	11 UNC	17/32	13.50	1/8	27 NPT	R	8.70
6	32 UNC	36	2.85		18 UNF	37/64	14.50	1/4	18 NPT	7/16	11.10
	40 UNF	33	2.95	3/4	10 UNC	21/32	16.50	3/8	18 NPT	37/64	14.50
8	32 UNC	29	3.50		16 UNF	11/16	17.50	1/2	14 NPT	23/32	18.00
	36 UNF	29	3.50	7/8	9 UNC	49/64	19.50	3/4	14 NPT	59/64	23.25
10	24 UNC	25	3.90		14 UNF	13/16	20.40	1	11 1/2 NPT	1 5/32	29.00
	32 UNF	21	4.10	1	8 UNC	7/8	22.25	1 1/4	11 1/2 NPT	1 1/2	38.00
12	24 UNC	16	4.50		12 UNF	59/64	23.25	1 1/2	11 1/2 NPT	1 47/64	44.00
	28 UNF	14	4.70		14 UNS	15/16	23.50	2	11 1/2 NPT	2 7/32	56.00
1/4	20 UNC	7	5.10	1 1/8	7 UNC	63/64	25.00	2 1/2	8 NPT	2 5/8	67.00
	28 UNF	3	5.50		12 UNF	1 3/64	26.50	3	8 NPT	3 1/4	82.50

Drill sizes based on approximately 72%–77% of full thread.

Table 16. Tap Drill Sizes, Metric

TAP SIZE mm	PITCH mm	TAP DRILL mm	ALT. TAP DRILL	TAP SIZE mm	PITCH mm	TAP DRILL mm	ALT. TAP DRILL	TAP SIZE mm	PITCH mm	TAP DRILL mm	ALT. TAP DRILL
ISO METRIC COARSE				24	3.00	21.00	53/64	11	1.00	10.00	X
1.6	0.35	1.25	3/64	27	3.00	24.00	61/64	12	1.00	11.00	7/16
1.7	0.35	1.35	55	30	3.50	26.50	1 3/64	12	1.25	10.75	27/64
1.8	0.35	1.45	54	33	3.50	29.50	15/32	12	1.50	10.50	Z
2	0.40	1.60	1/16	36	4.00	32.00	1 1/4	13	1.50	11.50	29/64
2.2	0.45	1.75	50	36	4.00	35.00	1 3/8	13	1.75	11.25	7/16
2.3	0.40	1.90	49	ISO METRIC FINE				14	1.25	12.75	1/2
2.5	0.45	2.05	46	3	0.35	2.65	37	14	1.50	12.50	31/64
2.6	0.45	2.15	44	4	0.35	3.65	27	15	1.50	13.50	17/32
3	0.50	2.50	40	4	0.50	3.50	29	16	1.00	15.00	19/32
3.5	0.60	2.90	33	4.5	0.45	4.05	21	16	1.25	14.75	37/64
4	0.70	3.30	30	5	0.50	4.50	16	16	1.50	14.50	9/16
4.5	0.75	3.70	27	5	0.70	4.30	18	18	1.00	17.00	43/64
5	0.80	4.20	19	5	0.75	4.25	18	18	1.25	16.75	21/32
5.5	0.90	4.60	15	5.5	0.50	5.00	9	18	1.50	16.50	41/64
6	1.00	5.00	9	6	0.50	5.50	7/32	18	2.00	16.00	5/8
7	1.00	6.00	15/64	6	0.75	5.25	5	20	1.00	19.00	3/4
8	1.25	6.80	H	7	0.75	6.25	D	20	1.50	18.50	47/64
9	1.25	7.80	5/16	8	0.50	7.50	M	20	2.00	18.00	45/64
10	1.50	8.50	O	8	1.00	7.00	J	22	1.00	21.00	53/64
11	1.50	9.50	3/8	9	0.50	8.50	Q	22	1.50	20.50	13/16
12	1.75	10.20	Y	9	1.00	8.00	O	22	2.00	20.00	25/32
14	2.00	12.00	15/32	10	0.50	9.50	3/8	24	1.00	23.00	29/32
16	2.00	14.00	35/64	10	0.75	9.25	U	24	1.50	22.50	7/8
18	2.50	15.50	39/64	10	1.00	9.00	T	24	2.00	22.00	55/64
20	2.50	17.50	11/16	10	1.25	8.75	11/32	24	2.50	21.50	27/32
22	2.50	19.50	49/64								

Table 17. Unified Screw Threads, Basic Data *(Courtesy of SKF & Dormer Tools)*

UNIFIED COARSE (U.N.C.)

THREAD FORM

r = Basic Radius = ·1443 p

hn = Basic Height of Internal Thread & Depth of Thread Engagement = ·54127 p

hs = Basic Height of External Thread = ·6495 p

p = Pitch = $\frac{1}{t.p.i.}$

nom. size	t.p.i.	basic major diameter inches	basic effective diameter inches	basic minor dia. of ext. threads inches	basic minor dia. of int. threads inches	recommended tapping drill size mm	clearance drill size mm
No.1	64	0·0730	0·0629	0·0538	0·0561	1.55	1.95
No.2	56	0·0860	0·0744	0·0641	0·0667	1.85	2.30
No.3	48	0·0990	0·0855	0·0734	0·0764	2.10	2.65
No.4	40	0·1120	0·0958	0·0813	0·0849	2.35	2.95
No.5	40	0·1250	0·1088	0·0943	0·0979	2.65	3.30
No.6	32	0·1380	0·1177	0·0997	0·1042	2.85	3.60
No.8	32	0·1640	0·1437	0·1257	0·1302	3.50	4.30
No.10	24	0·1900	0·1629	0·1389	0·1449	3.90	4.90
No.12	24	0·2160	0·1889	0·1649	0·1709	4.50	5.60
1/4	20	0·2500	0·2175	0·1887	0·1959	5.10	6.50
5/16	18	0·3125	0·2764	0·2443	0·2524	6.60	8.10
3/8	16	0·3750	0·3344	0·2983	0·3073	8.00	9.70
7/16	14	0·4375	0·3911	0·3499	0·3602	9.40	11.30
1/2	13	0·5000	0·4500	0·4056	0·4167	10.80	13.00
9/16	12	0·5625	0·5084	0·4603	0·4723	12.20	14.50
5/8	11	0·6250	0·5660	0·5135	0·5266	13.50	16.25
3/4	10	0·7500	0·6850	0·6273	0·6417	16.50	19.25
7/8	9	0·8750	0·8028	0·7387	0·7547	19.50	22.50
1	8	1·0000	0·9188	0·8466	0·8647	22.25	25.75
1,1/8	7	1·1250	1·0322	0·9497	0·9704	25.00	29.00
1,1/4	7	1·2500	1·1572	1·0747	1·0954	28.00	32.00
1,3/8	6	1·3750	1·2667	1·1705	1·1946	30.75	35.50
1,1/2	6	1·5000	1·3917	1·2955	1·3196	34.00	38.50
1,3/4	5	1·7500	1·6201	1·5046	1·5335	39.50	45.00
2	4,1/2	2·0000	1·8557	1·7274	1·7594	45.00	51.00
2,1/4	**4,1/2**	**2·2500**	**2·1057**	**1·9774**	**2·0094**	**52·00**	**58·00**

UNIFIED FINE (U.N.F.)

THREAD FORM

r = Basic Radius = ·1443 p

hn = Basic Height of Internal Thread & Depth of Thread Engagement = ·54127 p

hs = Basic Height of External Thread = ·6495 p

p = Pitch = $\frac{1}{t.p.i.}$

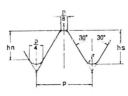

nom. size	t.p.i.	basic major diameter inches	basic effective diameter inches	basic minor dia. of external threads inches	basic minor dia. of internal threads inches	recommended tapping drill size mm	clearance drill size mm
No.0	80	0·0600	0·0519	0·0447	0·0465	1.25	1.60
No.1	72	0·0730	0·0640	0·0560	0·0580	1.55	1.95
No.2	64	0·0860	0·0759	0·0668	0·0691	1.90	2.30
No.3	56	0·0990	0·0874	0·0771	0·0797	2.15	2.65
No.4	48	0·1120	0·0985	0·0864	0·0894	2.40	2.95
No.5	44	0·1250	0·1102	0·0971	0·1004	2.70	3.30
No.6	40	0·1380	0·1218	0·1073	0·1109	2.95	3.60
No.8	36	0·1640	0·1460	0·1299	0·1339	3.50	4.30
No.10	32	0·1900	0·1697	0·1517	0·1562	4.10	4.90
No.12	28	0·2160	0·1928	0·1722	0·1773	4.70	5.60
1/4	28	0·2500	0·2268	0·2062	0·2113	5.50	6.50
5/16	24	0·3125	0·2854	0·2614	0·2674	6.90	8.10
3/8	24	0·3750	0·3479	0·3239	0·3299	8.50	9.70
7/16	20	0·4375	0·4050	0·3762	0·3834	9.90	11.30
1/2	20	0·5000	0·4675	0·4387	0·4459	11.50	13.00
9/16	18	0·5625	0·5264	0·4943	0·5024	12.90	14.50
5/8	18	0·6250	0·5889	0·5568	0·5649	14.50	16.25
3/4	16	0·7500	0·7094	0·6733	0·6823	17.50	19.25
7/8	14	0·8750	0·8286	0·7874	0·7977	20.40	22.50
1	12	1·0000	0·9459	0·8978	0·9098	23.25	25.75
1,1/8	12	1·1250	1·0709	1·0228	1·0348	26.50	29.00
1,1/4	12	1·2500	1·1959	1·1478	1·1598	29.50	32.00
1,3/8	12	1·3750	1·3209	1·2728	1·2848	32.75	35.50
1,1/2	12	1·5000	1·4459	1·3978	1·4098	36.00	38.50

Table 18. American National Pipe Threads, Basic Data *(Courtesy of SKF & Dormer Tools and The L.S. Starrett Company)*

AMERICAN NATIONAL TAPER PIPE (N.P.T.)

THREAD FORM

h = Basic Depth of Thread = ·8 p

$p = Pitch = \dfrac{1}{t.p.i.}$

Taper 1 in 16 on dia (shown exaggerated in diagram)

nom. size inches	t.p.i.	basic depth of thread inches	outside diameter of pipe inches	effective diameter at small end of external thread inches	effective diameter at large end of internal thread inches	recommended tapping drill size mm
1/16	27	0·0296	0·3125	0·2712	0·2812	6·30
1/8	27	0·0296	0·405	0·3635	0·3736	8·70
1/4	18	0·0444	0·540	0·4774	0·4916	11·10
3/8	18	0·0444	0·675	0·6120	0·6270	14·50
1/2	14	0·0571	0·840	0·7584	0·7784	18·00
3/4	14	0·0571	1·050	0·9677	0·9889	23·25
1	11, 1/2	0·0696	1·315	1·2136	1·2386	29·00
1, 1/4	11, 1/2	0·0696	1·660	1·5571	1·5834	38·00
1, 1/2	11, 1/2	0·0696	1·900	1·7961	1·8223	44·00
2	11, 1/2	0·0696	2·375	2·2690	2·2963	56·00
2, 1/2	8	0·1000	2·875	2·7195	2·7622	67·00

American Standard Pipe Thread and Tap Drill Sizes

Pipe Size Inches	Threads Per Inch	Root Diameter Small End of Pipe and Gage	Tap Drill	
			Taper NPT	Straight NPS
1/8	27	.3339"	Q	11/32"
1/4	18	.4329"	7/16"	7/16"
3/8	18	.5676"	9/16"	37/64"
1/2	14	.7013"	45/64"	23/32"
3/4	14	.9105"	29/32"	59/64"
1	11-1/2	1.1441"	1-9/64"	1-5/32"
1-1/4	11-1/2	1.4876"	1-31/64"	1-1/2"
1-1/2	11-1/2	1.7265"	1-47/64"	1-3/4"
2	11-1/2	2.1995"	2-13/64"	2-7/32"

Table 19. Acme Screw Threads, Basic Data *(Courtesy of The L.S. Starrett Company)*

American Standard Acme Screw Thread Dimensions

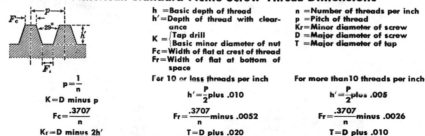

h = Basic depth of thread
h' = Depth of thread with clearance
K = $\begin{cases} \text{Tap drill} \\ \text{Basic minor diameter of nut} \end{cases}$
Fc = Width of flat at crest of thread
Fr = Width of flat at bottom of space

n = Number of threads per inch
p = Pitch of thread
Kr = Minor diameter of screw
D = Major diameter of screw
T = Major diameter of tap

$p = \dfrac{1}{n}$

K = D minus p

$Fc = \dfrac{.3707}{n}$

Kr = D minus 2h'

For 10 or less threads per inch

$h' = \dfrac{p}{2}$ plus .010

$Fr = \dfrac{.3707}{n}$ minus .0052

T = D plus .020

For more than 10 threads per inch

$h' = \dfrac{p}{2}$ plus .005

$Fr = \dfrac{.3707}{n}$ minus .0026

T = D plus .010

Threads per inch (n)	Depth of Thread with Clearance (h')	Flat at Top of Thread (Fc)	Flat at Bottom of Space (Fr)	Space at Top of Thread	Thickness at Root of Thread
1	.5100	.3707	.3655	.6293	.6345
1⅓	.3850	.2780	.2728	.4720	.4772
2	.2600	.1854	.1802	.3146	.3198
3	.1767	.1236	.1184	.2097	.2149
4	.1350	.0927	.0875	.1573	.1625
5	.1100	.0741	.0689	.1259	.1311
6	0933	.0618	.0566	.1049	.1101
7	.0814	.0530	.0478	.0899	.0951
8	.0725	.0463	.0411	.0787	.0839
9	.0655	.0412	.0360	.0699	.0751
10	.0600	.0371	.0319	.0629	.0681
12	.0467	.0309	.0283	.0524	.0550
14	.0407	.0265	.0239	.0449	.0475
16	.0363	.0232	.0206	.0393	.0419

Table 20. ISO Metric Threads, Basic Data *(Courtesy of SKF & Dormer Tools)*

I.S.O. METRIC COARSE

THREAD FORM

r = Basic Radius = ·1443 p

hn = Basic Height of Internal
Thread & Depth of
Thread Engagement
= ·54127 p

hs = Basic Height of External
Thread = .61344 P

p = Pitch

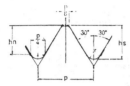

nom. dia.	pitch	basic major diameter	basic effective diameter	basic minor diameter of external threads	basic minor diameter of internal threads	recom- mended tapping drill size	clearance drill size
mm	mm	mm	mm	mm	mm	mm	mm
1	0.25	1.000	0.838	0.693	0.729	0.75	1.05
1.1	0.25	1.100	0.938	0.793	0.829	0.85	1.15
1.2	0.25	1.200	1.038	0.893	0.929	0.95	1.25
1.4	0.30	1.400	1.205	1.032	1.075	1.10	1.45
1.6	0.35	1.600	1.373	1.170	1.221	1.25	1.65
1.8	0.35	1.800	1.573	1.370	1.421	1.45	1.85
2	0.40	2.000	1.740	1.509	1.567	1.60	2.05
2.2	0.45	2.200	1.908	1.648	1.713	1.75	2.25
2.5	0.45	2.500	2.208	1.948	2.013	2.05	2.60
3	0.50	3.000	2.675	2.387	2.459	2.50	3.10
3.5	0.60	3.500	3.110	2.764	2.850	2.90	3.60
4	0.70	4.000	3.545	3.141	3.242	3.30	4.10
4.5	0.75	4.500	4.013	3.580	3.688	3.70	4.60
5	0.80	5.000	4.480	4.019	4.134	4.20	5.10
6	1.00	6.000	5.350	4.773	4.917	5.00	6.10
7	1.00	7.000	6.350	5.773	5.917	6.00	7.20
8	1.25	8.000	7.188	6.466	6.647	6.80	8.20
9	1.25	9.000	8.188	7.466	7.647	7.80	9.20
10	1.50	10.000	9.026	8.160	8.376	8.50	10.20
11	1.50	11.000	10.026	9.160	9.376	9.50	11.20
12	1.75	12.000	10.863	9.853	10.106	10.20	12.20
14	2.00	14.000	12.701	11.546	11.835	12.00	14.25
16	2.00	16.000	14.701	13.546	13.835	14.00	16.25
18	2.50	18.000	16.376	14.933	15.294	15.50	18.25
20	2.50	20.000	18.376	16.933	17.294	17.50	20.25
22	2.50	22.000	20.376	18.933	19.294	19.50	22.25
24	3.00	24.000	22.051	20.319	20.752	21.00	24.25
27	3.00	27.000	25.051	23.319	23.752	24.00	27.25
30	3.50	30.000	27.727	25.706	26.211	26.50	30.50
33	3.50	33.000	30.727	28.706	29.211	29.50	33.50
36	4.00	36.000	33.402	31.093	31.670	32.00	36.50
39	4.00	39.000	36.402	34.093	34.670	35.00	39.50
42	4.50	42.000	39.077	36.479	37.129	37.50	42.50
45	4.50	45.000	42.077	39.479	40.129	40.50	45.50
48	5.00	48.000	44.752	41.866	42.587	43.00	48.50
52	5.00	52.000	48.752	45.866	46.587	47.00	53.00
56	5.50	56.000	52.428	49.252	50.046	50.50	57.00

Table 21. Wire Gage Standards Used in the United States *(Courtesy of The L.S. Starrett Company)*

Number of Wire Gage	American or Brown & Sharpe	Birmingham or Stubs' Iron Wire	Washburn & Moen, Worcester, Mass.	W. & M. Steel Music Wire	American S. & W. Co's. Music Wire Gage	Stubs' Steel Wire	U.S. Standard Gage for Sheet and Plate Iron and Steel	Number of Wire Gage
00000000	....	....	....	.0083	....	....	....	00000000
0000000	....	....	....	.0087	....	....	....	0000000
000000	....	....	....	.0095	.004	....	.46875	000000
00000	....	....	....	.010	.005	....	.4375	00000
0000	.460	.454	.3938	.011	.006	....	.40625	0000
000	.40964	.425	.3625	.012	.007	....	.375	000
00	.3648	.380	.3310	.0133	.008	....	.34375	00
0	.32486	.340	.3065	.0144	.009	....	.3125	0
1	.2893	.300	.2830	.0156	.010	.227	.28125	1
2	.25763	.284	.2625	.0166	.011	.219	.265625	2
3	.22942	.259	.2437	.0178	.012	.212	.250	3
4	.20431	.238	.2253	.0188	.013	.207	.234375	4
5	.18194	.220	.2070	.0202	.014	.204	.21875	5
6	.16202	.203	.1920	.0215	.016	.201	.203125	6
7	.14428	.180	.1770	.023	.018	.199	.1875	7
8	.12849	.165	.1620	.0243	.020	.197	.171875	8
9	.11443	.148	.1483	.0256	.022	.194	.15625	9
10	.10189	.134	.1350	.027	.024	.191	.140625	10
11	.090742	.120	.1205	.0284	.026	.188	.125	11
12	.080808	.109	.1055	.0296	.029	.185	.109375	12
13	.071961	.095	.0915	.0314	.031	.182	.09375	13
14	.064084	.083	.0800	.0326	.033	.180	.078125	14
15	.057068	.072	.0720	.0345	.035	.178	.0703125	15
16	.05082	.065	.0625	.036	.037	.175	.0625	16
17	.045257	.058	.0540	.0377	.039	.172	.05625	17
18	.040303	.049	.0475	.0395	.041	.168	.050	18
19	.03589	.042	.0410	.0414	.043	.164	.04375	19
20	.031961	.035	.0348	.0434	.045	.161	.0375	20
21	.028462	.032	.03175	.046	.047	.157	.034375	21
22	.025347	.028	.0286	.0483	.049	.155	.03125	22
23	.022571	.025	.0258	.051	.051	.153	.028125	23
24	.0201	.022	.0230	.055	.055	.151	.025	24
25	.0179	.020	.0204	.0586	.059	.148	.021875	25
26	.01594	.018	.0181	.0626	.063	.146	.01875	26
27	.014195	.016	.0173	.0658	.067	.143	.0171875	27
28	.012641	.014	.0162	.072	.071	.139	.015625	28
29	.011257	.013	.0150	.076	.075	.134	.0140625	29
30	.010025	.012	.0140	.080	.080	.127	.0125	30
31	.008928	.010	.0132	....	.085	.120	.0109375	31
32	.00795	.009	.0128	....	.090	.115	.01015625	32
33	.00708	.008	.0118	....	.095	.112	.009375	33
34	.006304	.007	.0104	....	....	.110	.00859375	34
35	.005614	.005	.0095	....	....	.108	.0078125	35
36	.005	.004	.0090	....	....	.106	.00703125	36
37	.004453	....	....	....	....	.103	.006640625	37
38	.003965	....	....	....	....	.101	.00625	38
39	.003531	....	....	....	....	.099	....	39
40	.003144	....	....	....	....	.097	....	40

Table 22. Weights of Steel Bars *(Courtesy of A.J. Forsyth & Co. Ltd.)*

WEIGHTS OF STEEL BARS

Theoretical weights, based on .28334 pounds to the cubic inch or 489.6 pounds to the cubic foot.

ROUNDS　　Sectional Weight Factor = 1.000

SQUARES　　Sectional Weight Factor = 1.273
　　　　　　Distance across corners = 1.414 × diameter

HEXAGONS　Sectional Weight Factor = 1.103
　　　　　　Distance across corners = 1.155 × diameter

OCTAGONS　Sectional Weight Factor = 1.056
　　　　　　Distance across corners = 1.078 × diameter

Example: Require weight of 5 inch Octagon. Sectional weight factor of Octagon = 1.056 × weight of 5 inch round or 66.7 # equals 70.44 # theoretical weight.

WEIGHT PER FOOT IN POUNDS								
Size in Inches	Round	Square	Hexagon	Octagon	Size in Inches	Round	Square	Hexagon	Octagon
$1/16$	.010	.013	.012	.01	2	10.68	13.60	11.78	11.28
$1/8$	.042	.053	.046	.04	$2^1/16$	11.36	14.46	12.53	12.10
$3/16$	.094	.120	.104	.10	$2^1/8$	12.06	15.35	13.30	12.71
$1/4$	.167	.213	.184	.18	$2^3/16$	12.78	16.27	14.09	13.62
$5/16$	.261	.332	.288	.28	$2^1/4$	13.52	17.22	14.91	14.24
$3/8$	.376	.478	.414	.40	$2^5/16$	14.28	18.19	15.75	15.72
$7/16$	.511	.651	.564	.54	$2^3/8$	15.07	19.18	16.61	15.88
$1/2$	.668	.850	.736	.70	$2^7/16$	15.86	20.20	17.49	16.75
$9/16$	.845	1.076	.932	.89	$2^1/2$	16.69	21.25	18.40	17.65
$5/8$	1.043	1.328	1.150	1.10	$2^9/16$	17.53	22.33	19.34	18.20
$11/16$	1.262	1.607	1.392	1.33	$2^5/8$	18.40	23.43	20.29	19.45
$3/4$	1.502	1.913	1.656	1.58	$2^11/16$	19.29	24.56	21.27	20.62
$13/16$	1.763	2.245	1.944	1.83	$2^3/4$	20.20	25.00	22.27	21.78
$7/8$	2.045	2.603	2.254	2.16	$2^13/16$	21.12	26.90	23.29	22.53
$15/16$	2.347	2.988	2.588	2.48	$2^7/8$	22.07	28.10	24.34	23.28
					$2^15/16$	23.04	29.34	25.41	24.32
1	2.67	3.40	2.94	2.82	3	24.03	30.60	26.50	25.36
$1^1/16$	3.01	3.83	3.32	3.18	$3^1/16$	25.04	31.89	27.62	26.43
$1^1/8$	3.37	4.30	3.73	3.56	$3^1/8$	26.08	33.20	28.76	27.50
$1^3/16$	3.76	4.79	4.15	3.97	$3^3/16$	27.13	34.55	29.92	28.39
$1^1/4$	4.17	5.31	4.60	4.40	$3^1/4$	28.20	35.92	31.10	29.28
$1^5/16$	4.60	5.85	5.07	4.85	$3^5/16$	29.30	37.31	32.31	30.69
$1^3/8$	5.04	6.42	5.57	5.32	$3^3/8$	30.42	38.73	33.54	32.10
$1^7/16$	5.51	7.02	6.09	5.82	$3^7/16$	31.56	40.18	34.79	33.33
$1^1/2$	6.01	7.65	6.63	6.34	$3^1/2$	32.71	41.65	36.07	34.56
$1^9/16$	6.52	8.30	7.19	6.88	$3^9/16$	33.90	43.14	37.37	35.86
$1^5/8$	7.05	8.97	7.78	7.32	$3^5/8$	35.09	44.68	38.69	37.05
$1^11/16$	7.60	9.68	8.39	8.02	$3^11/16$	36.31	46.24	40.04	38.37
$1^3/4$	8.17	10.41	9.02	8.64	$3^3/4$	37.56	47.82	41.41	39.68
$1^13/16$	8.77	11.17	9.67	9.25	$3^13/16$	38.81	49.42	42.84	41.04
$1^7/8$	9.38	11.95	10.35	9.92	$3^7/8$	40.10	51.05	44.27	42.40
$1^15/16$	10.02	12.76	11.05	10.57	$3^15/16$	41.40	52.71	45.70	43.76
					4	42.73	54.40	47.13	45.12

Theoretical weights, based on .28334 pounds to the cubic inch or 489.6 pounds to the cubic foot.

Table 23. Weights of Steel Bars *(Courtesy of A.J. Forsyth & Co. Ltd.)*

WEIGHTS OF STEEL BARS
Rounds and Squares
WEIGHT PER FOOT IN POUNDS

Size in Inches	WEIGHT		Size in Inches	WEIGHT		Size in Inches	WEIGHT	
	Round	Square		Round	Square		Round	Square
4 1/16	44.0	56.1	6 3/4	121.7	154.9	9 7/16	237.9	302.8
4 1/8	45.4	57.8	6 13/16	123.9	157.8	9 1/2	241.0	306.8
4 3/16	46.8	59.6	6 7/8	126.2	160.8	9 9/16	244.2	310.9
4 1/4	48.2	61.4	6 15/16	128.5	163.6	9 5/8	247.4	315.0
4 5/16	49.6	63.2	7	130.9	166.6	9 11/16	250.6	319.1
4 3/8	51.1	65.0	7 1/16	133.2	169.6	9 3/4	253.9	323.2
4 7/16	52.5	66.9	7 1/8	135.6	172.6	9 13/16	257.1	327.4
4 1/2	54.0	68.8	7 3/16	137.9	175.6	9 7/8	260.4	331.6
4 9/16	55.5	70.7	7 1/4	140.4	178.7	9 15/16	263.7	335.8
4 5/8	57.1	72.7	7 5/16	142.8	181.8	10	267.0	340.0
4 11/16	58.6	74.7	7 3/8	145.3	184.9	10 1/16	270.4	344.3
4 3/4	60.2	76.7	7 7/16	147.7	188.1	10 1/8	273.8	348.5
4 13/16	61.8	78.7	7 1/2	150.2	191.3	10 3/16	277.1	352.9
4 7/8	63.4	80.8	7 9/16	152.7	194.4	10 1/4	280.6	357.2
4 15/16	65.1	82.8	7 5/8	155.2	197.7	10 5/16	284.0	361.6
5	66.7	85.0	7 11/16	157.8	200.9	10 3/8	287.4	366.0
5 1/16	68.4	87.1	7 3/4	160.3	204.2	10 7/16	290.9	370.4
5 1/8	70.1	89.3	7 13/16	163.0	207.6	10 1/2	294.4	374.9
5 3/16	71.8	91.4	7 7/8	165.6	210.8	10 9/16	297.9	379.4
5 1/4	73.6	93.7	7 15/16	168.2	214.2	10 5/8	301.4	383.8
5 5/16	75.3	95.9	8	171.0	217.6	10 11/16	305.0	388.3
5 3/8	77.1	98.2	8 1/16	173.6	221.0	10 3/4	308.6	392.9
5 7/16	78.9	100.5	8 1/8	176.3	224.5	10 13/16	312.2	397.5
5 1/2	80.7	102.8	8 3/16	179.0	228.0	10 7/8	315.8	402.1
5 9/16	82.6	105.2	8 1/4	181.8	231.4	10 15/16	319.5	406.8
5 5/8	84.4	107.6	8 5/16	184.5	234.9	11	323.1	411.4
5 11/16	86.3	110.0	8 3/8	187.3	238.5	11 1/16	326.8	416.1
5 3/4	88.2	112.4	8 7/16	190.1	242.0	11 1/8	330.5	420.9
5 13/16	90.2	114.9	8 1/2	193.0	245.6	11 3/16	334.3	425.5
5 7/8	92.1	117.4	8 9/16	195.7	249.3	11 1/4	337.9	430.3
5 15/16	94.1	119.9	8 5/8	198.7	252.9	11 5/16	341.7	435.1
6	96.1	122.4	8 11/16	201.0	256.6	11 3/8	345.5	439.3
6 1/16	98.1	125.0	8 3/4	204.4	260.3	11 7/16	349.4	444.8
6 1/8	100.2	127.6	8 13/16	207.4	264.1	11 1/2	353.1	449.6
6 3/16	102.2	130.2	8 7/8	210.3	267.9	11 9/16	357.0	454.5
6 1/4	104.3	132.8	8 15/16	213.3	271.6	11 5/8	360.9	459.5
6 5/16	106.4	135.5	9	216.3	275.4	11 11/16	364.8	464.4
6 3/8	108.5	138.2	9 1/16	219.3	279.3	11 3/4	368.6	469.4
6 7/16	110.7	140.9	9 1/8	222.4	283.2	11 13/16	372.6	474.4
6 1/2	112.8	143.6	9 3/16	225.4	287.0	11 7/8	376.6	479.5
6 9/16	114.9	146.5	9 1/4	228.5	290.9	11 15/16	380.6	484.5
6 5/8	117.2	149.2	9 5/16	231.5	294.9	12	384.5	489.6
6 11/16	119.4	152.1	9 3/8	234.7	298.9			

Theoretical weights, based on .28334 pounds to the cubic inch or 489.6 pounds to the cubic foot.

Table 24. Weights of Flat Bar Steel Per Linear Foot *(Courtesy of A.J. Forsyth & Co. Ltd.)*

	½	⅝	¾	⅞	1	1⅛	1¼	1⅜	1½	1¾	2	2¼	2½	2¾	3	3½	4	5	6
⅛	0.213	0.266	0.320	0.372	0.426	0.479	0.530	0.585	0.640	0.745	0.850	0.955	1.07	1.18	1.28	1.49	1.70	2.13	2.56
3/16	0.319	0.399	0.480	0.558	0.639	0.718	0.790	0.878	0.960	1.12	1.28	1.43	1.60	1.76	1.92	2.24	2.55	3.20	3.83
¼	0.425	0.533	0.640	0.743	0.852	0.958	1.06	1.17	1.28	1.49	1.70	1.91	2.13	2.34	2.56	2.98	3.40	4.26	5.11
5/16	0.531	0.665	0.800	0.929	1.06	1.20	1.33	1.46	1.60	1.86	2.13	2.39	2.66	2.92	3.19	3.72	4.25	5.32	6.38
⅜	0.638	0.798	0.960	1.12	1.28	1.43	1.59	1.75	1.91	2.23	2.55	2.87	3.20	3.51	3.83	4.46	5.10	6.40	7.66
7/16	0.744	0.931	1.12	1.30	1.49	1.67	1.86	2.05	2.23	2.60	2.98	3.35	3.72	4.09	4.46	5.21	5.95	7.44	8.92
½	....	1.07	1.28	1.49	1.70	1.91	2.13	2.34	2.55	2.98	3.40	3.83	4.26	4.68	5.10	5.96	6.80	8.52	10.20
9/16	....	1.20	1.44	1.67	1.91	2.15	2.39	2.63	2.87	3.35	3.83	4.30	4.78	5.26	5.74	6.69	7.65	9.56	11.50
⅝	....	....	1.60	1.86	2.12	2.39	2.66	2.92	3.19	3.72	4.26	4.79	5.32	5.86	6.39	7.44	8.52	10.64	12.78
11/16	....	....	1.76	2.04	2.34	2.63	2.92	3.22	3.51	4.09	4.68	5.26	5.84	6.43	7.01	8.18	9.35	11.70	14.00
¾	....	....	....	2.23	2.55	2.86	3.19	3.50	3.83	4.46	5.10	5.74	6.40	7.02	7.65	8.92	10.20	12.80	15.30
13/16	....	....	....	2.41	2.76	3.11	3.45	3.80	4.14	4.83	5.53	6.22	6.91	7.60	8.29	9.67	11.10	13.80	16.60
⅞	....	....	....	....	2.98	3.34	3.72	4.09	4.46	5.21	5.96	6.70	7.46	8.19	8.94	10.42	11.92	14.92	17.88
15/16	....	....	....	....	3.19	3.59	3.98	4.38	4.78	5.58	6.38	7.17	7.97	8.77	9.56	11.20	12.80	15.90	19.10
1	....	....	....	....	....	3.82	4.25	4.68	5.10	5.96	6.80	7.66	8.52	9.36	10.20	11.92	13.60	17.04	20.40
1⅛	....	....	....	....	....	....	4.78	5.27	5.74	6.71	7.65	8.61	9.59	10.54	11.48	13.41	15.30	19.17	22.95
1¼	....	....	....	....	....	....	....	5.85	6.38	7.45	8.50	9.57	10.65	11.71	12.76	14.90	17.00	21.30	25.61
1½	....	....	....	....	....	....	....	7.02	7.67	8.94	10.20	11.49	12.78	14.04	15.30	17.88	20.40	25.56	30.00

Table 25. Hardness Conversions *(Courtesy of SKF & Dormer Tools)*

Diamond Pyramid Hardness No.	Brinell Hardness No.	Rockwell C. Scale Hardness No.	Tensile Strength		Diamond Pyramid Hardness No.	Brinell Hardness No.	Rockwell C. Scale Hardness No.	Tensile Strength	
			Tons/sq.in.	Kilos/sq.mm				Tons/sq.in.	Kilos/sq.mm
940		68·0							
920		67·5			412	390	42	86	135
900		67·0			402	381	41	85	134
883		66·5			392	371	40	83	132
865		66·0			382	362	39	81	129
848		65·5			372	353	38	80	126
832		65·0	150	236	363	344	37	78	123
817		64·5	147	231	354	336	36	76	120
800		64·0	145	228	345	327	35	74	117
787		63·5	142	223	336	319	34	72	113
772		63·0	140	220	327	311	33	70	110
759		62·5	139	218	318	301	32	68	107
746		62·0	137	215	310	294	31	67	105
733		61·5	135	212	302	286	30	65	103
720		61	133	209	294	279	29	64	101
697		60	129	203	286	273	28	62	98
674		59	126	198	279	267	27	61	96
653		58	123	193	272	261	26	59	93
633		57	120	189	266	258	25	58	91
613		56	117	184	260	253	24	57	90
595		55	114	179	254	248	23	55	88
577		54	112	176	248	243	22	54	85
560	510	53	109	171	243	239	21	53	84
544	500	52	107	168	238	235	20	52	82
528	487	51	104	163	228	226		50	79
513	475	50	102	160	217	216		47	74
498	464	49	100	157	207	206		45	71
484	450	48	98	154	196	195		43	68
471	442	47	96	151	187	187		41	65
458	432	46	94	148	176	176		39	61
446	421	45	92	145	165	165		37	58
434	410	44	90	142	145	145		33	52
423	401	43	88	139	131	131		30	47

Left Brinell column (top range): Not a practicable method of testing in this range

Right Rockwell C. Scale column (lower range): Not a practicable method of testing in this range

Table 26. Classification of Steels *(Courtesy of A.J. Forsyth & Co. Ltd.)*

THE AISI, SAE SYSTEM OF STEEL IDENTIFICATION

In the AISI, SAE specifications, capital letter prefixes are used to indicate the manufacturing process. Numbers are used to indicate the chemical composition. The first two numbers of the series indicate the alloy class to which the steel belongs and the last two numbers indicate the carbon range and sometimes minor departure from the usual range of simple alloying elements such as manganese and sulphur.

THE CAPITAL LETTER PREFIXES

A   denotes Basic Open Hearth Alloy Steel.
B   denotes Acid Bessemer Carbon Steel.
C   denotes Basic Open Hearth Carbon Steel.
CB denotes either Acid Bessemer or Basic Open Hearth at option of manufacturer.
D   denotes Acid Open Hearth Carbon Steel.
E   denotes Electric Furnace Alloy Steel.

THE NUMBER SERIES

10XX series—Plain Carbon Steel.
11XX series—Free Cutting Sulphurized Carbon Steels.
12XX series—Phosphorized Carbon Steels.
13XX series—Manganese Alloy Steels.
23XX series—Nickel Alloy Steels.
31XX series—Nickel Chromium Alloy Steels.
33XX series—High Nickel, High Chromium Alloy Steels.
40XX series—Molybdenum Alloy Steels.
41XX series—Chromium Molybdenum Alloy Steel.
43XX series—Nickel Chromium, Molybdenum Alloy Steel.
46XX series—Nickel Molybdenum Alloy Steel.
51XX series—Chromium Alloy Steels
52XX series—Chromium Spring Steels.
61XX series—Chromium Vanadium Alloy Steels.
92XX series—Silicon Manganese Spring Steel.
NE—8XXX series are National Emergency Steels developed during WWII to conserve vital alloying elements.

SAE
(Society of Automotive and Aeronautical Engineers)

AISI (American Iron and Steel Institute)

Table 27A. Steel Compositions *(Courtesy of A.J. Forsyth & Co. Ltd.)*

AISI & SAE STANDARD STEEL COMPOSITIONS
CARBON STEELS (Revised 1976)
Basic Open Hearth and Acid Bessemer

AISI No.*	C	Mn	P Max.	S Max.	SAE No.
C 1008	0.10 max.	0.30-0.50	.040	.050	1008
C 1010	0.08-0.13	0.30-0.60	.040	.050	1010
C 1012	0.10-0.15	0.30-0.60	.040	.050	1012
C 1015	0.13-0.18	0.30-0.60	.040	.050	1015
C 1016	0.13-0.18	0.60-0.90	.040	.050	1016
C 1017	0.15-0.20	0.30-0.60	.040	.050	1017
C 1018	0.15-0.20	0.60-0.90	.040	.050	1018
C 1019	0.15-0.20	0.70-1.00	.040	.050	1019
C 1020	0.18-0.23	0.30-0.60	.040	.050	1020
M 1020	0.17-0.24	0.25-0.60	.040	.050	....
C 1021	0.18-0.23	0.60-0.90	.040	.050	1021
C 1022	0.18-0.23	0.70-1.00	.040	.050	1022
C 1023	0.20-0.25	0.30-0.60	.040	.050	1023
C 1024	0.19-0.25	1.35-1.65	.040	.050	1524
C 1025	0.22-0.28	0.30-0.60	.040	.050	1025
C 1026	0.22-0.28	0.60-0.90	.040	.050	1026
C 1027	0.22-0.29	1.20-1.50	.040	.050	1527
C 1029	0.25-0.31	0.60-0.90	.040	.050	....
C 1030	0.28-0.34	0.60-0.90	.040	.050	1030
C 1035	0.32-0.38	0.60-0.90	.040	.050	1035
C 1036	0.30-0.37	1.20-1.50	.040	.050	1536
C 1037	0.32-0.38	0.70-1.00	.040	.050	1037
C 1038	0.35-0.42	0.60-0.90	.040	.050	1038
C 1039	0.37-0.44	0.70-1.00	.040	.050	1039
C 1040	0.37-0.44	0.60-0.90	.040	.050	1040
C 1041	0.36-0.44	1.35-1.65	.040	.050	1541
C 1042	0.40-0.47	0.60-0.90	.040	.050	1042
C 1043	0.40-0.47	0.70-1.00	.040	.050	1043
C 1044	0.43-0.50	0.30-0.60	.040	.050	1044
C 1045	0.43-0.50	0.60-0.90	.040	.050	1045
C 1046	0.43-0.50	0.70-1.00	.040	.050	1046
C 1048	0.44-0.52	1.10-1.40	.040	.050	1548
C 1049	0.46-0.53	0.60-0.90	.040	.050	1049
C 1050	0.48-0.55	0.60-0.90	.040	.050	1551
C 1051	0.45-0.56	0.85-1.15	.040	.050	....
C 1052	0.47-0.55	1.20-1.50	.040	.050	1552
C 1053	0.48-0.55	0.70-1.00	.040	.050	....
C 1055	0.50-0.60	0.60-0.90	.040	.050	1055
C 1060	0.55-0.65	0.60-0.90	.040	.050	1060
C 1070	0.65-0.75	0.60-0.90	.040	.050	1070
C 1078	0.72-0.85	0.30-0.60	.040	.050	1078
C 1080	0.75-0.88	0.60-0.90	.040	.050	1080
C 1084	0.80-0.93	0.60-0.90	.040	.050	1084
C 1090	0.85-0.98	0.60-0.90	.040	.050	1090
C 1095	0.90-1.03	0.30-0.50	.040	.050	1095

* Prefix C denotes basic open-hearth carbon steels.
When Silicon is required, the following ranges and limits are commonly used for basic open-hearth steel grades:

Standard Steel Designation	Silicon Ranges or Limits
Up to C 1015 Excl.	0.10 max.
C 1015 to C 1025 Incl.	0.10 max., 0.10/0.20, or 0.15/0.30
Over C 1025	0.10/0.20, or 0.15/0.30

Copper is specified as an added element when required.

Table 27B. Steel Compositions *(Courtesy of A.J. Forsyth & Co. Ltd.)*

AISI & SAE STANDARD STEEL COMPOSITIONS
CARBON STEELS (Revised 1976) *(Continued)*
Basic Open Hearth and Acid Bessemer — Resulphurized

AISI No.*	C	Mn	P Max.	S Max.	SAE No.
C 1109	0.08-0.13	0.60-0.90	.040	0.08-0.13	1109
C 1110	0.08-0.13	0.30-0.60	.040	0.08-0.13	....
B 1111	0.13 max.	0.60-0.90	0.07 0.12	0.10-0.15	....
B 1112	0.13 max.	0.70-1.00	0.07-0.12	0.16-0.23	1112
B 1113	0.13 max.	0.70-1.00	0.07-0.12	0.24-0.33	1113
C 1116	0.14-0.20	1.10-1.40	.040	0.16-0.23	....
C 1117	0.14-0.20	1.00-1.30	.040	0.08-0.13	1117
C 1118	0.14-0.20	1.30-1.60	.040	0.08-0.13	1118
C 1119	0.14-0.20	1.00-1.30	.040	0.24-0.33	1119
C 1132	0.27-0.34	1.35-1.65	.040	0.08-0.13	1132
C 1137	0.32-0.39	1.35-1.65	.040	0.08-0.13	1137
C 1139	0.35-0.43	1.35-1.65	.040	0.13-0.20	....
C 1140	0.37-0.44	0.70-1.00	.040	0.08-0.13	1140
C 1141	0.37-0.45	1.35-1.65	.040	0.08-0.13	1141
C 1144	0.40-0.48	1.35-1.65	.040	0.24-0.33	1144
C 1145	0.42-0.49	0.70-1.00	.040	0.04-0.07	1145
C 1146	0.42-0.49	0.70-1.00	.040	0.08-0.13	1146
C 1151	0.48-0.55	0.70-1.00	.040	0.08-0.13	1151

Basic Open Hearth Rephosphorized,
Resulphurized and Lead Bearing

C 1211	0.13 max.	0.60-0.90	0.07-0.12	0.10-0.15	1111
C 1212	0.13 max.	0.07-1.00	0.07-0.12	0.16-0.23	1112
C 1213	0.13 max.	0.07-1.00	0.07-0.12	0.24-0.33	1113
C 1215	0.09 max.	0.75-1.05	0.04-0.09	0.26-0.35	....
†C12L14	0.15 max.	0.85-1.15	0.04-0.09	0.26-0.35	....

* Prefix C denotes basic open hearth carbon steels; B, acid-bessemer.
† 12L14 has 0.15 to 0.35% lead added.

When Silicon is required, the following ranges and limits are commonly used for basic open hearth steel grades:

Standard Steel Designation	Silicon Ranges or Limits
Up to C 1110 Excl.	0.10 max.
C 1116 and Over.	0.10 max., 0.10/0.20, or 0.15/0.30

It is common practice not to produce basic open hearth or acid bessemer rephosphorized and resulphurized carbon steels to specified limits of silicon.

MECHANICAL TUBING — Seamless and Welded
Low Carbon — Open Health

AISI No.	C**	Mn	P Max.	S Max.
MT 1010	0.05-0.15	0.30-0.60	.040	.050
MT 1015	0.10-0.20	0.30-0.60	.040	.050
MT X1015	0.10-0.20	0.60-0.90	.040	.050
MT 1020	0.15-0.25	0.30-0.60	.040	.050
MT X1020	0.15-0.25	0.70-1.00	.040	.050

**Carbon limits are minimum and maximum on check analysis, while the other elements are subject to AISI Stardard Variations for check analyses.

Table 27C. Steel Compositions *(Courtesy of A.J. Forsyth & Co. Ltd.)*

AISI & SAE STANDARD STEEL COMPOSITIONS
ALLOY STEELS (Revised 1976)
Open Hearth and Electric Furnace

AISI No.*	C	Mn	Ni	Cr	Mo	Other Elements	SAE No.
Manganese Steels							
1330	.28-0.33	1.60-1.90	....	....	....	a,b	1330
1335	.33-0.38	1.60-1.90	....	....	....	a,b	1335
1340	.38-0.43	1.60-1.90	....	....	....	a,b	1340
1345	.43-0.48	1.60-1.90	....	....	....	a,b	1345
Nickel — Chromium Steels							
3140	.38-0.43	0.70-0.90	1.10-1.40	0.55-0.75	....	a,b	3140
† 3310	.80-0.13	0.45-0.60	3.25-3.75	1.40-1.75	....	a,b	3310
Molybdenum Steels							
4012	.09-0.14	0.75-1.00	....	....	.15-.25	a,b	4012
4023	.20-0.25	0.70-0.90	....	....	.20-.30	a,b	4023
4024	.20-0.25	0.70-0.90	....	....	.20-.30	b,c	4024
4027	.25-0.30	0.70-0.90	....	....	.20-.30	a,b	4027
4028	.25-0.30	0.70-0.90	....	....	.20-.30	b,c	4028
4037	.35-0.40	0.70-0.90	....	....	.20-.30	a,b	4037
4042	.40-0.45	0.70-0.90	....	....	.20-.30	a,b	4042
4047	.45-0.50	0.70-0.90	....	....	.20-.30	a,b	4047
4063	.60-0.67	0.75-1.00	....	....	.20-.30	a,b	4063
Chromium — Molybdenum Steels							
4118	.18-0.23	0.70-0.90	....	0.40-0.60	.08-.15	a,b	4118
4130	.28-0.33	0.40-0.60	....	0.80-1.10	.15-.25	a,b	4130
4137	.35-0.40	0.70-0.90	....	0.80-1.10	.15-.25	a,b	4137
4140	.38-0.43	0.75-1.00	....	0.80-1.10	.15-.25	a,b	4140
4142	.40-0.45	0.75-1.00	....	0.80-1.10	.15-.25	a,b	4142
4145	.43-0.48	0.75-1.00	....	0.80-1.10	.15-.25	a,b	4145
4147	.45-0.50	0.75-1.00	....	0.80-1.10	.15-.25	a,b	4147
4150	.48-0.53	0.75-1.00	....	0.80-1.10	.15-.25	a,b	4150
4161	.56-0.64	0.75-1.00	....	0.70-0.90	.25-.35	a,b	4161
Nickel — Chromium — Molybdenum Steels							
4320	.17-0.22	0.45-0.65	1.65-2.00	0.40-0.60	.20-.30	a,b	4320
4340	.38-0.43	0.60-0.80	1.65-2.00	0.70-0.90	.20-.30	a,b	4340
E 4340	.38-0.43	0.65-0.85	1.65-2.00	0.70-0.90	.20-.30	a,b	E4340
Nickel 1.75% — Molybdenum 0.25% Steels							
4615	.13-0.18	0.45-0.65	1.65-2.00	....	.20-.30	a,b	4615
4620	.17-0.22	0.45-0.65	1.65-2.00	....	.20-.30	a,b	4620
4621	.18-0.23	0.70-0.90	1.65-2.00	....	.20-.30	a,b	4621
4626	.24-0.29	0.45-0.65	0.70-1.00	....	.15-.25	a,b	4626
Nickel 1.05% — Chromium 0.45% — Molybdenum 0.20%							
4718	.16-0.21	0.70-0.90	0.90-1.20	0.35-0.55	.30-.40	a,b	4718
4720	.17-0.22	0.50-0.70	0.90-1.20	0.35-0.55	.15-.25	a,b	4720

* Prefix E denotes electric furnace steel; all others are open hearth steel.
† For open hearth steel the manganese is 0.40 to 0.60%.
(a) Phosphorus and Sulphur content for basic open hearth steel is .040 max.;
 for basic electric furnace steel, .025 max.
(b) Silicon content is .20 to .35%.
(c) Phosphorus content is .040 max.; sulphur .035 to .050%.

Table 27D. Steel Compositions *(Courtesy of A.J. Forsyth & Co. Ltd.)*

AISI & SAE STANDARD STEEL COMPOSITIONS
ALLOY STEELS (Revised 1976)
Open Hearth and Electric Furnace

AISI No.*	C	Mn	Ni	Cr	Mo	Other Elements	SAE No.
Nickel 3.50% — Molybdenum 0.25%							
4815	.13-0.18	0.40-0.60	3.25-3.75	....	.20-.30	a,b	4815
4817	.15-0.20	0.40-0.60	3.25-3.75	....	.20-.30	a,b	4817
4820	.18-0.23	0.50-0.70	3.25-3.75	....	.20-.30	a,b	4820
Chromium Steels							
5015	.12-0.17	0.30-0.50	....	0.30-0.50	....	a,b	5015
5120	.17-0.22	0.70-0.90	....	0.70-0.90	....	a,b	5120
5130	.28-0.33	0.70-0.90	....	0.80-1.10	....	a,b	5130
5132	.30-0.35	0.60-0.80	....	0.75-1.00	....	a,b	5132
5135	.33-0.38	0.60-0.80	....	0.80-1.05	....	a,b	5135
5140	.38-0.43	0.70-0.90	....	0.70-0.90	....	a,b	5140
5145	.43-0.48	0.70-0.90	....	0.70-0.90	....	a,b	5145
5147	.45-0.51	0.70-0.95	....	0.85-1.15	....	a,b	5147
5150	.48-0.53	0.70-0.90	....	0.70-0.90	....	a,b	5150
5155	.51-0.59	0.70-0.90	....	0.70-0.90	....	a,b	5155
5160	.56-0.64	0.75-1.00	....	0.70-0.90	....	a,b	5160
E51100	.98-1.10	0.25-0.45	....	0.90-1.15	....	a,b	51100
E52100	.98-1.10	0.25-0.45	....	1.30-1.60	....	a,b	52100
Chromium — Vanadium Steels							
6118	.16-0.21	0.50-0.70	....	0.50-0.70	.10-.15V	a,b	6118
6150	.48-0.53	0.70-0.90	....	0.80-1.00	.15min.V	a,b	6150
Nickel 0.55% — Chromium 0.50% — Molybdenum 0.20%							
8615	.13-0.18	0.70-0.90	0.40-0.70	0.40-0.60	.15-.25	a,b	8615
8617	.15-0.20	0.70-0.90	0.40-0.70	0.40-0.60	.15-.25	a,b	8617
8620	.18-0.23	0.70-0.90	0.40-0.70	0.40-0.60	.15-.25	a,b	8620
8622	.20-0.25	0.70-0.90	0.40-0.70	0.40-0.60	.15-.25	a,b	8622
8625	.23-0.28	0.70-0.90	0.40-0.70	0.40-0.60	.15-.25	a,b	8625
8627	.25-0.30	0.70-0.90	0.40-0.70	0.40-0.60	.15-.25	a,b	8627
8630	.28-0.33	0.70-0.90	0.40-0.70	0.40-0.60	.15-.25	a,b	8630
8637	.35-0.40	0.75-1.00	0.40-0.70	0.40-0.60	.15-.25	a,b	8637
8640	.38-0.43	0.75-1.00	0.40-0.70	0.40-0.60	.15-.25	a,b	8640
8642	.40-0.45	0.75-1.00	0.40-0.70	0.40-0.60	.15-.25	a,b	8642
8645	.43-0.48	0.75-1.00	0.40-0.70	0.40-0.60	.15-.25	a,b	8645
8655	.51-0.59	0.75-1.00	0.40-0.70	0.40-0.60	.15-.25	a,b	8655

*Prefix E denotes electric furnace steel; all others are open hearth steel.
†For open hearth steel the manganese is 0.40 to 0.60%.
(a) Phosphorus and Sulphur content for basic open hearth steel is .040 max.; for basic electric furnace steel, .025 max.
(b) Silicon content is .20 to .35%.
(c) Phosphorus content is .040 max.; sulphur .035 to .050%.

Table 28. Spark Testing Metals

Mild Steel (SAE 1020)	Carbon Tool Steel	High-Speed Steel	Stainless Steel	Cast Iron	Carbide Tools

METALS	SPARK CHARACTERISTICS
Mild Steel (SAE 1020)	Moderate volume of sparks, white close to the wheel, white near the end of the stream, forked spurts, stream about 70 inches long.
Carbon Tool Steel	Large volume of sparks, white close to the wheel, white near the end of the stream, many fine spurts, stream about 50 inches long.
High-Speed Steel	Small volume of sparks, reddish close to the wheel, straw color near the end of the stream, a few forked spurts, stream about 60 inches long.
Stainless Steel	Moderate volume of sparks, straw color close to the wheel, white near the end of the stream, forked spurts, stream about 50 inches long.
Gray Cast Iron	Small volume of sparks, red close to the wheel, reddish straw color near the end of the stream, fine spurts, about 24 inches long.
Carbide Tools	Volume of sparks very small, light orange close to the wheel, light orange near the end of the stream, no spurts, stream about 2 inches long, if that.

APPENDIX 2

Three areas have been selected to present definitions of the terms — gearing, sawing, and grinding. Most of the actual terms used in each area are covered. In addition, general information is presented on the use of different grinding wheels. Specifications for the different grinding wheels needed for specific grinding operations are given in chart form. Most of the common grinding operations are covered by specifying the material being ground and by stating the proper wheel to be used. The specifications given are based on actual Norton Company field practice. The specifications and the terms of definition are both commonly used in the machine tool industry.

SPUR GEARING — DEFINITION OF TERMS

Addendum (a) — The height of the gear tooth from the pitch circle to the outside diameter or top of the tooth measured along the center line of the tooth.

Backlash — The play or looseness between mating gears. Backlash is required to prevent any unnecessary noise between mating gears in motion.

Base Circle — A circle with a diameter slightly less than that of the pitch circle. It is the circle from which the involute curve of the gear tooth actually starts.

Center Distance (C) — The distance from the center of one gear to the center of the mating gear, or the distance from the pitch line of a gear rack to the center of the mating gear. Center distance is sometimes called the center-to-center distance of mating gears or mating gear and gear rack.

Circular Pitch (c) — The distance from the center of one gear tooth to the center of the adjacent gear tooth, measured on the pitch circle. The circular pitch system is not used too much in gearing except in places like a pattern shop where patterns of large gear teeth are made to be cast in a foundry. When the circular pitch system is used, it is usually specified as such.

Clearance (c) — The distance between the top of the gear tooth and the bottom of the mating slot between gear teeth measured along the centerline of the gear tooth. The clearance is necessary to prevent the gear tooth from bottoming in the mating slot, which would cause the gear to growl and be noisy.

Chordal Addendum — The distance from the line made by the chordal thickness of the gear tooth to the top of the gear tooth measured along the center line of the tooth.

Chordal Thickness — The length of the chord formed between the sides of the gear tooth at the pitch circle.

Dedendum (b) — The distance from the pitch circle to the bottom of the tooth at the root circle, measured along the centerline of the gear tooth. The root circle which forms the root diameter of the gear is sometimes called the dedendum circle. The dedendum includes the clearance which means that the dedendum is larger than the addendum by the amount of the clearance.

Diametral Pitch (P) — The ratio between the number of gear teeth in the gear and each inch of pitch diameter. Diametral pitch is usually referred to as "Pitch." The diametral pitch system is the most common system used in gearing.

Length of Gear Rack (L) — The length of the gear rack measured along the pitch line. It is found by multiplying the number of teeth by the circular pitch.

Line of Action — A line perpendicular to the point of contact between gear teeth and the base circle. It is at this point that mating gear teeth exert their greatest pressure on each other. The angle formed by the line of action and the base circle is called the Pressure Angle of the gear tooth.

Number of Teeth in the Gear (N) — The total number of teeth in a gear or on a gear rack.

Outside Diameter (Do) — The outside diameter of a gear. It is also the diameter of the gear blank before the teeth are cut.

Pitch Diameter (D) — The diameter of the pitch circle. This is the most important diameter in gearing since all calculations are made with reference to the pitch circle.

Root Diameter (Dr) — The diameter of the circle formed by the bottom of the gear teeth.

Tooth Thickness (T) — The thickness of the gear tooth measured on the pitch circle. It is actually the chordal distance.

Whole Depth (ht) — The total depth of the gear slot. It is also equal to the full depth set and cut by the gear cutter.

Working Depth (hk) — The depth actually penetrated by a gear tooth into its mating slot.

BEVEL GEARS — DEFINITION OF TERMS

Addendum (As) — The addendum at the small end of the tooth is the distance from the top of the tooth to the center line of the tooth at the small end of the tooth.

Addendum Angle (Al) — The angle formed between the center line of the pitch angle and the top of the gear tooth.

Apex Distance (V) — The distance from the apex of the pitch cone to a line formed by the top of the large end of the gear tooth.

Apex Distance (v) — The distance from the apex of the pitch cone to the line formed by the top of the small end of the gear tooth.

Dedendum Angle (cl) — The angle formed between the center line of the pitch angle and the bottom of the gear tooth.

Face Angle (Fa) — The angle formed by the center line of the bevel gear and the top of the gear tooth.

Imaginary Number of Teeth (Nc) — The number of teeth of the imaginary spur gear for which the gear cutter is selected.

Root Angle (d) — The angle formed by the center line of the gear and the root of the gear tooth.

Set-over (n) — The calculated amount of set-over used to offset the milling machine from center when cutting an approximate bevel gear on a milling machine.

Tooth Thickness (TL) — The thickness of the tooth at the large end of the tooth measured on the center line of the tooth.

Tooth Thickness (Ts) — The thickness of the tooth at the small end of the tooth measured on the center line of the tooth.

Width of the Face (W) — The actual width of the face of the gear measured along the top of the gear tooth

SAWING — DEFINITION OF TERMS

Abrasive Band — A continuous fabric band coated with aluminum oxide or silicon carbide grit used for polishing parts previously sawed and file finished.

Additive — A chemical added in small quantities to cutting fluid to give specific properties to the cutting fluid.

Add Mixture, Admixture — A mixture of cutting fluid concentrate and water used to replenish the original cutting fluid mixture. It is usually mixed at one-half the strength of the original mixture.

AISI — American Iron and Steel Institute. The initials are used to prefix numbers designating iron and steel alloys by composition; also SAE, Society of Automotive Engineers.

Auxiliary Guard — A saw band guard introduced in the throat of the band machine when using only two carrier wheels on a three-wheeled model.

Back Clearance Angle — The angle at the back of a saw tooth measured in relation to the cutting edge of the band tool.

Backoff — Rapid withdrawal of the band tool from contact with the workpiece.

Backup Bearing — Thrust bearing mounted in the saw

guide to support the back of the saw band.

Bactericide — A substance that inhibits growth of bacteria; may be used to protect cutting fluid.

Band Machine — A machine employing continuous band tools for either sawing, slicing, filing, or polishing.

Band Tension — The relative tautness given to the band tool on the machine after the idler wheel has been adjusted to exert a pulling force against the opposite loop of the band anchored around the fixed drive wheel.

Band Tool — Any saw, wire, file, or other tool in band form for use on a band machine.

Band Wheels — Wheels around which the continuous band tool revolves on a band machine. One is the drive wheel, and the other is the idler wheel.

Beam Strength — The resistance a band tool has to back deflection when subjected to the edge thrust of feeding force.

Bond — In band welding, the juncture of joined band ends.

Built-up Edge — The deposit of metal that accumulates at or near the cutting edge of the saw band tooth as heat generated by the cutting action softens the workpiece enough so that particles of it weld themselves to the tooth face.

Burr — A turned over edge on work resulting from cutting.

Butt Welder — The built-in accessory on contour band machines in which two ends of a saw blade are butted together and welded to form a band.

Buttress — A tooth form consisting of a shallow gullet with widely spaced teeth to provide for ample chip clearance.

Camber — An arcing or bending of the back or cutting edge of the band tool. In positive camber, the cutting edge arcs backward; in negative camber, the cutting edge arcs forward.

Center Plate — A circular plate that fits into an opening on fixed table contour band machines at the point of cut. The band tool passes through a slot cut into the plate.

Chatter — A rumbling sound in the band machine caused by trying to take too heavy a cut. The sound comes from overloading the machine or transmission, or springing the saw guide post or worktable.

Chip Clearance — The gullet area between two saw teeth.

Chip Load — The average tooth advance or feed rate into the work.

Chip Welding — The fusing of a portion of the chip to a tooth face, as in a built-up edge. It is caused by extreme heat and pressure in the cutting area.

Claw Tooth — A tooth form consisting of a shallow gullet and widely spaced teeth having a positive rake angle.

Contour Band Machine — A machine with the ability to cut to a curved layout line using a band saw blade or other band tools. It is also informally called a band saw.

Contour Sawing Attachment — A device for a contour band machine that holds and guides the work when making contour cuts.

Cut-off — A process of cutting a piece of stock into usable pieces.

Cutting Rate — The speed at which the cross-sectional area of the workpiece is cut, expressed in square inches of cutting per minute.

Dart — A type of high carbon, spring-tempered-back saw blade, superior to standard carbon alloy blade in tooth hardness, abrasion resistance, and ability to be tensioned for faster, more accurate cutting.

Deburring — Removing burrs, sharp edges, or fins from metal parts by filing, grinding, or rolling the work in a barrel with abrasives suspended in a suitable liquid medium.

Diamond-edged Saw Band — Continuous saw band having industrial diamonds bonded to the cutting edge. It is used to cut superhard materials.

Disc Cutter — An attachment for making circular cuts on contour band machine.

Dish — The curved surface of the side wall of a cut produced when the band tool wanders from the natural course of cut in sawing round workpieces.

Electro Band — A straight, knife-edge band used to cut honeycomb and cellular material. An electric spark leaps from the knife edge of the band and disintegrates the material ahead of it in a narrow kerf.

Etching Pencil — A marking tool used in connection with the welder that burns a permanent mark in metal.

Extension Bar — Used in the T-slots in a contour band machine to provide support for workpieces larger than the worktable.

Feed Force — The pressure exerted by the work against the cutting edge of a saw band measured in pounds.

Feed Rate — The linear movement of the work into the saw band measured in inches per minute or meters per minute.

File Band — A continuous band tool made from a series of small file segments riveted to a flexible steel band.

File Card — A small, flat wire brush for cleaning the file band.

Filler Plate — A bar that slides into the long thin slot running from the center of fixed-table, contour band machines. It covers the slot through which saw bands are installed, to provide an unbroken table surface for small work.

Flash — A bulge of excess material left at the point of saw blade weld.

Flash Welding — A resistance butt welding process in which the weld covers the entire blade ends held in the welder jaws. It is done by pressure and heat: the heat produced by the electric arcs between the blade ends, and the pressure by the movement towards each other of the welder blades.

Friction Sawing — A method of sawing ferrous metals. It utilizes the frictional heat generated by the high saw band velocity and heavy feed force to soften the material. The saw band teeth then remove the softened material by scooping it out of the kerf.

Gage — The thickness of the back of the saw band, expressed in thousandths of an inch.

Gullet — The throat within the curved area between two saw band teeth.

Gullet Depth — The distance from the tooth tip to the bottom the gullet.

Heavy Work Holding Clamps — Holding devices for heavy work on fixed table, contour band machines.

Heavy Work Slide — Bars fitted with rows of ball bearings and placed on the worktable of a contour band machine to ease operator effort in feeding heavy material.

High Speed Sawing — Band machining at saw band velocities between 2000 and 6000 fpm (615–1846 m/min).

Imperial Bimetal Blade — Trademark of a DoAll bimetal band saw blade.

Job Selector — A dial located on some contour band machines that is used to select the correct band tool, band speed, feed force, and cutting fluid for the most commonly used materials.

Kerf — The slot made in the work by the cutting tool as it moves through the work. The width of the kerf is determined by the set of the teeth of the saw band.

Knife Band — A sharp-edged, continuous band used to slice soft and fibrous materials. It normally parts material without cutting a kerf.

Lead — The tendency of a saw band to wander from the natural course of cut.

Line Grind Band — A band tool with a continuous abrasive coating of aluminum oxide, diamond, or Borazon/CBN. It is used on all DoAll CNC contour wire band machines to cut electrodes for EDM machines and to saw carbides and hardened tool and die alloys.

Machine Speed Clamp — A "quick release" work holding device that fastens to the worktable T-slots of power table contour band machines or to those of movable fixtures on fixed table machines.

Mandrel Slotting Fixture — A plug over which the workpiece drops for proper positioning before cutting.

Mitering Accessories — Attachments for fixed table, contour band machines for making angular cuts.

90 Degree Saw Guide Bracket — An attachment for contour band machines that twists the saw bands 90 degrees, permitting the cutting of work longer than the throat depth will accommodate.

Notching — A sawing operation in which two straight cuts intersect to remove a single piece of metal.

Pitch — The number of teeth per inch, or 25.4 mm, in a saw blade.

Plate Fixtures — Work holding devices with locating pins, nests or other locators on a flat piece of metal used for fixtured production.

Precision Tooth — A tooth form consisting of a deep gullet with a smooth radius at the bottom.

Production Band Machining — A term applied to the process of producing a large number of identical parts in a continuous operation, usually of a highly automated nature.

Protractor Workstop and Aligning Gage — A device for expediting straight and angular cuts up to 45 degrees on production jobs on contour band machines.

Quenched Arc Cutting — The principle used in electroband machining. A flood of coolant limits or quenches the arc as it leaps off the knife-edge band tool to the workpiece, to prevent it from burning the workpiece.

Rake Angle — The angle the tooth face makes with respect to a perpendicular line from the back edge of the saw band. It is positive when the tooth angles forward in the direction of the cutting action, and negative when the tooth angles backward from the direction of the cutting action.

Raker Set — A saw tooth pattern in which one tooth is offset to the right, the next to the left, and the third is straight.

Ratchet Feed — A feeding device used on fixed-table contour band machines consisting of a rack gear and a ratchet lever.

Rip Fence — Fixture guide used in making long straight cuts.

Ripping — A term for dividing work into pieces. It usually refers to sawing sheet or plate material into strips.

Rocking Technique — The procedure of oscillating a workpiece up and down in friction sawing to permit sawing work thicker than 1 inch.

Roller Guide — Rollers mounted on a saw guide to resist lateral movement of the saw band. Usually used when employing fast band speeds because they entail less friction and wear on both the band tool and guides than do insert-type guides at high speeds.

Saw Band, Saw Band Tool — A blade or other tool in the form of a continuous loop or band used on a band machine.

Saw Band Velocity — The rate of travel of the saw band through the work, expressed in linear or surface feet per minute (sfpm) or meters per minute (m/min).

Saw Gage — A slotted piece of metal used to align saw guide inserts.

Saw Guide — A device mounted above and below the worktable on contour band machines that holds saw guide inserts (or rollers) and backup bearings for positioning the band tool properly in relation to the work.

Saw Guide Inserts — Metallic pieces, normally of HSS or carbide, mounted in saw guides to support or resist lateral movement of the saw band.

Saw Guide Post — A post extending down from the head of a contour band machine toward the worktable. It supports the upper saw guide, saw band guard, and coolant and air nozzles.

Sawing — Cutting a workpiece with a band blade, a circular blade, or hacksaw blade.

Scallop Edge — A type of knife-edge saw band with a sharp cutting edge characterized by scallop-like undulations of pointed crests and rounded troughs.

Scratch Pitch — The endless pattern of ridges and grooves on the side walls of a cut produced by the saw teeth under the combined actions of band speed and feeding force.

Sectioning — A word used to describe a straight cut that separates stock into usable parts.

Segmenting — A word for notching, a sawing operation in which two straight cuts intersect to remove a single piece of material.

Servo-contour Directional Control Feed — A device incorporated into hydraulic contour band machines that automatically increases or decreases feed force as the saw band encounters more or less pressure while the work is manipulated around corners.

Set — The amount of bend given to saw blade teeth to create side clearance for the back of the band when cutting through material. The amount of set is measured from the outside corner of a tooth bent to the left to the outside corner of a tooth bent to the right.

Shear — A cut-off device on a blade welder for cutting a saw blade square.

Side Clearance — The difference in dimension between the set of the teeth and the back of the band tool. It provides space for maneuvering the band in contour cuts, prevents "leading" when making straight cuts, and minimizes transfer of frictional heat to the work.

Sleeve Slotting Fixture — Tube holding device into which the workpiece is inserted for slotting.

Slicing — A cut-off operation usually performed with knife-edge saw bands.

Slotting — A sawing operation that produces a narrow slit in the work.

Slugging — The process of sawing blanks or slugs from bar stock or billets for subsequent machining operations. This is usually performed on cut-off band machines, but can be done on contour band machines.

Snips — A metal cutting scissor-like tool used for cutting saw blades.

Spiral Saw Band — A saw band resembling a wire with a continuous spiral cutting edge.

Splitting — A straight cut dividing a part made as a single piece.

Spray Manifold — A nozzle applicator for a coolant system mounted on the saw guard of a fixed table contour band machine.

Stack Sawing — A technique of making a pile of several pieces of work material, holding them together by some means, and sawing them all at the same time.

Starwheel — A star-shaped control wheel, such as the one on the back of fixed-table contour band machines, used to control movement of the saw guide post.

Straight Set — A saw tooth pattern in which all teeth are set symmetrically, one to the right followed by one to the left.

Super Dart — A hard back high carbon alloy blade similar to a Dart blade but with higher red hardness and abrasion resistance.

Swarf — In band machining, the intimate mixture of workpiece chips and fine particles of abrasive resulting from the grinding action of abrasive band tools used on CNC or diamond band machines.

T-Slot — Narrow T-shaped grooves in the worktable of a power table, contour band machine, or other machine tools. They are used for hold-down bolts or studs and to aid in fixturing. They may also be used on movable fixtures for fixed-table machines.

Tachometer — An instrument used for measuring revolutions of a shaft. It is used on contour band machines to indicate band speed in surface feet per minute (sfpm).

Tilt Lock Handwheel — A control on the base of contour band machines for holding the worktable in an angular position.

Tilt Lock Nut — A knob or starwheel on the hub of the upper idler band carrier wheel used to secure the tilt position of the wheel after properly tracking the band tool.

Tooth Face — The surface of the saw tooth on which the chip forms as it is cut away from the work.

Tooth Form — The shape of a tooth on a band tool designed to achieve specific results.

Tracking — The process of aligning a band tool on the band carrier wheels so that the band tool is in proper relation to the table, work, backup bearings, and saw guide inserts.

Trimming — Sawing off unwanted scrap elements of a workpiece such as the sprue of a casting, to make a usable part.

Trunnion — A pivoting device under the contour band machine worktable that enables the table to be tilted right or left.

Tungsten Carbide Saw Blade — A saw blade with tungsten carbide inserts in the tooth tips. They are used for cutting extremely hard material.

Twist — The tendency of a saw band to spiral after use.

Universal Calibrated Work Fixture — An optional attachment for the power table contour band machine. It holds and squares work for straight line cuts and angle sawing up to 45 degrees.

Universal Vise — A device for holding material for accurate, straight, and angular cuts up to 45 degrees on power table contour band machines.

Upset — The localized increase in volume resulting from the application of pressure during welding of the band tool. The merging of band end materials forms a stronger bond than with butt welding.

Upset Material — The flash or bulge of material left on the saw band after welding.

Vernier Control — An auxiliary control on a power table contour band machine used to make find adjustments in table feed force.

Vibration — Effect on a saw band caused by improper tracking, velocity, feed force, tension, tooth spacing, pitch, work thickness.

Warp — The rounded surface produced when lead occurs in straight cutting.

Wave Set — A saw tooth pattern having one group of teeth set progressively to the right and back to center, and the next group of teeth set progessively to the left and back to center. It has the appearance of a wave when viewed from the top of the band.

Wavy Edge — A type of knife-edge saw band having a sharp cutting edge of wavy undulations characterized by rounded crests and troughs.

Weight-type Power Feed — A device used on fixed-table contour band machines to provide a steady mechanical feeding force.

Width — The nominal dimension of a band tool as measured from the tip of the tooth to the back of the band.

Work, Workpiece — Pieces, parts, or objects that are to be or are being machined, processed, or treated.

Work Hardening, Work Hardness — Hardness that develops in metal as a result of cold working, dull saw band teeth, excessive band speed, or too light a feeding force.

Work Holding Jaw — A fixed-angle holding device for easier guiding of material while feeding it into the saw band on contour band machines.

Work Stop — Device used to limit table travel on a contour band machine

GRINDING — DEFINITION OF TERMS

Abrasive — A substance used in the process of grinding, polishing, and lapping of material. It is the actual cutting tool which does the cutting, abrading, or wearing away. Natural abrasives include corundum, emery, garnet, and diamond. Manufactured abrasives include aluminum oxide, silicon carbide, boron carbide, cubic boron nitride, and synthetic diamond.

Abrasive File — A special shaped sharpening stone used to reach into corners or rounded surfaces for special sharpening jobs.

Accuracy — Conformity in dimension to an exact standard of dimension.

Alumina — Unfused aluminum oxide.

Aluminum Oxide — An abrasive made by fusing the mineral bauxite (Al_2O_3).

Alundum — Norton Company's registered trademark for aluminum oxide abrasive.

Arbor — The spindle of the grinding machine on which the wheel is mounted.

Arbor Hole — The hole in a grinding wheel sized to fit the machine arbor.

Arc of Contact — That portion of the circumference of a grinding wheel touching the work being ground.

Area of Contact — The total area of the grinding surface of a grinding wheel in contact with the work being ground.

Arkansas Stone — An extremely hard quartzite quarried in the Ozark mountains, used as sharpening stones and abrasive files for producing the finest edges possible.

Armored Diamond — A Norton trademarked name for diamond abrasive with a coating of metal applied directly to the diamond. The coating provides a tenacious bond, preventing premature loss of the diamond.

Aztec — A Norton trademark for a special heat-and-wear retardant bond used in CBN grinding wheels.

Balance (Dynamic) — A piece in static balance is in dynamic balance if, upon rotating, there is no vibration or "whip" action due to unequal distribution of its weight throughout its length.

Balance (Static) — A grinding wheel is in static balance when, centered on a frictionless horizontal arbor, it remains at rest in any position.

Balancing — Testing a wheel for balance, adding or subtracting weight to put a piece into either static or dynamic balance.

Bauxite — A mineral ore high in aluminum oxide content from which Alundum abrasive is manufactured.

Bearing — Point of support. The part of a machine in which the spindle revolves.

Bench Stand — An offhand grinding machine with either one or two wheels mounted on a horizontal spindle, attached to a bench.

Blotter — A disc of compressible material usually of blotting paper stock, used between a wheel and flanges when mounting, to equalize pressure exerted by the flanges on the wheel.

Bond — The material in a grinding wheel which holds the abrasive grains together.

Bonded Abrasives — Grinding wheels, sharpening stones, and other abrasive products in which the abrasive is held together with bonding material.

Boron Carbide — An abrasive trademarked by Norton Company as Norbide.

Bort — Industrial grade natural or manufactured diamond suitable for use as an abrasive.

Brick — A block of bonded abrasive used for such purposes as rubbing down castings, scouring castings, general foundry and machine shop use, scouring chilled iron rolls, and polishing marble.

Burning (the work) — A change in the work being ground caused by the heat of grinding, usually accompanied by a surface discoloration.

Burr — A turned over edge of metal resulting from punching a sheet or from grinding or cutting-off operations.

Burring (Pulpstones) — Passing over the face of a pulpstone with a special tool to develop a pattern to provide a freer cutting surface.

Bushing — A material, usually lead, babbitt, or plastic, which sometimes serves as a lining for the hole in a grinding wheel.

Cam Wheel — An expression used to designate wheels used for grinding cam shafts.

CBN — Cubic boron nitride abrasive, a manufactured abrasive used for precision grinding of steel.

Centerhole — Tapered precision holes in the ends of workpieces to be ground between centers on a cylindrical grinder.

Centerhole Lapping — The cleaning or lapping of centerholes with a bonded or abrasive wheel cemented onto a steel mandrel.

Centerless Grinding — Grinding the outside or inside diameter of a round piece not mounted on centers.

Centers — Conical steel pins of a grinding machine upon which the work is centered and rotated during cylindrical grinding.

Ceramics — The science and art of clay working and various related industries. The use of vitrified bonds brings abrasive wheel manufacturing under this classification.

Chatter Marks — Surface imperfections on the work being ground, usually caused by vibrations between the wheel and the work.

Chuck — A device for holding grinding wheels of special shape or for holding the workpiece being ground.

Coated Abrasives — Abrasive products in which the abrasive is coated in a relatively thin layer on a backing of cloth, fiber, or paper. Coated abrasive products include sheets, rolls, belts, discs, and specialty shapes.

Cone Wheel — A type 17 wheel with a conical shape, used in portable grinding.

Controlled Structure — The Norton process of manufacturing grinding wheels whereby the relationship between the abrasive and bond is positively controlled.

Coping — Sawing stone with a grinding wheel.

Corner Wear — The tendency of a grinding wheel to wear on a corner so that it does not grind sharp corners without fillets.

Corundum — A natural abrasive of the aluminum oxide type, of higher purity than emery.

Crank Wheel — An expression used to designate wheels for grinding crankshafts.

Crush Truing (or Forming) — The process of using steel or tungsten carbide rolls to true or form grinding wheels to special face shapes.

Crystalline — Made up of crystals.

Crystallize — To convert into crystals.

Crystolon — Norton Company's registered trademark for silicon carbide abrasives.

Cubic Boron Nitride (CBN) — An extremely hard manufactured abrasive used for precision grinding of steels.

Cup Wheel — A Type 6 grinding wheel shaped like a cup or bowl.

Cut-off Wheel — A thin grinding wheel, often reinforced and usually made with an organic bond, used for cutting through workpiece material; an abrasive saw.

Cutters — The part of a mechanical dresser for grinding wheels which comes in contact with the wheel and actually does the dressing.

Cutting Rate — The amount of material removed by a grinding wheel per unit of time.

Cutting Surface — The surface or face of the wheel against which the material is ground.

Cylinder Wheel - A grinding wheel similar in characteristics to a straight wheel, with a large hole size in proportion to its diameter, and usually several inches in height. It is often reinforced with winding.

Cylindrical Grinding — Grinding the outside surface of a cylindrical part mounted on centers.

Deburring — The process of removing burrs from metal.

Diamond Dressing Tool — A tool for dressing or truing a wheel. It is made with a single diamond point or multiple diamond points.

Diamond Wheel — A grinding wheel in which the abrasive is mined or manufactured diamond.

Disc Grinding — A type of grinding used to produce flat surfaces on workpieces. Parallel opposing surfaces can be ground with disc grinders having double spindles.

Disc Wheel — A grinding wheel used in a disc grinder, with a shape similar to a Type 1 straight wheel. It is usually mounted on a plate for reinforcement. The side of the wheel is used for grinding.

Dish Wheel — A Type 12 wheel shaped like a dish.

Dog — A device like a lathe dog attached to the workpiece by means of which the work is revolved.

Dressers — Tools used for dressing a grinding wheel.

Dressing — The process of restoring, improving, or altering the cutting action of the face of a grinding wheel.

Dressing Stick — An abrasive stick used to dress the face of the grinding wheel.

Dry Grinding — Performing a grinding operation without the use of grinding fluids.

Emery — A natural abrasive of the aluminum oxide type.

External Grinding — Grinding on the outside surface

of an object as distinguished from internal grinding or cutting-off.

Face — The periphery of a grinding wheel. It is also the part of a grinding wheel brought into contact with the workpiece.

Face Shape — The physical shape of the periphery or face of a grinding wheel. It is designated by a letter in the wheel description.

Feed, Cross — In surface grinding, it is the distance of horizontal feed of the wheel across the table.

Feed, Down — In surface grinding, it is the rate at which the abrasive wheel is fed into the work.

Feed, Index — In cylindrical grinding, it is a measurement indicated by the index of the machine. On most machines, this measurement refers to the diameter of the work. On a few machines, this measurement refers to the radius of the work.

Feed Lines — A pattern on the work produced by grinding. The finer the finish, the finer and more evident are these lines. Some types of feed lines indicate an incorrect grinding condition.

Fin — A thin projection on a casting usually removed by snagging.

Finish — The surface quality or appearance of the work surface such as that produced by grinding or other machining operation.

Finishing — The final cuts taken with a grinding wheel to obtain accuracy and the surface desired.

Flanges — The circular metal plates on a grinding machine used to hold and drive the grinding wheel.

Flaring Cup — A Type 11 cup wheel with the rim extending from the back at an angle so that the diameter at the outer edge is greater than at the back.

Floorstand Grinders — An offhand grinder mounting either one or two wheels running on a horizontal spindle fixed to a metal base attached to the floor.

Fluting — Grinding the grooves of a twist drill or tap.

Form Grinding — A type of grinding used to produce or finish a special form or shape in a workpiece.

Gate — The part of a casting formed by the opening in the mold through which the molten metal enters the mold cavity. This is often removed by snagging.

Gemini — Norton's trademark name for pre-engineered diamond wheels and cut-off wheels.

Generated Heat — Heat resulting from the friction

between the workpiece and the running grinding wheel.

Glazing — An extreme condition of loading on a grinding wheel face caused by dulled abrasive and a build-up of swarf (the same as a built-up edge on any other cutting tool), resulting in sharply decreased cutting rates.

G-Ratio — A measurement for grinding efficiency made by dividing cubic inches of metal removed from a workpiece by the cubic inches of wheel wear.

Grade — The strength of bonding in a grinding wheel, also referred to as the hardness of a grinding wheel.

Grain (Grit) — The abrasive particles used in a grinding wheel or as loose abrasive materials.

Grain (Grit) Size — The size of the abrasive particles in a grinding wheel classified according to the number of linear holes in a screen through which the grits will pass.

Grinding — The process of removing material with a grinding wheel.

Grinding Action — The cutting ability of, and the finish produced, by a grinding wheel.

Grinding Fluid — A liquid or solution used in wet grinding to cool the workpiece, to lubricate the grinding wheel, to clean away the swarf, and to prevent rust.

Grinding Machine — Any machine on which a grinding wheel is operated.

Grinding Wheel — A cutting tool of circular shape made of abrasive grains bonded together.

Grindstone — A flat circular grinding wheel cut from natural sandstone sometimes used for sharpening tools.

Guard — A cover surrounding a grinding wheel, grinding machine table, or any moving part of the grinding machine. It is used to protect the operator in case of wheel breakage, to direct sparks, swarf, and cutting fluid away from the operator, and to provide machine safety in operation.

Hand Hone — A flat, metal stick with vitrified diamond abrasive on one or both ends, used for very fine touch-up jobs.

Hemming Machine — A machine or process for grinding flat surfaces, such as cutlery blades and ice skate blades.

Honing — An abrasive operation typically performed on internal cylindrical surfaces and employing bonded abrasive sticks in a special holder to remove stock and obtain surface accuracy.

Hoods — Metal guards used for protection against wheel breakage.

Huntington Dresser — A tool using star-shaped cutters for truing and dressing grinding wheels.

India — Registered trademark of the Norton Company for oilstones for producing keen cutting edges. They are made from Alundum abrasive and are oil impregnated.

Inserted Nut — Disc, segment or cylinder wheels having nuts embedded in the back surface for mounting on the machine.

Internal Grinding — Grinding the inside surface of a hole in a piece of work.

Lapping — A finishing process typically employing loose abrasive grains, but now often including similar types of operation with bonded abrasive wheels.

Loading — Filling of the pores of the grinding wheel surface with the material being ground, usually resulting in a decrease in production and poor finish.

Lubricant — The liquid or solution used to lubricate the wheel and promote a more efficient cutting action.

Magnetic Chuck — A workpiece holding device, used in surface grinding, which holds the workpiece by means of magnetic force.

Mandrel — A solid cylindrical piece of metal on one end of which a grinding wheel or abrasive is mounted. The other end is fixed in the machine chuck. It may be made of steel or a heavier metal to provide greater stability and resistance to whipping.

Maximum Safe Speed — The fastest speed at which a grinding wheel or mounted wheel can be operated safely. The speed may be expressed as RPM or SFPM.

Metal Bond — A bonding material used in diamond grinding wheels for work on nonmetallics such as germanium, stone, silicon, glass, asbestos, ceramics, and ferrite.

Mounted Wheels — Small grinding wheels, often with special shapes, that are mounted on metal mandrels.

Mounting — Putting the grinding wheel on the arbor or spindle of a grinding machine, and getting it ready to go to work.

Multipoint Dressing Tools — Dressing tools which use a large number of fractional-carat diamonds, embedded in a matrix.

Natural Abrasive — A hard mineral found in nature and used as the abrasive component of a grinding wheel.

Norbide — Norton's registered trademark name for boron carbide abrasive.

NorZon — Norton's registered trademark name for grinding wheels made with NZ Alundum abrasive. NorZon portable wheels and raised hub wheels are used for the offhand grinding of steel.

Notching — The grinding of relieving notches in the weld joints of pipelines. Usually done with raised hub wheels.

NZ Alundum — A zirconia alumina abrasive alloy specially formulated for use in NorZon grinding wheels.

Offhand Grinding — Grinding by holding the work against the grinding wheel by hand. In floorstand grinding, the operator holds the work. In portable grinding, the operator holds the machine mounted wheel against the workpiece.

Oilstone — A sharpening stone or abrasive file, usually prefilled with oil, which uses oil as a lubricant to flush away swarf in the operation.

Operating Speed — The speed of a grinding wheel expressed in either revolutions per minute or surface feet per minute.

Organic Bond — A bond made of organic materials such as synthetic resin, rubber, or shellac.

Peripheral Speed — The speed at which any point or particle on the face of the wheel is travelling when the wheel is revolved, expressed in surface feet per minute (SFPM). To find SFPM, multiply the wheel circumference in feet by the wheel revolutions per minute.

Periphery — The line bounding a rounded surface, which is equal to the circumference of the wheel.

Planer Type — A type of surface grinding machine built similar to an open side planer.

Plate Mounted — A disc, segment, or cylinder wheel cemented to a steel back plate having projecting studs or other means for mounting on the machine.

Polishing — An operation to smooth off roughness from a workpiece or put a high finish on metal by using a polishing wheel.

Polishing Wheel — A wheel made of one of several kinds of materials and coated with abrasive grain and glue.

Polypax Dressing Tools — Norton's trademark name for single-point diamond dressing tools which use a manufactured polycrystalline diamond.

Portable Grinder — A hand-held grinding machine which uses portable wheels, raised hub wheels, plugs, or cones.

Portable Wheel — A grinding wheel used for portable grinding and stock removal. These wheels usually have hard durable abrasives, coarse grit sizes, organic bonds, and may have molded-in bushings.

Precision Grinding — Those types of grinding which result in the workpiece being ground to exact measurements, finish, etc.

Pre-engineered Grinding Wheels — Grinding wheels made to an established specification for cost-efficiency in handling a specific range of jobs.

Process E — Norton's trademark name for diamond and CBN grinding wheels and mandrels made by electroforming the abrasive directly to a preformed shape.

Production — The quantity of workpieces turned out, or the amount of work done in a given time. Also, the amount of work done during the life of a grinding wheel.

Profilometer — An instrument for measuring the degree of surface roughness in microinches.

Raised Hub Wheels — Type 27 or Type 28 grinding wheels with the area near the arbor hole offset to accommodate the mounting nut and flange. Used for light to medium portable grinding jobs.

Recessed Wheels — Type 5 or Type 7 wheels made with a recessed section on one or both sides to fit special types of flanges or sleeves on grinding machines.

Reinforcing — One or more layers of fiberglass material molded into the grinding wheel to add strength and stability in operations. Cut-off wheels, raised hub wheels, floorstand snagging wheels, and portable wheels use reinforcing. Cylinder wheels often have several bands of wound reinforcing on their periphery to add stability and protection.

Resinoid Bond — An organic bonding material made up of synthetic resin.

Rest — A part of the grinding machine used to support the workpiece being ground, or a dresser or truing tool when applied to the wheel. Also called a work rest.

Roll Grinding — A specialized type of cylindrical grinding used to finish or refinish the long, large diameter rolls used in metal rolling mills, paper mills, rubber and plastic processing and calendering plants.

Rough Grinding — A type of grinding used for stock removal, or where a precision dimension or finish is not the primary requirement.

RPM — The revolutions per minute or speed of a revolving object such as a grinding wheel.

Rubber Bond — An organic bonding material composed of natural or synthetic rubber.

Rubber Wheels — Grinding wheels made with rubber bond.

Safety Devices — Devices made to provide protective safety for a grinding machine operator. Also, devices attached to the grinding machine to provide safe operation and protection in case of malfunction.

Safety Flanges — Special types of flanges designed to hold together a grinding wheel in case it breaks, thus protecting the operator.

Saucer Wheel — A Type 13 wheel with a shallow saucer shape.

Saw Gummer — A grinding wheel used for gumming and sharpening saw blades.

Saw Gumming — A grinding operation for removing gum resins from wood-cutting saw blades and sharpening the blades.

Scleroscope — An instrument for determining the relative hardness of materials by a drop and rebound method.

Scratches — Marks left on a ground surface caused by a dirty coolant or a grinding wheel unsuited for the operation.

Scythe Stone — A long, narrow sharpening stone for sharpening or whetting scythes by hand.

Segments — Bonded abrasive sections of various shapes to be assembled to form a continuous or intermittent grinding wheel.

SFPM — Surface feet per minute, the distance moved by a point or grain on the circumference or face of a grinding wheel in one minute. Multiply the circumference in feet by the wheel revolutions per minute.

Sharpening Stone — A natural or manufactured abrasive product used for sharpening, honing and whetting tools and sharp edges.

Sharpening Stone Oil — A special mineral-base oil used to lubricate the sharpening stone and flush swarf from its surface.

Shellac Bond — An organic bonding material composed of shellac.

Silica — Silicon oxide found in silica sand. One of the raw material components of silicon carbide abrasives.

Silicon Carbide — An abrasive made from silica sand and coke, known as Norton Crystolon abrasive.

Snagging — A type of stock removal grinding used in foundries to remove gates, fins, and sprues from castings.

Steel Conditioning — A type of stock removal grinding used to condition the surfaces of steel billets after they have been poured and have cooled.

Stock Removal — The grinding or abrading away of material from a workpiece. Also general type of grinding in which stock removal is the primary requirement, as in rough grinding.

Straight Wheel — A Type 1 grinding wheel made with straight parallel sides, a straight face, and a straight or tapered arbor. Straight wheels contain no recesses, grooves, bevels, or dovetails.

Structure — A term designating the relative grain spacing in a grinding wheel, expressed as a number in the wheel marking. Dense structures have low numbers, while open structures have higher numbers.

Stub — That portion of a grinding wheel left after it has been used and worn down to the discarding diameter.

Surface Grinding — A type of precision grinding used to produce flat plane surfaces.

Swarf — Residue from grinding usually consisting of small chips from the workpiece, spent abrasive grains, and particles of bond.

Swing Frame Grinder — A grinding machine suspended from a chain at the center point so that it may be turned and swung in any direction for the grinding of billets, large castings, or other heavy work.

Table — That part of the grinding machine which directly or indirectly supports the work being ground.

Table Traverse — Reciprocating movement of the table of a grinding machine.

Tapered Wheel — A Type 4 grinding wheel shaped similar to a straight wheel, but having a taper from the hub of the wheel to the face, and thus being thicker at the hub than at the face.

Thread Grinding — A type of precision grinding used to generate screw-type threads.

Tool and Cutter Grinding — A type of precision grinding used to restore cutting edges and sharpen tooling and cutters.

Treatment — A material used to impregnate a grinding wheel so as to improve its grinding action, often by reducing the tendency for loading and glazing.

Truing — The process of shaping a grinding wheel to an accurate form. This is done to make the wheel face run absolutely true and grind without making chatter marks. A diamond tool is usually used for truing.

Universal Grinding Machine — A machine which can be set up to do several types of grinding, usually cylindrical, internal, and surface grinding. Genrally it is used for toolroom work.

Vitrified Bond — An organic bonding material, usually composed of clay. The bonding becomes vitrified when the wheel is fired in a kiln.

Wet Grinding — A kind of grinding where a flow of grinding fluid is directed over the wheel and workpiece.

Wheel Sleeve — A form of flange used on precision grinding machines where the wheel hole is larger than the wheel arbor. Usually the sleeve is so designed that the wheel and sleeve are assembled as one unit.

Wheel Speed — The speed at which a grinding wheel is revolving, measured in either RPM or SFPM.

Wheel Traverse — The rate of movement of the wheel across the work during grinding.

Wheel Type — A description of the overall shape and dimension of a grinding wheel, designated as a number. A Type 1 wheel is a straight wheel.

Work, Workpiece — The material, part, component, or piece to be processed by grinding.

Work Rest Blade — A blade used on a centerless grinding machine to support the workpiece between the two wheels during grinding.

Work Speed — In cylindrical, centerless, and internal grinding, the rate at which the work revolves, measured in RPM or SFPM. In surface grinding, the rate of table traverse measured in feet per minute.

ZF Alundum — A zirconia alumina abrasive alloy specially formulated for use in floorstand and swing frame grinding wheels.

Zirconia Alumina — An abrasive made by a proprietary Norton process where aluminum oxide and zirconium oxide are fused to produce a true abrasive alloy.

ZS Alundum — A zirconia alumina abrasive alloy specially formulated for use in steel conditioning and other high pressure grinding applications.

GRINDING WHEEL RECOMMENDATIONS

The most common grinding operations are listed below under the material being ground. The grinding wheel recommended for the particular grinding operation is listed at the right.

Alnico

Offhand	23AC36-N5B5
Cylindrical	23A54-L5VBE
Surfacing (straight wheel)	32A60-H8VBE
Cutting-off (wet) up to 7500 SFPM	23A80-L4R55
Cutting-off (dry)	23A60-L8B2
Disc – wet	23A60-HB14
Centerless	23A54-L5VBE
Internal	32A60-J8VBE

Aluminum
   Cylindrical   32A46-18VBE
   Centerless   32A46-L7VBE
   Surfacing (straight wheel)   37C36-J8V
   Portable Cut-off   A24-PBNA

General Reinforced Cut-off
   Less than 16 inches   A24-PBNA
   16 inches and over   A24-RBNAX286
   Surfacing (Discs)   37C24-JB14
   Internal   37C36-K5V
   Floor Stands   AC20-PBSX4
                                  #12 Treated
   Portable Grinders   A24-OBX4
                                  #12 Treated or
                                  A20-04B5
                                  #12 Treated

Bars
   Cutting-off (dry)
      9000–16000 SFPM   A244-Q6B9
                            #12 Treated
   Cutting-off (wet)
      7500–12000 SFPM   A46-P4R55

Aluminum Castings
   Disc
      Small – light work   37C24-IB14
      Large – heavy work   37C16-MB14
   Surfacing
      Cylinders, Cups, Segments   32A30-G12VBEP

Aluminum Oxide (Ceramic)
   Cylindrical – roughing   MD120S-N100M
           – finishing   SD320-R100B56
   Surfacing (wheels)
           – roughing   MD100S-N100M
           – finishing   SD320-R100B56
   Cutting   MD150-N25M

Aluminum Plate
   Surfacing – cylinders, cups,
      segments   37C30-H8V

Armature (Laminations)
   Cylindrical – roughing   32A100-18VBE
           – finishing   37C320-19E
   Internal   23A46-J5VBE or
                            23A46-I12VBEP

Balls for Bearings
   Semi-finishing, (soft – small) A100-Z69V
   Semi-finishing, (hard – large) A180-Z69V
   Final Finishing (hard – small) 37C240F-Z38V
   Final Finishing   37C400-Z38V

Ball Bearing Races
   Disc surfacing outer/inner
      races – soft   53A46-JB14
   Disc surfacing outer/inner
      races – hard   53A801-FB14
   Grind O.D. cups – roughing
      and finishing   57A80-M5VBE
   Grind outer race (oscillating)
      Wheel diameter
         1/4–7/8 inch   A901-T14R34
         Wheel diameter 1–3 inch   A150-R12R3 or
                              A901-T8R34
         Wheel diameter 3 inch
           and up   A1001-P4R30 or
                              A1001-R4R3
   Grind inner race (oscillating) A1801-R8R3
   Internal grind bore   23A80-K5VBE

Bars (Centerless)
   Hard or soft steel
      3/4 inch and smaller   23A54-SB17
      3/4–2 1/2 inch   23A54-RB17
      over 2 1/2 inch   23A54-QB17
   Aluminum, some tool and
      stainless steel   23AC54-QB17
   Steel tubing   23A60-0B17
   To achieve commercial or
      better finish   23A80-0B17

Bearings (Centerless)
   Pins (straight) up to
      1/8 inch O.D.
      roughing   23AC100-U9BH
      finishing   23AC100-TB17
   Pins (straight) over
      1/8 inch O.D.
      roughing   23A60-QB17
      finishing   A100-RB17
   Races O.D.
      roughing   32A60-K5VBE
      finishing   57A100-M5VBA
   Rollers
      roughing   23A60-M5VBE or
                              23A60-R2R3 or
                              2360-P6R30 or
                              23A801-R4R30
      finishing   A80-T6R34 or
                              A120-P4R30

Bolts (Screws and Studs)
   Cylindrical   57A60-N5VBE
   Centerless   23A60-06VBE

Brake Drums (Automotive)
  Regrinding   23A46-K5VBE

Brass
  Centerless   37C36-LVK
  Cylindrical   37C36-KVK
  Internal   37C60-L7V
  Surfacing
    Straight wheel   37C36-J8V
    Cylinders, cups, segments   37C24-H8V
  Surfacing (Discs)   37C24-JB14
  Snagging (Floor Stands)
    Up to 12,500 SFPM   AC20-PBSX4
        #12 Treated
  Portable Cut-off   A24-PBNA
  General Reinforced Cut-off
    Less than 16 inch   A24-RBNA
    16 inch and over   A24-TBNA
  Cutting-off Rod (Dry)
    9000–16000 SFPM   A241-T6B9
        R sides
  Cutting-off Rod (Wet)
    7500–12000 SFPM   A46I-N4R55
  Cutting-off Tubing (Wet)
    9000-16000 SFPM   A120-P6R55
  Cutting-off Tubing (Dry)
    7500–12000 SFPM   A80-P6R55

Brass Rod
  Cutting-off (Dry)   A241-T6B9 D sides
  Cutting-off (Wet)   A46-N4R55

Brass Tubing
  Cutting-off (Dry)   A80-P6R55
  Cutting-off (Wet)   A120-P6R55

Broaches
  Sharpening   38A46-K8VBE or
        32A60-K8VG

Bronze (Soft)
  Use same wheels as for brass.

Bronze (Hard)
  Centerless   32A46-M5VBE
  Cylindrical   23A46-M5VBE
  Internal   23A60-L5VBE
  Snagging (Floor Stands) to
    12000 SFPM   AC20-PBSX4
  Cutting -off (Dry)
    9000–16000 SFPM   23A30-P6B9
  Cutting-off gates and risers
    (Dry, Tabor R sides Mach.
    16000 SFPM)   A241-SBMA

Reinforced Resinoid
  Wheels   A24-RBNA
  Surfacing – cylinders, cups,
    segments   32A30-G12VBEP
  Surfacing (Straight Wheels)   32A36-K8VBE or
        23A36-J5VBE
  Surfacing (Discs)   37C24-JB14

Bushings (Hardened Steel)
  Centerless   23A60-M5VBE
  Cylindrical   23A60-L5VBE
  Internal   32A60-K8VBE or
        32A60-K5VBE

Bushings (Bronze)
  Centerless   37C46-OVK

Bushings (Cast Iron)
  Cylindrical   37C46-KVK
  Internal   37C46-J5V or
        32A60-K5VBE

Camshaft Bearings (Automotive)
  Cylindrical   23A46-N5VBE

Carbon
  Surfacing   37C16-LVK
  Surfacing (Discs)   37C36-JB14
  Surfacing Fine Finish   37C320-I7B

Carbon (Metallic)
  Cutting-off (dry)   37C24-O8B2

Carbon (Soft)
  Surfacing   BD60-N100M
  Cutting-off (dry)   37C60-N8B2
        D sides
  Cutting-off (wet)   37C46-K8B2
        D sides
  Cutting-off   BD60-L50M
        1A1R

Carbon (Hard)
  Centerless   37C36-NVK
  Cutting-off (dry)   37C60-N8B2
        D sides
  Cutting-off (wet)   37C46-K8B2
        D sides
  Cutting-off   BD60-L50M
        1A1R

Cast Iron
  Centerless   37C46-LVK
  Cutting-off (dry)
    9000–16000 SFPM   A30-T6B21
        R sides
  Cutting-off (wet)   A46-P4R55
  Cylindrical   37C36-JVK

Portable Cut-off	A24-TBNAX142	Note: Carbide only – Use ASDC or A2D in above	
General Reinforced Cut-off		specifications.	
Less than 16 inch	A24-RBNA	Surface Grinding Straight	
16 inch and over	A24-VBNA	Wheels	
Internal	37C46-J5V or	Roughing (wet)	ASD100-R100B56
	32A60-K5VBE	Finishing (wet)	ASD220-R100B56
		General Purpose (wet)	Gemini GTM 104,
Surfacing – cylinders, cups,			105, 106
segments		Cylindrical Grinding	
Ductile	23A30-F12VBEP	Roughing (wet)	ASD100-R75B56
Gray	37C30-H8V	Finishing (wet)	ASD220-N75B56
Chilled	37C30-H8V	Internal	
Ni Hard	32A36-F12VBEP	Roughing	RMD150-N100V5
Surfacing – disc	37C20-KB14	Finishing	RMD220-N100V5
Snagging (Floorstand) to		Centerless	ASD120-R50B56
9500 SFPM	ZF14-R5B23S	Hand Honing or Stoning	
Snagging (Swing Frame) to		Rectangular Hone	SD320-V5 or
9500 SFPM	ZF14-Q5B23S		37C280-P10V
Snagging (Portable Grinder)		Lapping	
7000–9500 SFPM	FP4310	Cup Wheel (wet)	SD400S-L50B56
Cemented Carbides		Norbide Abrasive Grain	
Single-point tools, offhand		and Cast Iron Disc	320F
Cup or plate mounted		Cutting-off	ASD100S-R100B56
wheels		Chain Saw Blades	
Roughing (wet and dry)	39C60-18VK or	Disc	57A36-LB14
	MD120-Process E	Chisels (Woodworking)	
Semi-finishing	39C100-H8VK	Sharpening	57A60-K5VBE
Finishing (Diamond		Clutch Plates	
Wheel)	RMD220-P50V5 or	Disc	23A36-IB14
	MD220-Process E	Commutators (Copper)	
General Purpose	Gemini GTM109	Roughing and Finishing	
Straight Wheels		(Wheel)	37C60-M4E
Roughing (wet and dry)	39C60-18VK	Copper	
Semi-finishing (wet and		Cylindrical	37C60-KVK
dry)	39C100-H8VK	Surfacing (Cups and	
Chip Breaker Grinding		Cylinders)	37C16-JVK
Straight Wheels		Surfacing (Discs)	37C36-HB14
Less than 1/4 inch thick	SD150-R100B11	Cutting-off	
1/4 inch and thicker	RMD150-N100V5 1/8	Rod (dry)	A46-P4R55
Milling Cutters, Reamers, etc.		Rod (wet)	A60-N4R55
Backing-off Cup Wheel		Tubing (dry)	A80-Q8R55
Roughing (Type 11V9)	ASD100-R75B76	Tubing (wet)	A120-P6R55
Finishing (Type 11V9)	ASD180-R75B76	Gates and Risers (Dry,	
Combination	ASD150-R75B76	Tabor Mach. 16000	
General Purpose		SFPM)	A24-VBNA
(Type 11V9)	Gemini GTM 101,	Cutlery	
	102, 103, 108	Knives – Butcher	
(Type 12A2)	Gemini GTM 107		

Hemming and Klotz
Machines
Carbon Steel and Stainless
Steel   57A801-D12VBEP
                           #22 Treated
Offhand
Surfacing Sides   A150-IVBA
Sharpening (Production)   A100-NVBA
Knives – Kitchen
Hemming Machine
Carbon Steel and Stainless
Steel   57A120-G8VBE
                           #22 Treated
Hollow Grinding   A60-F2RR
Knives – Pocket
Hemming Machine
Carbon Steel and Stainless
Steel   57A120-G8VBE
                           #22 Treated
Knives – Table
Hemming Machine
Stainless Steel   32A801-E12VBEP
                           #22 Treated
                           23A601-F4RR

Cutters
Sharpening (Machine)   32A46-K8VBE
Cutters (Molding)
Sharpening (Offhand)   57A60-N5VBE
Cylinders, Automotive
(Cast Iron) Internal
Regrinding (Wheels)   37C36-H8V or
                           37C46-G12VP
Honing (New Cylinders,
Sticks)
Commercial Finish   37C120-P8V
Mirror Finish   37C500-IV
Cylinders (Aircraft) Internal
Molybdenum Steel
Roughing   23A46-J5VBE
Finishing   32A60-I5VBE
Regrinding   23A54-I5VBE
Nitrided
Before nitriding   37C60-I5V
After nitriding   23A54-J5VBE
Regrinding   37C60-J5V
Drills (Manufacturing)
Cutting-off Soft (wet)   A46-R4R55

Cutting-off Soft (dry)   A241-V6B21 R sides
Cutting-off Hard (wet)   23A60-N4R55
Cutting-off Hard (dry)   23A46-P6B21 R sides
Cylindrical   23A54-N5VBE
Centerless (Soft)   23A60-N5VBE
Centerless (Hard)   32A80-N6VBE
Precision Sharpening   23A46-L5VBE
Fluting   32A1001-S8BH
Pointing
1/2 inch and smaller   38A60-L8VBE
Larger than 1/2 inch   23A36-L5VBE
Grinding Relief
1/2 inch and smaller   23A54-M5VBE
Larger than 1/2 inch   23A36-L5VBE
Drills (Salvaging)
Cutting-off – small   A60-OE7 or
                         A60-M8B2
Cutting-off – large   23A46-P6B21 R sides
Drills (Resharpening)
1/4 inch and smaller
– Machine   23A100-18VBE
1/4 inch and smaller
– Offhand   23A80-L6VBE
1/4 inch to 1 inch
– Machine   32A46-L5VBE
1/4 inch to 1 inch
– Offhand   23A60-L5VBE
Ductile Iron
Portable Cut-off   A24-TBNAX142
General Reinforced Cut-off
Less than 16 inch   A24-TBNA
16 inch and over   A24-VBNAX286
Duralumin (Tubing)
Cutting-off (dry)   A60-V10R20
Cutting-off (wet)   A80-V10R20
Ferrite
Cut-off   MD150-N25M or
                         PD240-Process E
Surface   MD150-N75M9
Internal   MD150-N75M9
Cylindrical   39C100-IVK or
                         MD120-N75M9 or
                         PD80-Process E
Centerless   MD120-N75M9
Disc   MD120-N50M9
Form Grinding Motor Arc
Segments   PD46-Process E

Ferrotic

Cylindrical	23A60-K5VBE
Surface	32A60-18VBE
Internal	32A80-J8VBE
Hardened State	ASD100-R100B56

Forgings

Centerless	23A60-N5VBE
Cylindrical	A46-M5VBE
Surfacing (discs)	23A16-JB14

Gages (Plug)

Cylindrical	32A80-K8VBE
Cylindrical, High Finish	37C500-J9E

Gages (Thread)

Grinding Threads 12 pitch and coarser	32A90-K8VBE
Grinding Threads 13–20 pitch	32A120-L8VBE
Grinding Threads 24 pitch and finer	32A220-M9VBE

Gages (Ring)

Internal

Roughing	32A60-L7VBE
Finishing	38A100-18VBE
Fine Finishing	37C320-J9E

Gears (Cast Iron)

Cleaning Between Teeth (Offhand)	37C24-T6R30

Gears (Hardened Steel)

Teeth – Form Precision	32A60-L8VBE or 32A80-J12VBEP
Teeth – Generative Precision Grinding	32A60-K8VBE or 32A80-J12VBEP
Internal	19A60-K5VBE
Surfacing (Cups and Cylinders)	32A36-18VBE
Surfacing (Segments)	23A36-J8VBE
Surfacing (Discs, Remove Burrs)	53A36-IB14
Surfacing (Straight Wheels)	32A46-J8VBE

Gun Barrels

Spotting and O.D.

Cylindrical	57A60-M5VBE

Internal

Grind Contours (Cartridge Chamber)	32A60-L7VBE

Hastalloy

Surface

Straight Wheel	32A46-H8VBE

Vertical Spindle	32A46-F12VBEP
Internal	23A54-L5VBE
Cylindrical	32A54-K5VBE
Centerless	32A54-L5VBE

Inconel (with H.D. soluble)

Surface

Straight Wheel	32A60-18VBE
Vertical Spindle	32A36-F12VBEP
Internal	32A60-J5VBE
Cylindrical	32A601-K5VBE
Centerless	32A601-L5VBE

Inconel X (with H.D. soluble)

Surface

Straight Wheel	32A60-H8VBE
Vertical Spindle	32A46-F12VBEP
Internal	32A60-J5VBE
Cylindrical	32A601-J5VBE
Centerless	32A601-L5VBE

Knives (Machine)

Cutting-off (dry)	A46-Q6B21 R sides
Cutting-off (wet)	A60-M4R55
Chipper and Barker, Sharpening	23A36-18VBE
Hog, Sharpening	A36-J5VBE
Leather Fleshing, Sharpening (Bricks)	A36-P5VBE
Leather Shaving, Sharpening – Cyl.	A54-P7VBE
Leather Splitting, Sharpening	A30-J3E
Molding, Offhand Sharpening	23A46-M5VBE
Machine, Sharpening	23A46-18VBE
Paper, Sharpening	23A70-18VBE
Section, Beveling	23A46-M5VBE
Surfacing Backs	23A46-18VBE

Lapping (General Purpose)

Aluminum	39C280-J9V
Brass	37C190-J9V
Cast Iron	37C180-J9V
Copper	39C320-J9V
Stainless	39C280-J9V
Steel	39C220-J9V

Lawn Mowers

Resharpening	23A60-M8VBE

Malleable Castings

Portable Cut-off	A24-TBNAX142
General Reinforced Cut-off	
Less than 16 inch	A20-SRN
16 inch and over	23A24-RBNAX286

Malleable Castings (Annealed)	
Floorstands to 12500 SFPM	A14-05B23S or
	4ZF14-Q5B23S
Swing Frames	
5000–6500 SFPM	A14-R5VBE
Swing Frames to 12500 SFPM	4ZF14-P5B23S
Disc	
Portable Grinders to	
9500 SFPM	A16-Q4BH or
	FP3410
Malleable Castings (Unannealed)	
Floorstands (to 12500 SFPM)	ZF14-P5B23S
Molybdenum	
Cylindrical	23A60-J8VBE
Surfacing	23A46-I8VBE
Monel Metal	
Portable Cut-off	23A30-RBNA
General Reinforced Cut-off	
Less than 16 inch	23A30-RBNA
16 inch and over	23A30-RBNA
Internal	37C60-J8V
Cutting-off (dry)	A301-R6B9 R sides
Cutting-off (wet)	A46-P4R55
Floorstands to 9500 SFPM	23A20-P5B23
Cylindrical	37C60-KVK
Nickel Rods and Bars	
Cutting-off (dry)	A46-Q6B21
Cutting-off (wet)	23A461-N4R55
Portable Cut-off	23A46-PBNA
General Reinforced Cut-off	
Less than 16 inch	23A46-PBNA
16 inch and over	23A36-PBNAX163
Ni Hard	
Cylindrical	23A80-K5VBE
Internal	32A80-K8VBE
Surfacing	
Wheels	32A46-I8VBE
Segments	32A36-F12VBEP
Pipe (Steel)	
Cutting-off (dry)	A301-T6B9 R sides
Cutting-off (wet)	A461-P4R55
Pipe (Soft Steel)	
Cutting-off (dry) Minimum	
Burr	A60-W10R20
Cutting-off (wet)	A80-Q8R55
Pistons (Aluminum)	
Cylindrical	32A46-I8VBE or
	23A46-J5VBE

Centerless	37C46-LVK
Regrinding	23A46-I8VBE
Pistons (Cast Iron)	
Cylindrical	37C46-J8VKP or
	37C36-KVK
Centerless	37C46-LVK
Regrinding	23A46-I8VBE
Piston Pins	
Centerless Machine	
Roughing	23A60-O7VBE or
	23A54-PB17
Semi-finishing	23A60-NB17
	57A100-M7VBE
Finishing	37C320-N8E
Surfacing Ends (Discs)	53A60-IB17
Lapping (Norton Hyprolap)	37CXF-K6E
Platon Rings (Cast Iron)	
Surfacing Rough (Cylinders)	32A30-H8VBE
Surfacing (Discs)	
Roughing	37C24-KB14
Semi-finishing	39C46-IB17
Finishing	39C80-GB17
Surfacing (Straight Wheels)	32A80-K8VBE
Lapping (Norton Hyprolap)	37C400-J9V
Internal Snagging	37C36-VVK
Pulleys (Cast Iron)	
Cylindrical	37C36-JVK
Roughing (Cylinder Wheel)	37C24-JVK
Finishing (Cylinder Wheel)	A54-P0R30
Reamers	
Backing-off	32A80-H12VBEP or
	32A60-K5VBE
Cylindrical	23A60-L8VBE
Rods (Centerless)	
Miscellaneous Steel	23A60-N5VBE or
	23A54-QB17
300 Series Stainless	37C54-NVK or
	32A54-M5VBE or
	23AC54-RB17
Nitralloy (Before Nitriding)	23A60-M5VBE
Silichrome Steel	23A60-O6VBE
	23AC54-RB17
Brass and Bronze	37C60-KVK
Hard Rubber	37C30-KVK
Carbon	37C36-NVK
Plastic	32A80-N7VBE
Rollers For Bearings	
Centerless – Roughing	23A80-M6VBE

– Finishing	A120-P4R30
Surface Ends (Discs)	23A100-JB14

Scissors and Shears (Cast Iron)

Surfacing Sides of Blades	37C100-S8V

Scissors and Shears (Steel)

Surfacing Sides of Blades (Cyl.)	A150-O7E3
Grinding Flash from Bows	23A46-Q5VBE
Pointing and Shaping	23A80-P7VBE
Grinding Neck or Corner	A120-N5E
Strike Cutting Edges	23A120-O9VBE
Resharpening	
(Small Wheels)	32A100-M7VBE
(Large Wheels)	57A901-MVBA

Shafts (Centerless)

Pinion	23A60-M5VBE or 23A601-QB17
Spline	23A60-O5VBE

Shear Blades (Power Metal Shears)

Sharpening (Segments)	23A30-H8VBE
Sharpening (Cylinders)	23A30-G8VBE

Skate Blades

Resharpening	23A100-L8VBE

Spline Shafts

Centerless	23A60-O5VBE
Cylindrical	23A60-N5VBE
Grinding Splines	23A60-L5VBE

Sprayed Materials (Wet Grinding)

General Nickel, Chrome, Cobalt, and Stainless Materials

Surface	39C46-H8VK or 32A46-H8VBE
Cylindrical	39C60-J8VK or 32A60-J8VBE
Centerless	39C60-J8VK or 32A60-J8VBE
Internal	39C80-L5VK or 32A80-L5VBE

Carbide, Zirconia, Tungsten, Alumina, Zirconate, High Cobalt, and Titania Materials

Surface	39C80-F8VK or ASD120-R75B56
Cylindrical	39C80-G8VK or ASD120-R75B56
Centerless	39C80-G8VK or ASD120-R75B56
Internal	39C80-J5VK or SD150-R100B56 or RMD180-N100V5

For dry grinding of sprayed materials it may be necessary to specify a grade 1 to 2 grades softer than those above.

Sprayed Metal

Internal	37C46-J5V

Steel Castings (Low Carbon)

Swing Frames to 9500 SFPM	A16-R5B23S or AZF16-O5B23S
Floorstands to 9500 SFPM	
Floorstands – 9500–12500 SFPM	4ZF14-P5B23S
Portable Grinders – 7000–9500 SFPM	A16-Q4BH or FP3410

Steel (High Speed, Tool)

	Wet	Dry
Cutter Sharpening		
Straight		
Wheel	CB100-TBB1/16″	CB100-TBB1/16″
Cup Wheel	CB100-WBB1/16″	AZTEC100-T1/16″
Roughing	XB150-Process E	XB150-Process E
Finishing	XB240-Process E	XB240-Process E
Dish Wheel	CB100-TBB1/16″	CB100-TBB1/16″
Surfacing		
Straight		
Wheel	CB100-TBB1/16″	CB100-TBD1/16″
Internal		
Mounted		
Points		
or	CB120-TBB	CB120-TBB
Straight		
Wheel		

Steel (Hard)

Centerless (Fine Finish)	A120-P4R30
Centerless (Com. Finish)	23A60-L5VBE
Centerless (Feed Wheel)	A80-RR51
Cylindrical (Smaller Wheels)	23A60-L5VBE
Cylindrical (Larger Wheels)	23A54-L5VBE
Internal	32A80-L8VBE or 23A80-K5VBE
Surfacing (Straight Wheels)	32A46-I8VBE or 32A60-G12VBEP
Surfacing – Cyl., Cups, Segments	
Broad Contact	32A36-E12VBEP
Narrow Contact	32A46-H8VBE
Surfacing (Discs – rough)	23A24-IB14

Surfacing
  (Discs – finish)   23A46-HB14
Cutting-off (wet)
  9000–12000 SFPM   A60-M4R55
Cutting-off (dry)
  9000–12000 SFPM   23A46-P6B21 R sides

Steel (Mild)
  Portable Cut-off   A24-TBNAX142
  General Reinforced Cut-off
    Less than 16 inch   A24-RBNA
    16 inch and over   A24-TBNA

Steel (Soft)
  Cylindrical
    Less than 1″ dia.   23A60-M5VBE
    Over 1″ dia.   23A54-L5VBE
  Disc
    Roughing   23A20-KB14
    Finishing   23A36-JB14
  Surfacing
    Straight Wheel   32A36-K8VBE
    Cylinders, Cups, Segments   32A24-G12VBEP

Steel (Stainless – 300 series)
  Centerless   32A46-K8VBE
  Centerless (Feed Wheel)   A80-RR51
  Cylindrical   32A46-J8VBE
  Internal   32A54-I8VBE
  Surfacing (Straight Wheels)   32A46-J8VBE
  Surfacing (Cups and
    Cylinders)   32A30-H8VBE
  Surfacing (Segments)   32A46-F12VBEP
  Cutting-off (dry)
    12000–16000 SFPM   23A36-R6B21 R sides

Steel (Stainless)
  Portable Cut-off   23A30-TBNA
  General Reinforced Cut-off
    Less than 16 inch   32A30-RBNA
    16 inch and over   23A30-TBNAX163 or
                         23A30-PBNAX286

Steel (High Speed)
  Centerless (Com. Finish)   23A60-L5VBE
  Centerless (Fine Finish)   A120-P4R30
  Centerless (Feed Wheel)   A80-RR51
  Cylindrical (14″ and smaller)   32A60-L5VBE
  Cylindrical (16″ and larger)   23A54-L5VBE
  Internal   32A80-K8VBE
  Surfacing (Straight Wheels)   32A46-H8VBE
                         32A60-F12VBEP

Surfacing (Cups and
  Cylinders)   38A46-G8VBE
Surfacing (Segments)   32A46-H8VBE
Cutting-off (soft-dry)
  12000–16000 SFPM   A301-T6B9 R sides
Cutting-off (soft-wet)
  9000–12000 SFPM   A46-Q4R55
Cutting-off (hard-dry)
  9000–12000 SFPM   23A46-P6B21 R sides
Cutting-off (hard-wet)
  9000–12000 SFPM   A60-M4R55
Cutting-off (wet)
  9000–12000 SFPM   A46-P4R55
Cutting-off (Tubing-dry)
  12000 SFPM or less   A80-V10R20
Cutting-off (Tubing-wet)
  9000–12000 SFPM   A80-Q8R55

Steel (Stainless – 400 Series Hardened)
  Centerless (Com. Finish)   23A60-L5VBE
  Centerless (Fine Finish)   A120-P4R30
  Centerless (Feed Wheel)   A80-RR51
  Cylindrical (Smaller Wheels)   23A60-L5VBE
  Cylindrical (Large Wheels)   23A54-K5VBE
  Internal   23A60-K5VBE
  Surfacing (Straight Wheels)   32A46-18VBE
  Surfacing (Cups and
    Cylinders)   32A36-G8VBE
  Surfacing (Segments)   32A24-G12VBEP
  Cutting-off (dry)   23A46-P6B21 R sides
  Cutting-off (wet)
    9000–12000 SFPM   A60-M4R55

Stellite (also Rexalloy, Tantung)
  Cylindrical   23A46-M5VBE
  Cutter Grinding   32A46-J8VBE
  Surfacing (Cups and
    Cylinders)   32A46-G8VBE
  Surfacing (Discs)   19A36-K5VBE
  Surfacing (Straight Wheels)   32A46-H8VBE
  Cutting-off (dry)   23A46-P6B21
  Cutting-off (wet)   A60-M4R55
  Tools
    Offhand   A46-N5VBE
    Machine   32A46-L8VBE

Taps
  Fluting (Small Taps)   A1001-V8R20
    (Large Taps)   A1001-W10R20
  Grinding Relief   32A80-K8VBE
  Squaring Ends   A60-N5VBE

Threading
  Precut
    2 through 6*   32A80-J8VBE
    8 through 12*   32A100-K8VBE
    14 through 24*   32A150-L9VBE
Taps
  Threading
    Solid
      4 through 12*   32A100-K8VBE
      14 through 20*   32A120-L8VBE
      24 through 36*   32A180-N9VBE
      40 and Finer*   32A220-N9VBE
    Shanks (Cylindrical)   23A80-M6VBE
Tools – Single Point
  Carbon and High Speed Steel
    Offhand Grinding
      Bench and Pedestal
        Grinders
          Coarse   A36-O5VBE
          Fine   A60-M5VBE
    Offhand Grinding
      Wet Tool Grinders
        12 to 14″ dia. wheels   A36-O5VBE
        Over 24″ dia. wheels   A24-M5VBE
    Machine Grinding
      Straight Wheels
        15″ dia. wheels   23A36-L5VBE
        Over 24″ dia. wheels   23A24-M5VBE
      Cup or Cylinder wheels 23A24-L8VBE
Tubing, Steel
  Centerless   23A60-N7VBE
  Cutting-off (dry)
    12000 SFPM or less
    Steel, finish unimportant A461-T6B9
    Steel   A80-W10R20
    Stainless Steel   A80-W10R20

    Chrome-Molybdenum A80-W10R20
  Cutting-off (wet)
    9000–12000 SFPM
    Steel   A80-Q8R55
    Stainless Steel   A80-Q8R55
    Chrome-Molybdenum A80-Q8R55
Tubing (Cut-off)
  Brass-Bronze
    Dry   A80-P6R55
    Wet   A100-P6R55
  Low Carbon, Minimum Burr
    Dry   A120-W10R20
    Wet   A801-Q8R55
  Low Carbon, Finish
    Unimportant
    Dry   A461-T6B9
    Wet   A461-R4R55
Tungsten
  Cylindrical
    Rolled Tungsten   32A54-L5VBE
    Sintered Tungsten   37C601-JVK
  Centerless
    Rolled Tungsten   32A46-N5VBE
    Sintered Tungsten   37C601-KVK
  Surfacing – 2000 SFPM 23A46-J8VBE
          – 6000 SFPM 37C46-J8V
Welds
  Carbon Alloy Steels
    Portable Grinders to
    9500 SFPM   A16-Q4BH or
                    FP3410

  Stainless Steel
    Portable Grinders to
    9500 SFPM   A16-R5B7 or
                    FP3610

APPENDIX 3

Whenever a single part or a complete project is to be made, a plan of some kind is necessary. The plan can be a simple hand sketch or an elaborate working drawing. Without a plan, too many errors can be made. These mistakes would require the use of extra stock, as well as the loss of time and the resulting frustration. It is much less frustrating and is much cheaper to make mistakes on paper where they are much easier to correct. Having a good plan to start with should naturally lead to the making of a reasonable order of operations, keeping in mind that a job can probably be done in more than one way. Projects 1 and 2 illustrate the use of project plans and show the value of listing and using an order of operations.

PROJECT 1

Figure 1 is a small spirit or bubble level. The sketch calls for three parts: a hexagon steel body, a steel plug, and a replacement bubble which can usually be purchased at most local building supply stores, or wherever carpentry tools are sold.

A typical order of operations is as follows:

Cutting the Stock

1. Cut the 5/8″ hexagon stock for the body, allowing at least 1/16″ extra in length to allow for facing or squaring the ends.
2. Select a scrap length of 1/2″ mild steel round stock for the plug and face, and square one end on the lathe. Also cut a 30-degree bevel to the single thread depth on the faced end.

Making the Body

1. Set the 5/8″ hexagon stock in a three-jaw chuck and make sure the stock is running true. If this can't be accomplished with the three-jaw chuck, switch to a four-jaw chuck. Make the work run true.
2. Face one end and partially round the end with a file. It would be safer to file left-handed.
3. Turn the job end-for-end, set it running true, and square the other end so that the total length of the job is 4″, as called for in the sketch.
4. Center drill the end and using a 29/64″ drill (the tap drill for 1/2″-20 UNF), drill a hole 3 1/2″ deep.
5. Select a 1/2″-20 UNF starting tap, insert it in the 29/64″ hole and align the tap accurately with a lathe center set in the lathe tailstock, by setting the lathe center in the small center hole in the end of the tap.
6. Using a wrench that fits snugly on the square end of the tap, turn the tap in about three turns. Make sure that the tap stays in alignment by turning the tailstock handle so that the lathe center moves with the tap. Use the proper cutting fluid for the job. Back off the lathe center and back the tap out of the hole in order that the chips can be cleaned out. Re-insert the tap in the hole and continue until the tap has penetrated 9/16″ deep. Next, use the plug tap and then the bottoming tap,

until the threaded hole is full depth for the entire 9/16″ length.

Making the Plug

1. The length of 1/2″ mild steel round stock can be placed in the bench vise and the 1/2″-20 UNF thread can be cut with a hand threading die. Allow at least 1/8″ extra in thread length and do not cut off the plug.

   This thread may also be cut on a threading machine. The more advanced student may wish to set up the job and cut the thread on a lathe. Whichever method is used, make sure the thread runs smoothly in the threaded hole of the body.

2. Screw the plug in to the bottom of the hole and cut off the excess stock. Face the end square and partially round the end with a file (the same as was done with the other end).

3. Cut the 1/16″ × 1/16″ slot in the plug, being careful not to cut the body of the level. This can be done carefully with a hacksaw, or the

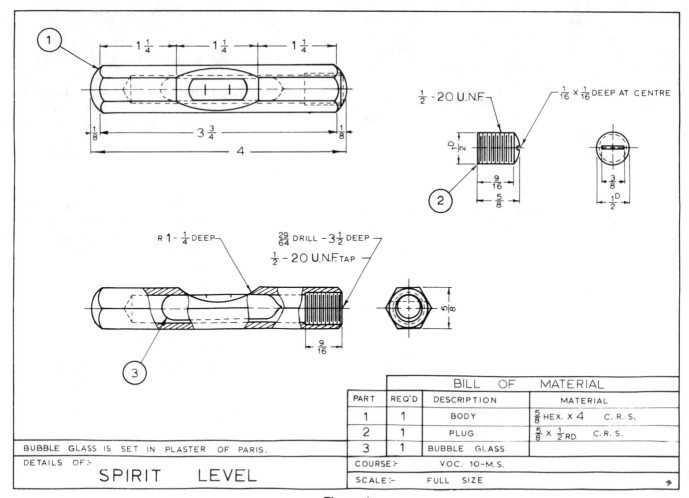

Figure 1

work can be set up on a milling machine where the slot can be milled.

Completing the Body

1. The curved recess can now be cut either by hand with a half-round file, or the work can be clamped to a horizontal milling machine worktable, where a 2″ diameter cutter, 3/4″ wide will quickly make the necessary cut.
2. The job should now be cleaned and polished, either by hand or by finishing on a surface grinder.

Setting the Bubble

1. To set the bubble in place, it is best to work on a surface plate and use a precision bubble level for comparison. Make a solution of plaster of paris and pour a little into the top of the body. Quickly place the bubble in through the curved recess. Level the bubble by comparing it with the precision bubble level, and let it set into position. If the bubble will not pass in through the curved recess, the plug can be removed, and the bubble can be inserted through the hole from the end of the body.
2. To complete the job, once the bubble has set into position, use a thin solution of plaster of paris to fill in any gaps. Make sure to use cold water when mixing the plaster of paris solution. If hot water is used, the plaster will set immediately, not allowing any time to place the bubble into position. After solidification and drying, scrape away excess plaster and do any final finish polishing which may be required.

Individual instructors can use this order of operations as a guide or they can set up their own order of operations. Variations in stock material are also optional. What is important is that a plan should be required for each project. In addition, some manner of order of operations is desirable. Students should be taught a proper order of doing a particular job.

PROJECT 2

Another project is the Plumb Bob, Figure 2, which consists of three parts: a brass body, a hexagon aluminum plug, and a carbon tool steel tip. There are three different sizes of plumb bob shown in the figure, but the medium one will do for our purposes. We will use the sketch and set up a suggested order of operations

Selecting and Cutting the Stock

1. Cut a piece of 3/4″ diameter brass about 4 1/8″ long for the body of the plumb bob.
2. For the plug select a random length of hexagon aluminum, 7/8″ across the flats.
3. Select a random length of 1/2″ diameter annealed carbon tool steel for the tip.

Making the Body

1. Set up the brass body in either a three- or four-jaw chuck, making it run true. Only about 1/2″ is to project out of the chuck.
2. Face the end of the body square and center drill.
3. Select a 1/2″ drill and drill 2″ deep into the body. Select a 33/64″ drill (the tap size drill for a 9/16″-18 UNF), and drill 1/2″ deep.
4. Select a 9/16″-UNF starting tap. Insert it in the 33/64″ hole and align the tap with a lathe center. Using a snug fitting wrench on the square end of the tap, turn the tap about three turns, keeping the tap in alignment with the lathe center. Back the lathe center out and unscrew the tap to clean out chips. Return the tap to the hole and continue tapping until the tap penetrates 1/2″ in depth, keeping the tap aligned with the lathe center. Do the same with the plug tap and the bottoming tap until a full depth thread is completed for the entire 1/2″ thread length. Use the appropriate cutting fluids where needed.
5. Next, set up to make a medium knurl. Pull the body of the plumb bob out of the chuck.

Grip it on the end, supporting the threaded hole with a revolving center. Figure 2 calls for a 2 1/8″ long knurl to within 3/4″ of the end. Run a medium knurl with a full diamond pattern. Remve the body of the plumb bob.

6. Switch the body end-for-end in the chuck, protect the knurl, set it running true, and face the body to the finished length of 4″. Center drill, then drill with a #3 drill (the tap drill

for 1/4″-28 UNF) to a depth of 3/8″. Thread the hole, keeping the tap in alignment with the tailstock.

Making the Plug

1. Set the 7/8″ hexagon aluminum in a three- or four-jaw chuck. Leave enough length for the total length of the job (1 3/16″), plus material for parting off. Face the end square.

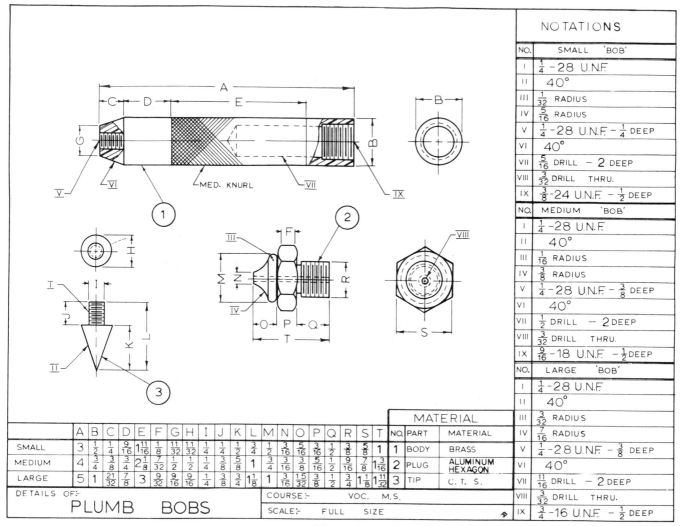

Figure 2

2. Turn the 9/16″ diameter (the thread diameter), 1/2″ long, and cut the neck for threading. This can be threaded on the lathe with a hand die, or the thread can be run easily on the lathe by setting up for threading. Cut the thread. Center drill the end and drill a 3/32″ diameter hole for the total length of the job plus the width of the parting tool. Use the appropriate cutting fluids where needed. Countersink the end of the hole very slightly to remove the sharp edge.

3. Set up the parting tool and cut straight in to the top of the bead, a diameter of 3/4″. Set the parting tool over to the full length of the plug (1 3/16″ measured to the right side of the parting tool). Cut straight in to a diameter of about 5/16″.

4. Set the angles (30 degrees) and cut the bevels on each side of the hexagon.

5. Set up and cut the radius of 3/8″, using a 3/8″ radius tool.

6. Cut the bead using a small bead forming cutter.

7. Finish parting off the plug.

8. Set a finished hexagon nut, 9/16″-18 UNF, (or the brass body) into the chuck and set it running true. Insert the plug into the threaded hole and finish face the end. Countersink the end of the 3/32″ hole slightly or just enough to remove the sharp edge. Remove the plug and hexagon nut or the brass body.

Making the Point

1. Set up the 1/2″ diameter carbon tool steel in the lathe chuck. Make sure that it is running true. Turn the 1/4″ diameter 3/8″ long. Cut the 1/4″-28 UNF thread, either with a hand die (leaving the job set in the chuck), or by running the thread with the lathe. Fit the thread to the threaded hole in the brass body.

2. Screw the tip into the brass body to the full depth of the thread. Cut off the excess carbon tool steel, leaving enough for the full length of the point which is 5/8″.

3. Set up the body in the lathe chuck with the point end out and running true.

4. Set up and cut the 40 degree point angle using the compound tool rest method.

5. Finish as desired and harden and temper the point.

Select about 6 feet of good twine (or other length as needed) and run it through the 3/32″ hole of the plug. Knot each end to prevent the twine from unraveling. When using the plumb bob, leave out enough twine for the job at hand. Tie a slip knot at the inner end of the plug, leaving the balance of twine inside the plumb bob out of sight and out of the way. Again, individual instructors can use this order of operations or set up their own, as desired.

PROJECT 3

The Dent Puller, Figure 3, is a slightly more complex project in that it involves five parts and some fitting. Some assembly is also required. All of the stock required is standard in size. Just the main dimensions are given in the sketch.

Cutting Off Stock

1. Part #1 is a standard sheet metal screw, available at any hardware store. The sketch calls for a 1″ long, #10, pan head screw. A round head screw is also satisfactory.

2. Part #2, the hexagon retainer, is made from 3/4″ hexagon steel. Cut a piece 1 3/8″ long.

3. Part #3, the slide hammer, is 1 1/2″ diameter mild steel. Cut a piece 4 1/8″ long.

4. Part #4, the slide, is 5/8″ diameter mild steel. Cut a piece 12 1/8″ long.

5. Part #5, the handle, is 1″ diameter mild steel. Cut a piece 3 1/4″ long. The extra length will provide a chucking grip for knurling.

Making the Hexagon Retainer

1. Set the stock in a three-jaw chuck, running true. Face the end square. Turn the taper

using the compound tool rest method.

2. Turn the stock in the chuck and face the work to full length, 1 1/4″ long.

3. Center drill the end, then drill 13/64″ (a clearance size for the #10 sheet metal screw); drill 29/64″ (the tap drill for 1/2″-20 UNF) 15/16″ deep. Set up and tap in the lathe, keeping the taps in alignment with the lathe center set in the tailstock. A machine tap can also be used directly in the tailstock spindle. Use the appropriate cutting fluids.

Making the Slide Hammer

1. Set the stock in a three-jaw chuck, running true. Face the end square. Reverse the work in the chuck and face the other end square and to the full length of 4″.

2. Center drill the end and drill 39/64″ (1/64″ smaller than 5/8″). The hole must be slightly larger than the slide for a sliding fit. This can be accomplished with an adjustable reamer, turning it through by hand and keeping it in alignment with a lathe center in the tailstock, just as is done with a tap. Use the appropriate cutting fluids.

3. Set up for knurling by gripping the work on the end in the three-jaw chuck, supporting the outer end with a revolving center. Set the knurling tool and knurl with a medium knurl. Use the appropriate cutting fluids.

4. The bevel and curved recess on each end are

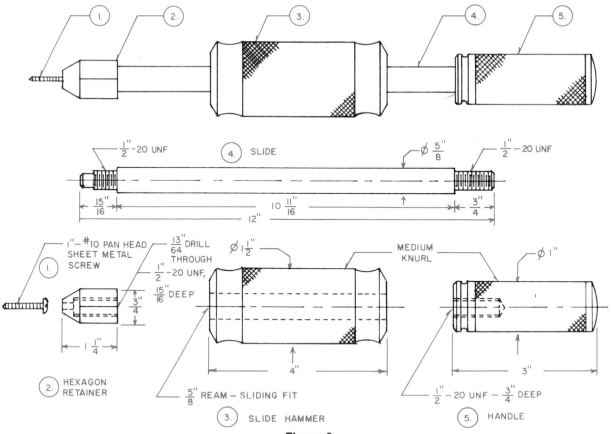

Figure 3

entirely optional since they are for design purposes only. Cut the bevel with the compound tool rest and the curve with a forming tool.

Making the Slide

1. The 5/8″ diameter slide requires the ends to be turned for the 1/2″-20 UNF threads. This can be done using a three- or four-jaw chuck, with the work running true. Turn the ends to the required length. Cut the threads either with a screw threading die or by running the threads with the lathe.
2. Threading to a shoulder is good practice if the thread is run on a lathe since the cross slide has to be pulled outwards at the same time as the threading lever is opened to stop the carriage movement.

Making the Handle

1. Set up the handle in a three-jaw chuck and face the end square. Center drill the end. Drill a 29/64″ hole (the tap drill for 1/2″-20 UNF) 3/4″ deep. Thread the hole, keeping the taps in alignment with the lathe center in the tailstock. Cut the half-round groove and bevel the end. These operations are optional since they are for design purposes only. Use the appropriate cutting fluids.
2. Pull the handle out of the chuck and grip it on the end. Support the threaded hole with a revolving center. Set up the knurling tool and knurl with a medium knurl. Use the appropriate cutting fluid.
3. Reverse the handle in the chuck taking care to protect the knurl from any damage. Face the end square and to the full length of 3″.

Assembly of the Dent Puller

1. Finish all parts as required.
2. The dent puller can now be assembled. It can be left assembled or can be dismantled and cleaned after each use. This is up to the individual.

In all of the projects presented in Appendix 3, it is also an option to use the metric measuring system if desired. It is an easy matter to substitute metric sizes for any of the fractional sizes.

PROJECT 4

Another simple and very common type of job done in a machine shop is the making of a bushing or bearing or sleeve, Figure 4. The main problem with this job, and with many other similar types of work, is to keep the outside diameter and the bore concentric. This means that bore and outside diameter must remain on the same center. Neither the bore nor the outside diameter can be even slightly off center. The same applies to the turning of a shaft with more than one step in it. There is a definite procedure for ensuring that all setps are on the same center. An order or sequence of operations is required.

Cutting or Selecting the Stock

1. Select a bronze casting which, in this case, is specially made for the bushing. It measures 1 3/4″ in diameter, is 2″ long, and has a cored hole 3/4″ in diameter.
2. As is often the case in a machine shop, the patternmaker makes the casting deliberately

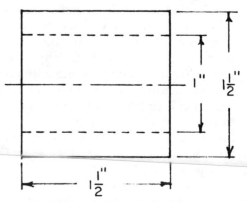

Figure 4 Bronze Bushing

oversize in outside diameter, undersize in the bore, and oversize in length. The reason for this is to provide the machinist with a certain amount of material that can be cut off so that a finished bushing will result. This material is called the machining allowance.

Making the Bushing

1. Set up the bushing material, preferably in a four-jaw chuck for the strongest grip. A large revolving center may also be required to support the bushing material for the rough turning operation.
2. Set up the cutting tool and rough turn the outside diameter, leaving the outside diameter at least 1/16″ oversize (or 1 9/16″ diameter). Do not turn a finished size at this time.
3. Rough face the end square with not too heavy a cut. Clean up the outer end.
4. Set up the boring tool and rough bore the inside diameter, leaving it at least 1/16″ undersize (or 15/16″ diameter). Do not bore a finished size at this time.
5. Now the bushing is rough turned, rough faced, and rough bored. It will be found that because of the rough cutting and the pressures involved, the bushing will have moved slightly despite the seemingly tight grip. Neither the outside diameter nor the inside diameter will be concentric with each other. Also, the face will not be square with either diameter at this time. This is the reason for allowing for the light finishing cuts to be taken. Not too much pressure will be exerted on the bushing during the light finish cuts.
6. Set up and finish turn the outside diameter to the exact 1 1/2″ diameter.
7. Set up and finish face the outer end of the bushing.
8. Set up and finish bore the inside diameter to the exact 1″ diameter.
9. Set up the parting tool, set the length of 1 1/2″

by measuring from the outside end of the bushing to the right side of the parting tool, and proceed with the parting cut. Go straight in part way and remove the tool from the cut. Lightly file the two outer corners of the bushing to remove the sharp edges. Do the same for the outside end of the bore with either a file or scraper. Finish parting off the bushing, allowing the slight leading edge of the parting tool to break through first to leave the ragged edge on the stock rather than on the bushing. This will leave the inner end of the bore clean but still sharp. Remove the sharpness with a file, or scraper, or emery cloth, as desired.

If this procedure is followed, the bushing produced will be concentric outside and inside, and the faces will be square with the outside and inside diameters. This is a sound reason for establishing and using a good order of operations.

PROJECT 5

The steel sleeve in Figure 5 has two steps and a hole through the center. This means there are three surfaces that must be concentric with each other. In addition, the two end faces must be at right angles with all three surfaces. The order of operations is similar to that given for the bushing in Figure 4.

Cutting the Stock

1. Select and cut off the steel stock for the work, 2 1/4″ diameter by 1 3/4″ long. These sizes include the machining allowance, and also provide for chucking the work.

Making the Steel Sleeve

1. Set up the work in a three-jaw chuck. Face the end square and center drill the end. Use the appropriate cutting fluid.
2. Select a small drill, approximately 3/16″ (or the size of a 15/16″ drill point), to act as a pilot

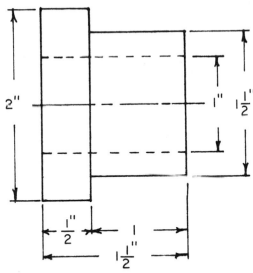

Figure 5 Steel Sleeve

operations, the sleeve has moved slightly despite the chucking and support of the revolving center.

6. Only light cutting is required now to finish the sleeve. Set up and finish the shoulder square between the two steps.
7. Finish face the end square using a light cut from the center outwards.
8. Set up and finish turn the small step to 1 1/2″ diameter, 1″ long.
9. Finish turn the large step to 2″ diameter, 1/2″ long.
10. Set up and finish bore to 1″ diameter.
11. Remove the sharpness of both outside corners of the two steps and also the inside corner of the bore at the outer end.
12. Reverse the work in the chuck so that the shoulder is against the chuck jaws. Set the work running true. Protect the outside diameter of the small step from being damaged by the chuck jaws.
13. Set up and finish face the end square so that the large step is 1/2″ long.
14. Remove the sharpness of the outer corner of the step and the inner corner of the bore. Finish all surfaces as required during finish cuts.

The sleeve is now complete. The outer diameters are concentric with the inner diameters and the ends are square with all diameters. This project again proves the value of using a proper order or sequence of operations.

drill. Drill the pilot hole through the sleeve, using the appropriate cutting fluid. Follow this with the 15/16″ drill. Drill through the sleeve using the appropriate cutting fluid.

3. Use a large revolving center to support the end of the sleeve with the tailstock.
4. Set up and rough turn the small step to 1 9/16″ diameter. Make the shoulder length 15/16″ long. Rough turn the large step to 2 1/16″ diameter.
5. The sleeve is now roughly machined to size. It will be found that because of the pressures involved during the drilling and turning

GLOSSARY

Accuracy: correctness, exactness.

Accurate: correct, exact.

Acme Thread: one of the power or translation screw threads with an included angle of 29 degrees.

Acme 29-Degree Thread Gage: a gage used to grind and set up an Acme screw threading tool.

Adapter: a tool that will fit or adapt another tool to a machine; e.g., fit a milling cutter or cutters to a milling machine spindle.

Addendum: "to be added"; in gearing, that part of the gear tooth above the pitch circle to the outside diameter.

Alloy: a metal made by mixing and fusing two or more metals together.

Aluminum Oxide: a manufactured electric furnace abrasive, probably the most widely used abrasive.

Angle Plate: used to support work in a vertical position.

Angular: having angles between two lines that are joined at one end.

Angular Indexing: the indexing of angles that is performed on a dividing or index head on a milling machine.

Angular Milling: the milling or cutting of any inclined surface on a milling machine.

Annealing: the process of heating and slow cooling steel for the purpose of softening and refining grain structure.

Approximate: nearly correct.

Arbor: the main shaft of a machine; in milling, the shaft on which the rotating milling cutter is held.

Arc of Contact: in grinding, the portion of the grinding wheel that is in contact with the work.

Babbitt: a soft bearing metal made essentially from lead, tin, and antimony.

Ball Pein Hammer: the most common striking tool used in a machine shop; the rounded top is called the pein and the flat bottom is called the face.

Band Machining: cutting with a rotating bandsaw blade; also band filing with rotating band files and band polishing with emery belts.

Bands (Fringe Patterns): a series of alternate light and dark bands or fringes that appears on a metal surface when a monochromatic light source shines through an optical flat to the work surface reflecting the light back up through the optical flat to the eye.

Base Line: in layout work, the line (a machined edge or scribed line) from which measuring and layout are done.

Bearing: a part of a machine on which another part will rotate or slide.

Bessemer Converter: a container for converting iron into steel by the Bessemer process which was the first widely used commercial method for producing steel in quantity.

Blast Furnace: a tall, cylindrical furnace used to convert iron ore into pig iron by blasting air from the bottom up through the ore, producing extremely high heat.

Bond: in grinding, the material that binds or holds the emery grits together.

Boring: to enlarge a hole by means of a cutting tool that revolves, or by means of a cutting tool that is held in a boring bar.

Bur: a rotary cutting tool used in a hand grinder.

Burr: a ragged edge left on the work or stock after a cut has been made.

Butt Welder: in band machining, a device to hold and weld saw bands.

Caliper: a measuring tool used to measure the diameter or thickness of the work.

Carbide: a blend of tungsten and carbon powders bound together with powdered cobalt to produce a cemented carbide cutting tool.

Carbon: a common nonmetallic element that is added to iron to give iron and steel their various characteristics.

Carbon Tool Steel: a group of carbon steels containing from 0.5% to 1.6% carbon that are usually used for making cutting tools for various uses.

Carriage: in lathe work, a sliding part of the machine that carries the tool bit into the rotating work.

Case Hardening: increasing the hardness of a steel surface by changing the composition of the surface by adding carbon (C) or nitrogen (N) or by adding both carbon and nitrogen (CN). The processes are called carburizing, nitriding, and cyaniding.

Cast Iron (gray, white, malleable, nodular): usually made from pig iron and cast into its final shape.

Cathead: an outer support bearing for supporting a shaft when using a steady (center) rest. The support arms of the steady rest run on the cathead rather than directly on the shaft.

Cemented Carbide: powdered tungsten carbide blended with powdered cobalt, compacted into shape, and heated in an oxygen-free furnace to cement the particles together.

Cementite: combined carbon and iron, which has great strength, is very brittle, and is very hard.

Center Drilling: drilling a center hole in the end of a shaft to act as a support for that shaft, or to act as a starting hole for another drill when drilling that shaft.

Center Gage: a small gage used to grind a thread cutting tool to shape and also used to set the cutting tool accurately to the work.

Center Head: a V-shaped instrument with a scale through the middle of the V.

Center Punch: a pointed punch used to mark the center for a hole to be drilled or punched.

Ceramic Cutting Tools: made primarily from aluminum oxide, ceramics are the hardest of all cutting tool materials currently used in industry, except for titanium carbide and diamond.

Chatter: vibration of the cutting edge of a metal cutting tool.

Chip Breaker: notches or shoulders used to break a chip into smaller units.

Chip Curler: a curved rake used to tightly curl a chip.

Chips: the cuttings made by any cutting tool.

Chord: a straight line between two points on the circumference of a circle.

Chuck: a holding device used for holding either the work, as in a lathe chuck, or the cutting tool, as in a drill chuck.

Circular Pitch: the distance from the center of one tooth to the center of the adjacent tooth measured on the pitch circle.

Circular Pitch System: a system for measuring gear teeth.

Clearance: a relief ground under the cutting edge of a cutting tool to permit the tool to cut without rubbing.

Clearance Angle: the angle at which the relief is ground.

Climb Milling: when a rotary milling cutter tends to cut down on the work as it moves into the work.

Clutch: a mechanism for engaging or disengaging a mechanical movement.

CNC (Computer Numerical Control): the operating of a machine tool by computer after the taped data has been transferred to the computer memory.

Cold Chisel: a very hard hand cutting tool for cutting metal.

Combination Square: a square with 90 degrees on one side and 45 degrees on the other, and an adjustable scale.

Combination Wrench: a combined open-end and box-end wrench.

Comparator: a gage for testing repetitive sizes of duplicate parts.

Concave: a surface such as a hollow surface.

Conventional Milling: when a rotary milling cutter tends to cut upward as it cuts into the work.

Convex: a surface that is curved outward like the outside of a circle.

Coolant Feeding Drill: a heavy-duty drill with oil holes running through the lands to the cutting lips.

Counter Bore: a parallel enlargement of an existing hole, forming a flat-bottomed recess for the head of a cap screw.

Countersink: an angular enlargement of an existing hole that forms an angled recess for a flat-head screw or bolt.

Cupola Furnace: a tall cylindrical furnace like a small blast furnace used specifically for the making of cast iron.

Cutting Fluid: a fluid used to improve the cutting action of a cutting tool, sometimes by cooling, sometimes by lubricating, or by a combination of both.

Cutting Tool: any tool in a machine shop that does the actual cutting, e.g., tool bits, carbide tools, grinding wheels, drills, milling cutters, etc.

Dedendum: that portion of the gear tooth from the pitch circle to the root diameter.

Dial: a graduated surface on which a pointer moves, usually in a rotary movement, for the purpose of accurate measurement, or graduated dials as found on the handwheels of machine tools.

Dial Indicator: an extremely sensitive, high precision measuring dial used principally for accurate setup of work or for comparing the size of duplicate parts.

Diametral Pitch System: a system for measuring gear teeth; it is the number of teeth for each inch of pitch diameter; e.g., a 30-tooth gear with a 3-inch pitch diameter would have a diametral pitch of 10.

Die: one of a pair of cutting or shaping tools operated by being pressed or driven toward each other; the smaller tool that enters the other is called the punch, the larger tool is a die. A die can also be a hollow, internally threaded tool used for cutting screw threads on bolts or studs.

Dielectric: a fluid in electrical discharge machining.

It acts as an insulator between tool and work, providing a path for the spark, acts as a coolant, and flushes away the chips.

Die Stocks: the handles that hold and rotate the threading die.

Dimension: a measurement of length, width, or thickness.

Dividers: a two-legged instrument for measuring distances. Both legs are pointed.

Division(s): If an inch is divided into 16 parts, there would be 16 divisions, or each part would be a division.

Dressing Wheels: in grinding, it exposes new cutting grits, squares the grinding wheel, makes the grinding wheel run true or concentric with the center.

Drill Chuck: a holding device for holding and rotating straight shank twist drills on a drill press.

Drill Drift: a tapered tool used to remove a taper shank twist drill or drill chuck from a sleeve or from a drill press spindle.

Drill Point Gage: a gage used to measure point angle and length of cutting lips.

Drill Press: a machine tool used to hold, rotate, and press a twist drill into the work.

Drill Press Vise: a tool used to hold the work in position to prevent the work from rotating while it is being drilled.

Drill Sleeve: a tool used to enlarge the taper shank of a twist drill, reamer, or drill chuck.

Drill Socket: a tool used to adapt a large taper shank to a smaller taper socket.

EDM (Electrical Discharge Machining) (conventional and wire cut): the cutting of metal with electric sparks, either with an electrode moving down into the work, or with a copper wire (through the work) cutting sideways into the work.

Electric Furnace: in steel making, an electric arc through carbon electrodes that generates heat.

Face Milling: the cutting face of the rotary cutter is perpendicular to the centerline of the spindle which can be either in a horizontal or vertical position.

Face Plate: in lathework, a circular, slotted plate to

which the work can be clamped in order to rotate it for cutting.

Facing: in lathework, cutting perpendicular to the centerline of the main spindle. Facing is sometimes called "squaring."

Feed: the movement of a cutter into the work. In lathework, feed is measured in inches per revolution and in milling, it is measured in inches per minute.

Feed Clutch: a mechanism for starting or stopping the feeding movement.

Feed Rod: in lathework, the rod connecting the quick-change gear box to the apron.

Ferrite: free iron.

Ferrous Metals: any combination of metals containing iron as one of the major elements.

File: a hand cutting tool for flattening and smoothing a surface.

File Card: a wire brush for cleaning a file.

Fit: the closeness with which mating parts fit together.

Fixture: a mechanical device that is firmly attached or fastened to a machine to become a permanent part of that machine.

Flame Spraying (metallizing): the spraying of material onto a prepared surface for the purpose of building up or coating that surface.

Flat Plane: any flat or level surface.

Float Lock Vise: a safety vise attached to a drill press table that is used to hold the work and prevent it from revolving.

Flute: the grooves in a cutting tool, as in a drill, reamer, or tap.

Follower Rest: a support for turning long shafts; clamped to the saddle, it follows the tool bit.

Form Milling: the cutting of a surface that has a regular or irregular outline.

Friction Sawing: cutting on a band machine by friction.

Friction Thimble: a thimble that will slip and revolve when the correct tension is reached with a micrometer reading.

Fringe Patterns: *see* Bands.

Gage: a standard measure or a table of standard measurements or a measuring instrument used to show measurements.

Gage Block: a block made to exacting measurements, accurate to plus or minus 0.000001 in (plus or minus 0.00003 mm), one of the uses being to test other precision instruments for accuracy.

Gear: a disc or wheel with teeth on its periphery (outer diameter) that will mate with another gear to transmit movement from one to the other.

Gib Key: a driving key, tapered, with a head on one or both ends.

Graphite: a soft, black, greasy form of carbon.

Grinding: cutting by abrasive grits that are bound together in the form of a wheel called a grinding wheel.

Hacksaw: a hand saw used for cutting metal.

Hammer: a striking tool with a handle.

Hand Scraping: removing a small amount of metal from a surface with a hand scraper.

Hardening: increasing the hardness of metal, usually by heating and quick cooling (quenching), followed immediately by tempering or drawing.

Headstock: that part of the lathe that contains all of the speed mechanisms and the main spindle.

Helical Milling: a type of gear cutting in which the gear blank is fed into the cutter while at the same time the gear blank is revolved slowly to attain a curved tooth.

Hermaphrodite Caliper: a caliper with one leg curved at the end and the other leg pointed with a scriber point.

High-Speed Sawing: a type of bandsawing that uses speeds of 2000 to 6000 sfpm.

High-Speed Steel: an alloy steel used principally for making cutting tools.

Increment(s): the amount by which something increases.

Indexing: in milling work, the dividing and rotating of a gear blank; e.g., a fixed amount for each cut (or tooth) so that on completion of all cuts, all gear teeth are equal in size.

Indiscriminate: no fixed pattern or rules of order.

Interference Bands: alternate light and dark bands that appear when a light beam (single wavelength) is directed through an optical flat to a steel surface.

Involute Gear: a gear with a tooth that has an involute curve on its flank so that each tooth actually rolls into its mating tooth.

Jig: a mechanical device that is usually associated with drilling. It usually holds the work in position and also guides the drill into the work.

Jig Bore: a machine that is similar in concept to a vertical milling machine but much more sensitive and much more accurate. The essential operations are drilling, reaming, and boring.

Kerf: the cut made by a saw blade.

Keyseat Clamp: a clamp used to hold a rule parallel to the centerline of a round shaft.

Knurl: a pattern formed on the outside of a shaft in the form of a diamond or a straight pattern, not by cutting but by impressing the rollers into the surface.

Knurling Tool: an impression-type tool with patterned rollers used to form a knurl.

Land: the area of the drill body between the flutes, containing both the margin and the body clearance.

Lathe: the basic machine tool in a machine shop. The lathe is used essentially for revolving the work so that it can be cut.

Lathe Bed: the main support or backbone supporting the headstock, tailstock, and carriage of the lathe.

Lathe Center: a 60-degree center used mainly for supporting and revolving work on centers.

Lathe Dog: a rotating clamp used to drive or rotate a shaft while it is being supported between centers.

Layout: a plan or design drawn on the work to act as a guide for the machinist to work to.

Leadscrew: on a lathe, the threaded shaft connecting the quick-change gear box to the apron to provide the driving power for cutting screw threads.

Light Wave Measurement: the most accurate method of measuring the flatness of a surface by the use of light waves.

Limit(s): the minimum or maximum sizes that the work can be cut to.

Linear: a straight line measure.

Machine Tool: a machine such as a lathe that does the work of machining.

Machining: the work of cutting or forming the work into its final shape.

Machinist: the tradesman who does the job of machining.

Maintenance: the work of maintaining or keeping a machine in good working order.

Mandrel: a shaft used to support and rotate the work between centers.

Measure: to find out the size of the work by using a standard of measure.

Measurement: the actual size as compared to a standard of measure.

Metallurgy: the study of the composition and manufacture of metals from the ore in the ground to the finished product.

Micrometer: a precision measuring tool for the accurate measurement of diameter or thickness of materials.

Mike: a term for measuring in millionths of an inch. One millionth of an inch is one mike.

Milling: a type of cutting that makes use of a revolving type of cutter with one or more teeth.

Module Gearing: in gearing, a standard of measure that is an actual measurement either in English (U.S. Customary) or metric.

Monel: a nickel alloy containing nickel and copper.

Monochromatic Light: a light consisting of only one wavelength used in light wave measurement.

N/C (Numerical Control): the operating of machine tools by means of a punched tape or card inside a control mechanism where actions are caused by numerical data punched on the tape or card.

N/C Control: the heart of numerical control machining; it collects the numerical data from the tape, interprets the data, and distributes the interpretations to the various control mechanisms on the machine tool to activate or stop the movements.

Nonferrous Metals: any combination of metals containing little or no iron.

Normalizing: returning tool steel to a normal condition after forging has changed the structure.

Open Hearth Furnace: a type of furnace used in the production of steel.

Optical Flat: a highly polished, fused, quartz, optical glass used in light wave measurement to test the flatness of surfaces.

Outside Diameter: any diameter that forms the maximum diameter of an object.

Parting: cutting off material in a lathe with a blade called a parting tool.

Pearlite: one of the cast-iron group of metals composed chiefly of ferrite with some cementite.

Pig Iron: the product of the blast furnace that resembles cast iron. It is sent to the iron foundry to be refined into cast iron, or it is sent to the steel plant to be refined into steel.

Pilot and Pilot Bushing: tools used to keep a boring bar running true.

Pilot Hole: a small hole equal in size to the web of a drill. Its purpose is to remove the metal normally encountered by the point of the drill.

Pitch: in sawing, the number of teeth per inch on a saw blade; in threading, the distance from point to point of adjacent threads; in gearing, usually the diametral pitch (the number of gear teeth per inch of pitch diameter).

Pitch Circle: the circle of a gear on which all calculations and measurements are made.

Pitch Diameter: the diameter of the pitch circle.

Plain Milling: the cutting of a flat surface parallel to the axis of the cutter. In some cases, this is called slab milling.

Pliers: a tool used to hold, turn, or bend objects.

Point of Contact: where the point of a cutting tool actually contacts the work.

Point Thinning: thinning the point of a drill to relieve pressure on the drill.

Powder Metallurgy: the forming of metal objects by blending metal powders, placing those powders in a mold, exerting tons of pressure to form the object, ejecting the object from the mold, and heating the object in a furnace to weld the powders together.

Precision: the accuracy or exactness with which a machine tool or machine part is made.

Prick Punch: a layout punch with a longer point than usual, used chiefly for marking lines.

Protractor: a precision measuring tool for measuring angles.

Rake: the slope of the top of the cutting edge of a cutting tool which can be a positive, zero, or negative slope depending on the application and on the type of cutting tool.

Ratchet: a releasing type of driving mechanism that will slip after a certain pressure is reached, thereby protecting the tool involved.

Reamer: a cutting tool used to enlarge a hole to the correct size.

Reference Surface: a flat surface such as a surface plate that provides a surface to measure from or to layout from. All measurements start from this flat surface or reference surface.

Refraction: the bending of a light ray as it passes from air through something more dense like a glass prism.

Relative: to relate to or compare to; e.g., "accuracy is relative to the work." You would not use a micrometer to measure for cutting off stock. A scale would be close enough.

Rotary Table: a rotating table used for circular and rotary machining operation.

Scraper: a hand cutting tool used to flatten or decorate a surface by means of a scraping action that removes a very small amount of metal.

Screw Pitch Gage: used to measure the number of threads per inch or the pitch in millimeters.

Screw Thread: a groove of specific shape or form cut or rolled on the outside of a cylinder or on the inside walls of a hole in the form of a parallel or tapered helix.

Scriber: a marking or scratching tool used for scratching a line on a surface when laying out.

Scribing: the act of using a scriber.

Sensitive Drill Press: a drill press with hand feeds only.

Set: the amount that the saw teeth are offset from side to side to give clearance for the saw blade in the kerf or cut.

Silicon Carbide: an abrasive made from coke (C) and silica sand (SI); e.g., Norton's Crystolon (SiC).

Sine Bar: used with gage blocks for the accurate testing or setting up of angles.

Sine Plate: an adjustable plate that will permit angular setups to be made precisely and easily; both single and compound angles can be set.

Sledgehammer: a large, two-faced, striking hammer with a long handle.

Spacer Collar: any plain collar that takes up space; e.g., spacers on a milling machine arbor that take up space on both sides of the milling cutter.

Spade Drill: a flat, replaceable drill with two sharp cutting lips mounted in a holder.

Speed: revolutions per minute of the chuck and work on a lathe, or of the milling cutter on a milling machine.

Spider: three or more adjustable arms radiating out from a center hole and used to support the end of pipe or tubing on a lathe.

Spigot: a live center gripped in a lathe chuck, turned with a maximum taper of 3 degrees.

Spindle: on a lathe, the main spindle supports and drives the chucks and face plates that rotate the work; in a milling machine, the main spindle drives the arbor and milling cutters.

Spline: a long, round shaft with multiple, evenly spaced cuts running lengthwise along a part or all of its entire length such as long gear teeth.

Split Nut: a long nut in two halves attached to the back of the lathe apron with the leadscrew passing through it.

Spot Drilling: drilling a hole exactly on center using a proper hole layout as a guide.

Spot Face: machining the face or faces perpendicular to the hole so that a bolt head and/or nut has a square face to be tightened to.

Square: a precision tool for testing the trueness of faces at right angles to each other or for squaring lines perpendicular to a straight edge or face.

Steady (Center) Rest: a support attachment for a lathe for supporting a shaft that is too far out of the chuck.

Stock: the material from which a project is made.

Straight Shank: as in a twist drill, a parallel shank to be gripped in a drill chuck.

Surface Gage: a gage used in layout for scribing lines parallel to a reference surface (surface plate), for setting work level in vises, and for setting revolving shafts to run true on lathes.

Surface Plate: a precision, flat plate usually made of cast iron or granite that is used in layout and testing. It provides an accurate, flat surface to measure from.

Swing: the maximum diameter of work that can be swung or revolved over the ways of a lathe.

Tailstock: a support mechanism on the ways of a lathe that is used for supporting a long shaft or for drilling and reaming; also used on a milling machine as a support and called a footstock or tailstock.

Tap: a cutting tool used for cutting a screw thread in a hole of specific size called a tap drill hole.

Tap Drill: a standard twist drill of a specific size for a particular size and coarseness of screw thread.

Tape: in N/C and CNC, the punched tape that feeds through the control designating the various movements to the machine tool.

Taper: a difference in diameters over a unit length.

Tapering: the act of cutting a taper on a machine tool.

Taper Shank: in drilling, a Morse taper shank that can be jammed in a socket or spindle to drive a drill.

Tap Wrench: used specifically for rotating a tap to cut a screw thread; also for revolving a square shank reamer for enlarging a hole.

Technique: a method of doing a particular operation; e.g., grinding a tool bit is an individual technique.

Temper: to draw some of the hardness out so as to have the correct hardness for a particular steel.

Tempering (Drawing): the act of tempering or drawing some of the hardness out of hardened steel.

Thermoplastics: plastics that become soft when

exposed to heat. They reharden when cooled and can be reshaped many times.

Thermosets: plastics that set or cure to permanent shape by heat.

Thimble: the rotating part of the micrometer that is attached to the spindle.

Thread Dial: a chasing dial connected to the leadscrew of a lathe that indicates when to close the split nut so that the thread cutting tool starts in the right place for each cut.

Threading: the actual cutting of a screw thread, whether by hand or by machine.

Tool Bit: a single-point cutting tool used on a lathe, and as a fly-cutter on a milling machine.

Tool Post: a device for holding the tool holder in position on a machine tool.

Tool Steel: a steel with 0.5% to 1.6% carbon that is capable of being hardened and tempered for use as a cutting tool.

Tooth Form: the shape of a tooth, as in a gear cutter or saw tooth.

Trammels: a type of dividers or calipers with a beam of metal or wood to connect the two legs together.

Transfer Calipers: a type of caliper with one of the two legs being capable of movement so that a groove can be calipered; the leg can be moved and replaced in position so that the calipers can be removed from the groove and the leg replaced in position to retain the calipered size.

True: an exact or accurate formation, position, or adjustment.

Truing: the act of making an exact or accurate formation, position, or adjustment.

Turned Center: a piece of round steel placed in a chuck and used for accurate turning between centers.

Turning: the act of cutting an outside diameter on a lathe.

Turret Tool Post: a type of post capable of being revolved and having more than one station for mounting more than one cutting tool at a time.

Universal Milling Machine: a horizontal milling machine with a table that can be swiveled in either direction so that helical milling can be done.

V-Block: a block used to hold a round shaft parallel to a surface plate or worktable.

Vernier: an extra scale on a measuring tool with sliding or rotating action; it is extremely accurate.

Vise: a holding device to hold the work firmly in place while being worked on.

Ways: on a lathe, the slides on top of the bed on which the carriage sits and moves.

Whole Depth: the whole depth of the gear tooth of a gear which includes the addition of the clearance. This is the depth to which the gear cutter is set.

Working Depth: the depth to which two mating gear teeth roll into each other. This does not include the clearance.

Wrench: a tool for loosening or tightening nuts and bolts.

Wringing: when two gage blocks are rubbed together in a specific way, they are so accurate in flatness that they will stick together. This is called "wringing gage blocks."

INDEX

Abrasive cutting-off wheels, 123
Abrasive grits, 461, 462–63
Absolute programming, 500–501, 501–2 *fig.*
Accuracy, degree of, 12
Acme thread, 304, 305 *fig.*
Acrylic plastics, 609
Adjustable reamers, 200
Air wedge, 66, 64 *fig.*
Algorithms, 519–20
Allowance, 50–51
Alloy steels, 575–76
Aluminum, 600–602, 601 *fig.*
Aluminum oxide, 462
American pattern files, 111
American Standard Code for Information Interchange
    (ASCII), 499, 500
Amino plastics, 608
Angle plates, 93, 94 *fig.*
Angular cutters, 371, 374 *fig.*
Angular indexing, 402, 404, 406
Angular milling, 369, 394, 396 *fig.*
Annealing, 584, 610–11
    saw blade weld, 137–38, 138–39 *fig.*
Aqueous cutting fluids, 361
Artificial machine vision, 592–94, 592–94 *fig.*
Auto mode, 521
Automated guided vehicles (AGV), 196–97, 196–97 *fig.*
Automatic center punch, 79, 80 *fig.*
Automatic Speech Recognition (ASR), 524
Axes, 388
Axis orientation, 502–3, 502–5 *fig.*

Babbitt, 363
Babbitt bearings, 604–5
Backlash eliminator, 395–96
Ball pein hammer, 77, 99, 99 *fig.*
Ball penetrator, 587, 590 *fig.*
Band machining
    band machines, 123, 124 *fig.*

contour band machining, 123
cutting off stock, 123
cutting off with band machines, 146
feed systems, 123, 125 *fig.*
operation recommendations, 150, 150–52 *tab.*
principles of, 123–24, 126–27 *fig.*
Band tools, 144–46, 144–47 *fig.*, 147–48 *tab.*
Base line, 74
Basic oxygen converter, 573, 575 *fig.*
Bearing bronze, 603
Bed-type milling machine, 383, 386, 386 *fig.*
Bench grinder, 461, 462 *fig.*
Benchwork
    chisels, 105–7, 106–8 *fig.*
    definition of, 97
    files, 110–12, 117, 110–19 *fig.*, 117 *tab.*
    hacksaws, 103–5, 103–5 *fig.*
    hammers, 98–99, 99 *fig.*
    hand tool safety, 95–96
    scrapers, 107–8, 108–9 *fig.*
    screwdrivers, 99–100, 99–100 *fig.*
    tool care, 97
    wrenches and pliers, 100–101, 101–2 *fig.*
Bessemer, Henry, 573
Bessemer process, 573, 575 *fig.*
Bevel gears, 413, 414 *fig.*, 440–41, 446–47, 446 *fig.*,
    448–51 *fig.*, 452 *tab.*
Binary code, 500
Bituminous coal, 567
Blast furnace, 567, 567–68 *fig.*
Blotters, 475
Body clearance, 175
Bond, 464–65
Boring, 199, 199 *fig.*, 203, 204 *fig.*
    center drilling, 266–77, 267–77 *fig.*, 268–69 *tab.*
    commercial tool holders, 261–62, 263–64 *fig.*
    drilling and reaming, 264, 265 *fig.*
    lathe milling operations, 266, 266 *fig.*
    methods of, 264, 265 *fig.*

Boring bars, 346, 349–50 *fig.*
Bottoming tap, 204, 205 *fig.*
Brass, 602–4, 604 *tab.*
Brass wire, 602–3
Brinell hardness test, 587, 588–90 *fig.*
Bronze, 602
Brown and Sharpe taper, 282
Built-up edge, 361

Calipers, 18, 19–22 *fig.*
Cape chisels, 105, 106 *fig.*
Carbide burs, 117, 119 *fig.*
Carbide cutting tools, 553–54, 555 *fig.*, 570, 571, 571 *fig.*
Carbide tooling
    basic tool types, 343, 343 *fig.*
    boring bars, 346, 349–50 *fig.*
    Kennametal grades, 345, 351 *tab.*
    machining recommendations, 346, 349–50, 352–53 *tab.*
    standard tool types, 343, 344–45 *fig.*
    tool holder identification system, 346, 347–48 *tab.*
    tool nomenclature, 342, 342 *fig.*
    tool selection, 343, 346
    tool wear, 354 *fig.*
    troubleshooting tools, 353–54
Carbon, 568, 573, 576
Carbon steels, 575–76
Carbon tool steel cutting tools, 586
Carbon tool steel tool bits, 326
Carburizing, 585
Case hardening, 585, 586
Casein plastic, 608
Casite, 362–63
Cast-alloy cutting tools, 570–71
Cast iron, 567–68, 569 *fig.*
    classes of, 568–69
    effect of other chemicals on, 570
    effect of silicon, sulfur, phosphorus, and manganese on, 569–70
    machining of, 570–71, 571 *tab.*, 571 *fig.*
Cast material alloys, 328
Casting alloys, 601
Catheads, 274–75, 275 *fig.*
Cellulose plastics, 609
Cemented carbide cutting tools, 328
Cementite, 568
Center drilling, 266–77, 267–77 *fig.*, 268–69 *tab.*
Center punches, 79, 79–80 *fig.*
Center rests, 240–41, 243–45 *fig.*
Centerless grinding, 484, 485 *fig.*

Ceramic materials, 556
Ceramic-oxide cutting tools, 570
Ceramic tools, 328
Chatter, 179, 186 *fig.*, 201
Chemical cutting fluids, 362
Chip breaker, 188–89, 190 *fig.*, 333, 336 *fig.*, 581–82, 581 *fig.*
Chip curler, 324–25, 325 *fig.*, 333, 336 *fig.*
Chip formation, 188–89
Chipping, 105
Chisel edge, 175
Chisels, 105–7, 106–8 *fig.*
Chlorinated oils, 362
Chromium, 576
Cinmill control, 511, 512–13 *fig.*, 517 *fig.*
Circular dividing table, 391, 393 *fig.*
Circular Pitch System, 416
Clamping, 511, 514 *fig.*
Clamps, 377–78, 381 *fig.*
Clearance, 328–32, 329–33 *fig.*, 416
    for free machining alloys, 604, 604 *tab.*
Clearance angles, 333–34
Climb milling, 395–96, 397–98 *fig.*
Coated abrasives, 461
Coating materials (powder process), 555–56
Cobalt, 576
Coil springs, 305, 305 *fig.*
Cold forming taps, 204–5, 206 *fig.*
Collet, 211
Column-and-knee-type milling machines, 382–83, 383 *fig.*
Combination back and side rake, 580, 579 *fig.*
Combination chuck, 238, 241 *fig.*
Commercial tool holders (lathe), 233–34, 233–35 *fig.*
Compound tool rest method, 283, 283 *fig.*
Computed tomography (CT), 534–35, 534–35 *fig.*
Computer, 509
Computer Assisted Design (CAD), 90, 90 *fig.*
Computer Assisted Manufacturing (CAM), 90, 90 *fig.*
Computer graphics, 511–13, 513–16 *fig.*
Computer numerical control (CNC), 497, 509–11, 510–13 *fig.*
Continue function, 521
Continuous path algorithm, 519–20
Contour, 325, 326 *fig.*
Contour band machining, 121, 123
Control, 499, 499 *fig.*
Control algorithms, 519–20
Controlled-path algorithm, 520
Controlled pulse circuits (CPC), 544, 544 *fig.*

Conventional EDM, 541–42, 541–42 *fig.*
Conventional milling, 397
Conventional-type thinning, 190, 191 *fig.*
Coolant feeding drills, 175, 177 *fig.*
Coordinate system, 388, 390–91 *fig.*
Copper, 602–4, 603 *fig.*, 604 *tab.*
Copper-lead based babbitt, 604
Copper-lead-tin based babbitt, 604–5
Corrected addendum, 436–37, 437 *fig.*
Counterboring, 199, 199 *fig.*, 203, 205 *fig.*
Countersinking, 199, 199 *fig.*, 203–4, 205 *fig.*
Crater, 349
Cross filing, 111, 116 *fig.*
CRT (cathode ray tube), 511
Cubic boron nitride (CBN), 463
Cutting a worm, 438, 439–41 *fig.*
Cutting edge, 320–21, 322–23 *fig.*
Cutting fluids, 361–64
    for copper alloys, 604
    in sawing, 132, 132 *tab.*
Cutting lips, 175, 176 *fig.*
Cutting off, 255–58, 257–59 *fig.*
Cutting speed (CS), 191, 192–94 *tab.*, 356–57, 357 *tab.*
Cutting threads, 305–6
Cutting tools
    carbide tooling, 342–43, 346, 349–50, 342–45 *fig.*,
        347–48 *tab.*, 349–50 *fig.*, 352–53 *tab.*, 354 *fig.*
    cutting fluids, 361–64
    cutting principles, 319–21, 320–23 *fig.*
    cutting tools, 324–29, 331–36, 325–27 *fig.*, 329–37 *fig.*
    safe use of, 317–18, 318 *fig.*
    speed, feed, and depth of cut, 356–59, 357 *tab.*, 359
        tab., 360 *fig.*
Cyaniding, 585
Cylindrical grinding, 483–84, 483–85 *fig.*
Cylindrical plug gage, 51, 55 *fig.*
Cylindrical ring gage, 51, 54 *fig.*

Dead center, 272
Deionization, 543
Depth of cut, 357
Dial bore gage, 50, 54 *fig.*, 55 *tab.*
Dial indicators, 50, 51 *fig.*, 53 *fig.*, 387, 387 *fig.*
Dial-type hardness testers, 587–88, 596–97 *fig.*
Diametral Pitch System, 416–17
Diamond abrasives, 462
Diamond-edge bands, 144, 145 *fig.*
Diamond penetrator, 587, 590 *fig.*
Diamond-point chisels, 106, 106 *fig.*

Diamond tools, 328
Die, 386
Dielectric constant, 542
Dielectrics, 542
Differential indexing, 402, 406–7, 408–11 *tab.*
Digital readouts, 71, 142–43, 142–43 *fig.*
Direct indexing, 402, 402 *fig.*
Discharge, 543
Dividers, 82, 82 *fig.*
Dormer fast lock holder, 377, 380 *fig.*
Double-cut files, 110, 111 *fig.*
Dragging, 249
Draw filing, 111, 116 *fig.*
Dressing and trueing the wheel, 475–77, 480–81 *fig.*
Drill, 164
Drill drift, 167, 170 *fig.*
Drill point angles and clearances, 178–79, 182–83 *fig.*
Drill point gage, 179, 185 *fig.*
Drill point thinning, 190
Drill press, 164, 165 *fig.*
Drill press vise, 167, 171 *fig.*
Drill socket, 166, 170 *fig.*
Drills and drilling operations, 161
    drilling machines and tool holding methods, 164–68,
        164–73 *fig.*
    drilling operations, 188–91, 194, 188–91 *fig.*, 192–94
        tab.
    drills, 174–79, 175–78 *fig.*, 180–81 *tab.*, 182–86 *fig.*
    other drilling operations, 198–205, 207–12, 199–202
        fig., 203 *tab.*, 204–12 *fig.*, 213–15 *tab.*, 216 *fig.*
    safety practices, 162–63

Electric arc spraying, 549
Electric furnace process, 573, 577 *fig.*
Electrical discharge machining (EDM), 537, 541–45,
    541–47 *fig.*, 560–62, 560–62 *fig.*
Electrodes, 543, 544
Electronics Industry Association (EIA) Code, 499, 500
Emergency drills, 211–12, 216 *fig.*
Employment opportunities
    career selection, 4–5
    job application, 5–6
    job titles, 3–4, 5 *fig.*
    machine trades, 1–3, 2 *fig.*
End mills, 373–75, 377–80 *fig.*
End of block (EB), 499
English (U.S. Customary) measurement system, 13, 14 *tab.*
English (U.S. Customary) to Metric conversions, 619 *tab.*
Epoxy plastics, 608

Expansion mandrel, 277, 277 *fig.*
Expansion reamers, 200
Eye protection, 8

Face milling, 369, 394, 395 *fig.*
Face milling cutters, 372–73, 376–77 *fig.*
Face plate, 240, 243 *fig.*
Facing, 248–49, 249 *fig.*
Fast finishing steels, 327
Fast taper, 281–82
Feed, 194, 194 *fig.*, 357–59, 359 *tab.*, 360 *fig.*, 398, 400 *tab.*
Feed systems (band machines), 123, 125 *fig.*
Feed systems (lathe), 224–26, 226–28 *fig.*
Ferrite, 568
Ferromanganese, 570
Ferrous metallurgy, 565, 566
  blast furnace, 567, 567–68 *fig.*
  cast iron, 567–71, 568–69 *fig.*, 571 *fig.*, 571 *tab.*
File bands, 145, 146 *fig.*
Files, 110–12, 117, 110–19 *fig.*, 117 *tab.*
Filing, 110
Firm joint calipers, 18, 20 *fig.*
Fit, 50
Fitting, 1–2, 2 *fig.*
Fixed gages, 50, 54–56 *fig.*
Fixture, 386
Flame spraying
  power process, 554–56, 558, 555–58 *fig.*
  wire process, 549–50, 552–54, 549–55 *fig.*
Flash, 137, 138 *fig.*
Flat chisels, 105, 106 *fig.*
Flat files, 111
Flattening the lips, 190, 191 *fig.*
Flexible Manufacturing Cell (FMC), 45, 44–45 *fig.*
Flexible Manufacturing System (FMS), 44–45, 44–45 *fig.*, 90–91, 91 *fig.*
Float-lock vise, 168, 171 *fig.*
Fluorocarbon plastic, 609
Flutes, 117, 174
Follower rest, 241, 243
Form milling, 369, 394, 396 *fig.*
Form relieved tooth, 369–70
Formed cutters, 371–72, 375–76 *fig.*
Four-jaw chuck, 236–37, 238 *fig.*
Frequency, effect on surface finish, 543, 543 *fig.*
Friction sawing, 145
Fringe pattern formations, 64–66, 64–66 *fig.*
Frustrum, 440
Full-nose cutting tool, 580–81, 581 *fig.*

Gage, 386
Gage (saw blade), 128, 131 *fig.*
Gage blocks, 57, 57–62 *fig.*
  care of, 61
  use of, 58, 62 *fig.*
  using a sine bar with, 60–61, 63 *fig.*
Gang drilling machine, 165
Gang milling, 369, 369 *fig.*, 395
Gash-type thinning, 190, 191 *fig.*
Gear teeth measurement
  bevel gears, 440–41, 446–47, 446 *fig.*, 448–51 *fig.*, 452 *tab.*
  corrected addendum, 437, 437 *fig.*
  gear tooth vernier, 436–37, 437 *fig.*
  involute gear cutters, 447, 454 *tab.*, 454 *fig.*
  module system, 439, 446 *tab.*
  tooth sizes, 447, 454 *fig.*
  worm gearing, 437–38, 438–41 *fig.*, 442 *tab.*, 444 *tab.*
Gear tooth vernier, 35, 40 *fig.*, 436, 437 *fig.*
Gearing
  cutting a spur gear, 417, 420–21, 417 *fig.*, 420–21 *fig.*, 421–22 *tab.*
  gear diameters, 414–16, 415–16 *fig.*
  gear tooth parts, 416–17, 416–17 *fig.*
  type of gears, 413–14, 413–15 *fig.*
Glass dust, 608
"Go" and "no-go" gage, 51
Grade, 465, 465 *fig.*
Granite plates, 77, 77–78 *fig.*
Graphite, 568
Gray cast iron, 568, 570
Grinding
  cylindrical grinding, 483–84, 483–85 *fig.*
  grinding wheel, 461–66, 468, 471–73, 462–71 *fig.*
  precision grinding, 474–77, 480–82, 475–82 *fig.*
  safety practices, 459–60
  threading tool, 304–5
  tool and cutter grinding, 484–85, 485 *fig.*, 486–87 *tab.*
  tool bits, 335–36, 336–37 *fig.*, 340 *tab.*
Grinding wheel, 461–66, 468, 471–73, 462–71 *fig.*
Grit blasting, 558
Grit size, 463–64, 464 *fig.*
Grooving, 255, 256 *fig.*

Hacksaws, 103–5, 103–5 *fig.*
Half round files, 111, 112 *fig.*
Hammers, 77, 98–99, 99 *fig.*
Hand taps, 204, 205 *fig.*
Hard spots, 570

Hardening, 584
Hardness conversion table, 587, 595 *fig.*
Hardness testing, 587–88, 588–91 *fig.*, 595–97 *fig.*
Headstocks (lathe), 227
Heat-affected zone, 543–44, 543 *fig.*
Heat treating furnaces, 9
Helical fluted reamers, 201–2
Helical gearing
    calculating change gears for required lead, 425–26,
       427–31 *tab.*, 432–33 *fig.*, 434–35 *tab.*
    cutting a helix, 424, 425 *fig.*
    determining the direction to set the table, 424–25
    terminology, 423–24
Helical teeth, 413, 414 *fig.*
Helix (def.), 423–24
Helix angle, 175, 178–79, 183 *fig.*, 288
Hermaphrodite calipers, 82, 83 *fig.*
Herringbone gears, 414, 415 *fig.*
High helix twist drill, 175, 177 *fig.*
High-speed sawing, 145
High-speed steel, 145, 327–28
High-speed steel cutting tools, 586
High-speed tool bits, 601, 601 *fig.*
Home position, 503
Honing, 326
Hook rule, 15, 16–17 *fig.*

Incremental measuring, 501, 502 *fig.*
Indexable insert tools, 343
Indexing
    direct indexing, 402, 402 *fig.*
    simple indexing, 403–4, 405 *tab.*
    universal dividing head, 403, 403 *fig.*
Industrial robot, 518–19, 520–22 *fig.*
Inside micrometers, 27, 30, 30–32 *fig.*
Instrument mechanic, 3, 5 *fig.*
Interfacing, 521, 522
Interference bands, 63
Involute gear cutters, 447, 454 *tab.*, 454 *fig.*
Iron, 566, 573
Iron-carbide diagram, 576, 578 *fig.*

Jarno taper, 282
Jig, 386
Jig boring
    machines, 386, 386 *fig.*
    setting up work, 387–88, 388–90 *fig.*
    tools used in, 387, 388 *fig.*
    using vertical milling machine for, 389, 391, 391–93 *fig.*

Job selector, 129, 131 *fig.*

K Monel, 582, 582 *tab.*
Kennametal grades, 346, 351 *tab.*
Kerf, 105
Keyless chucks, 166, 168 *fig.*
Keyseat clamps, 92, 93 *fig.*
Knife blades, 144, 144 *fig.*
Knife files, 111, 113 *fig.*
Knurling, 258–59, 260 *fig.*

Land, 175
Lard oil, 362
Laser assisted machining (LAM), 419
Laser cladding, 338–39, 338–39 *fig.*
Laser cutting and machining, 418–19, 418–19 *fig.*
Laser lathe machining, 488–89, 488–89 *fig.*
Laser surface alloying, 339, 339 *fig.*
Laser surface hardening, 384–85, 384–85 *fig.*
Lathe and lathe operations
    boring, 261–77, 262–77 *fig.*, 268–69 *tab.*
    facing, 248–49, 249 *fig.*
    lathe, 223–29, 223–29 *fig.*
    lathe tools, 233–38, 240–41, 243, 233–47 *fig.*
    safety practices, 221–22
    tapering, 281–83, 282–84 *fig.*, 285–86 *tab.*
    thread cutting, 301–6, 301–5 *fig.*, 306–8 *tab.*
    threading, 287–90, 292–95, 288–96 *fig.*, 296 *tab.*
    threading tools (commercial), 297–98, 298–300 *fig.*
    turning, 250–59, 251–60 *fig.*
Lathe chuck, 236–38, 240, 238 *fig.*, 241–42 *fig.*
Lathe filing, 111, 116 *fig.*
Lathe tool bits, 319, 320 *fig.*
Lathe tool holders, 233–36, 233–37 *fig.*
Layout
    angle plates, 93, 94 *fig.*
    dividers, 82, 82 *fig.*
    hammers, 77
    hermaphrodite calipers, 82, 83 *fig.*
    keyseat clamps, 92, 93 *fig.*
    laying out, 74–75, 75 *fig.*
    protractor and depth gage, 85, 87 *fig.*
    punches, 79, 79–81 *fig.*
    safe use of tools, 73–74
    squares, 85, 85–87 *fig.*
    surface gage, 85, 87–88 *fig.*
    surface plates, 76–77, 77–78 *fig.*
    surface preparation, 75, 75 *fig.*
    trammels, 92, 93 *fig.*

Layout *(continued)*
    V-blocks, 92, 93 *fig.*
    vernier height gage, 89, 89 *fig.*
Lead, 288
Lead angle, 288
Lead-based babbitt, 604
Letting-down process, 586
Limestone, 567
Linear measurement, 12–13
Liquid bath carburizing, 586–87
Live center, 272
Locating microscope, 387, 388 *fig.*
Lock joint calipers, 18, 21 *fig.*
Low helix twist drill, 175, 177 *fig.*

Machine operator, 3
Machine work, 2–3
Machinist, 3
Machinist's file, 111
Magnetic base indicator holder, 50, 52 *fig.*
Maintaining the gap, 543–44
Major diameter, 290
Malleable cast iron, 569, 570
Mandrel, 275–77, 275–77 *fig.*
Manganese, 569, 575, 576–77
Manganese bronze, 603
Manufacturing Automation Protocol (MAP), 613–14,
    614 *fig.*
Margin, 174
Marking off. *See* Layout
Measurement. *See* Precision measurement
Measuring systems, 13, 14 *tab.*
Mechanical comparator, 50, 52 *fig.*
Mechanical technician, 3–4
Mechanical technologist, 4
Melting furnaces, 9
Metallizing, 549, 549 *fig.*
Metallurgy, 565
    ferrous metals, 566–71, 567–79 *fig.*, 571 *tab.*
    heat treatment of steels, 584–88, 588–91 *fig.*
    nonferrous metals, 600–605, 601 *fig.*, 603 *fig.*, 604 *tab.*
    steel, 573, 575–82, 574–75 *fig.*, 577–81 *fig.*, 582 *tab.*
Metric rule, 15, 16 *fig.*
Metric system, 13, 14 *tab.*
Metric threading, 294–95, 295–96 *fig.*, 296 *fig.*
Micrometer height gage, 57, 62 *fig.*
Micrometers
    care of, 30–31, 32 *fig.*
    holding and using, 31, 33, 33–34 *fig.*

inside micrometers, 27, 30, 30–32 *fig.*
    reading of, 23, 25–26, 26–29 *fig.*
Mill file, 111, 113 *fig.*
Milling and milling operations
    basic operations, 369–75, 377–78, 369–81 *fig.*
    gear teeth measurement, 436–41, 446–47, 437–41 *fig.*,
        442–46 *tab.*, 446–51 *fig.*, 452–54 *tab.*, 454 *fig.*
    gearing, 413–17, 420–21, 413–17 *fig.*, 420–22 *fig.*,
        423 *tab.*
    helical gearing, 423–26, 425 *fig.*, 427–31 *tab.*, 432–33
        fig., 434–35 *tab.*
    indexing, 402–4, 406–7, 402–4 *fig.*, 405 *tab.*, 406 *fig.*,
        408–11 *tab.*
    machine safety, 367–68
    milling machines, 382–83, 386–89, 391, 383 *fig.*,
        386–93 *fig.*
    milling operations, 394–98, 400, 395–98 *fig.*, 399–400
        tab.
Milling cutters, 369–75, 371–80 *fig.*
Milling operations, 266, 266 *fig.*
Mineral-lard oils, 362
Mineral oils, 362
Minor diameter, 290
Miter gears, 440
Modified-nose cutting tool, 580, 580 *fig.*
Module system (gearing), 439, 446 *tab.*
Molybdenum, 577
Monel metal, 582, 582 *tab.*
Morse taper, 166, 166–67 *fig.*, 282, 285 *tab.*
Mounted points (or wheels), 465, 470–71 *fig.*
Multi-spindle drilling machine, 165
Mushet, Robert, 573

Naval brass, 603
Needle file, 111, 114 *fig.*
Negative rake tools, 343
Nickel, 576
Nickel alloys, 582, 582 *tab.*
Nitriding, 585
Nodular cast iron, 569, 570
Nonferrous metallurgy, 565
Nonferrous metals
    aluminum and its alloys, 600–602, 601 *fig.*
    babbitt bearings, 604–5
    copper and its alloys, 602–4, 603 *fig.*, 604 *tab.*
Normalizing, 584
Norton wheel marking systems, 465, 467 *fig.*
Nose radius, 334–35

Numerical control (N/C), 497, 498
  axis orientation, 502–3, 502–5 *fig.*
  machining with, 503–5, 507–8, 504–8 *fig.*
  programming, 500–501, 501–2 *fig.*
  tapes and codes, 499–500, 501 *fig.*

Offhand grinding, 461
Offset, calculation of, 283
Open hearth process, 573, 577 *fig.*
Optical comparators, 67–69, 68–71 *fig.*
Optical flats, 62–66, 64–66 *fig.*
Outside diameter (Do), 416
Overcut, 544
Oxidation-resistant alloys, 556

Pack carburizing, 586
Parting, 256–58
Phenolic plastics, 608
Phosphor-bronze, 603
Phosphorus, 569, 573, 575
Pillar file, 111, 113 *fig.*
Pilot holes, 189–90, 191 *fig.*
Pinion gear, 440
Pitch (P), 103, 417
Pitch (band machining), 128, 129 *fig.*
Pitch (screw threads), 288
Pitch circle, 415, 415 *fig.*
Pitch diameter (Pd.), 290, 415
Plain metal slitting saw, 371, 372–73 *fig.*
Plain milling, 369, 394, 395 *fig.*
Plain milling machine, 382, 383 *fig.*
Plasma spraying, 549
Plastics, 607
  machining of, 609–12, 610–11 *tab.*
  types of, 608–9
Pliers, 101, 102 *fig.*
  safety using, 96
Plunge cutting, 255
Plunge grinding, 483, 484 *fig.*
Point (drill), 175
Point angles, 178–79, 182–83 *fig.*
Pointed cutting tool, 320, 322 *fig.*
Point-to-point algorithm, 519
Point-to-point controls, 143
Polishing band guides, 145, 147 *fig.*
Polishing bands, 145, 146 *fig.*
Polyamide plastic, 609
Polyester plastics, 608
Polyolefin plastics, 609

Powder metallurgy, 537, 538–40, 539 *fig.*
Powder process (flame spraying), 554–56, 558, 555–58 *fig.*
Power threads, 293, 293–94 *fig.*
  cutting of, 304, 305 *fig.*
Precision grinding, 462
  dressing and trueing the wheel, 475–77, 480–81 *fig.*
  mounting and dressing wheels, 475, 479 *fig.*
  setting the work, 477, 480–82
  surface grinding, 474–75, 475–77 *fig.*
  testing wheels before mounting, 475, 478 *fig.*
Precision measurement, 11–12
  measuring systems, 13, 14 *tab.*
  optical comparators and digital readouts, 67–71, 68–71 *fig.*
  tool care, 13
  using calipers, 18, 19–22 *fig.*
  using dial indicators, dial bore gages, and fixed gages, 50–51, 51–56 *fig.*
  using gage blocks, sine bars and tables, and optical flats, 57–66, 57–66 *fig.*
  using micrometers, 23–33, 24–36 *fig.*
  using rules (scales), 15, 16–17 *fig.*
  using telescoping gages, protractors, and thickness gages, 42, 46, 43–49 *fig.*
  using vernier calipers, 33, 35, 36–40 *fig.*
Pressing, 539
Prick punches, 79, 80–81 *fig.*
Primary cutting edge, 321
Programming, 500–501, 501–2 *fig.*
Protractor and depth gage, 85, 87 *fig.*

Rack and pinion, 413, 413 *fig.*
Radial arm drill, 164–65, 165 *fig.*
Rake, 332–33, 335–36 *fig.*
  for nonferrous metals, 604 *tab.*
Rake angles, 334
Rapid cooling, 568
Rasp-cut file, 110, 110 *fig.*
Reamers, 200–203, 201–2 *fig.*, 203 *tab.*
Reaming, 198–99, 199 *fig.*, 264, 265 *fig.*
Red brass, 602–3
Red hardness, 327–28, 570
Red shortness, 569
Retang sleeve, 166, 169 *fig.*
Riffler file, 111, 115 *fig.*
Right-hand cutter with chip curler, 580, 579–80 *fig.*
Right-hand rule, 321, 323 *fig.*
Rigid threading tool holder, 297, 298 *fig.*

Robotics, 518
    applications for, 521–22, 526–28 *fig.*
    control algorithms, 519–20
    control system, 519, 523–24 *fig.*
    industrial robot design, 519, 520–22 *fig.*
    modes of operation, 521, 523 *fig.*, 525 *fig.*
    new developments in, 523–24
Robots, 230–31, 278–79, 497
Rockwell hardness test, 587, 588–90 *fig.*
Root diameter (Dr), 416
Rose-type chucking reamer, 200
Rotary bur, 117, 119
Rotary file, 117, 117–18
Rotary impulse generators (RIG), 544, 544 *fig.*
Round file, 111, 113 *fig.*
Round-nose chisel, 105, 106 *fig.*
Round-nose tool bit, 320, 321 *fig.*
Rules (scales), 15, 16–17 *fig.*

Safe edges, 110
Safety
    eye protection, 8
    general rules of, 7–8
    with melting and heat treating furnaces, 9
Sand pocket, 570
Saw blades
    cutting fluids, 132, 132 *tab.*
    job selector, 129, 131 *fig.*
    selection of, 128, 129–31 *fig.*
Sawing
    band machining, 123–24, 124–27 *fig.*
    other band tools, 144–46, 144–48 *fig.*
    recommendations for, 149–50, 150–52 *tab.*
    safety in using, 121–22
    saw blade, 128–29, 132, 129–32 *fig.*
    troubleshooting, 152–54
    welding saw blades, 133–41, 133–40 *fig.*
Scrapers, 107–8, 108–9 *fig.*
    safety using, 96
Screw threads, 204–5
    forms of, 292–93, 293 *fig.*
    measurement of, 288–90, 292, 291–92 *fig.*
    metric threading, 294–95, 295–96 *fig.*, 296 *tab.*
    power threads, 293, 293–94 *fig.*
    single and multiple, 288, 291 *fig.*
    unified screw threads, 294, 294–95 *fig.*
Screwdrivers, 99–100, 99–100 *fig.*
    safety using, 95
Scribers, 78, 78 *fig.*

Second-cut grade (file), 111
Self-bonding coating, 556
Self-fluxing alloys, 555–56
Sensitive drill press, 164, 164 *fig.*
Set, 128, 131 *fig.*
Setting the compound, 297–98, 299–300 *fig.*
Shape of cutting tool, 325, 326 *fig.*
Shaped edge cutting tool, 321, 322 *fig.*
Sharp-nose tool bit, 320, 321 *fig.*
Sharpening (drills), 179, 183–85 *fig.*
Shell end mill adaptor, 377, 380 *fig.*
Shell reamers, 200
Shoulder treatment (flame spraying), 550, 551–53 *fig.*
Side milling cutters, 371, 372–73 *fig.*
Side rake, 333, 336 *fig.*, 579, 579 *fig.*
Silicon, 569, 575, 577
Silicon carbide, 462
Simple indexing, 402, 403–4, 405 *tab.*
Sine bar, 60–61, 63 *fig.*
Sine tables, 61
Single-cut file, 110, 110 *fig.*
Single-point boring bar, 387, 387 *fig.*
Sintering, 539
Sledgehammer, 99, 99 *fig.*
Slow cooling, 568
Slow taper, 281
Small hole gage, 42, 46 *fig.*
Small round-nose tool bit, 320, 321 *fig.*
Smooth-cut grade (file), 111
Soft-face hammer, 99, 99 *fig.*
Spade drills, 176–77, 177 *fig.*
Spark erosion, 541
Special-purpose processes, 537
    electrical discharge machining (EDM), 541–45, 541–47 *fig.*
    powder metallurgy, 538–40, 539 *fig.*
    thermal spray processes, 548–50, 552–56, 558, 549–58 *fig.*
Speeds (milling), 397–98, 399–400 *tab.*
Spider, 274, 274–75 *fig.*
Spiegeleisen, 570
Spigot, 273–74, 274 *fig.*
Spindle noses, 228–29, 229 *fig.*, 232 *fig.*
Spiral (def.), 424
Spiral-edge band, 144
Spiral fluted taps, 205, 206–7 *fig.*
Spiral point taps, 204, 206 *fig.*
Split point thinning, 190, 191 *fig.*
Spot drilling, 188, 188–89 *fig.*

Spot-facing, 199, 199 *fig.*, 203
Sprayed metal, 549
Spring calipers, 18, 19 *fig.*
Spring tempered rule, 15, 16 *fig.*
Spring-type threading tool holder, 297
Spur gear, 413, 413 *fig.*
    cutting of, 417, 420–21, 417 *fig.*, 420–21 *fig.*,
        421–22 *tab.*
Square thread, 293, 293 *fig.*
Squares, 85, 85–87 *fig.*
Standard brazed tools, 343
Standard drill press vise, 167, 171 *fig.*
Standard drill sleeve, 166, 168 *fig.*
Standard tapers, 282
Standardization, screw threads, 294–95, 295–96 *fig.*,
    296 *tab.*
Steady (center) rest, 240–41, 244–45 *fig.*
Steel, 573
    alloying elements in, 576–78
    carbon content and use of, 576, 578 *fig.*
    classification of, 575–76
    heat treatment of, 584–88, 588–91 *fig.*
    machining of, 578–82, 579–81 *fig.*
    metallurgy of, 573, 575
Step block, 168, 172 *fig.*
Straddle milling, 395, 396 *fig.*
Straight cut controls, 143
Straight edge cutting tool, 320–21, 322 *fig.*
Straight fluted drill, 175, 177 *fig.*
Straight helix, 287
Straight or spiral tooth, 369–70, 371 *fig.*
Straight shank, coolant feeding drill, 175, 177 *fig.*
Straight shank drill, 166, 167 *fig.*
Strap clamps, 168, 172 *fig.*
Style "E" holder, 377, 381 *fig.*
Styrene plastics, 609
Sulfur, 569, 575
Sulfurized oils, 362
Surface gage, 85, 87–88 *fig.*
Surface grinding, 474–75, 475–77 *fig.*
Surface plates, 76–77, 77–78 *fig.*
Swing, 227
Swiss-pattern file, 111, 113 *fig.*

Tailstock set-over method, 283, 284 *fig.*
Tang, 110, 166
Taper attachment, 283, 284 *fig.*
Taper pin reamers, 200, 202 *fig.*
Taper pins, 282, 283 *fig.*

Taper reamers, 200
Taper shank, high-speed coolant inducer drill, 175, 177 *fig.*
Tapered helix, 287
Tapering, 281–83, 282–84 *fig.*, 285–86 *tab.*
Tapers, 281–82, 282 *fig.*
Tapping, 199, 200 *fig.*
Taps, 204–10, 205–10 *fig.*, 213–15 *tab.*
Teach mode, 521
Teach pendant, 521, 525 *fig.*
Telescoping gage, 42, 43 *fig.*
Temper mills, 584
Tempering, 584
Thermal spray processes, 548
    electric arc spraying, 549
    flame spraying, wire process, 550, 552–54, 550–54 *fig.*
    metallizing, 549, 549 *fig.*
    plasma spraying, 549
Thermoplastics, 609
Thermosets, 608
Thermospray gun, 555, 557 *fig.*
Thickness gage, 46, 49 *fig.*
Thread cutting, 301, 301 *fig.*
    cutting a metric thread, 305–6
    cutting power threads, 304, 305 *fig.*
    cutting routine, 301–2, 302 *fig.*
    cutting tapered threads, 306
    grinding threading tool bits, 304, 304 *fig.*
    making coil springs, 305
    picking up a thread, 303–4, 303 *fig.*
    resource tables for, 306–8 *tab.*
    threading dial use, 302–3, 302 *fig.*
Thread mechanism (lathe), 226–27, 226 *fig.*, 229 *fig.*
Thread plug gage, 56 *fig.*
Thread ring gage, 55 *fig.*
Threaded spindle, 228, 229 *fig.*
Threading, 287, 288 *fig.*
    screw threads, 288–90, 292–93, 289–93 *fig.*
    tools, commercial, 297–98, 298–300 *fig.*
Threading dies, 210–11, 210–12 *fig.*
Three-jaw chuck, 238, 241–42 *fig.*
Three square file, 111, 113 *fig.*
Tin, 570
Tin-based babbitt, 604
Titanium, 570
Tobin bronze, 603
Tolerance, 51
Tool and cutter grinding, 484–85, 485 *fig.*, 486–87 *tab.*
Tool bits, 325, 326 *fig.*, 335–36, 336–37 *fig.*, 340 *tab.*
Tool center point (TCP), 519

Tool posts, 234–35, 235–36 *fig.*
Tool steels, heat treatment of, 585–86
Tooth sizes, 447, 454 *fig.*
Tooth thickness, 416, 417 *fig.*
Tracking function, 521
Trammels, 92, 93 *fig.*
Traverse grinding, 483, 484 *fig.*
Triangular saw file, 111, 113 *fig.*
Troubleshooting
    carbide cutting tools, 353–54
    poor welds, 139–41, 139–40 *fig.*
    sawing operations, 152–54
Tungsten, 577
Turning, 250–51, 251 *fig.*, 311–15
    cutting off, 255–58, 257–59 *fig.*
    cutting position, 254, 254 *fig.*
    grooving, 255, 255–56 *fig.*
    knurling, 258–59, 260 *fig.*
    methods of, 252–53, 253 *fig.*
    tool position, 251–52, 252–53 *fig.*
    turning a shoulder, 254, 255 *fig.*
Turret tool holders, 235–36, 235–37 *fig.*
Twist drills, 174–75, 177, 175–77 *fig.*, 180–81 *tab.*

Undercutting, 550, 551 *fig.*
Unified screw threads, 294, 294–95 *fig.*
Uniform chip, 124, 127 *fig.*
Universal bevel protractor, 42, 47–48 *fig.*
Universal dividing head, 403, 403 *fig.*
Universal heavy-duty dial test indicator, 50, 51 *fig.*
Universal milling machine, 382, 383 *fig.*
Universal rotary table, 168, 171 *fig.*
Urethane plastic, 608

Vanadium, 570, 577
V-blocks, 92, 93 *fig.*, 168

Vernier calipers, 33, 35, 36–40 *fig.*
Vernier height gage, 89, 89 *fig.*
Vertical milling machine, 382, 383 *fig.*
    for jig boring, 389, 391, 391–93 *fig.*
Vinyl plastics, 609
Vises, 377–78

Web, 175, 176 *fig.*
Welding saw blades, 133–34, 133 *fig.*
    making the weld, 136–38, 138–39 *fig.*
    preparation of blade and welder, 134–36, 134–37 *fig.*
    troubleshooting poor welds, 139–41, 139–40 *fig.*
White cast iron, 569, 570
Whole depth, 416, 417 *fig.*
Width (blade), 128, 130 *tab.*
Wire-cut EDM, 541, 544–45, 545–47 *fig.*
Wire process (flame spraying), 550, 552–54, 551–53 *fig.*
Woodruff keyseat cutters, 375, 379 *fig.*
Working depth, 416, 417 *fig.*
Worm and worm gear, 414, 415 *fig.*
Worm gearing, 437–38, 438–41 *fig.*, 442 *tab.*, 444 *tab.*
Wrenches, 100–101, 101 *fig.*
    safety using, 95–96
Wringing, 57
Wrought alloys, 601

X-axis, 502–3

Yellow brass, 602, 604

Z-axis, 502–3
Zero back rake, 603 *fig.*